Deformation and Fracture in Materials

This book provides information on the basics of deformation and fracture in materials and on current, state-of-the-art experimental and numerical/theoretical methods, including data-driven approaches in the deformation and fracture study of materials. The blend of experimental test methods and numerical techniques to study deformation and fracture in materials is discussed. In addition, the application of data-driven approaches in predicting material performance in different types of loading and loading environments is illustrated.

Features:

- Includes clear insights on deformation and fracture in materials, with clear explanations of mechanics and defects relating to them
- Provides effective treatments of modern numerical simulation methods
- Explores applications of data-driven approaches such as artificial intelligence, machine learning, and computer vision
- Reviews simple and basic experimental techniques to understand the concepts of deformation and fracture in materials
- Details modeling and simulation strategies of mechanics of materials at different scales

This book is aimed at researchers and graduate students in fracture mechanics, finite element methods, and materials science.

Deformation and Fracture in Materials

Advances in Experimental and Numerical Studies

Edited by

Anoop Kumar Mukhopadhyay and Dhaneshwar Mishra

CRC Press is an imprint of the
Taylor & Francis Group, an informa business

First edition published 2025
by CRC Press
2385 NW Executive Center Drive, Suite 320, Boca Raton FL 33431

and by CRC Press
4 Park Square, Milton Park, Abingdon, Oxon, OX14 4RN

CRC Press is an imprint of Taylor & Francis Group, LLC

ISBN: 9781032417066 (hbk)
ISBN: 9781032417073 (pbk)
ISBN: 9781003359364 (ebk)

DOI: 10.1201/9781003359364

Typeset in Times
by Deanta Global Publishing Services, Chennai, India

DM would like to dedicate this book to his parents for their constant encouragement and support throughout his life.

AKM would like to dedicate this book to Roopkatha, Sanjhbati, Manjulekha and Ahaan.

Contents

PART 3 Recent Advances in Modeling of Deformation and Fracture in Materials

PART 4 Progress in Experimental Approaches

PART 5 Future Research Directions

Preface

The relationship between materials and humans is as old as human civilization itself. However, the importance is of maximum significance today as we live in an age of technology that is extended to nano dimensions and, hence, to nanoscience and nanotechnology. Almost all devices and daily accessories are made of materials either natural or synthesized. We also have engineered materials developed for specific applications, especially in the strategic sector. However, no matter how good a material is, it eventually fails under the governance of external forces. It may happen instantaneously or over time. It is easy to conceptualize that deformation precedes fracture. In other words, there is a limiting condition up to which a given material can sustain under an externally applied force. Beyond this limiting condition, the material and the related structure fail by either catastrophic or gradual fracture. For example, the global healthcare system suffers a cost of $400 billion per annum due to osteoporotic fracture. Thus, if all other materials and fracture incidents of machinery, power plant boilers, etc. are included in the global costs of fracture the total could be well over $1 trillion per annum. This is why it is important to study the deformation and fracture of materials. This is also what defines the scope of the current book, *Deformation and Fracture in Materials: Advances in Experimental and Numerical Studies*. The book is intended for undergraduate, postgraduate, and PhD research scholars as well as the professional research community studying deformation and fracture.

The book is comprised of 18 chapters, distributed over five parts. Part 1 includes Chapters 1, 2, and 3 that focus on important theoretical approaches necessary to be understood to have a grasp of basic and advanced knowledge of deformation and fracture of different types of materials. Chapters 4, 5, 6, and 7 of Part 2 focus on the standard, conventional, and advanced techniques being utilized today for conducting experimental studies of deformation and fracture in metals, ceramics, polymers, and composites.

Part 3, Chapters 8, 9, 10, 11, 12, 13, and 14, focuses on recently emerging modeling approaches being utilized to comprehend deformation and fracture, e.g., path-independent integrals (Chapter 8), crack growth modeling (Chapter 9), nanoindentation using finite element method (Chapter 10) and creep (Chapter 11), channelling induced electric field in hybrid polymer nanocomposites (Chapter 12), and atomistic, MD simulations in nanocomposites (Chapter 13), and aluminum–graphene composites (Chapter 14).

The chapters of Part 4 focus mainly on advanced experimental techniques being utilized for deformation and fracture characterization of nickel-based superalloys at the macroscopic length scale (Chapter 15) and of metals (Chapter 16) and glasses (Chapter 17) at microstructural length scales. Finally, the concluding Chapter 18 focuses on how the recently emerging disciplines of AI and machine learning and their many variants can be utilized to comprehend nanoscale deformations in a given material. The chapters are enriched with a comprehensive survey of current literature and future research directions that need to be pursued.

This book sums up the basic theoretical and experimental techniques to advanced modeling approaches and evolving experimental methodologies for almost all materials, including metals, ceramics, polymers, and a wide variety of composite materials. The editors sincerely hope this book will be very useful for its intended readership and even beyond that.

The editors express their sincerest gratitude to the entire CRC team and reviewers, and especially Dr. Gagadeep Singh for invaluable help and support at each and every step of the uphill journey. The editors thank all the contributing authors, colleagues, students, institute, and university authorities, and especially family members, whom without this mammoth dream could not have seen the light of day. Thanks to all concerned.

Editors

Anoop Kumar Mukhopadhyay, born in 1958, is currently Director of Research and Education at VCentMedia, Kolkata, India; and Director of Materials Research at a Kolkata-based start-up. Prior to this he was Chief Scientist and Head of the Advanced Mechanical and Materials Characterization Division at CSIR–Central Glass and Ceramic Research, Kolkata (1986–2018); Professor in the Department of Physics with the Faculty of Science and Dean with the Faculty of Science at Manipal University Jaipur, Jaipur, Rajasthan, India (2018–2021); and Professor in the Department of Physics at the Sharda School of Basic Sciences and Research, Sharda University, Greater Noida, Uttar Pradesh, India (2022–2023). This book was written while he was at Sharda University. He earned his MA in physics and PhD in science from Jadavpur University, Kolkata, India and conducted postdoctoral studies at the University of Sydney, Australia (1990–1992), and Forschungszentrum Jülich, Germany (2000). He is the recipient of many awards and accolades including the Sir C V Raman Award of the Acoustical Society of India and the MRSI Medal from the Materials Research Society of India. He has handled national and international projects worth of few tens of crores of INR sponsored by DST, DST-SERB, DAE-BRNS, ISRO-RESPOND, and CSIR. His major areas of research include synthesis and characterization of nanomaterials, nano-biomaterials, nanoindentation, deformation and fracture of bulk glass, structural ceramics, functional ceramics, thin films, smart materials, armour materials, TBC and bio-coatings, multilayer thin films, synthetic biomaterials, and natural biomaterials. He also has a keen interest in high-strain-rate deformation of materials. His recent research interests also include the application of artificial intelligence and machine learning to the deformation and fracture of glass, ceramics, and other materials.
Google Scholar: https://scholar.google.co.in/citations?user=tcFVCpAAAAAJ&hl=en; total SCI/SCOPUS indexed peer-reviewed journal publications, over 200; patents, 8-plus; citations, 5194; h-index, 38; i-10 index, 123; 3 books, CRC Press, USA; PhD guidance, 10-plus; postdoc guidance, 3-plus.

Dhaneshwar Mishra is Associate Professor at the Department of Mechanical Engineering, Manipal University Jaipur, Rajasthan, India. He completed his PhD in mechanical engineering from Ajou University, Suwon, South Korea, in 2011. Mishra received the prestigious Brain Korea 21 (BK-21) scholarship from the Government of Korea for his PhD study and postdoctoral research at Ajou University. His domain of research is mechanics and materials in general, and defect mechanics of electronic materials in particular. He has published many journal articles in high-impact peer-reviewed journals, patents with the Indian Patent Office, book chapters, and conference proceedings, and presented his research output at numerous international conferences. Mishra completed his Master of Engineering with a specialization in engineering design in 2003 from Anna University Chennai, India, and his Bachelor of Engineering in Mechanical Engineering in 1998 from Motilal Nehru Regional Engineering College (currently known as MNNIT) Allahabad as an international candidate selected by the Government of India.

Ingrained with an exceptionally strong and diverse teaching experience, Mishra started his academic career in 1998 after completing his undergraduate in mechanical engineering as a lecturer at Kathmandu University, Dhulikhel, Nepal. He worked as an assistant professor at Ajou University, South Korea as well. In his 11 years of stay in South Korea, Mishra also worked as a senior researcher at Advanced Institutes of Convergence Technology, Seoul National University,

Suwon, and a postdoctoral researcher at Ajou University, Suwon. After joining Manipal University Jaipur in 2018, Mishra was instrumental in setting up the Multiscale Simulation Research Center (MSRC), a high-end computing facility for research scholars at Manipal University Jaipur. In addition, the undergraduate and graduate students mentored by Mishra at MSRC published high-quality papers and received lucrative research offers from reputed international universities based in the U.S. and Germany.

Contributors

Payel Bandyopadhyay
Department of Physics and Nanotechnology
College of Engineering and Technology
SRM Institute of Science and Technology
Kattankulathur, Tamil Nadu, India

Raman Bedi
Department of Mechanical Engineering
Dr B R Ambedkar National Institute of Technology
Jalandhar, Punjab, India

Manjima Bhattacharya
Advanced Mechanical and Materials Characterization Division
CSIR–Central Glass and Ceramic Research Institute
Kolkata, India

Anamitra Datta
Department of Metallurgy and Materials Engineering
Indian Institute of Engineering Science and Technology
Shibpur, Howrah, India

Shubhada S. Garnaik
Department of Mechanical Engineering
SUNY Korea
Incheon, Republic of Korea

Mohit Jain
Department of Metallurgy and Materials Engineering
Indian Institute of Engineering Science and Technology
Shibpur, Howrah, India

Jash Rana
Department of Mechanical and Aerospace Engineering
Institute of Infrastructure, Technology, Research and Management (IITRAM)
Ahmadabad, Gujrat, India

Mithilesh K. Dikshit
Department of Mechanical and Aerospace Engineering
Institute of Infrastructure, Technology, Research and Management (IITRAM)
Ahmadabad, Gujrat, India

Akhil Khajuria
Department of Mechanical Engineering
Dr B R Ambedkar National Institute of Technology Jalandhar
Punjab, India

Saleem Khan
R&D Sensor Division
Multi Nano Sense Technologies
Nagpur, Maharashtra, India

Reinhold Kienzler
Department of Production Engineering
University of Bremen
Bremen, Germany

Ashok Kumar
Department of Physics
School of Basic Sciences & Research
Sharda University
Greater Noida, Uttar Pradesh, India

Rajnish Kumar
Department of Mechanical Engineering
Dr B R Ambedkar National Institute of Technology
Jalandhar, Punjab, India

Manoj Kumar
Department of Mechanical Engineering
Dr B R Ambedkar National Institute of Technology
Jalandhar, Punjab, India

Amitava Basu Mallick
Department of Metallurgy and Materials Engineering
Indian Institute of Engineering Science and Technology
Shibpur, Howrah, India

Payel Maiti
Department of Applied Mechanics
Indian Institute of Technology Madras
Chennai, India

Dhaneshwar Mishra
Department of Mechanical Engineering
Manipal University Jaipur
Jaipur, Rajasthan, India

Anoop Kumar Mukhopadhyay
Department of Physics
Sharda School of Basic Sciences
Sharda University
Greater Noida, Uttar Pradesh, India

Neetu Chaudhary
Department of Mechanical and Aerospace Engineering
Institute of Infrastructure, Technology, Research and Management (IITRAM)
Ahmadabad, Gujrat, India

Anh Tay Nguyen
Department of Mechanical Engineering
Northwestern University
Evanston, Illinois, USA

Y. Eugene Pak
Department of Mechanical Engineering
SUNY Korea
Incheon, Republic of Korea

Harsha Pandey
Department of Mechanical Engineering
Manipal University Jaipur
Jaipur, Rajasthan, India

Vimal Kumar Pathak
Department of Mechanical Engineering
Manipal University Jaipur
Jaipur, Rajasthan, India

Sangeeta Rawal
Department of Physics
Sharda School of Basic Sciences and Research
Sharda University
Greater Noida, Uttar Pradesh, India

Abhishek Rizal
Department of Civil Engineering
Manipal University Jaipur
Jaipur, Rajasthan, India

Aniruddha Samanta
Electro Mineral Division
Carborundum Universal Limited
Kalamassery, Ernakulam, India

Arpana Pal Sharma
Department of Physics
Manipal University Jaipur
Jaipur, Rajasthan, India

Charanjeet Singh Tumrate
Department of Civil Engineering
Manipal University Jaipur
Jaipur, Rajasthan, India

Kulwant Singh
Skill Faculty of Engineering & Technology
Shri Vishwakarma Skill University
Palwal, Haryana, India

Ramanpreet Singh
Department of Mechanical Engineering
Manipal University Jaipur
Jaipur, Rajasthan, India

Ashish Kumar Srivastava
Department of Mechanical Engineering
Manipal University Jaipur
Jaipur, Rajasthan, India

Monika Srivastava
Department of Physics
Sharda School of Basic Sciences and Research
Sharda University
Greater Noida, Uttar Pradesh, India

Aman Taylor
Department of Mechanical Engineering
Manipal University Jaipur
Jaipur, Rajasthan, India

Sujal Laxmikant Vajire
Department of Mechanical Engineering
Manipal University Jaipur
Jaipur, Rajasthan, India

Uvais Valiyaneerilakkal
Department of Physics
Manipal University Jaipur
Jaipur, Rajasthan, India

Part 1

Mechanics and Physics of Deformation and Fracture in Materials

Theoretical Approaches

1 Theoretical Approaches in Deformation and Fracture of Materials

Dhaneshwar Mishra, Aman Taylor, and Anoop Kumar Mukhopadhyay

1.1 INTRODUCTION

Deformation refers to the change in size or shape of an object made up of various classes of materials. The deformation pattern in various classes of materials can be different. The deformation in materials that can be recovered after releasing the external load is termed elastic deformation. Elastic deformation in material is governed by Hook's law. The elastic deformation in material is followed by yielding and plastic deformation which is beyond the scope of Hook's law and mostly nonlinear in nature. Plastic deformation, which is also termed permanent deformation (i.e., not recoverable after releasing the load as well), is mostly observed in ductile materials such as metals.

Failure and fracture in material are the result of excessive deformation (beyond the permissible limit). Deformation in materials and structures beyond the permissible limit can also be termed as failure, which signifies the non-compliance of the structures or components to attain certain designed functions. Fracture is the separation of structure/components into two or more pieces. It starts with bond breaking at the atomic level, which grows leading to fracture at continuum scale. The mechanism of fracture is also different in different classes of materials. The brittle materials, in which plastic deformation is negligible, fractures into pieces catastrophically once the stress concentration reaches beyond their elastic limit. Ductile materials, such as mild steel rods, give enough signs before they fracture. The concepts of fracture and failure in materials and related mechanics improved considerably after the Second World War when the catastrophic failure of Liberty ships was observed.

Love [1] in 1892 described the condition of rupture for the first time. Leonardo da Vinci gave some explanations on material failure by measuring the strength of iron wires and concluded that the strength of the wire varies with the wire length. That was interpreted as the defect or flaw length affects the strength. These attempts to link flaw size and material strength were qualitative in nature.

Griffith [2] in 1920 published the relationship between the flaw size and the fracture stress to develop the quantitative connections between them. This result was derived based on the experiment conducted by Inglis [3] of stress analysis of an elliptical hole to the unstable crack propagation. Griffith used the first law of thermodynamics, which is basically on energy balance to formulate fracture theory.

According to Griffith's energy balance theory, the strain energy change due to the external critical loading is sufficient to overcome the surface energy of the material. This theory is basically applicable to perfect brittle material, which leads to catastrophic fracture. This theory is not applicable to metals where there is considerable plastic deformation before rupture. Therefore, G. R. Irwin [4] in 1948 modified Griffith's theory to make it applicable to metals by adding energy dissipated by local plastic flow. Orwan [5] also modified the original Griffith theory to accommodate

DOI: 10.1201/9781003359364-2

plastic deformation. Mott [6] extended Griffith's energy criterion of fracture to include rapidly propagating cracks.

The stress-based approach near the vicinity of the crack tip is another useful method to quantify the flaw size, the material parameter, and the stress intensity. This approach was developed by Westergaard [7], Irwin [8], and Williams [9] by developing a closed-form solution for the stresses around the crack tip in a linear, elastic, isotropic material. There has been much research carried out since then to explain the fracture mechanism in different classes of materials [10–24].

As explained earlier, fracture mechanics can be broadly classified into two broad categories, namely, linear elastic fracture mechanics (LEFM) and elastic–plastic fracture mechanics (EPFM). These two theories are applied to understand fracture mechanisms in ductile and brittle materials. Cotterell [25] in 2002 summarized the evolution of fracture mechanics very well in his review article "The Past Present and Future of Fracture Mechanics".

As material development is leading toward 2-D materials, understanding the mechanism of fracture is another important aspect of current research on fracture mechanics. In this regard, another review article on the development of fracture mechanics by Kiener et al. [26] in 2022 has beautifully summarized a century of development of fracture mechanics. Readers are referred to that reference for better understanding.

The scope of this chapter is to introduce the basic concepts of deformation and fracture mechanics. Therefore, it is divided into two broad parts. The first part of this chapter is dedicated to explaining the deformation mechanism, which includes both elastic and plastic deformations in various classes of materials along with the concepts of stress and strain tensors, stress–strain diagram, etc. The second part of the chapter is focused on explaining the basics of fracture mechanics. Both LEFM and EPFM are explained along with the resistance to fracture in materials and the brief on experimental evaluation of the fracture toughness.

1.2 MECHANICS OF DEFORMATION

This section is well studied and already part of different texts and reference books. Still, it is presented here for completeness and better understanding of the readers. Mechanics of materials books [27, 28] are referred to for the content presented here.

1.2.1 Concept of Stress

Let us assume that an external force (F_i) is applied to a given body. Deformation of any given body is the material's response to the external loading (Figure 1.1) and the loading environment. Every material intrinsically has the character to resist any external force applied to it. This resistance is expressed as the internal, resistive force that acts per unit area of the material's cross-section (A_i). This is the conceptual idea of stress. So, for any arbitrary force F_i applied to an area A_i this is termed as stress (σ_i). It is defined as

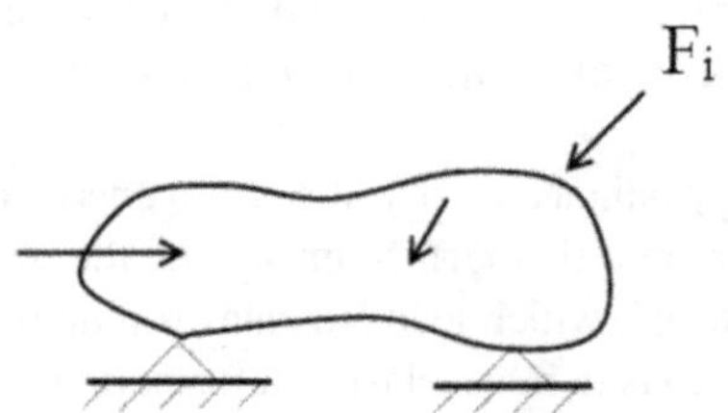

FIGURE 1.1 Schematic representation of body under external load.

$$\sigma_i = \frac{F_i}{A_i} \tag{1.1}$$

1.2.2 Concept of Stress Tensor

Let us assume that there is a solid body. For the sake of convenience and ease of understanding, let us represent the body in a Cartesian coordinate system (Figure 1.2). For the sake of generality, it is assumed that a number of forces are acting on this body. Since the force directions are arbitrarily chosen, their net resultant vanishes. Let $\Delta\, F = \Delta\, F_x + \Delta\, F_y + \Delta\, F_z = \Delta\, F_x\, \hat{x} + \Delta\, F_y\, \hat{y} + \Delta\, F_z \hat{z}$ be the force that acts on an elementary area small area $\Delta A_x = \Delta A_x \hat{x}$ parallel to the yz-plane.

Accordingly, we have the normal and shear stress magnitudes defined as

$$\sigma_{xx} = \lim_{\Delta A_x \to 0} \frac{\Delta F_x}{A_x} \quad \tau_{xy} = \lim_{\Delta A_x \to 0} \frac{\Delta F_y}{A_x} \quad \tau_{xz} = \lim_{\Delta A_x \to 0} \frac{\Delta F_z}{A_x} \tag{1.2}$$

In all these expressions, the first subscript represents the plane in which the stress is active. In a similar manner, the second subscript represents the force direction. In a similar fashion, we have the stresses for the other two corresponding Cartesian systems as

$$\tau_{yx} = \lim_{\Delta A_y \to 0} \frac{\Delta F_x}{A_y} \quad \sigma_{yy} = \lim_{\Delta A_y \to 0} \frac{\Delta F_y}{A_y} \quad \tau_{yz} = \lim_{\Delta A_y \to 0} \frac{\Delta F_z}{A_y}$$

and (1.3)

$$\tau_{zx} = \lim_{\Delta A_z \to 0} \frac{\Delta F_x}{A_z} \quad \tau_{zy} = \lim_{\Delta A_z \to 0} \frac{\Delta F_y}{A_z} \quad \sigma_{zz} = \lim_{\Delta A_z \to 0} \frac{\Delta F_z}{A_z}$$

The condition of static force equilibrium requires the following condition to be fulfilled, i.e., $\sigma xy = \sigma yx$, $\sigma yz = \sigma zy$, and $\sigma xz = \sigma zx$. This equivalence reduces the number of actual independent scalar quantities from 9 to 6. These are arranged in an ordered 3×3 matrix form. This is called stress tensor (σ_{ij}) as

$$\sigma_{ij} = \begin{bmatrix} \sigma_{xx} & \tau_{xy} & \tau_{xz} \\ \tau_{yx} & \sigma_{yy} & \tau_{yz} \\ \tau_{zx} & \tau_{zy} & \sigma_{zz} \end{bmatrix} \tag{1.4}$$

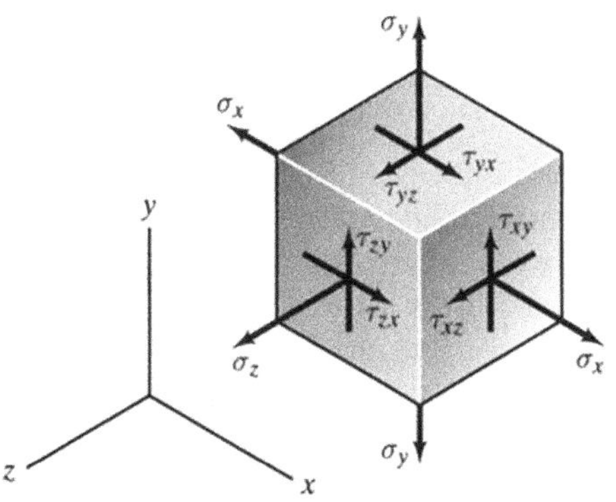

FIGURE 1.2 Body under 3-D state of stress [28].

1.2.3 Concept of Strain Tensor

The result of very small stress application is deformation at the initial stage. This is measured as the amount of deformation per unit dimension. So, deformation in length can be measured as the change in length divided by the original length where the change in length is measured in the direction in which the force is applied. This is called the normal strain or, equivalently speaking, longitudinal strain. Thus, if the force is applied in the X-direction and longitudinal strain is measured also along the X-direction, it is denoted as $_{xx}$. Here also the first suffix reflects the direction in which the strain is measured. Similarly, the second suffix reflects the direction of force application, per se.

In addition, there can be dimension changes that occur in the other two directions, i.e., Y and Z, which are mutually perpendicular to the direction of force application, and X as per the present discussion. The corresponding strains are represented as $_{yx}$ and $_{zx}$. Here again, the first suffix reflects the direction in which the strain is measured. Similarly, the second suffix reflects the direction of force application, per se. However, if Y is the direction of force application, the strains are represented as $_{yy}$, $_{zy}$, and $_{xy}$. It follows that if Z is the direction of force application, the strains are represented as $_{zz}$, $_{yz}$, and $_{xz}$. This points to the fact that application of any force to a solid object can simultaneously cause, at least in principle, translation, rotation, and deformation of the solid body. Hence, it becomes convenient to define a common vector function $u\ (x, y, z)$ that can describe the relative changes between the initial and final states of each point of the given solid body. This function is defined as

$$\boldsymbol{u}(x,y,z)=\begin{bmatrix} u(x,y,z) \\ v(x,y,z) \\ w(x,y,z) \end{bmatrix} \tag{1.5}$$

The spatial rates of all corresponding and partial changes in $u\ (x, y, z)$ can be represented as

$$\nabla u=\begin{bmatrix} \frac{\partial u}{\partial x} & \frac{\partial u}{\partial y} & \frac{\partial u}{\partial z} \\ \frac{\partial v}{\partial x} & \frac{\partial v}{\partial y} & \frac{\partial v}{\partial z} \\ \frac{\partial w}{\partial x} & \frac{\partial w}{\partial y} & \frac{\partial w}{\partial z} \end{bmatrix} \tag{1.6}$$

It is trivial to show that $\nabla\ u + \nabla\ u\ T = 2e$, where

$$e = {}_{kl} = \begin{bmatrix} {}_{xx} & {}_{xy} & {}_{xz} \\ {}_{yx} & {}_{yy} & {}_{yz} \\ {}_{zx} & {}_{zy} & {}_{zz} \end{bmatrix} \tag{1.7}$$

The tensor $_{kl}$ is defined as the strain tensor.

As mentioned earlier the normal longitudinal strain is E_{xx} for the X-direction. Similarly, the normal longitudinal strain is ε_{yy} for the Y-direction. Further, the normal longitudinal strain is $_{zz}$ for the Z-direction. All the other terms $_{ij}$ $(i = x, y, z; j = x, y, z; i \neq j)$ are the shear terms as explained earlier. The off-diagonal terms are equal to one-half of the corresponding engineering shear strain, e.g., $\epsilon_{xy} = \gamma_{xy}/2$. These are related to the actual deformations as

$$_{xy}=\frac{1}{2}\left(\frac{\partial u}{\partial y}+\frac{\partial v}{\partial x}\right) \quad _{xz}=\frac{1}{2}\left(\frac{\partial u}{\partial z}+\frac{\partial w}{\partial x}\right) \quad _{yz}=\frac{1}{2}\left(\frac{\partial v}{\partial z}+\frac{\partial w}{\partial y}\right) \tag{1.8}$$

For the strain tensor $_{kl}$, $_{xy} = {}_{yx}$, $_{xz} = {}_{zx}$, and $_{yz} = {}_{zy}$.

1.2.4 Generalized Strain Dependency of Stress

Each term of $_{kl}$ gives rise to the corresponding stress σ_{pq} in the stress tensor σ_{ij}. The generalized dependency of stress on strain can be expressed as, e.g., $\sigma_{ij} = C_{ijkl}\ _{kl}$, where C_{ijkl} are the material constants. Thus, as a typical example, σ_{xx} can be explicitly expressed as

$$\sigma_{xx} = C_{xxxx}\ _{xx} + C_{xxxy}\ _{xy} + C_{xxxz}\ _{xz} + C_{xxyy}\ _{yy} + C_{xxyz}\ _{yz} + C_{xxzx}\ _{zx} + C_{xxzy}\ _{zy} + C_{xxzz}\ _{zz} \quad (1.9)$$

It is well known that there are (3^4), i.e., 81 terms possible for C_{ijkl}. However, out of these only 21 terms are there that are uniquely different from each other. Conversion of the off-diagonal strain terms to engineering shear strains leads to a contracted notation relationship between stress and strains as

$$\begin{bmatrix} _{xx} & 2\ _{xy} & 2\ _{xz} \\ 2\ _{yx} & _{yy} & 2\ _{yz} \\ 2\ _{zx} & 2\ _{zy} & _{zz} \end{bmatrix} = \begin{bmatrix} _{xx} & \gamma_{xy} & \gamma_{xz} \\ \gamma_{yx} & _{yy} & \gamma_{yz} \\ \gamma_{zx} & \gamma_{zy} & _{zz} \end{bmatrix} \quad (1.10)$$

The resulting matrix is no longer a tensor. The reason is that it can no longer follow the coordinate transformation rules.

1.2.5 Generalized Stress Dependency of Strain

For linear elastic isotropic materials (LEIM), the properties are direction invariant. So, they are characterized by three unique properties: Young's modulus E, Poisson's ratio v, and the shear modulus G [={(E)}/{2(1+υ)}]. Thus, for LEIM we have the inverse dependency, i.e., stress dependencies of strain as

$$\begin{aligned} _{xx} &= \frac{1}{E}\left[\sigma_{xx} - v\left(\sigma_{yy} + \sigma_{zz}\right)\right] \\ _{yy} &= \frac{1}{E}\left[\sigma_{yy} - v\left(\sigma_{zz} + \sigma_{xx}\right)\right] \\ _{zz} &= \frac{1}{E}\left[\sigma_{zz} - v\left(\sigma_{xx} + \sigma_{yy}\right)\right] \end{aligned} \quad (1.11)$$

Here, Poisson's ratio (v) is defined as the ratio of lateral to longitudinal strains:

$$(v) = (\ _{lateral})/(\ _{longitudinal}) \quad (1.12)$$

1.3 DEFORMATION

There can be two types of deformation possible in a material subjected to external force. The first one is elastic deformation. If the given material regains all of its original macroscopic dimensions when the externally applied force is removed, it is called elastic deformation. The second is plastic deformation. If the given material does not regain all of its original macroscopic dimensions when the externally applied force is removed, it is called plastic deformation. This is a permanent deformation that stays in the material even when the externally applied force is removed.

1.3.1 Elastic Deformation

As mentioned earlier, the original shape and dimensions are recovered in elastic deformation. This happens when the externally applied force is withdrawn. Thus, elastic deformation can be of two types. The first one is linear. This occurs mainly in ceramics and to a great extent in metals as well. The limit up to which the materials show the linear elastic behavior is called the elastic limit. The corresponding stress and strain are termed linear stress and linear strain. In other words, the linear elastic deformation is reversible when the externally applied force is withdrawn. Hence, such deformation is non-permanent. Therefore, it needs to be recalled that the classical Hooke's law is valid only within the linear elastic limit. As a result, in the region of validity of Hooke's law, the rate of deformation is proportional to the rate of force application. From a physics point of view, such deformation is a result of only slight stretching of elastic bonds between the relevant atoms. In this limit of small deformation, the bond lengths regain their equilibrium values as soon as the externally applied force is withdrawn. The second one is non-linear in nature. It occurs mainly in polymers and also to some extent in materials such as gray cast iron and cementitious materials such as concrete. Thus, Hooke's law is defined as

$$\sigma = E\varepsilon \tag{1.13}$$

Here $\sigma$is stress, E is Young's modulus and ε is strain. The higher the E value, the stiffer the material. So, it is more difficult to plastically deform such a stiffer material. Similarly, we have

$$\tau = G\gamma \tag{1.14}$$

Here, G is the shear modulus, τ is the shear stress, and γ is the shear strain. Thus, E and G are related as $E = 2G(1 + \upsilon\ )$. Ceramic materials have mostly ionic bonding. So, they possess strongly bonded atoms. Therefore, generally speaking, they are rather difficult to deform. So, they have very small (e.g., 10^{-3}) strain values since they have very limited (almost zero) ability to deform plastically. Consequently, they have a high Young's modulus as the stress is very high. These modulus values are much higher than those of metals and especially those of polymers. Generally, E is anisotropic for single crystals but isotropic for polycrystalline materials, as expected. Further, the modulus of resilience is defined as

$$U_r = \frac{1}{2}\sigma_y\varepsilon_y \tag{1.15}$$

Here, σ_y is yield stress and ε_y is yield strain. On the other hand, if the elastic recovery process becomes time-dependent, it is called an anelastic deformation behavior. In a similar fashion, the time dependence of elastic deformation in especially polymers and amorphous materials (e.g., glass) is termed as viscoelastic deformation.

1.3.2 Plastic Deformation

Plastic deformation is a permanent deformation. So, it is irreversible in nature. In this case, bond breakage occurs. Hence, the elastic recovery fails to happen. It happens mostly in metals as well as many polymers but hardly ever in ceramics. Only under very high hydrostatic stress state conditions under special situations, e.g., as in the material just below the tip of a nanoindenter, limited localized plasticity occurs at even room temperature in both ceramics and metallic glasses. The yield stress values for linear and non-linear metals are taken at 0.2% and 0.5% strain beyond which the occurrence of ductile necking leads to tensile failure in metals and sharp brittle fracture leads to brittle failure in other brittle solids, e.g., mainly ceramics and glasses, which exhibit high engineering stress but very little engineering strain. But just the opposite is true for mainly metals. Therefore, they can absorb much higher energy and hence exhibit the highest toughness (≈50–300 MPa.m$^{0.5}$). On the contrary, the least toughness is exhibited by mostly the unreinforced polymers, e.g., ≈0.5 to 1 MPa.m$^{0.5}$. As expected, due to its brittleness, the ceramics exhibit intermediate toughness of rather small magnitude, e.g., ≈1–3 MPa.m$^{0.5}$. It is shown in the next section that toughness (K) is related to fracture strength (σ) and defect dimension (a) as

$$K = Y \sigma a^{0.5} \quad (1.16)$$

where Y is a loading geometry-dependent term.

1.4 FRACTURE MECHANICS

Man-made structures exhibiting fracture and failure is a very common phenomenon, as technological advancement has made things complex. For example, major airplane crashes these days are a result of their complex technology and working environment. Failures occur for many reasons, including uncertainties in the loading or environment, defects in the materials, inadequacies in design, and deficiencies in construction or maintenance. Designing against fracture has its own technology, and this is a very active field of study. The basic challenge in designing against fracture in high-strength materials is that the existence of local cracks or voids might vary the local stresses to such an amount that the designers' meticulous stress assessments are insufficient. When a fracture reaches a critical length, it can spread catastrophically across the structure, despite the fact that the gross stress is significantly lower than what would ordinarily produce yield or failure in a tensile specimen. The term "fracture mechanics" refers to an important subfield of solid mechanics in which we assume the presence of cracks/voids/defects intrinsic to the material and would like to develop quantitative relationships between the crack length, the material's inherent resistance to crack growth, and the stress at which the crack propagates rapidly enough to cause structural failure.

The aim of this section is to provide the basics of fracture mechanics to the readers so that those interested can follow the developments of this subject through recent literature. Hence, mostly Anderson [29] and Ewalds and Wanhill [30] were the references used in the following sections.

1.4.1 Linear Elastic Fracture Mechanics (LEFM)

LEFM is the fundamental theory of fracture, established by Griffith (1921–1924) and developed in its essential form by Irwin (1957, 1958) and Rice (1968) [3–7, 20, 21]. LEFM is a highly simplified but profound theory that deals with severe cracks in elastic bodies. If specific requirements are satisfied, LEFM applies to any material. These requirements rely on the presence of all basic ideal conditions examined in LEFM, in which all materials are elastic except for a vanishingly small region (a point) at the crack tip. In fact, the stress around the fracture tip is so high that some type of inelasticity must occur in the immediate region of the crack tip.

1.4.1.1 Griffith's Energy Balance Approach

Griffith proposed that there is a simple energy balance consisting of a loss in potential energy within the stressed body owing to fracture extension, which is balanced by an increase in surface energy due to increasing crack surface.

Griffith's theory establishes the theoretical strength of brittle material and the relationship between fracture strength and flaw size "a". The initial strain energy for the uncracked plate per thickness is given as

$$U_i = \int_A \frac{\sigma^2}{2E} dA \quad (1.17)$$

When a crack of size 2a is formed, the tensile force on an element "ds" on an elliptic hole is reduced to zero. The elastic strain energy released per unit width as a result of the increase in the length of the crack is given by

$$U_a = -4\int_0^a \frac{1}{2}\sigma.dx.v \tag{1.18}$$

where displacement $v = \left(\sigma/_E\right) a\,sin\theta$. Using $x = a\cos\theta$, $U_a = \frac{\pi\sigma^2 a^2}{E}$. The external work = $U_w = \int_\delta Fdy$, where F is resultant force = σ × Area, and δ is total relative displacement. The internal energy of the body is $U_p = U_i + U_a - U_w$. The surface energy increase due to creation of a new surface can be given as $U_\gamma = 4a\gamma_s$. The total elastic energy of the cracked plate can therefore be given as

$$U_t = \int_A \frac{\sigma^2}{2E}dA + \frac{\pi\sigma^2 a^2}{E} - \int_\delta Fdy + 4a\gamma_s \tag{1.19}$$

The variation of crack extension should be minimum

$$\frac{dU_t}{da} = 0 \rightarrow -\frac{2\pi\sigma^2 a}{E} + 4\gamma_s = 0 \tag{1.20}$$

Denoting σ as σ_f during fracture can be expressed as $\sigma_f = \left(\frac{2E\gamma_s}{\pi a}\right)^{1/2}$ for plane stress and $\sigma_f = \left(\frac{2E\gamma_s}{\pi a\left(1-v^2\right)}\right)^{1/2}$ for plane strain. Glass, mica, diamond, and refractory metals all obey the Griffith principle, which is brittle in nature.

1.4.1.2 Stress Intensity Approach and Modes of Fracture

In theory, the stress ahead of a sharp crack tip becomes limitless and cannot be utilized to explain the situation around a fracture. Fracture mechanics is used to define the stresses on a fracture, often utilizing a single parameter to represent the entire loading condition at the crack tip. Many parameters have been established. Since the plastic zone at the crack tip is tiny in comparison to the crack length, the stress state at the crack tip is the consequence of elastic forces within the material and in the domain of LEFM and may be defined using the stress intensity factor, *K*. Although a crack's load might be arbitrary, Irwin [4] discovered that every condition could be reduced to a combination of three independent stress intensity factors:

Mode I – Opening mode: The phenomenon where the two surfaces of a crack move away from each other symmetrically with respect to the plane occupied by the crack prior to the deformation is known as mode I (Figure 1.3). This mode of crack propagation is caused by normal stresses that are perpendicular to the crack plane. In mode I, the state of stress ahead of the crack tip (Figure 4 1.4) can be expressed as

$$\begin{aligned}
\sigma_x &= \frac{K_I}{\sqrt{2\pi r}}\cos\left(\frac{\theta}{2}\right)\left[1-\sin\left(\frac{\theta}{2}\right)\sin\left(\frac{3\theta}{2}\right)\right] \\
\sigma_y &= \frac{K_I}{\sqrt{2\pi r}}\cos\left(\frac{\theta}{2}\right)\left[1+\sin\left(\frac{\theta}{2}\right)\sin\left(\frac{3\theta}{2}\right)\right] \\
\tau_{xy} &= \frac{K_I}{\sqrt{2\pi r}}\sin\left(\frac{\theta}{2}\right)\left[\cos\left(\frac{\theta}{2}\right)\cos\left(\frac{3\theta}{2}\right)\right]
\end{aligned} \tag{1.21}$$

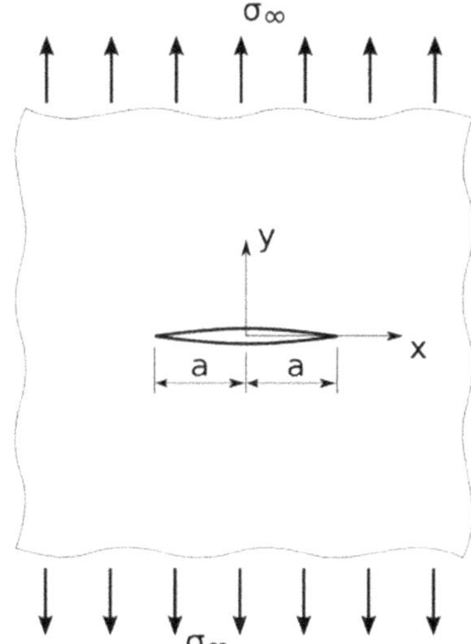

FIGURE 1.3 Infinite plate with central crack under far-field loading [29].

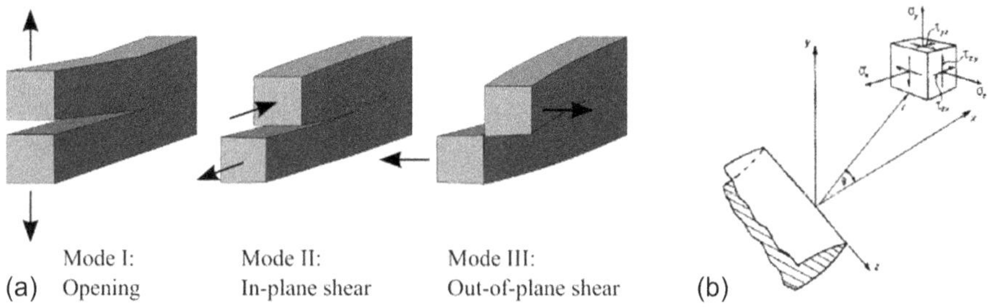

FIGURE 1.4 (a) Modes of fracture, (b) state of stresses at the crack tip [29].

where σ_x and σ_y are the normal stresses in the x- and y-directions, and τ_{xy} is the in-plane shear stress. K_I is the mode I stress intensity factor.

The displacement field ahead of the crack tip for linear elastic material can be expressed as

$$\begin{aligned} u &= \frac{2(1-v)}{E} K_1 \sqrt{\frac{r}{2\pi}} \cos\left(\frac{\theta}{2}\right)\left[\frac{\kappa-1}{2} + \sin^2\left(\frac{\theta}{2}\right)\right] \\ v &= \frac{2(1-v)}{E} K_1 \sqrt{\frac{r}{2\pi}} \sin\left(\frac{\theta}{2}\right)\left[\frac{\kappa-1}{2} + \cos^2\left(\frac{\theta}{2}\right)\right] \end{aligned} \tag{1.22}$$

where u and v are the displacements in the x- and y-directions; and E and υ are the material's elastic modulus and Poisson's ratio, respectively. The bulk modulus is $k = (3-4\upsilon)$ for the plane stress condition, and $k = \frac{(3-\upsilon)}{(1+\upsilon)}$ for the plane strain condition.

Mode II – Sliding mode: In the sliding mode, the crack surfaces slide over each other in opposite directions but within the same plane. This type of deformation results from shear stresses applied in-plane (Figure 1.5). The state of stress ahead of the crack tip in mode II can be expressed as

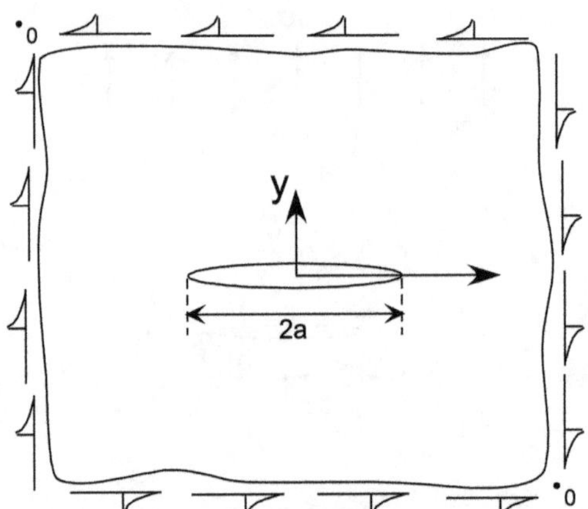

FIGURE 1.5 Infinite plate with central crack under far-field in-plane shear load [29].

$$\sigma_x = \frac{K_{II}}{\sqrt{2\pi r}} \cos\left(\frac{\theta}{2}\right)\left[\cos\left(\frac{\theta}{2}\right)\left(2+\cos\left(\frac{\theta}{2}\right)\cos\left(\frac{3\theta}{2}\right)\right)\right]$$

$$\sigma_y = \frac{K_{II}}{\sqrt{2\pi r}} \cos\left(\frac{\theta}{2}\right)\left[\sin\left(\frac{\theta}{2}\right)\cos\left(\frac{3\theta}{2}\right)\right] \tag{1.23}$$

$$\tau_{xy} = \frac{K_{II}}{\sqrt{2\pi r}} \cos\left(\frac{\theta}{2}\right)\left[1-\sin\left(\frac{\theta}{2}\right)\sin\left(\frac{3\theta}{2}\right)\right]$$

The displacement fields can be expressed as

$$u = \frac{K_{II}}{E}\sqrt{\frac{r}{2\pi}}(1+\nu)sin\left(\frac{\theta}{2}\right)\left[\kappa+2+cos(\theta)\right]$$

$$v = \frac{K_{II}}{E}\sqrt{\frac{r}{2\pi}}(1+\nu)cos\left(\frac{\theta}{2}\right)\left[\kappa+2+cos(\theta)\right] \tag{1.24}$$

Mode III – Tearing mode: In the tearing mode, the crack surfaces move parallel to the crack front within the crack plane, resulting from shear stresses acting parallel to the crack plane, which is perpendicular to the direction of the applied load. The stresses ahead of the crack tip in mode III are given by

$$\tau_{xz} = \frac{K_{III}}{\sqrt{2\pi r}} sin\left(\frac{\theta}{2}\right)$$

$$\tau_{yz} = \frac{K_{III}}{\sqrt{2\pi r}} cos\left(\frac{\theta}{2}\right) \tag{1.25}$$

$$\sigma_x = \sigma_y = \tau_{xy} = 0$$

and displacement fields are given by

$$w = \frac{K_{III}}{G}\sqrt{\frac{2r}{\pi}} sin\left(\frac{\theta}{2}\right)$$

$$\mathrm{u} = \mathrm{v} = 0 \tag{1.26}$$

$$K_{III} = \tau_o\sqrt{\pi a}$$

Stress Intensity Factor

At θ = 0, stress field on the crack plane in mode I can be expressed as

$$\sigma_{xx} = \sigma_{yy} = \frac{K_I}{\sqrt{2\pi r}}, \tag{1.27}$$

and the shear stress τ_{xy} turns out to be zero. At the crack tip, the far-field normal load causes a 1/r singularity. The distant boundary conditions influence stresses far from the fracture tip.

The amplitude of the crack tip singularity is defined by the stress intensity factor. This indicates that the strains around the fracture tip rise proportionally to *K*. The crack tip conditions are defined by the stress intensity factor, *K*; that is, given *K* is known, corresponding stresses, strains, and displacements may be calculated as a function of *r*. With *r* = *a* (half-length of the fracture) and θ= 0, the mode I stress intensity factor, $K_I = \sigma\sqrt{\pi a}$.

1.4.1.3 Relation between Energy Release Rate (*G*) and Stress Intensity Factor (*K*)

To explain fracture behavior, two parameters – the energy release rate (*G*) and the stress intensity factor (*K*) – are commonly used. The energy release rate quantifies the net change in potential energy and fracture extension, whereas the stress intensity factor correlates stress, strain, and displacement near the crack tip with crack extension. The energy release rate defines the global behavior, whereas *K* is a local characteristic. For linear elastic materials, *K* and *G* for mode I can be correlated as

$$G = \frac{K_I^2}{E} \text{ for plane stress and}$$
$$G_I = \frac{K_I^2\left(1-v^2\right)}{E} \text{ for plane strain} \tag{1.28}$$

1.4.1.4 Crack Tip Plasticity

In theory, the tension at the crack tip is limitless, which is referred to as stress singularity. This is not feasible in practice. In actual materials, the stress concentration at the crack tip is shown to be limited yet greater than the nominal stress applied to the specimen. This is due to plastic deformation at the fracture tip (blunt crack tip). Its deformation is principally determined by the applied stress in the appropriate direction, the fracture length, and the specimen shape. Irwin [4] calculated the plastic zone size at the fracture tip by correlating the material's yield strength to the far-field stress field normal to the crack plane:

$$\sigma_{ys} = \frac{K_I}{\sqrt{2\pi r_y}}, r_y = \frac{1}{2\pi}\left(\frac{K_I}{\sigma_{ys}}\right)^2, or\ r_y = \frac{a}{2}\left(\frac{\sigma}{\sigma_{ys}}\right)^2 \tag{1.29}$$

Stresses are redistributed to maintain balance in the presence of yielding. The elastic analysis represents forces that cannot be represented in an elastoplastic analysis because stresses cannot exceed the yield strength. The redistribution of excessive force necessitates an expansion of the plastic zone. This can happen if the material directly in front of the plastic zone is under increased stress. Irwin postulated that the fracture behaves as if it were greater than its true physical size, which may be stated as

$$a_{eff} = a + r_y \tag{1.30}$$

The effective stress intensity can be obtained by inserting a_{eff} into the *K* expression as

$$K_{eff} = \lambda\left(a_{eff}\right)\sigma\sqrt{\pi a_{eff}} \tag{1.31}$$

Equation 1.31 accommodates the small-scale yielding near the crack tip. This cannot take care of large-scale plastic deformation, which is the scope of elastic–plastic fracture mechanics.

1.4.2 Elastic–Plastic Fracture Mechanics (EPFM)

Under high load working circumstances, most engineering materials exhibit some nonlinear elastic and inelastic behavior. The assumptions of linear elastic fracture mechanics may not hold true in such cases, namely:

- the plastic zone at a crack tip may have the same order of magnitude as the crack size, and
- the size and form of the plastic zone may vary as the applied load and crack length increase.

Therefore, a more general theory of crack growth is needed for elastic–plastic materials that can account for:

- the local conditions for initial crack growth include the nucleation, growth, and coalescence of voids (decohesion) at a crack tip; and
- a global energy balance criterion for further crack growth and unstable fracture.

The notion of LEFM cannot be used to calculate fracture behavior for elastoplastic materials. In such instances, the route-independent integrals J-, M-, and L- are employed to investigate defect development. Another essential approach in this scenario is crack tip opening displacement (CTOD), which is used to measure the fracture characteristics. Information on these strategies can be found in Chapter 9 of this book.

1.4.3 Crack Growth Resistance in Materials (R-Curve)

Irwin's crack extension resistance curve, crack growth resistance curve, or R-curve was an early attempt in the direction of elastic–plastic fracture mechanics. This curve recognizes the fact that the resistance to fracture rises with crack size in elastic–plastic materials. The R-curve is a plot of the total energy dissipation rate as a function of crack size that may be used to investigate the processes of slow stable crack formation and unstable fracture (Figure 1.6). Unfortunately, the R-curve was not widely employed in applications until the early 1970s. The major reasons appear to be that the R-curve is dependent on the geometry of the specimen and the crack producing force may be difficult to determine. Details on the R-curve can be read in any standard fracture mechanics book. Readers can also follow specific literature to understand the R-curve behavior of any specific material under specific boundary conditions.

1.5 CONCLUSION

To summarize, this chapter provides, first, the conceptual foundation of deformation. Next, it provides the basic concepts of fracture and especially fracture mechanics. The fundamental concepts of stress, strain, deformation, elastic deformation, linear elastic deformation, non-linear elastic deformation, plastic deformation, inelastic deformation, and viscoelastic deformation are explained. The conceptual developments and mathematical formulations of both stress tensor and strain tensor are shown. Their interdependencies are explained. Further, the basic concepts and definitions of elastic modulus, modulus of resilience, and shear modulus are given. Furthermore, how the different moduli of a solid are related is explained. Finally, how these deformation features correlate

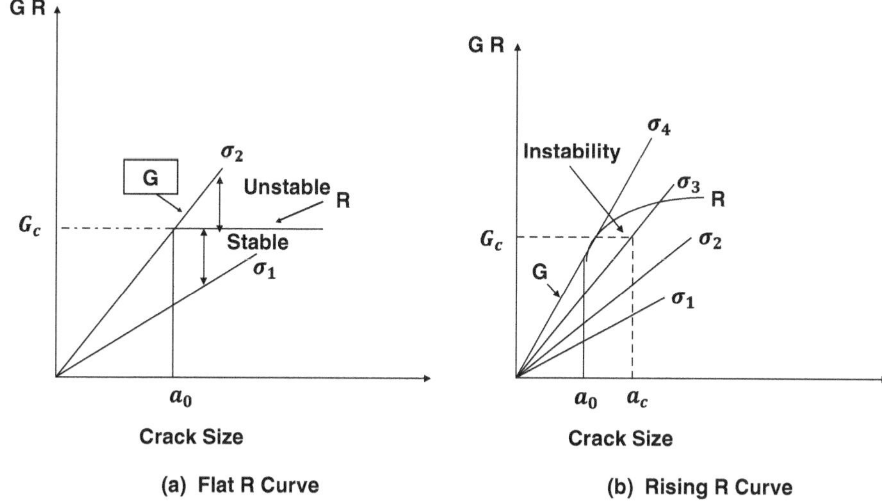

FIGURE 1.6 Schematic driving force/R-curve diagram [29].

to tensile strength, fracture strength, and toughness is explained regarding deformation. A huge amount of information is systematically developed to establish the fundamental concepts of fracture mechanics and its subsequent historical developments through the last several decades. The evaluation of stress intensity factors for mode I, mode II, and mode III loading is emphasized. The corresponding experimental techniques are also elaborated. Finally, why fracture mechanics is so important for materials engineering is hinted at.

ACKNOWLEDGEMENTS

The author AKM acknowledges the kind support and encouragement received from various directors at the Council of Scientific and Industrial Research (CSIR)–Central Glass and Ceramic Research Institute (CGCRI), Kolkata, India, when parts of the research work was carried out. The authors gratefully acknowledge the various infrastructural support received from various divisions of CSIR–CGCRI. The authors also acknowledge the financial support received from Indian sponsoring agencies, including the CSIR, UGC, DST(SERB), DAE-BRNS, and IPR, of the Government of India (GOI) in terms of various project sponsorships. Finally, AKM acknowledges the kind and continued support received from the authorities of Sharda University, Greater Noida, Uttar Pradesh, India.

DM acknowledges the support provided by the Multiscale Simulation Research Center (MSRC) at Manipal University Jaipur.

REFERENCES

1. Love A.E.H., *A Treatise on Mathematical Theory of Elasticity*. Dover Publications, New York, 1944.
2. Griffith A.A., The Phenomena of Rupture and Flow in Solids, *Philosophical Transactions Series A*, 221, 163–198, 1920.
3. Inglis C.E., Stresses in a Plate due to Presence of Cracks, and Sharp Corners, *Transactions of Institute of Naval Architects*, 55, 219–241, 1913.
4. Irwin G.R., *Fracture Dynamics, Fracturing in Metals*. American Society of Metals, Cleveland, 147–166, 1948.
5. Orwan E., Fracture and Strength of Solids, *Reports on Progress in Physics*, XII, 185–232, 1948.
6. Mott N.F., Fracture of Metals, Theoretical Considerations, *Engineering*, 165, 16–18, 1948.
7. Westergaard H.M., Bearing Pressure and Cracks, *Journal of Applied Mechanics*, 6, 49–53, 1939.

8. Irwin G.R., Analysis of Stress and Strain Near the Crack Traversing a Plate, *Journal of Applied Mechanics*, 24, 361–364, 1957.
9. Williams M.L., On the Stress Distribution at the Base of a Stationary Crack, *Journal of Applied Mechanics*, 24, 109–114, 1957.
10. Rice J.R., A Path Independent Integral and the Approximate Analysis of Strain Concentration by Notches and Cracks, *Journal of Applied Mechanics*, 35(2), 379–386, 1968, doi:10.1115/1.3601206.
11. Barenblatt G.I., The Mathematical Theory of Equilibrium Cracks in Brittle Fracture, (PDF), *Advances in Applied Mechanics*, 7, 55–129, 1962, doi:10.1016/s0065-2156(08)70121-2, ISBN 9780120020072.
12. Dugdale D.S., Yielding of Steel Sheets Containing Slits, *Journal of the Mechanics and Physics of Solids*, 8(2), 100–104, 1960, doi:10.1016/0022-5096(60)90013-2.
13. Willis J.R., A Comparison of the Fracture Criteria of Griffith and Barenblatt, *Journal of the Mechanics and Physics of Solids*, 15(3), 151–162, 1967, doi:10.1016/0022-5096(67)90029-4.
14. Liu, M., et al., An Improved Semi-Analytical Solution for Stress at Round-Tip Notches (PDF), *Engineering Fracture Mechanics*, 149, 134–143, 2015.
15. Crack Tip Plastic Zone Size. *Handbook for Damage Tolerant Design*. LexTech, Inc. Retrieved 20 November 2016.
16. Mishra D., Park C.Y., Yoo S.H., Pak Y.E., Closed-form Solution for Elliptical Inclusion Problem in Antiplane Piezoelectricity with Far-Field Loading at an Arbitrary Angle, *European Journal of Mechanics – A/Solids*, 40, 186–197, 2013.
17. Pak Y.E., Mishra D., Yoo S.H., Closed-form Solution for a Coated Circular Inclusion under Uniaxial Tension, *Acta Mechanica*, 223, 937–951, 2012.
18. Seo S.Y., Mishra D., Park C.Y., Yoo S.-H., Pak Y.E., Energy Release Rates for a Misfitted Spherical Inclusion under Far-Field Mechanical and Uniform Thermal Loads, *European Journal of Mechanics – A/Solids*, 49, 169–182, 2015.
19. Eugene Pak Y., Mishra D., Yoo S.H., Energy Release Rates for Various Defects under Different Loading Conditions, *Journal of Mechanical Science and Technology*, 26, 3549–3554, 2012.
20. Mishra D., Yoo S.H., Park C.Y., et al., Elliptical Inclusion Problem in Antiplane Piezoelectricity: Stress Concentrations and Energy Release Rates, *International Journal of Fracture*, 179, 213–220, 2013.
21. Pandey H., Fernandes D., Tumrate C.S., Mishra D., Numerical Investigation on Fracture Behavior in Novel Material Structure Inspired from First Hierarchy of Human Bone, *Materials Today: Proceedings*, 62–63, 1500–1504, 2022.
22. Pandey H., Fernandes D., Khandelwal A., Mishra D., Finite Element Analysis of Fatigue and Fracture Behavior in Idealized Airplane Wing Model with Embedded Crack Under Wind Load, *IOP Conference Series: Earth and Environmental Science*, 796, 012071, 2021.
23. Mishra D., Tumrate C.S., Mukhopadhyay A.K., Failure Behavior and Residual Strength of the Composite Sandwich Panels Subjected to Compression after Impact Testing, in *Sandwich Composites: Fabrication and Characterization*, CRC Press, Taylor and Francis, 75–97, 2022.
24. Singh A.P., Tailor A., Tumrate C.S., Mishra D., Crack Growth Simulation in a Functionally Graded Material Plate with Uniformly Distributed Pores using Extended Finite Element Method, *Materials Today: Proceedings*, 60, 602–607, 2022.
25. Kiener D., Han S.M., 100 Years after Griffith: From Brittle Bulk Fracture to Failure in 2D Materials, *MRS Bulletin*, 47(8), 792–799, 2022.
26. Cotterell B., The Past, Present, and Future of Fracture Mechanics, *Engineering Fracture Mechanics*, 69(5), 533–553, 2002.
27. Hibbeler R.C., *Statics and Mechanics of Materials in SI Units*. Pearson Higher Ed, 2018.
28. Beer F.P., Johnston E.R., DeWolf J.T., Mazurek D.F., *Mechanics of Materials*. 7th edition. McGraw-Hill Education Ltd., New York, 2015.
29. Anderson T.L., *Fracture Mechanics: Fundamentals and Applications*. CRC Press, 2017.
30. Ewalds H.L., Wanhill R.J., *Fracture Mechanics*, Edward Arnold, Delftse Uitgevers Maatschappij, 1984.

2 The Physics of Deformation Behavior in Nanoindentation Studies of Materials

Payel Maiti, Manjima Bhattacharya, and Anoop Kumar Mukhopadhyay

2.1 INTRODUCTION

The basics of nanoindentation have been described in our earlier publications [1–3]. Hence, they will not be repeated here. However, a great deal of the relevant details are incorporated in Chapter 6.

Nanoindentation is a process in which a nanoindenter with a tip radius of about 10–300 nm is pushed into a given surface. The resulting force (P) on the surface and displacement (h) of the surface are continuously recorded by appropriate transducers (e.g., capacitive or otherwise). By using the depth data, the area of contact is estimated. Then, the hardness of the surface is evaluated as the ratio of the maximum load/contact area. Since the tip radius of the nanoindenter is of nanometer order, the contact area is small. As the length scale involved in the entire process is also very small, the hardness evaluated is given the name "nanohardness (H)". This hardness value is likely to be representative of the microstructural length scale of a given material. In addition, using the experimentally measured stiffness and the fact that the reduced modulus is actually linked to the series combination of the nanoindenter and the sample, it is easy to find the experimentally measured value of Young's modulus (E) of the sample [1–3]. Further, using the same nanoindentation technique, the fracture toughness (K_{Ic}) can be also evaluated at the microstructural length scale [2]. Furthermore, it is possible to utilize the same nanoindentation technique for the evaluation of residual stress in a given material [3]. Moreover, the technique can be easily exploited to derive the nanomechanical characteristics of natural, hierarchical and synthetic biomaterials [1–3]. However, depending on the material the physics of deformation at the nanoscale changes by significant aspect. In spite of the wealth of literature [1–3], these aspects are yet to be fully understood, as the knowledge base is still evolving. This evolutionary nature of the knowledge base drives the urge to present the current understanding in this chapter.

In almost all cases, barring a few exceptions, diamond indenters are used as nanoindenters. The tips are artificially made narrow as desired. They could be sharp with a triangular pyramidal frontage, e.g., a Berkovich nanoindenter. They could be sharp with a square pyramidal frontage e.g., a Vicker's nanoindenter. They could as well be ultrasharp e.g., a cube corner nanoindenter. On the other hand, to study the elastoplastic transition in material deformation, a blunt nanoindenter, e.g., a spherical nanoindenter is used. Between these two ends of a sharp contact and a blunt contact, conical nanoindenters with relatively large tip radius are used. They help in simulating intermediate stress scenarios between an elastic contact and an elastoplastic contact.

Today in many domains, starting from mobile phones to laptops to strategic space shuttles and biomedical prosthetic applications, very small, e.g., micron size or nanometer size, functional materials are used. Such small functional structures are exposed to external stress. To make their usage repeatable, it is required that such materials, such as a carbon nanotube, preserve their structural integrity and mechanical stability. The problem is that here the deformation and/or fracture happens

DOI: 10.1201/9781003359364-3

at the microstructural length scale of approximately a micrometer or nanometer. For such small structures, it is impossible to apply the conventional fracture testing methodologies that involve large structures such as a universal testing machine (UTM). Thus, here a probe length scale compatible with the microstructural length scale is needed. These are the scenarios where the nanoindentation (NI) technique provides us with a solution. It can map the nanomechanical properties as a function of both position and depth. In addition, it can measure the properties of the interface in nanocomposites. Further, it can also map the nanomechanical properties across the layers of multilayer structures which are very common for all electronic components utilized in large-scale electronic devices as well as in microelectromechanical systems (MEMS), nanoelectromechanical systems (NEMS), etc. Furthermore, it can assess the nanomechanical properties of soft biomaterials and polymers at the microstructural length scale [1–3].

One of the most popular methods of evaluation of nanohardness (H) and Young's modulus (E) owes its origin to the seminal research of Oliver and Pharr [4]. This method is almost universally utilized in nanoindentation, although there are some other approaches too [1–3]. But these are yet to be as popular as the Oliver and Pharr (O-P) method. One very important aspect involved in the physics of deformation in the nanoindentation of various materials, especially metals and ceramics, is the indentation size effect (ISE). In short, it means that when measured at ultralow loads used in the nanoindentation experiments, often the magnitude of (H) becomes larger than what could be expected, especially at smaller depths of indentation. As the nanoindentation load is increased, the depth of penetration (h) greatly increases. It contributes to a relatively larger contact area. As a result, the measured magnitude of (H) drops at a higher depth of penetration. Thus, plotted as a function of depth from low to high depth of penetration, the magnitude of (H) drops from a relatively higher value to a relatively lower value. This is what is termed as ISE. A similar picture remains valid if the (H) values are plotted as a function of loads which can start from an ultralow value and then go up to a relatively higher value. There are indeed numerous models that proliferate the literature to explain the ISE. However, one of the most popular and successful models is the one proposed by Nix and Gao [5]. This model proposes that (over and above the statistically stored dislocations) the genesis of ISE is linked to the generation of more dislocations that are geometrically necessary to accommodate the strain due to nanoindentation. The gradient of strain change is highest at the ultrasmall depths generated at the ultralow loads. This change in strain gradient helps in enhancing the nanohardness at the depth of penetration. However, at a larger depth of penetration, the strain change is reduced and hence it does not enhance the nanohardness that much. This is how the ISE is created. Many other models are also reported to explain the ISE in various materials [6].

2.2 VERY RECENT WORKS ON NANOINDENTATION OF VARIOUS MATERIALS

Today the domain of nanoindentation spans a really wide zone. Recent work reports the application of nanoindentation in the twisted bilayers of MoS_2 [7], in multilayer thin films of ZrN and ZrCN [8], measurement of local nanomechanical properties in the skull [9], TaN-incorporated armor materials (e.g., ZrB_2 and SiC-ZrB_2) [10], additively manufactured Al alloy [11] and ultrasoft contact lens materials [12]. Further attempts are directed to propose a new method of "zero point" identification in continuous stiffness measurement by the nanoindentation technique [13]. The stress–strain relationships extracted from the nanoindentation P vs. h plots often propose a problem.

A very recent approach confirms that using a combination of nanoindentation experiments along with molecular dynamic (MD) simulation and finite element (FE) simulation could be very fruitful in solving this problem for elastomeric materials [14]. It is indeed shown that even as the indentation depth increases, the plastic zone underneath expands, but the shape remains hemispherical. Such a statement holds both at the microscale and at the nanoscale as long as a self-similar indenter, e.g., a Berkovich nanoindenter, is used [14].

In another very recent approach [15], both sphero-conical and Berkovich nanoindentation along with an artificial neural network model (ANN) are employed to dig out the mechanical properties of thin film metallic glass (TFMG). Thus, $Zr_{55}Cu_{30}Ag_{15}$ TFMG deposited on Si and soda lime silicate glass substrates are examined. Here, the ANN model is trained with a fixed database. Systematic finite element analysis is used to generate this data set. The most unique feature is that there is only a mere difference of 3% between the plastic energy values experimentally measured and predicted by the finite element-based approach using ANN.

It is interesting to note that nanoindentation is applied in nondestructive health monitoring of asphalt due to aging [16]. Thus, the recent work confirms that in the case of artificially aged asphalt and styrene-butadiene-styrene–modified asphalt, nanoindentation-based creep compliance is in good agreement with creep compliance measured by the dynamic shear rheometer using the frequency sweep tests [16].

Two-dimensional materials are now emerging as the materials for next-generation nanodevices. A significant approach to understanding their nanoscale mechanical properties is now proposed [17] to be a simulation of their nanoindentation responses by MD simulation. It is thus noted in a recent review that the more correct information on the chemical bonding in 2-D materials is available, the more accurately their nanomechanical response could be predicted by MD simulation of nanoindentation [17].

Recent work [18] also demonstrates the indentation strain rate sensitivity of three different metals, e.g., copper (FCC), alpha phase Fe (BCC) and alpha phase Ti (HCP). It is shown that after elimination of the ISE, a quantity W_{grad} can be defined to represent the local deformation energy scenario as depicted by the respective load (P) vs. h (depth) plots of the different metals with various crystal structures (e.g., FCC, BCC or HP). Systematic investigation reveals two different trends. The first one is as the indentation strain rate increases, the ratio of local deformation energy to total energy spent in the nanoindentation process decreases for metals with the FCC and BCC structure. On the other hand, for alpha Ti with the HCP structure the ratio of local deformation energy to total energy spent in the nanoindentation process increases as the strain rate increases. This happens as the physics of deformation changes with the change in the respective crystal structures [18].

Another recent work [19] shows that the nanoindentation technique can be applied to study the structural change and related stress relaxation mechanisms in artificially Pb-ion–implanted $ZrSiO_4$. The experimentally measured Young's modulus (E) and nanohardness (H) are affected due to swelling that happens due to nuclear irradiation damage and the resulting residual stress development in the material. Further, in a significant attempt, a finite element model (FEM) is developed. The FEM could predict the (H) and (E) values as a function of depth in both undamaged and irradiation damage-affected $ZrSiO_4$. The predicted (E) values are reported to be in close match with the experimentally measured data [19].

It is interesting to note that MD simulation [20] of nanoindentation in different directions in nanotwinned (nt) cubic boron nitride (c-BN) illustrates that mechanical properties in all these directions of loading are improved as compared to those of untwinned single crystal (sc)-c-BN. This improvement is linked to two significant changes in the physics of deformation. The first one is that a pile-up of dislocations happens within the boundary layer of the nt c-BN. The second one is that a multijunction of dislocations is generated near the boundary layer of the nt c-BN. These multijunctions are rather immovable. Thus, they resist expansion of the dislocations above them. It leads to enhancement of the back stress and, hence, the yield strength. The enhancement in the yield strength, in turn, enhances the nanohardness [20]. What is even more interesting to note is that when the dislocations cross over the boundaries of the twinned crystals, the nt c-BN crystals again start to generate dislocations. These dislocations once again pile up. This process finally leads once more to multidislocation junction formation. These multijunctions again resist dislocation expansion. As a result, again the back stress and, hence, the yield strength are enhanced. This process again repeats the further enhancement of the nanohardness [20].

A significant new approach [21] applies a new modeling technique to reproduce (E) data experimentally measured by the nanoindentation technique in the cases of ceramic composites (e.g., TaC-HfC, TaB_2-HfB_2, TaC-HfB_2, TaB_2-HfC) and metal matrix composites, (e.g., Al-1100-hBN

(1 volume%) and Al-6061-WS_2 (1 wt.%)). This model is the generative adversarial network (GAN) model. This model not only reproduces the experimentally measured (E) data but also reduces the standard deviation by as high as about ~50% for the TaB_2-HfB_2 composites. However, the reduction is also reported to be as high as more than 300% for other composites, e.g., TaC-HfB_2. Further, for the Al-1100-hBN and Al-6061-WS_2 composites, reductions of about 30% and 10%, respectively, are reported to be achieved [21]. Further, for advanced bulk metallic glasses, e.g., $Zr_{55}Cu_{30}Ag_{15}$ and $Zr_{65}Cu_{15}Al_{10}Ni_{10}$, the application of the ANN model and FEM reproduces the nanoindentation responses measured experimentally by a spheroconical nanoindenter and a Berkovich nanoindenter [22].

Furthermore, recent nanoindentation experiments [23] confirm that for a Ni-base superalloy, e.g., GH4169, the load, loading rate and ISE have significant effects on the microscale deformation behavior. Most importantly, the nanoindentation data measured experimentally are in good match with those predicted by the crystal plasticity finite element model [23].

Further, it is interesting to note that enhancement of gas pressure from 0.1 to 1.0 KPa enhances the sp^3 bonding, which in turn leads to nanohardness improvement from about 76 to 87 GPa in diamond thin films deposited by the hot filament chemical vapor deposition (CVD) technique [24]. In addition, based on the experimental data, it is proposed that rather than the conventional practice that it should be $\leq$10% of the film thickness, and the final depth of penetration should be restricted to $\leq$5% of the film thickness to obtain the valid nanohardness data free from the substrate deformation effect. However, in the case of hip fractures, the application of the nanoindentation technique to tissue-level nanomechanical properties assessment confirms that a significant correlation exists between the tissue-level strains and the corresponding organic content [25].

Recent [26] applications of the nanoindentation creep experiments to the fused silica glass and soda lime silica glass illustrate that an increase in holding time (t_h) at the peak load degrades both (H) and (E) values, while it enhances depth recovery as well as the densified volume and shear flow volume. In another recent work [27], high-speed nanoindentation mapping is done on polycrystalline cubic BN. Thus, assuming the microstructure as an assemblage of hard particles, binder regions, and interphase regions, the statistical analysis matches well with the experimental data and, hence, provides a better representation of the nanomechanical response spectrum [27].

A recent review [28] summarizes the results obtained from the application of the nanoindentation technique to bone, cartilage and muscle. Thereby, the versatility of this technique to the physics of deformation at a sub-micrometer to nanometer regime is well established [28]. In a very recent work [29], very soft microstructurally engineered PEG hydrogels that span the (E) range from a single kilopascal (kPa) to hundreds of kilopascal are exposed to the spherical nanoindentation with a non-cantilever-based system. This work thus successfully utilized the nanoindentation technique to assess the Young's modulus spectrum of soft biomaterials. Thus, recent works demonstrate the suitability of the nanoindentation technique to assess the nanomechanical properties of soft biomaterials [28, 29]. Another area of emerging importance is high-temperature nanoindentation [30]. It reveals further that in the range of room temperature to 500°C, a combination of high-temperature nanoindentation along with the application of the crystal plasticity theory and FEM can successfully develop the constitutive relations for reduced activation ferritic/martensitic steels. These are the materials thought of as the materials for next-generation reactors to conduct fusion experiments [30].

2.3 BRIEF LITERATURE REVIEW

The developments in nanoindentation physics begin with the early 1990s work of Oliver and Pharr [31]. This seminal work is followed by their further work [4] and our earlier attempts [1–3]. It needs to be appreciated that as per current literature [1–31], the physics of deformation at the microstructural length scale is complex. It is a domain of evolving knowledge. More often than not, numerous observations [1–31] note the presence of two particular scenarios. The first and possibly the most important one is the presence of stochastic discontinuities or irregularities in the (P-h) plots recorded during the loading cycle and unloading cycle. These are termed, respectively, as micro-pop-in and micro-pop-out events. The other is that an apparently higher value of (H) is recorded at the ultralow depth, while a relatively lower value of (H) is recorded at a relatively higher depth. This is called the indentation size effect. These aspects have been addressed in a number of very good earlier reviews.

The signature of such mechanical instability is the local serrations present in the (P-h) plots. If they are present in the loading part of the (P-h) plots, they are termed micro-pop-in events. On the other hand, if they are present in the unloading part of the (P-h) plots they are termed as micro-pop-out events. Generally, these are also called nanoscale plasticity (NSP) events. In the case of micro-pop-in events, the displacement increases at certain critical loads (P_c). In the immediate next situation, the load is enhanced at a constant depth. This sequence continues to repeat. As a result, a steplike shape prevails in the (P-h) plots. Thus, a serration appears in all such (P-h) plots during the loading cycle.

In a similar manner, during the unloading cycle, the load decreases at certain critical loads (P_c) at a constant depth. These are called micro-pop-out events. In the immediate next situation, the load is enhanced at a constant depth. This sequence continues to repeat. As a result, a steplike shape prevails in the (P-h) plots. Thus, a serration appears in all such (P-h) plots during the unloading cycle. For instance, the review by Schuh et al. in 2006 [32] focuses on the mechanical instability in (P-h) plots of bulk metallic glass and metallic alloys.

A later review by Pharr et al. in 2010 [33] links the genesis of the ISE to instrumental error and its propagation as well as to the tip-blunting effect. The physics of deformation involved in a depth-dependent nanoindentation response of polymers was comprehensively reviewed in 2015 [34]. Another review [35] addresses the ISE in the nanoindentation of metals and links it to an increase in hardness at low depth. This is attributed to the increase in geometrically necessary dislocations at ultralow depth. This review identifies the need for more extensive nanoindentation experiments and atomistic simulations to better understand the genesis of ISE in metals [35].

Further, the experimental, theoretical and computational studies linked to the genesis of pop-in events in mainly metals and bulk metallic glasses are comprehensively reviewed in 2021 [36]. This review suggests that the genesis of pop-in could be linked to the micro yielding behavior. However, the knowledge base regarding the spot at which the micro yielding initiates, the local stress condition that governs their initiation and the evolutionary steps that lead to the macroscopic yielding is yet to be fully proved. Further, it is suggested that very high localized stress close to the theoretical strength is required for the pop-in to initiate. Furthermore, the pop-in generation is suggested to involve dislocation generation/multiplication within the ultrasmall volume of the nanoindentation cavity.

In addition, it points out that the reason that leads to the generation of heterogeneous or homogeneous dislocation in crystals is yet to be comprehensively understood. It also suggests that the pop-in phenomenon may be linked in addition to the genesis of highly localized unstable plasticity due possibly to the dislocation avalanche formation [36]. The review stresses the need to conduct more definitive experiments along with computational and theoretical modeling in various materials [36]. Using the nano-electrical contact resistance (nano-ECR) method of nanoindentation response evaluation, a very recent review [37] comprehensively discusses the occurrence of pop-in events in a number of materials, including metals, ceramics, polymers and nanotubes. However, the reasons that lead to the genesis of these micro-pop-in events are yet to be comprehensively understood [37].

2.4 RECENT NEW OBSERVATIONS

Thus, the aforesaid brief review of the literature points out that the issues of micro-pop-in, micro-pop-out and nanoscale plasticity events in all materials, especially ceramics, are far from well understood. It is interesting to note that some very recent works [38, 39, 40, 41] also attempt to provide further insight into this domain.

In these recent works, the alumina ceramics are reported [38, 39, 40, 41] to be pressureless sintered with various sintering aids, e.g., (1, 3 and 5 wt.% MgO), (1, 3 and 5 wt.%) TiO_2 and a combination of both MgO and TiO_2. Thus, the MgO-densified alumina ceramics with 1, 3 and 5 wt.% MgO used as sintering aids are termed MDA1, MDA3 and MDA5. Similarly, the TiO_2-densified alumina ceramics with 1, 3, and 5 wt.% TiO_2 used as sintering aids are termed as TDA1, TDA3 and TDA 5 [40, 41]. Further, the 1 wt% MgO and 4 wt% TiO_2 co-densified pressureless sintered alumina samples are termed as AMT1. Similarly, the 2 wt% MgO and 3 wt% TiO_2 co-densified pressureless sintered alumina samples are termed as AMT2. In a similar manner, the 3 wt% MgO and 2 wt% TiO_2 co-densified pressureless sintered alumina samples are termed AMT3. In an analogous manner, 4 wt% MgO and 1 wt% TiO2 co-densified pressureless sintered alumina samples are termed as AMT4.

2.5 NANOINDENTATION RESPONSE OF ALUMINA DENSIFIED WITH VARIOUS SINTERING AIDS CASE STUDIES ON MGO-DENSIFIED ALUMINA CERAMICS

The relative densities of the MDA1, MDA3 and MDA5 ceramics are reported to be 96%, 96.6% and 97.3% of theoretical density (e.g., 3.98 $g.cc^{-1}$). The nanomechanical behavior of the MDA ceramics is studied under the applied nanoindentation load range of 10–1000 mN. The results obtained in this recent work [38, 41] for MDA ceramics conclusively prove the presence of both nanoindentation size effect and localized plasticity events in these ceramics (Figures 2.1–2.4). As a typical illustrative example, the microstructure of the MDA1 ceramics is shown in Figure 2.1a. The corresponding (P-h) plots are shown in Figure 2.1b.

The presence of ISE is shown in Figure 2.2a. Further, the comparison of experimentally measured nanohardness (H_E) and theoretically predicted nanohardness (H_T) is shown in Figure 2.2b as a function of the deformation resistance (df_r), which scales with the inverse of nanoindentation depth (h).

It is reported [38, 41] that the Huang model [6] provides a much better fit to the (H_E) value in comparison to the prediction made by the most famous Nix and Gao model [5]. A similar situation holds good for all the MDA ceramics. In other words, the Huang model [6] provides a much better description of the ISE present in the MDA ceramics. It needs to be noted that the nanoindentation size effect in the MDA ceramics is reported [38] to be determined by the load-dependent spatial variations in both the dislocation loop interaction zone size and the deformation resistance.

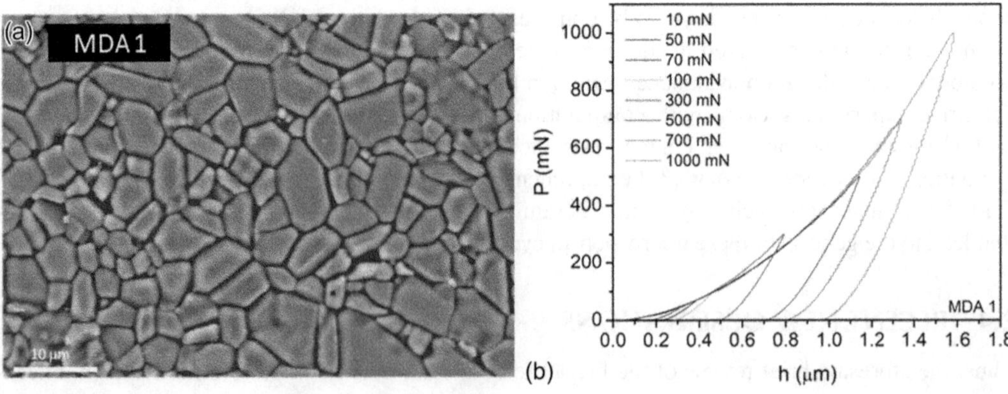

FIGURE 2.1 MDA 1 ceramics. (a) FESEM photomicrograph of microstructure, (b) (P-h) plots (reprinted with permission from [38]).

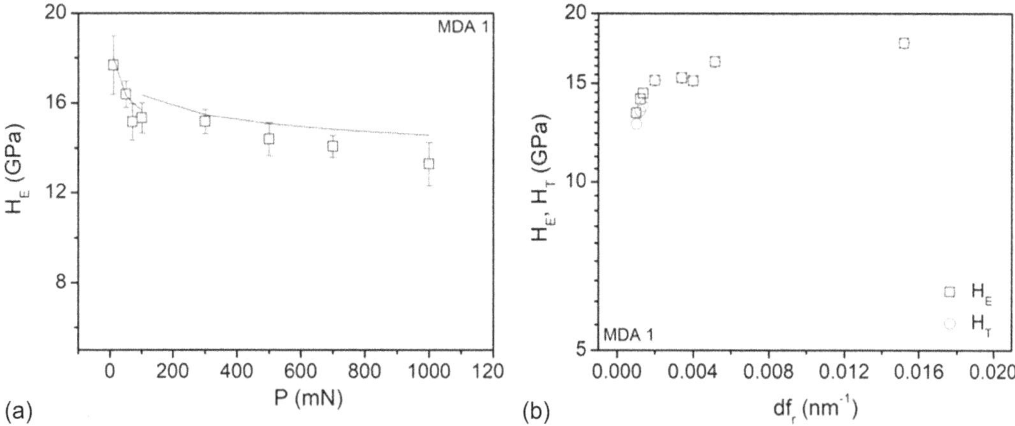

FIGURE 2.2 Nanohardness of MDA 1 ceramics (a) as a function of (P). (b) Experimental and predicted nanohardness of MDA1 ceramics as a function of deformation resistance (reprinted with permission from [38]).

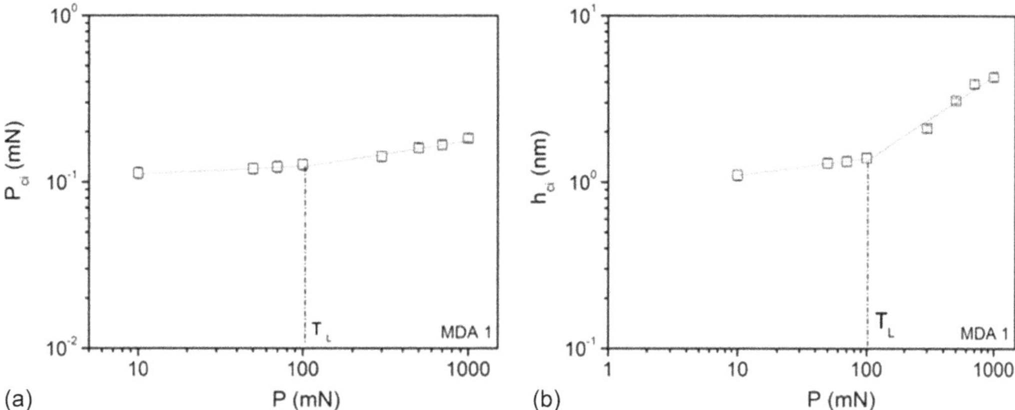

FIGURE 2.3 Variation of (a) critical load and (b) critical depth as a function of (P) (reprinted with permission from [38]).

It is noted that the localized plasticity events at nanoscale contact deformation induced by the present nanoindentation experiments happen in the MDA ceramics [38, 41]. As a typical illustrative example, the variation of critical load (P_c) at which the nanoscale plasticity (NSP) events happen and the corresponding nanoindentation depth (h_c) is shown as a function of the applied nanoindentation load (P) in Figure 2.3a and Figure 2.3b, respectively.

Both (P_c) and (h_c) register power law dependencies on (P) with a positive exponent with a high degree of goodness of fit (e.g., $\geqslant$90%) (Figure 2.3a and b) [38, 41]. The most notable physical new significance here is that the nanoscale contact deformation resistance (P_c) of these MDA ceramics is enhanced, as the applied nanoindentation load (P) applied to deform it at the nanoscale is enhanced. It is suggested [38, 41] that these micro-pop-in–induced NSP events happen in the loading curves due to the simultaneous contributions from dislocation nucleation, localized shear deformation band formation (Figure 2.4a) and microcracking formations (Figure 2.4b).

Further, it is reported [38, 41] that the critical loads at which localized plasticity events are initiated in the MDA ceramics increase with the amount of MgO dopant used as the sintering aid. It happens due to the simultaneous increase in the relative density, increase in the amount (i.e., wt.%)

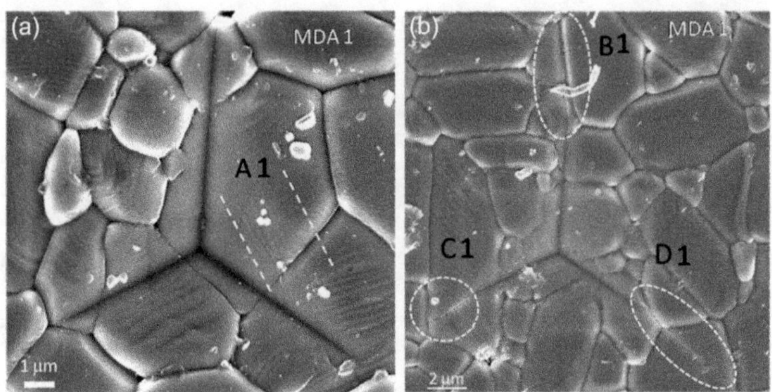

FIGURE 2.4 FESEM photomicrographs of nanoindentation cavity of MDA1 ceramics. (a) Presence of shear deformation bands, (b) presence of microcracking (reprinted with permission from [38]).

of the spinel phase, increase in the relative amounts of fine grains and decrease in the average size of the fine grains. Thus, these recent results [38, 41] imply a possibility that depending on the application perspective, the microstructure of alumina ceramics can be tuned to possess the desired contact deformation resistance at the nanoscale. Therefore, these new findings bring the concept of microstructurally engineered ceramics (MIEC) to the forefront. The MIEC can be developed with a tuned microstructure that could be suited to a given end application.

Further progress in this direction establishes that [39, 41] the nanohardness of all the MDA ceramics as mentioned earlier shows an increasing trend with the loading rate. With the increase in loading rate ($\dot{P}$) from ~1 to 1000 mN.s^{-1}, the nanohardness (H) values increase to 14.17, 13.98 and 13.92 GPa from the values of 12.26, 11.88 and 11.33 GPa for MDA1, MDA3 and MDA5 ceramics, respectively. The maximum shear stress is reported [39, 41] to play a significant role in micro-pop-in issues in these MDA ceramics. Further, here also, the intrinsic contact deformation resistance increases with the amounts of sintering aids in all the MDA ceramics. So, it is found out in these novel recent works [38, 39, 41] that these MDA ceramics developed with the MIEC philosophy have the interesting ability to adjust to the change in energy transfer rate from the loading train and accordingly also enhance its intrinsic resistance against contact-induced deformation at the nanoscale.

2.5.1 Case Studies on TiO_2-Densified Alumina Ceramics

Further application of the MIEC philosophy is carried out in the TDA ceramics as mentioned earlier [40, 41]. This recent work provides confirmatory experimental evidence that loading rate ($\dot{P} \approx$ 1–1000 mN.s^{-1}) dependent NSP events do occur in these TDA ceramics. As such, the TDA ceramics TDA1, TDA3, and TDA5 are reported [40, 41] to have a relative density of 98.17%, 92.03% and 93.42% of theoretical density (e.g., 3.98 g.cc^{-1}). These newly found [40, 41] NSP events are characterized by profuse serration and micro-pop-ins in the loading parts of the corresponding load–depth plots obtained during the nanoindentation experiments conducted with a diamond Berkovich nanoindenter.

As a typical illustrative example, the microstructures of TDA1, TDA3 and TDA5 ceramics are shown in Figure 2.5a, b and c, respectively.

The corresponding (P-h) plots obtained from the nanoindentation experiments conducted at various loading rates (e.g., $\dot{P} \approx 1$–1000 mN.s^{-1}) are shown, respectively, for TDA1, TDA3 and TDA5 ceramics in Figure 2.6a, b and c.

Further, the illustrative examples of the enhanced views of the (P-h) plots are shown in Figure 2.7a, b and c for TDA1, TDA3 and TDA5 ceramics in correspondence.

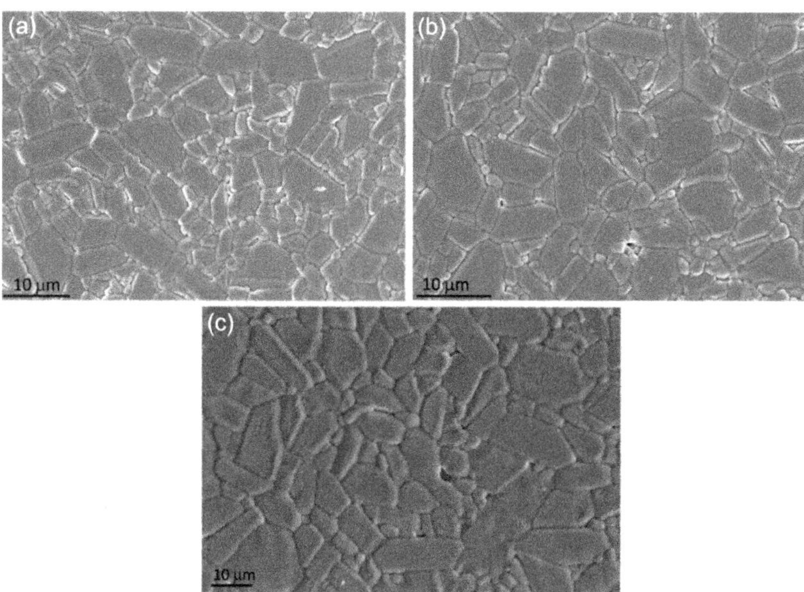

FIGURE 2.5 FESEM photomicrographs of microstructures of TDA ceramics. (a) TDA1, (b) TDA2 and (c) TDA3 (reprinted with permission from [40]).

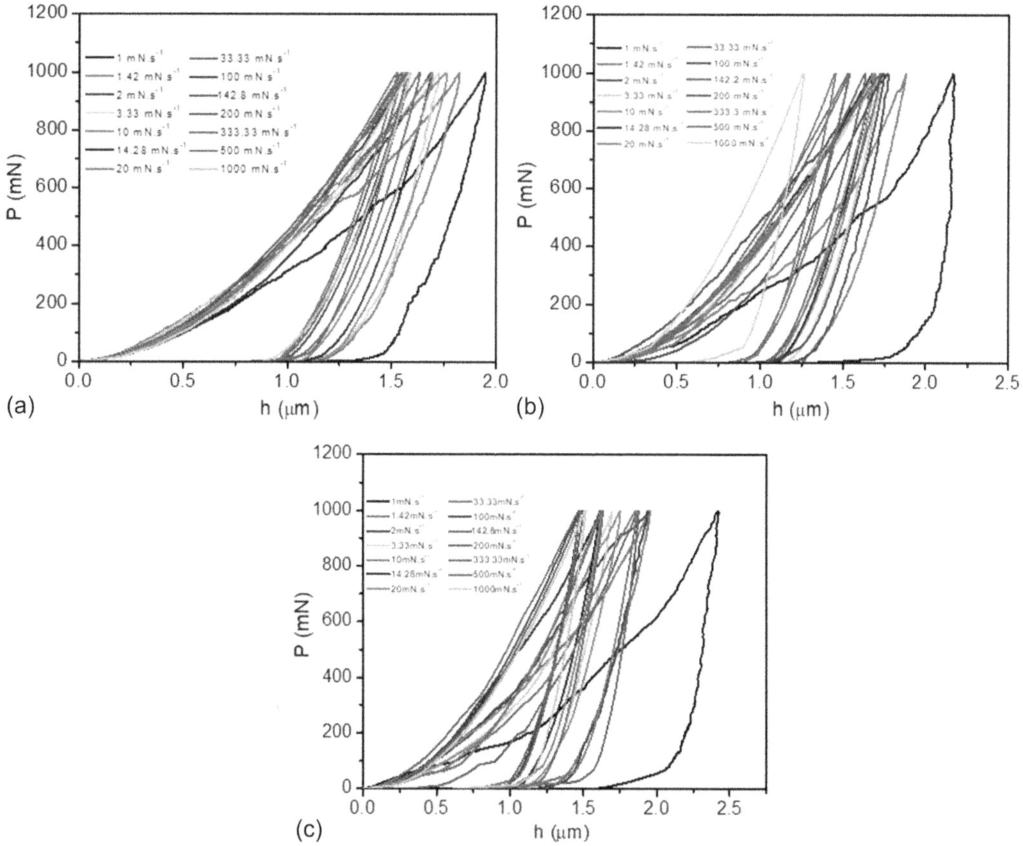

FIGURE 2.6 Nanoindentation experiment-based (P-h) plots of TDA ceramics. (a) TDA1, (b) TDA2 and (c) TDA3 (reprinted with permission from [40]).

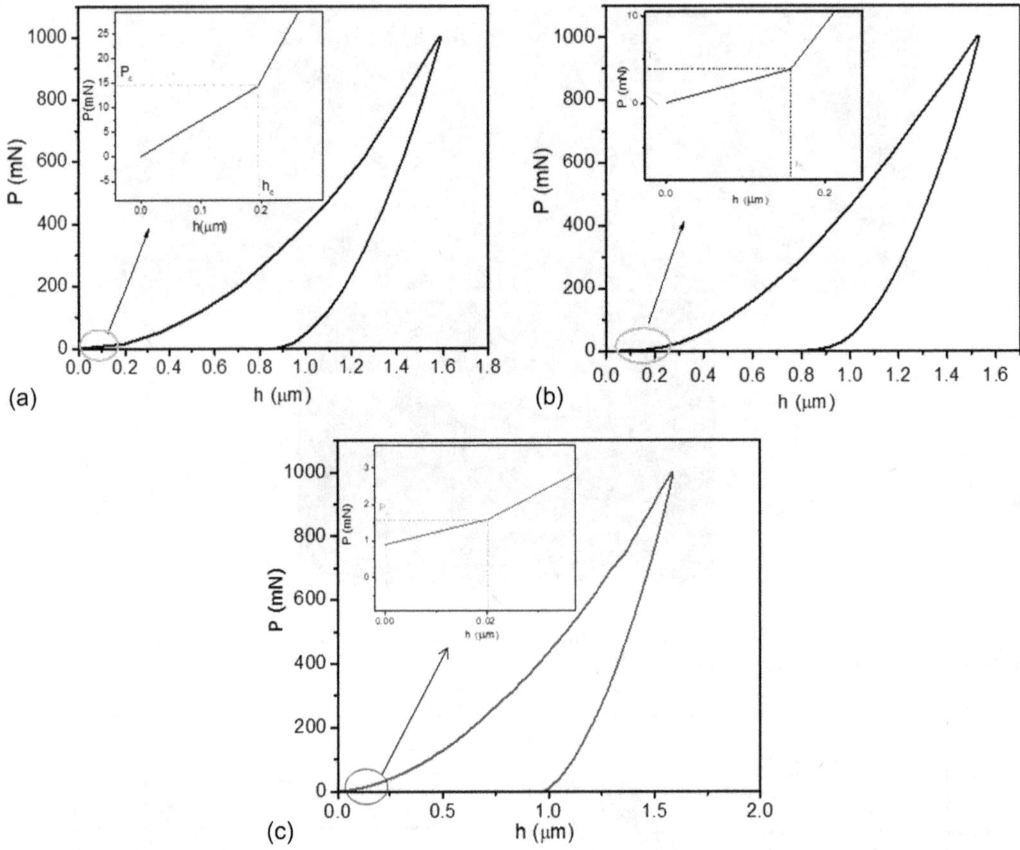

FIGURE 2.7 Views of (P-h) plots of TDA ceramics. (a) TDA1, (b) TDA2 and (c) TDA 3. Insets: Identification of critical load and depth (reprinted with permission from [40]).

Furthermore, the insets of these figures illustrate the identification of the corresponding critical load (P_c) and critical depth (h_c) values for TDA1, TDA3 and TDA5 ceramics.

In addition, it is reported [40, 41] that both [(H), Figure 2.8a; and (E), Figure 2.8b] values of the TDA ceramics enhance as loading rates ($\dot{P}$) are enhanced. It is worth noting that the increases in nanohardness values are about 8%, 30% and 35% for the TDA1, TDA3 and TDA5 ceramics, respectively (Figure 2.8a).

These data also indicate that the enhancement in nanohardness with loading rates is maximum in the TDA5 ceramics and minimum in TDA1 ceramics. Clearly, the amount of sintering aid emerges as an important parameter related to the enhancement in nanohardness. *However, further research work is desired to be done in the future to understand the role played by the enhanced amount of TiO_2 sintering aid in enhancing the nanohardness as a function of loading rates.*

A similar situation prevails for Young's modulus at the nanoscale. For instance, the average magnitude of Young's modulus is reported [40, 41] to be 370.04 ± 24.04 GPa for the loading rate range of 1 $mN.s^{-1}$ to 1000 $mN.s^{-1}$ applied to TDA1 ceramics (Figure 2.8b). Similarly, the average magnitude of Young's modulus is reported to be 353.94 ± 28.48 GPa for the loading rate range of 1 $mN.s^{-1}$ to 1000 $mN.s^{-1}$ applied to TDA3 ceramics (Figure 2.8b). In a similar manner, the average magnitude of Young's modulus is reported to be 360.44 ± 22.98 GPa for the loading rate range of 1 $mN.s^{-1}$ to 1000 $mN.s^{-1}$ applied to TDA5 ceramics (Figure 2.8c). These data match with those reported for alumina ceramics [42]. Thus, with the increase in loading rates the average values of Young's modulus increase by about 15%, 30 % and 35% for the TDA1, TDA3 and TDA3 ceramics,

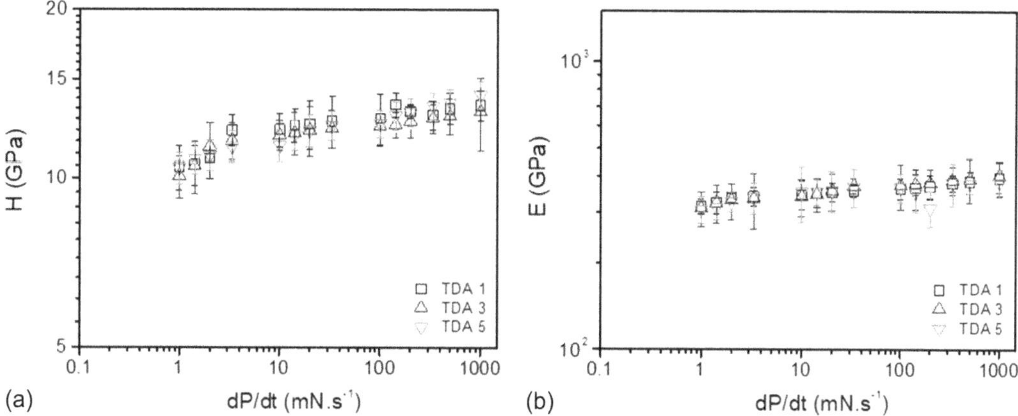

FIGURE 2.8 Variation of (a) nanohardness and (b) Young's modulus of TDA ceramics (TDA1, TDA2, TDA3) as functions of loading rate (dP/dt) (reprinted with permission from [40]).

respectively. These data also indicate that the enhancement in Young's modulus with loading rates is maximum in TDA5 ceramics and minimum in TDA1 ceramics. Clearly, the amount of sintering aid emerges as an important parameter related to the enhancement in Young's modulus and nanohardness. However, further research work is needed to understand the role played by the enhanced amount of TiO_2 sintering aid in enhancing the Young's modulus and nanohardness as a function of loading rates.

It needs to be noted from these recent works [38–41] that there are several parameters linked to NSP events expressed through the occurrence of micro pop in signatures. These are the critical load (P_c), the critical depth (h_c), the critical load (ΔP) difference between any two consecutive NSP events, the critical load difference (Δh) between any two consecutive NSP events, the maximum shear stress (τ_{max}) that generates just underneath the nanoindenter, the dislocation density (ρ_D) and the critical resolved shear stress (τ_{CRSS}). Recent work on TDA ceramics [40, 41] confirms that almost all the micro-pop-in related parameters, i.e., (P_c), (h_c), (ΔP), (Δh), ($_{\tau max}$; Figure 2.9), (ρ_D) and (τ_{CRSS}) also increase with the loading rate ($\dot{P}$).

For instance, in the case of TDA3 ceramics depending on the loading rate range 1 $mN.s^{-1}$ to 1000 $mN.s^{-1}$, the critical load P_c to initiate nanoscale plasticity is reported [40, 41] to vary from 0.1 mN to 14.44 mN. In correspondence, the critical depth h_c is reported [40, 41] to vary from 0.0003 μm to 0.16 μm. In tandem, the critical load difference ΔP is reported [40, 41] to vary from 0.002 mN to 98.94 mN. Further, the corresponding critical depth difference Δh varies from 0.0006 μm to 0.138 μm.

In a similar manner, in the case of TDA5 ceramics depending on the loading rate range 1 $mN.s^{-1}$ to 1000 $mN.s^{-1}$, the critical load P_c to initiate nanoscale plasticity is reported [40, 41] to vary from 0.1 mN to 14.54 mN. In correspondence, the critical depth h_c is reported [40, 41] to vary from 0.00008 μm to 0.195 μm. In tandem, the critical load difference ΔP is reported [40, 41] to vary from 0.003 mN to 99.1 mN. Further, the corresponding critical depth difference Δh is reported [40, 41] to vary from 0.0003 μm to 0.173 μm. All these parameters are reported to exhibit the empirical power law dependencies on ($\dot{P}$) with a positive exponent. These observations are justified by both qualitative arguments and theoretical simplistic modeling [40, 41].

It is also reported [40, 41] that the critical loads at which the first micro-pop-in happens show a Gaussian distribution. This happens at all the loading rates. This confirms one interesting fact. It reflects the intrinsically stochastic nature of the NSP events. It is also noted [40, 41] that the range of critical loads increases slightly with ($\dot{P}$). Further, the range of the critical loads also increases slightly with the increase in the amount of TiO_2 used as the sintering aid.

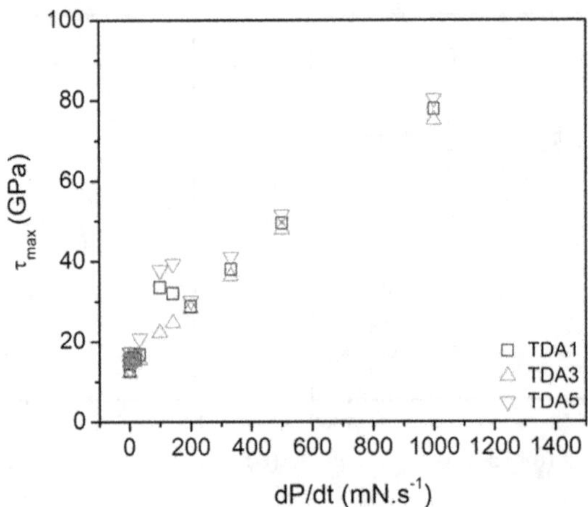

FIGURE 2.9 Variation of maximum shear stress of TDA ceramics (TDA1, TDA2, TDA3) as functions of loading rate (dP/dt) (reprinted with permission from [41]).

Based on reported works [40, 41], it seems plausible to argue that this slight increase in the range of (P_c) may be linked to the simultaneous reduction in the fine grain size and enhancement of the relative density in the TDA ceramics. Additional studies conducted on the nanoindentation cavity by the FESEM (field-emission scanning electron microscopy) technique [40, 41] confirm that three factors can contribute to the occurrence of NSP events in TDA ceramics. The first and probably the most important factor is the nucleation of homogeneous dislocations under the penetrating nanoindenter. The second most important factor is the formation of shear bands. This happens as the maximum shear stress active just underneath the nanoindenter is estimated to be much higher than the theoretical shear strength (Figure 2.9). Finally, the formation of microcracking along with the shear deformation band formation [40, 41] appears to be the other important factor.

Thus, the utilization of TiO_2 as a sintering aid provides a unique opportunity for the MIEC concept. These recent works [40, 41] provide undeniable evidence that both nanohardness and nanoscale contact deformation resistance, i.e., the critical load can be enhanced in tune with the enhancement in loading rate, which equivalently transfers to the rate of energy transfer from the loading train to the material. Using the MIEC concept, it may be possible to design advanced alumina ceramics for many important applications that involve resistance against high strain rate contact-induced deformation. All these works [38–41] also explore new insight into the physics of deformation at the microstructural length scale as well.

2.5.2 Case Studies with MgO-TiO_2 Co-Densified Alumina Ceramics

Recall that the 1 wt% MgO and 4 wt% TiO_2 co-densified pressureless sintered alumina samples are termed AMT1. Similarly, the 2 wt% MgO and 3 wt% TiO_2 co-densified pressureless sintered alumina samples are termed as AMT2. In a similar manner, the 3 wt% MgO and 2 wt% TiO_2 co-densified pressureless sintered alumina samples are termed AMT3, and 4 wt% MgO and 1 wt% TiO2 co-densified pressureless sintered alumina samples are termed as AMT4. The relative densities of the AMT1, AMT2, AMT3 and AMT4 ceramics are reported [41] to be about 93.79%, 92.47%, 92.86% and 92.20% of the theoretical density.

It is interesting to note the effect of load on the nanohardness of this new variant of MIEC materials, i.e., the AMT ceramics [41]. The typical x-ray diffraction (XRD) patterns of the co-doped MIEC materials, i.e., the AMT1, AMT2, AMT3 and AMT4 samples, are shown in Figure 2.10.

These XRD patterns confirm the presence of the MgO phase (ICSD code 01-087-0652), TiO_2 phase (ICSD code 01-089-4921) and alumina phase (ICSD code 01-073–1512), as expected. In this

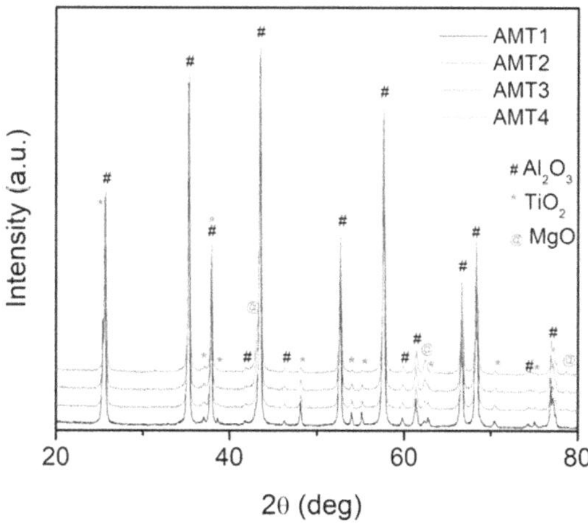

FIGURE 2.10 The XRD patterns of the AMT1, AMT2, AMT3 and AMT4 ceramics (reprinted with permission from [41]).

case, the load-controlled nanoindentation experiments are conducted at five different ultralow load values: 1000, 3000, 5000, 7000 and 10000 μN.

The typical load displacement (P-h) curves obtained at 1000, 3000, 5000, 7000, and 10000 μN applied during the nanoindentation experiments conducted on the mirror-polished AMT1, AMT2, AMT3 and AMT4 ceramics are shown in Figure 2.11a, b, c and d, respectively.

In all the AMT ceramics, the depth of penetration increases with the increase in the applied loads, as expected. All plots exhibit the expected elastoplastic nature of deformation. Further, the presence of serrations in the loading as well as unloading parts of the P-h plots is very prominent (Figure 2.12). These serrations are termed as micro-pop-in, as discussed earlier [38–41].

Recall that the load at which the micro pop-in initiates is termed the critical load (P_c) and the corresponding depth is termed the critical depth (h_c). As a typical illustrative example, the P-h plot obtained during the nanoindentation experiment conducted at an applied load of 3000 μN on the AMT1 sample is shown in Figure 2.12a. Thereafter, the corresponding critical load (P_c) and the critical depth (h_c) values measured for the AMT1 samples are shown in Figure 2.11b. The corresponding values of both ΔP and Δh are shown in Figure 2.13 for the applied load of 3000 μN.

Further, for the same AMT1 ceramics, the variations of all the nanoindentation-related parameters, i.e., P_c, h_c, ΔP and Δh, as a function of the applied loads are presented in Figure 2.14a, b, c and d, respectively.

Similarly, for AMT2 ceramics, the variations of P_c, h_c, ΔP and Δh, as a function of the applied loads, are presented in Figure 2.15 (a), (b), (c) and (d), respectively.

For AMT3 ceramics, the variations of P_c, h_c, ΔP and Δh, as a function of the applied loads, are presented in Figure 2.16a, b, c and d, respectively.

Finally, for AMT4 ceramics, the variations of P_c, h_c, ΔP and Δh, as a function of the applied loads, are presented in Figure 2.17a, b, c and d, respectively. The generic trends are similar. Hence, it suffices to discuss the traits of a given ceramic, e.g., the AMT1 samples.

Interestingly, both the critical load (P_c) (Figure 2.14a) and the critical depth (h_c) Figure 2.14b) show load-independent behavior. *It is a unique observation which means that the critical load to initiate nanoscale plasticity in these AMT1 ceramics does not depend on the applied loads.*

These new observations confirm that by appropriate choice of sintering aids the deformation physics at the nanoscale can be tailored to avoid sharp dependency on the applied nanoindentation load (P) while still preserving the occurrence of the NSP events as is noted for MDA and TDA ceramics.

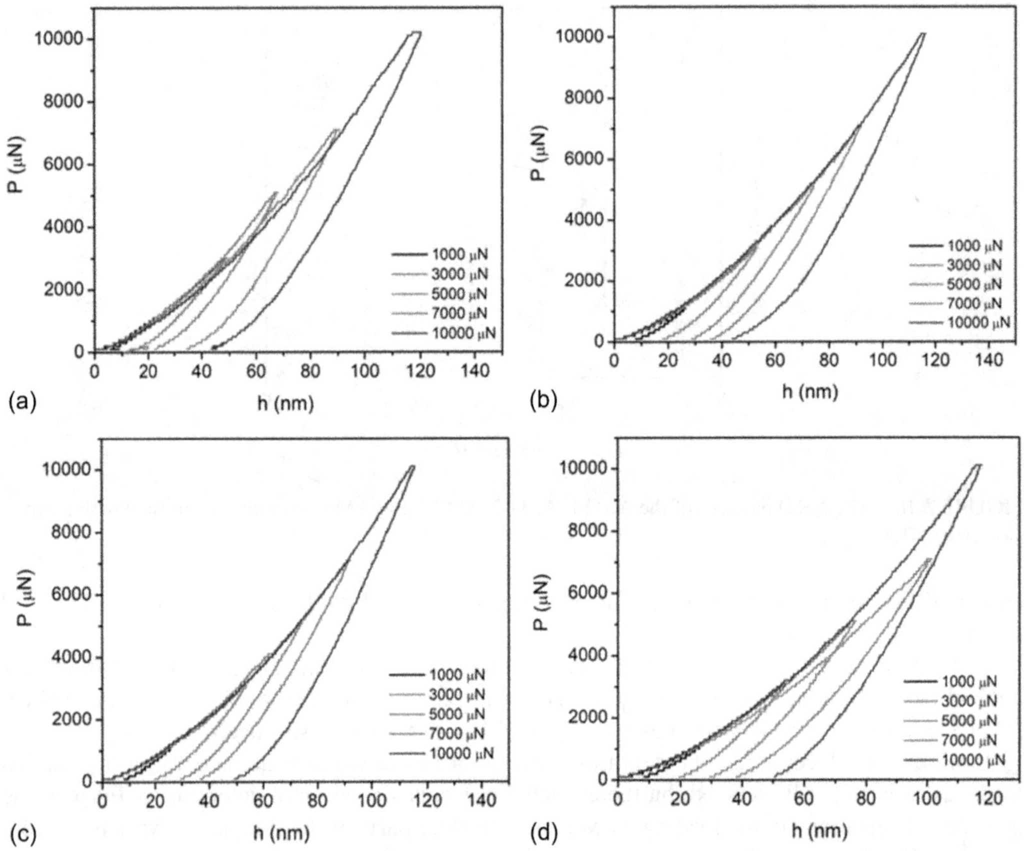

FIGURE 2.11 Typical P-h plots of (a) AMT1, (b) AMT2, (c) AMT3 and (d) AMT4 for different applied loads (reprinted with permission from [41]).

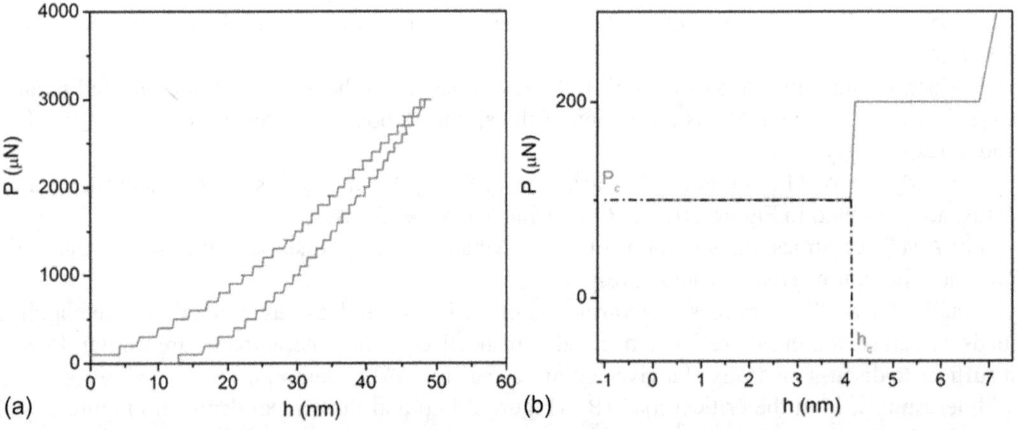

FIGURE 2.12 Nanoindentation results for AMT1 ceramics. (a) The typical P-h plot for the applied load of 3000 μN. (b) Identification of the critical load (P_c) and critical depth (h_c) values from the data plot shown in Figure 2.11a (reprinted with permission from [41].

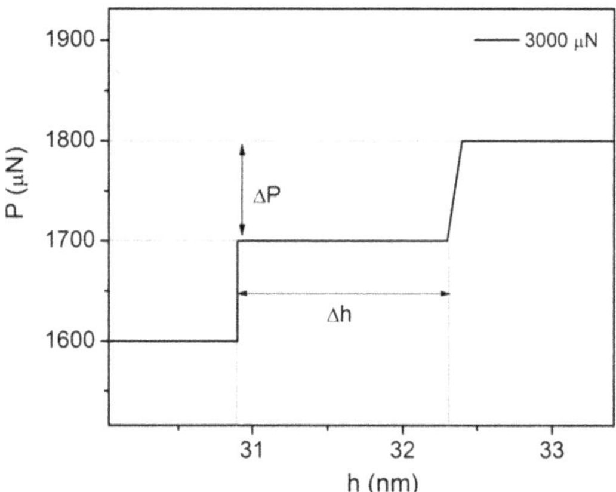

FIGURE 2.13 View of the P-h plot showing serrations as well as identification of ΔP and Δh for AMT1 ceramics during nanoindentation experiment conducted at the applied load of 3000 μN (reprinted with permission from [41]).

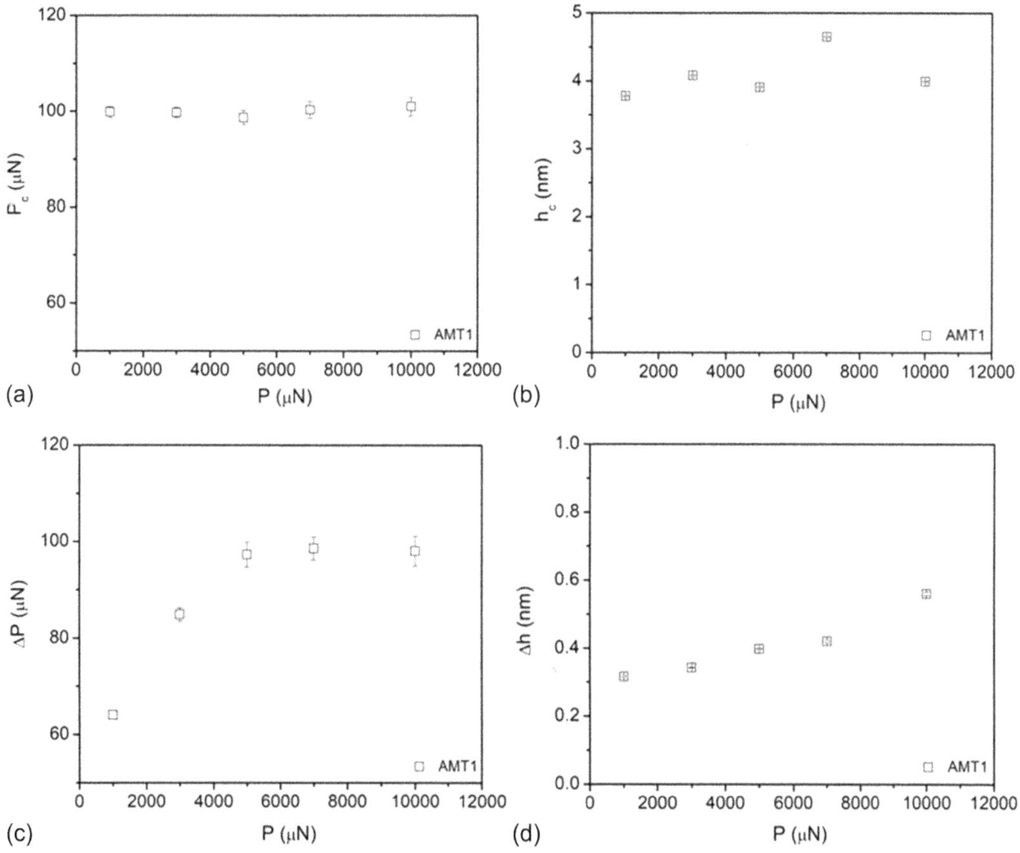

FIGURE 2.14 The variations of (a) P_c, (b) h_c, (c) ΔP and (d) Δh as a function of the applied nanoindentation loads in AMT1 ceramics (reprinted with permission from [41]).

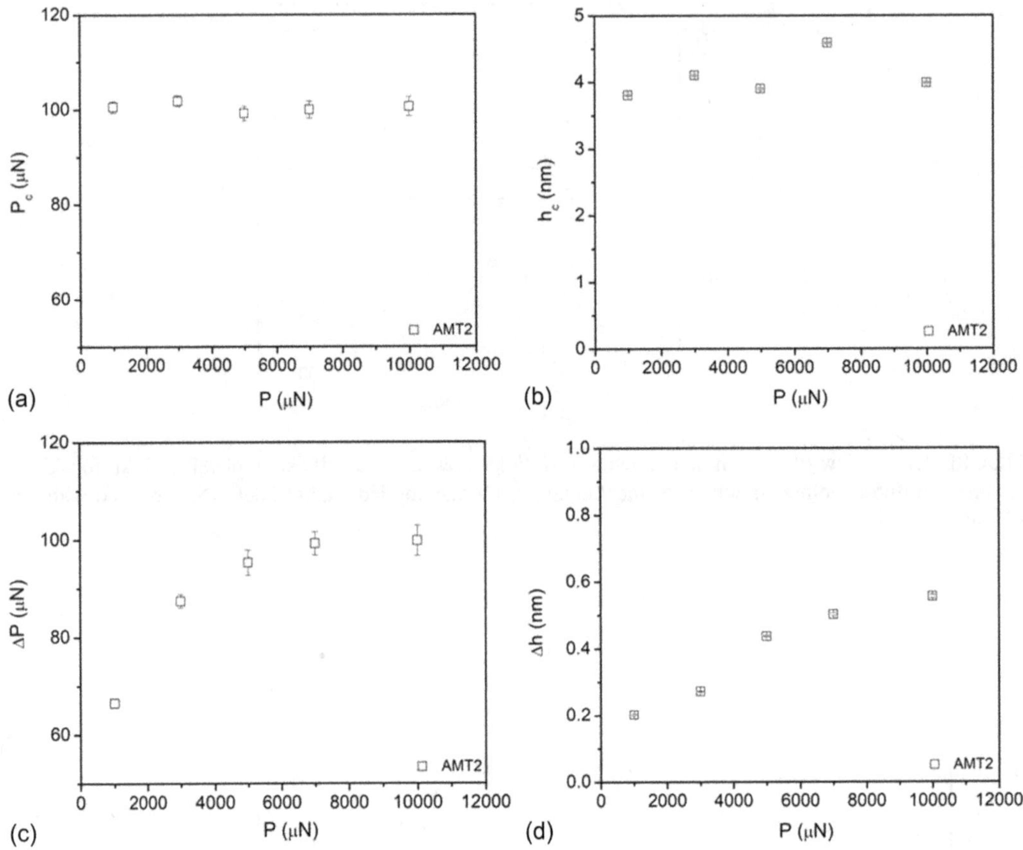

FIGURE 2.15 The variations of (a) P_c, (b) h_c, (c) ΔP and (d) Δh as a function of the applied nanoindentation loads in AMT2 ceramics (reprinted with permission from [41]).

Therefore, it is evident that further research work is necessary to understand if this is linked to the roles played by co-doping of sintering aids/grain size variation/extent of densification/reduction in coarse-grain size/enhancement in amounts of fine grain size or a simultaneous contribution of all such and other related factors, e.g., the structure of the grain boundary.

However, the ΔP values of AMT1, AMT2, AMT3 and AMT4 ceramics exhibit mildly increasing trends with an increase in the applied loads, as shown, respectively, in Figures 2.14c, 2.15c, 2.16c and 2.17c. As a result, the Δh values of AMT1, AMT2, AMT3 and AMT4 ceramics also exhibit a slightly increasing trend with an increase in the applied loads, as shown respectively in Figures 2.14c, 2.15c, 2.16c and 2.17c. In view of these present observations, it may only be cursorily mentioned here that this behavior is similar to what has been observed in other MIEC materials, e.g., MDA and TDA ceramics. *Further, the ΔP and Δh values of both MDA and TDA ceramics are reported [38–41] to exhibit an increasing trend with the applied loading rates.*

It is very interesting to note that both nanohardness (H) (Figure 2.18a) and Young's modulus (E) (Figure 2.18b) are almost totally insensitive to variations in the applied nanoindentation loads (P).

But both H and E are slightly sensitive to the variations in the amounts of sintering aids, although the variations are not very systematic. It is still interesting to note that at almost all the applied loads, AMT4 ceramics exhibit the highest magnitude of nanohardness.

On the other hand, it is also a fact that it has the lowest relative density. Therefore, these new observations are not in accordance with the conventional notion that higher density ensures higher nanohardness. In other words, there could be factors other than the relative density alone and,

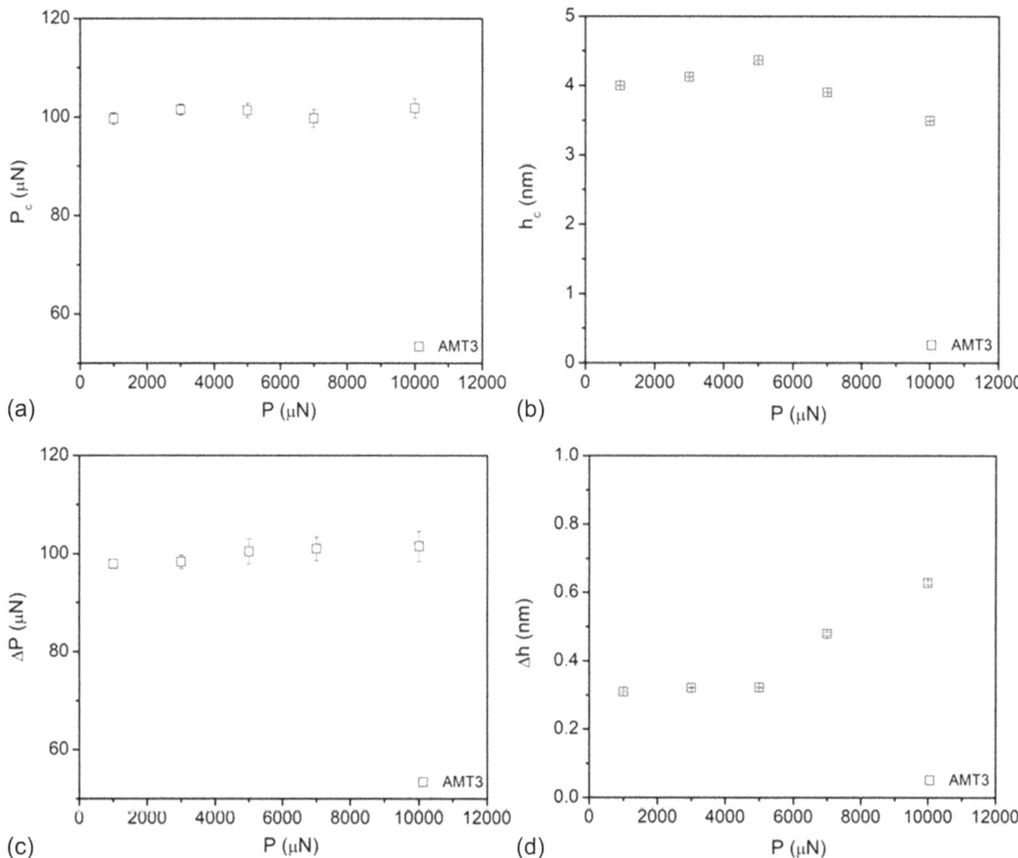

FIGURE 2.16 The variations of (a) P_c, (b) h_c, (c) ΔP and (d) Δh as a function of the applied nanoindentation loads in AMT3 ceramics (reprinted with permission from [41]).

hence, further research is desirable to be conducted in the future in this direction. Nevertheless, the overall grand average value of nanohardness is 33.69 ± 1.1 GPa as shown by the dotted line in Figure 2.17a. Similarly, it is also very interesting to note that at almost all the applied loads, AMT1 ceramics exhibit the highest values of Young's modulus. It may be recalled here that AMT1 ceramics exhibit the second highest relative density value. *These data may also reflect that at the nanostructural length scale, factors other than relative density may control Young's modulus of MIEC materials.* However, the overall grand average value of Young's modulus is 360.51 ± 10.3 GPa as shown by the dotted line in Figure 2.18b. *The reasons for such behavior need to be understood in the future to produce such or even better MIEC materials.*

Nevertheless, a load-independent nanohardness (H) is very important in the design aspects of structural load-bearing applications of alumina ceramics. Thus, these results point to a unique microstructural design possibility that may be fine-tuned further depending on the need for a given application. *However, the reasons for which these AMT ceramics are showing these load-independent nanomechanical properties are yet to be known and, hence, should provide a scope for future research.*

The variations of maximum shear stress (τ_{max}) with the applied nanoindentation loads for AMT1, AMT2, AMT3 and AMT4 ceramics are shown in Figure 2.19. Most interestingly, the magnitudes of the maximum shear stress values (τ_{max}) active just underneath the nanoindenter increase with an increase in the applied loads. Thus, the estimated magnitudes of (τ_{max}) are ~12–16 GPa. These (τ_{max}) values are much higher than the typical theoretical shear strength (i.e., ~3 GPa) of alumina ceramics [38–41].

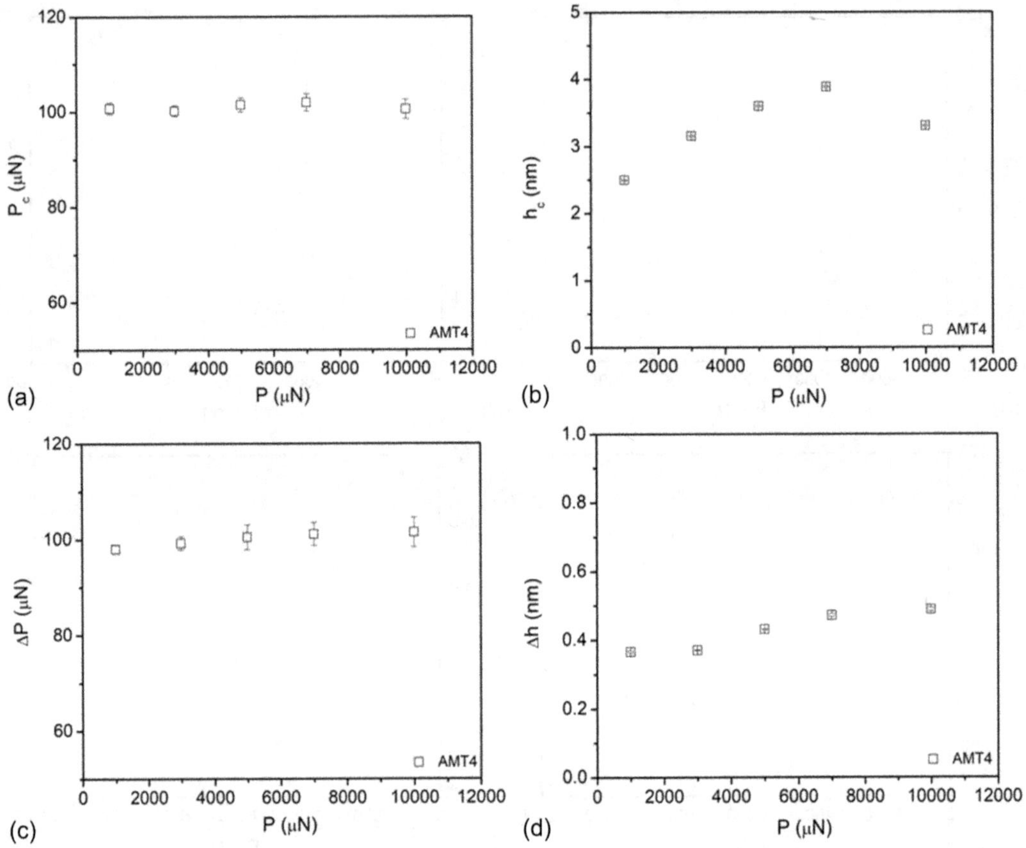

FIGURE 2.17 The variations of (a) P_c, (b) h_c, (c) ΔP and (d) Δh as a function of the applied nanoindentation loads in AMT4 ceramics (reprinted with permission from [41]).

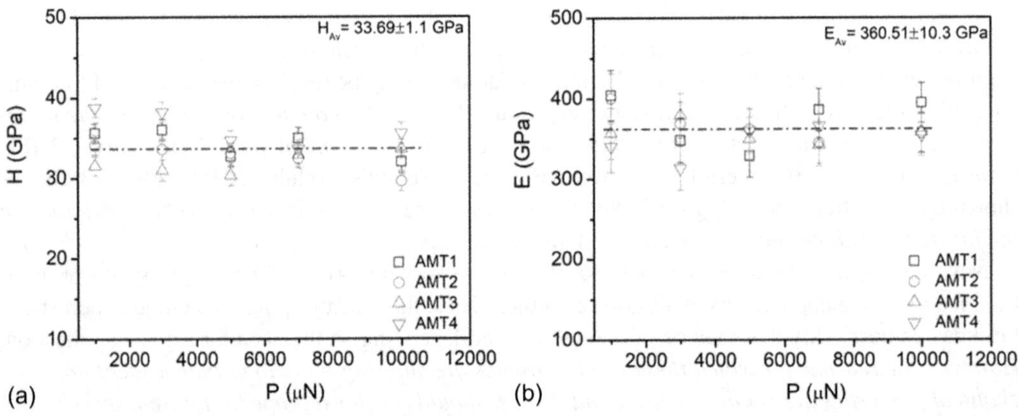

FIGURE 2.18 The variations in (a) nanohardness (H) and (b) Young's modulus (E) of AMT1, AMT2, AMT3 and AMT4 ceramics as a function of the loads applied during the nanoindentation experiments (reprinted with permission from [41]).

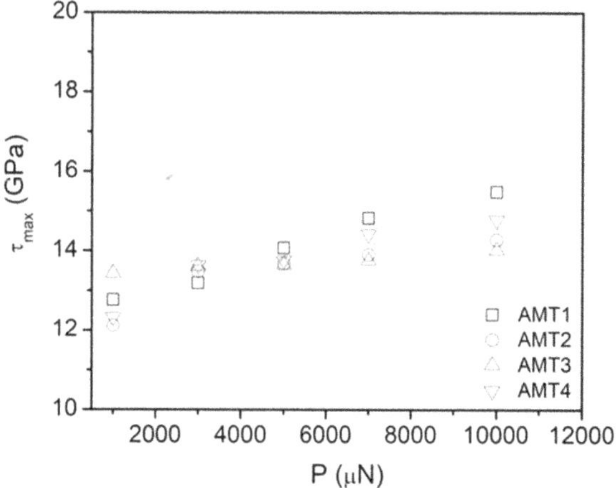

FIGURE 2.19 The variations of maximum shear stress with the applied nanoindentation loads in the AMT1, AMT2, AMT3 and AMT4 ceramics (reprinted with permission from [41]).

TABLE 2.1
Comparative Nanomechanical Properties of ZTA, MDA, TDA and AMT Ceramics Evaluated at an Applied Load of 10 mN

Sample	P (mN)	P_c (mN)	h_c (nm)	H (GPa)	E (GPa)	τ_{max} (GPa)	τ_{CRSS} (GPa)
MDA1	10	0.113	1.1	17.68 ± 1.3	346.1 ± 28.94	14.73 ± 0.69	0.26
MDA3	10	0.112	0.49	19.18 + 0.63	329.4 ± 20.6	14.71 ± 0.64	0.41
MDA5	10	0.11	1.2	19.57 ± 0.8	348.43 ± 18.93	13.67 ± 1.2	0.31
TDA1	10	0.14	1.08	20.43 ± 0.75	404.3 ± 23.56	14.40 ± 0.61	0.041
TDA3	10	0.103	0.6	25.31 ± 0.72	345.71 ± 24.4	13.63 ± 0.84	0.072
TDA5	10	0.11	1.11	20.05 ± 0.92	385.85 ± 23.95	14.55 ± 0.81	0.21
AMT1	10	0.101	3.99	31.99 ± 1.2	394.45 ± 24.87	15.48 ± 1.1	0.52
AMT2	10	0.1	3.98	29.58 ± 1.27	386.5 ± 24.1	14.27 ± 0.98	0.46
AMT3	10	0.101	3.49	30.56 ± 1.17	387.82 ± 24.64	13.64 ± 1.3	0.53
AMT4	10	0.1	3.30	33.52 ± 1.34	363.96 ± 25.93	14.78 ± 0.87	0.56

Therefore, it seems plausible to argue that as τ_{max} is much greater than τ_{theo} for all AMT ceramics; the probability of more dislocation nucleation inside the nanoindentation cavity is very much enhanced. These considerations suggest that shear stress-induced highly localized deformation is connected to the genesis of micro-pop-in events and. hence, nanoscale plasticity events in all AMT ceramics. However, detailed FESEM-based investigations will be required to confirm the validity of this suggestion.

2.5.3 The Final Comparative Overview

This section provides a final comparative overview of all the important nanomechanical properties of the materials developed in the current work. The research spanned a wide load range of 1000 μN to 1000 mN and a similarly wide range of loading rate of 1000 μN.s^{-1} (i.e., 1 mN.s^{-1}) to 1000 mN.s^{-1}. Therefore, to provide only a comparative picture, an applied nanoindentation load of 10 mN (i.e., 10000 μN) is chosen, and the comparative data evaluated at 10 mN load on P_c, h_c, H, E, τ_{max} and τ_{CRSS} are given in Table 2.1.

It is very interesting to note from the data presented in Table 2.1 that by the MIEC development approach, the nanohardness (H) values could be improved from 17 to 20 GPa in MDA ceramics to about 20 to 25 GPa in TDA ceramics to the highest value of about 30 to 35 GPa in AMT ceramics. These facts lend very credible support to the philosophical concept of MIEC material development.

It is also very interesting to note that by the MIEC development approach, Young's modulus (E) values could be improved from about 330 to 350 GPa in MDA ceramics to about 340 to 400 GPa in TDA ceramics to the highest value of about 386 to 400 GPa in AMT ceramics. These facts again lend very credible support to the philosophical concept of MIEC material development.

For ease of comparison, the nanohardness values of the ZTA (zirconia toughened alumina), MDA, TDA and AMT ceramics [38–41] are presented in Figure 2.20a, b, c and d, respectively.

Similarly, for the purpose of comparison, the Young's modulus values of the ZTA, MDA, TDA and AMT ceramics are presented in Figure 2.21a, b, c and d, respectively.

Finally, the overall grand comparisons of all nanohardness values of these MIEC materials are shown in Figure 2.22a. Similarly, the overall grand comparisons of all Young's modulus values of these MIEC materials are shown in Figure 2.22b.

In addition, it is very interesting to note from the data presented in Table 2.1 that the critical load values are almost certainly centered at about 100 to 110 nN with the singular exception of

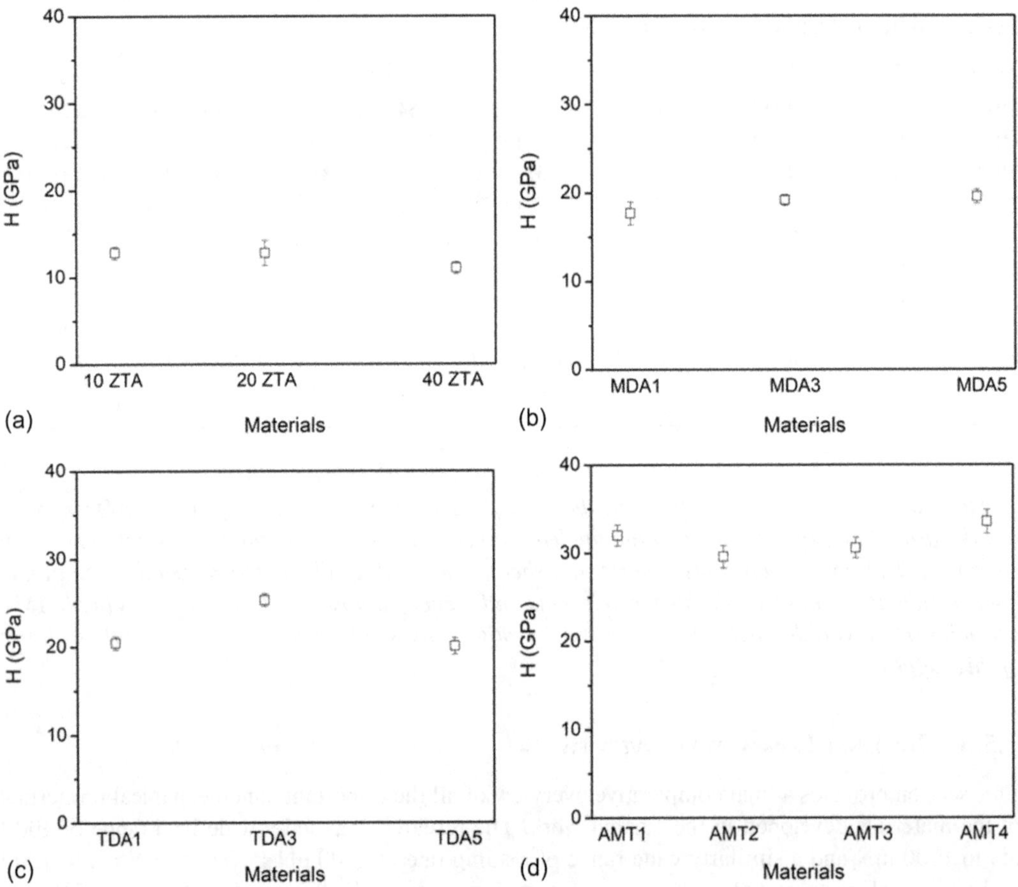

FIGURE 2.20 The variations of nanohardness (H) for the different MIEC materials (a) ZTA, (b) MDA, (c) TDA, and (d) AMT (reprinted with permission from [41]).

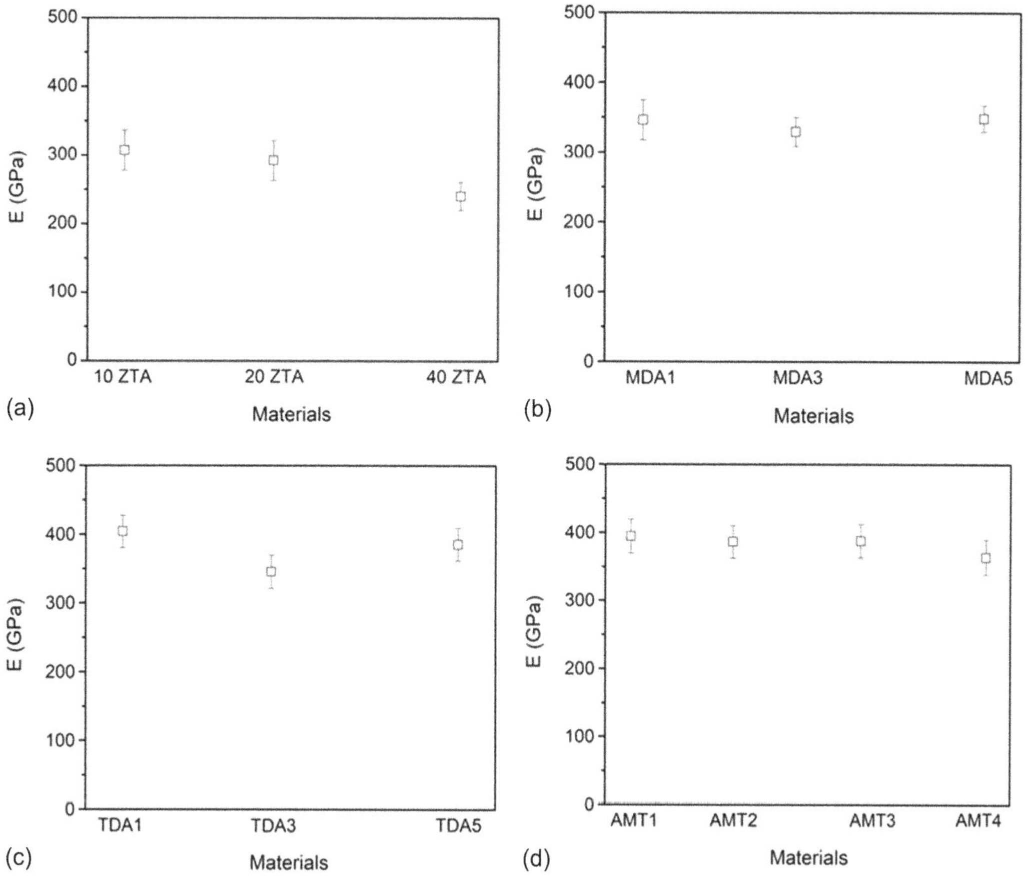

FIGURE 2.21 The variations of Young's modulus for the different MIEC materials (a) ZTA, (b) MDA, (c) TDA, and (d) AMT (reprinted with permission from [41]).

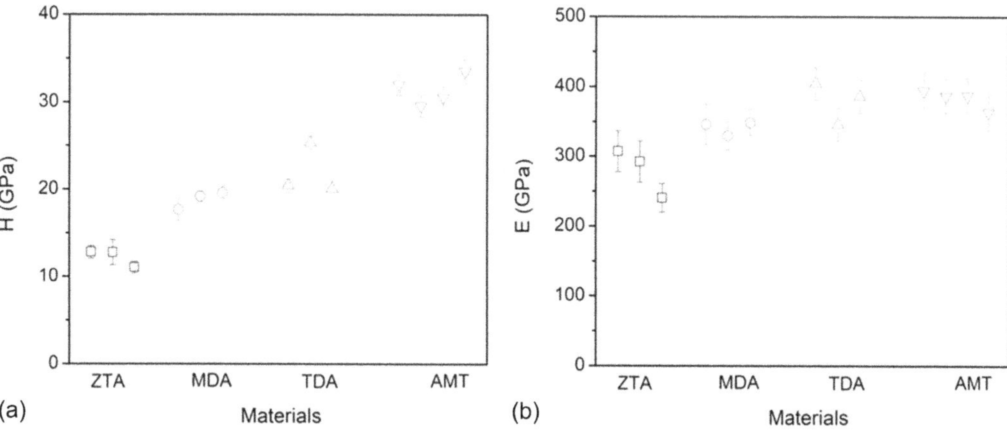

FIGURE 2.22 The grand comparisons of (a) nanohardness (H), and (b) Young's modulus (E) for the ZTA, MDA, TDA and AMT ceramics (reprinted with permission from [41]).

about 140 nN in the case of TDA1 ceramics. This points to an almost characteristic critical load value of about 100 to 110 nN for all the MIEC concept-based alumina ceramics developed in the present work.

In a similar manner, in almost all the cases of MDA and TDA ceramics, the critical depth values are centered at about 1 nm ± 10%. However, an exception happens in the case of AMT ceramics that show a critical depth value in the range of about 3 to 4 nm.

The data presented in Table 2.1 reveals that for MDA, TDA and AMT ceramics the maximum shear stress values are centered at about 14 GPa with an overall variation of about ±10%. The individual data scatters of these ceramics for maximum shear stress values are much lower though.

A similar situation pertains to most of the critical resolve shear stress values of the MDA and AMT ceramics. These stress values are centered at about 0.2 to 0.5 GPa with an average of about 0.35 GPa or 350 MPa. It is very interesting to note this magnitude matches with the typical flexural strength of alumina ceramics that often fail by the development of small grain boundary cracks that incubate and grow to final critical size prior to failure through catastrophic crack propagation. The only exception to this scenario is TDA ceramics, which exhibit critical resolved shear stress values of very low magnitude, e.g., 0.04 to 0.2 GPa.

These results (Table 2.1) therefore appear to be very interesting from the viewpoint of not only better structural ceramic development but also better understanding about the physics of deformation at the microstructural length. The reasons behind the genesis of such behaviors by the presently developed MIEC materials are therefore very important to be known. These situations depict both the scope and the need for future work in this arena so that better alumina ceramics with much enhanced nanoscale contact damage resistance can be developed through microstructural engineering in the future.

2.6 NANOINDENTATION RESPONSE OF DIFFERENT TYPES OF ALUMINA CERAMICS

Recent works also focus on microstructural aspects of the nanoindentation response of various alumina ceramics as well [43–50]. These significant results report on the mechanical response at the micro-/nanoscale level of different varieties of alumina, e.g., from a single crystal to a single grain to a polycrystalline alumina with fine grain sizes (e.g., about 2 μm), polycrystalline alumina with intermediate grain sizes (e.g., about 10 μm) and polycrystalline alumina with coarse grain sizes (e.g., as high as about 30 μm).

These results [43–50] strongly suggest that the mechanical response was a complex yet sensitive function of the microstructural parameters, i.e., grain size, grain size distribution, relative density, residual porosity, frozen in residual stress, presence or absence of texture on one hand and the rate of energy transfer from the loading train to the microstructural level of various alumina samples on the other hand. Just to support this proposed viewpoint, the basic trend of percentage enhancement ($\Delta H\%$) in nanohardness due to loading rate variations is shown in Figure 2.23 as a function of the variety of alumina microstructure utilized in the present work. It is interesting to note from the data presented in Figure 2.23 that ($\Delta H\%$) varies from about 7% to about 66% in going from the 20 μm grain-sized alumina to the single crystal alumina.

However, the loading rate effect is found in recent works [38–50] to be a generic mechanism that pertains to brittle solids. Therefore, for brittle solids in general and alumina in particular this issue needs to be better understood. Once a better understanding of the scientific issues involved in deformation mechanisms at the microstructural length scale is developed, it will certainly help in better materials engineering in the days to come. The lack of a knowledge base on the loading rate behavior of textured alumina has led to gathering deeper knowledge on the same at an ultralow length scale. What is presented next is just a glimpse of some specific issues of interest that are thought to be worthy of scientific attention in our future research.

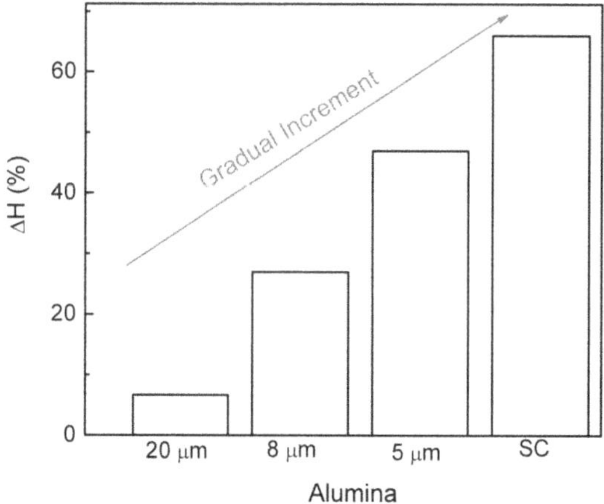

FIGURE 2.23 Gradual increase in the percentage increment of nanohardness (H) with the loading rates from coarse to intermediate to textured alumina with varying grain size to single crystal alumina (reprinted with permission from [43]).

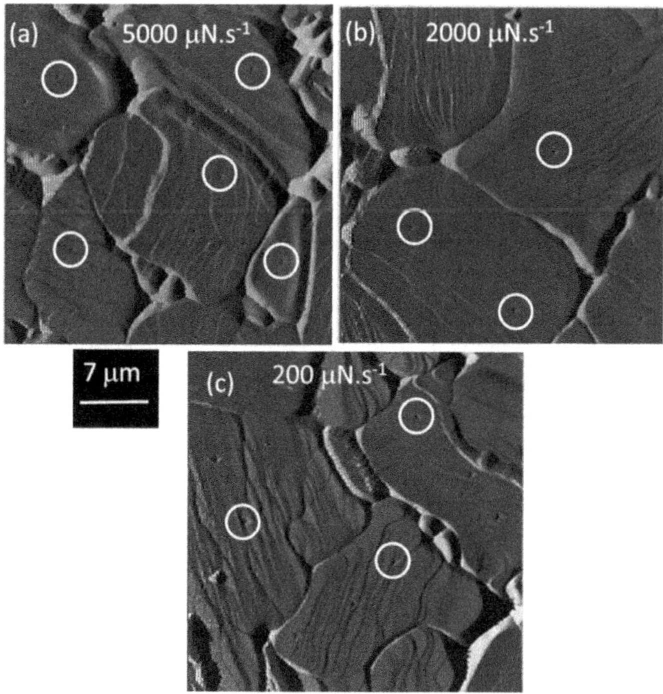

FIGURE 2.24 For a given peak load of 10000 μN, the SPM images of typical nanoindents at different loading rates of (a) 5000 $\mu N.s^{-1}$, (b) 2000 $\mu N.s^{-1}$ and (c) 200 $\mu N.s^{-1}$ (reprinted with permission from [43]).

2.6.1 Influence of Loading Rate on the Nanohardness of Textured Alumina at Ultralow Length Scale

The influence of loading rate on the nanohardness (H) of textured alumina is examined [43] at different loading rates (e.g., ~10–10000 $\mu N.s^{-1}$) at a peak load of 10000 μN. The scanning probe microscope (SPM) images of typical nanoindents at different grains of textured alumina at different loading rates are shown in Figure 2.24.

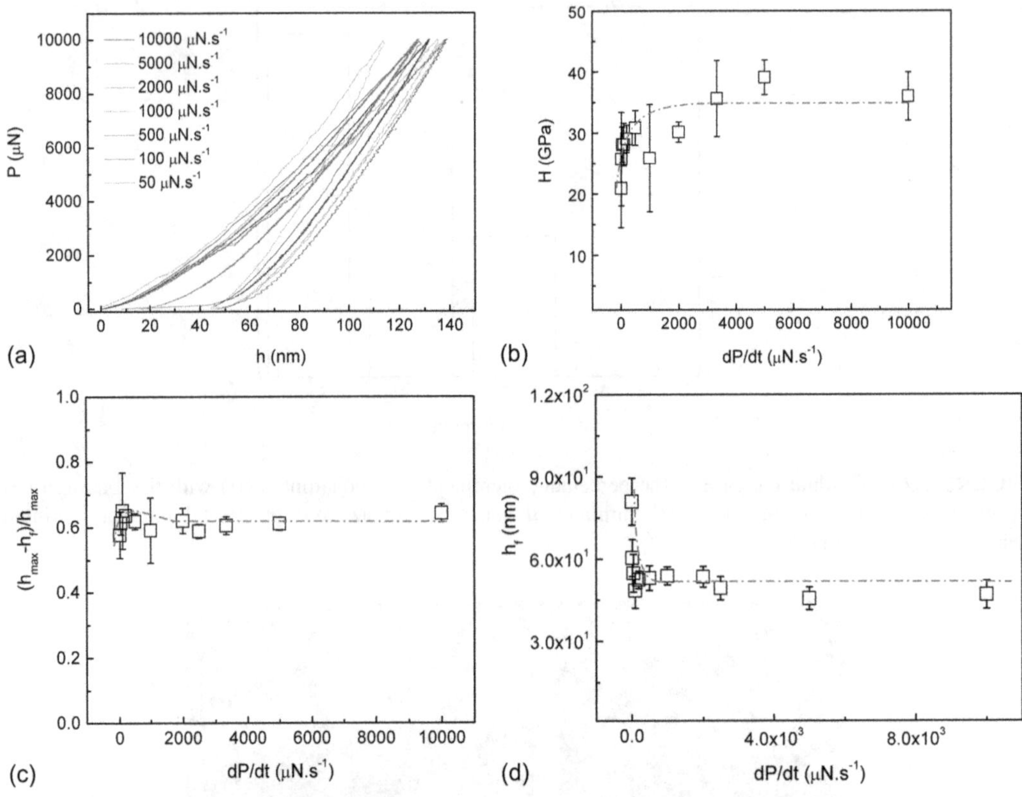

FIGURE 2.25 Loading rate effects in nanoindentation. (a) P-h plots, (b) H, (c) elastic recovery and (d) final depth as functions of dP/dt (reprinted with permission from [43]).

The nanoindents are marked by white hollow circles. The load–depth (P-h) plots obtained from the nanoindentation experiment at different loading rates are given in Figure 2.25. The peak load is kept constant at 10000 μN.

The nanohardness (H) of the present textured alumina initially increased with the loading rates until it saturated at higher loading rates, Figure 2.26.

However, the magnitude of nanohardness is relatively higher (~35 GPa). The increase in nanohardness with the loading rates corroborates with the variation of elastic recovery of the present textured alumina with the loading rates (Figure 2.27).

Elastic recovery is defined as the ratio of the difference between the maximum and final depth of penetrations ($h_{max} - h_f$) and the final depth of penetration (h_f) [1–6]. The elastic recovery also initially increases with the loading rates followed by gradual saturation at higher loading rates, Figure 2.27. The initial greater elastic recovery of the present alumina with the loading rates resulted in greater nanohardness of the present alumina with the loading rates (Figure 2.26). The increasing trend of nanohardness (H) with the loading rates complemented the corresponding trend of variation of the final depth of penetration (h_f) with the loading rates, as shown in Figure 2.28.

The presence of micro pop-in events is confirmed in the loading part of P-h plots (Figure 2.25). The number of multiple micro pop-in events found to occur at lower loading rates (e.g., ~10–100 $\mu N.s^{-1}$) is relatively more than the number of multiple micro pop-in events found to occur at higher loading rates (e.g., ~10000 $\mu N.s^{-1}$), Figure 2.25. These micro-pop-in events are also observed by other researchers [38–41] but the present observation is possibly the very first one in textured alumina under the present experimental condition as detailed above [43].

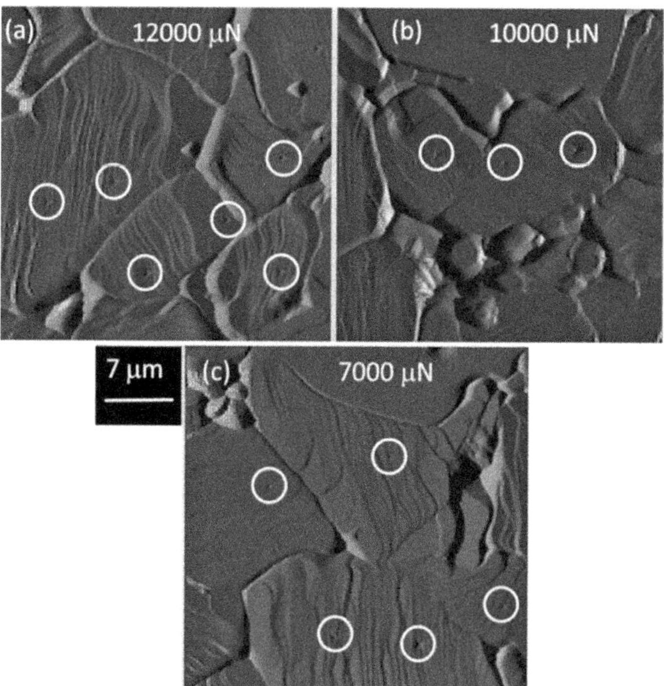

FIGURE 2.26 SPM image showing typical nanoindents marked by white hollow circles at different loads of (a) 12000μN, (b) 10000 μN and (c) 7000 μN (reprinted with permission from [43]).

2.6.2 Effect of Load (P) on the Nanohardness of Alumina

The recent work [43] also investigates the effect of load on the nanohardness (H) of the textured alumina. The SPM images showing typical impressions of the nano indents are shown in Figure 2.29a–c.

The load–depth (P-h) plots obtained from the partial load–unload experiments at different applied peak loads of 5000, 7000, 10000 and 12000 μN are shown in Figure 2.30.

Interestingly, the nanohardness (H) of the present textured alumina gradually decreases with the depth of penetration (h), as shown in Figure 2.31.

The gradual decrement of the nanohardness (H) with the applied load (P) and hence depth (h) signifies the presence of a strong ISE in the present textured alumina. This is further confirmed by the fitting using the well-known Nix and Gao model [5]. The characteristic form for the depth dependence of the hardness is thus modeled using Equation 2.1 [5]:

$$\frac{H}{H_O} = \sqrt{1 + \frac{h^*}{h}} \tag{2.1}$$

where H is the nanohardness for a given depth of penetration (h); H_O is the hardness at the limit of infinite depth; and h^* is a characteristic length that depends on the shape of the indenter, the shear modulus and H_O [5].

The nanohardness values obtained from the fitting of the Nix and Gao model match quite well with that of the present experimental data. This is shown in Figure 2.32. The magnitude of H_0 (~30 GPa) typically matches (Figure 2.32) with the depth-independent nanohardness data regime experimentally measured in the current work reported in Ref. [43].

The trend of variation of the nanohardness (H) of the present textured alumina also corroborates well with the corresponding variation of the final depth of penetration (h_f) with the applied loads

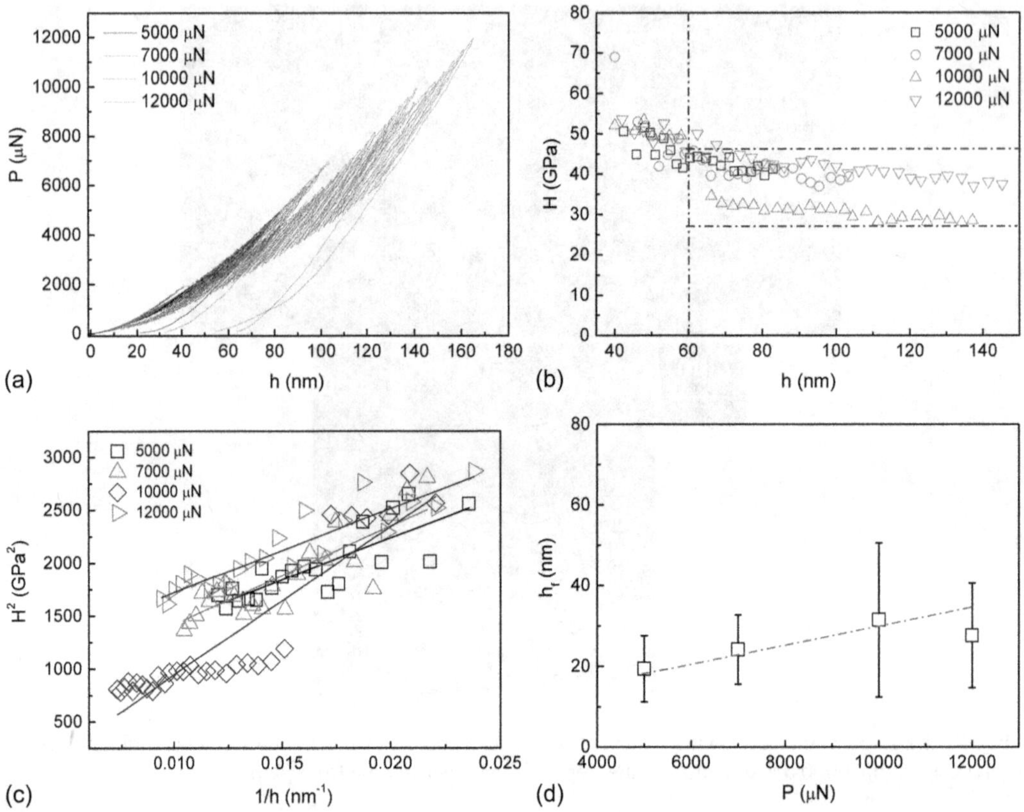

FIGURE 2.27 Partial load–unload nanoindentation experiments. (a) P-h plots, (b) H vs, h, (c) Nix and Gao model fitting and (d) final depth vs. P (reprinted with permission from [43]).

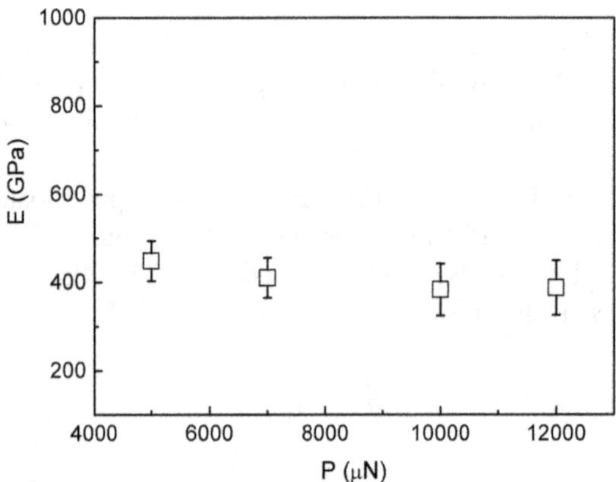

FIGURE 2.28 Variation of Young's modulus (E) with the applied loads (P) (reprinted with permission from [43]).

(P). This is shown in Figure 2.33. The average magnitudes of the final depth of penetration encounter gradual increments with the applied loads Figure 2.33).

This results in an increase in the corresponding projected contact area (A_P) of the present [43] textured alumina with the applied loads. The greater the projected contact area, the lesser the nanohardness (H), as nanohardness is given by the ratio of the applied load (P) and the projected contact

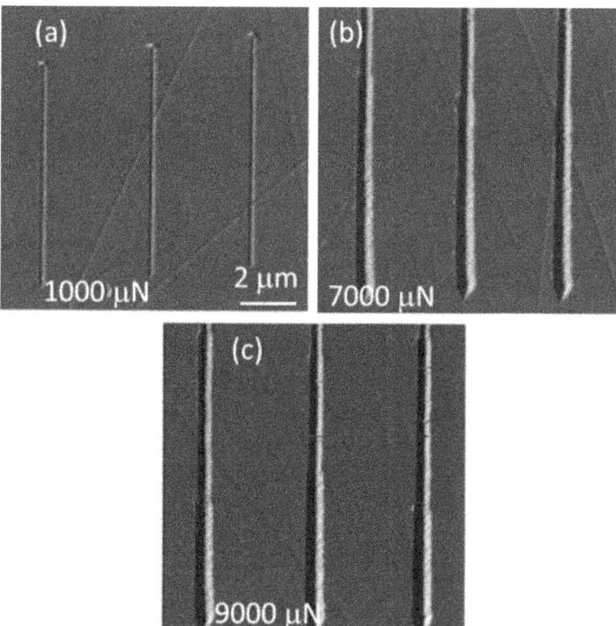

FIGURE 2.29 SPM images showing nanoscratches in single-crystal alumina (SC) at different normal loads of (a) 1000 μN, (b) 7000 μN and (c) 9000 μN (reprinted with permission from [43]).

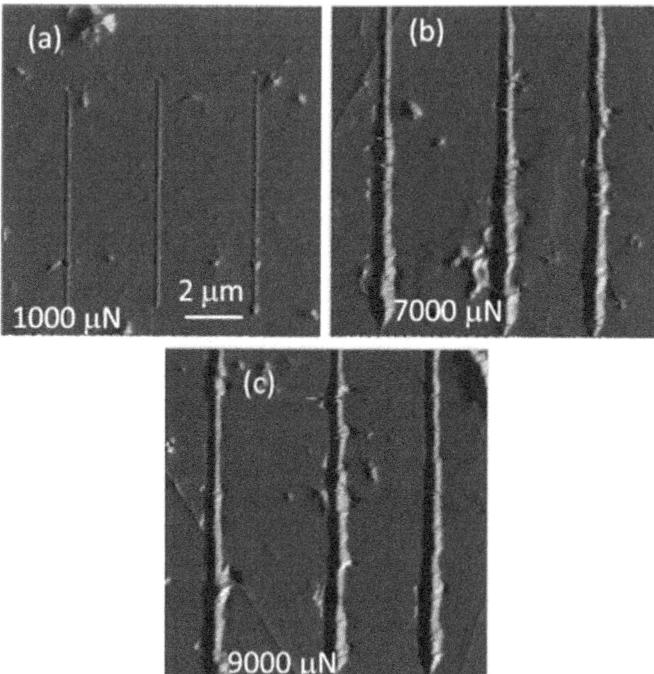

FIGURE 2.30 SPM images showing nanoscratches in polycrystalline fine-grain alumina (FG) at different normal loads of (a) 1000 μN, (b) 7000 μN and (c) 9000 μN (reprinted with permission from [43]).

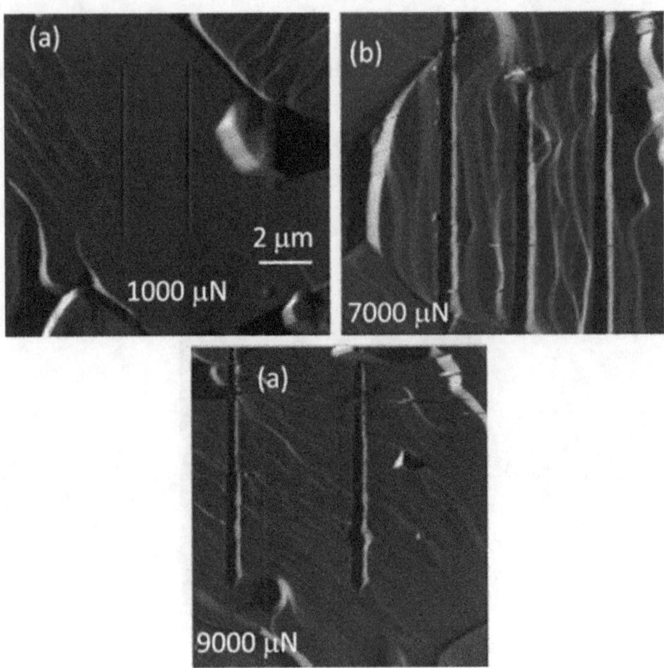

FIGURE 2.31 SPM images showing nanoscratches in polycrystalline textured alumina (MG) at different normal loads of (a) 1000 μN, (b) 7000 μN and (c) 9000 μN (reprinted with permission from [43]).

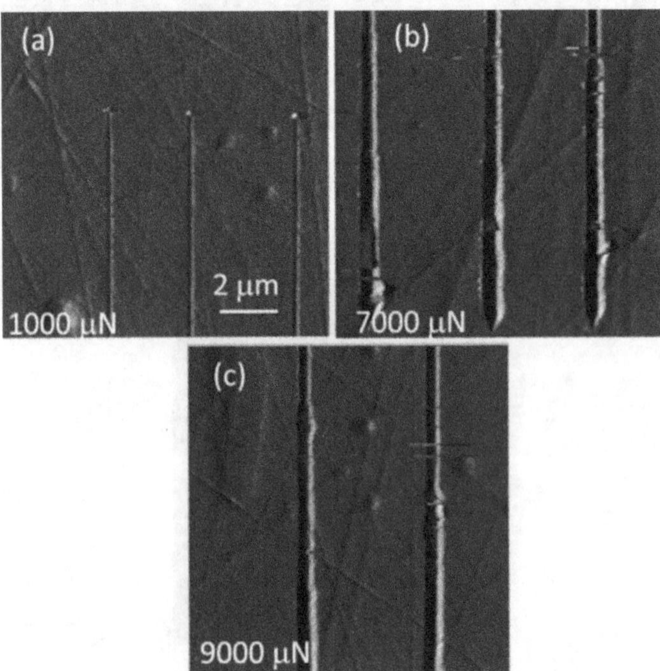

FIGURE 2.32 SPM images showing nanoscratches in polycrystalline coarse-grain alumina (CrG) at different normal loads of (a) 1000 μN, (b) 7000 μN and (c) 9000 μN (reprinted with permission from [43]).

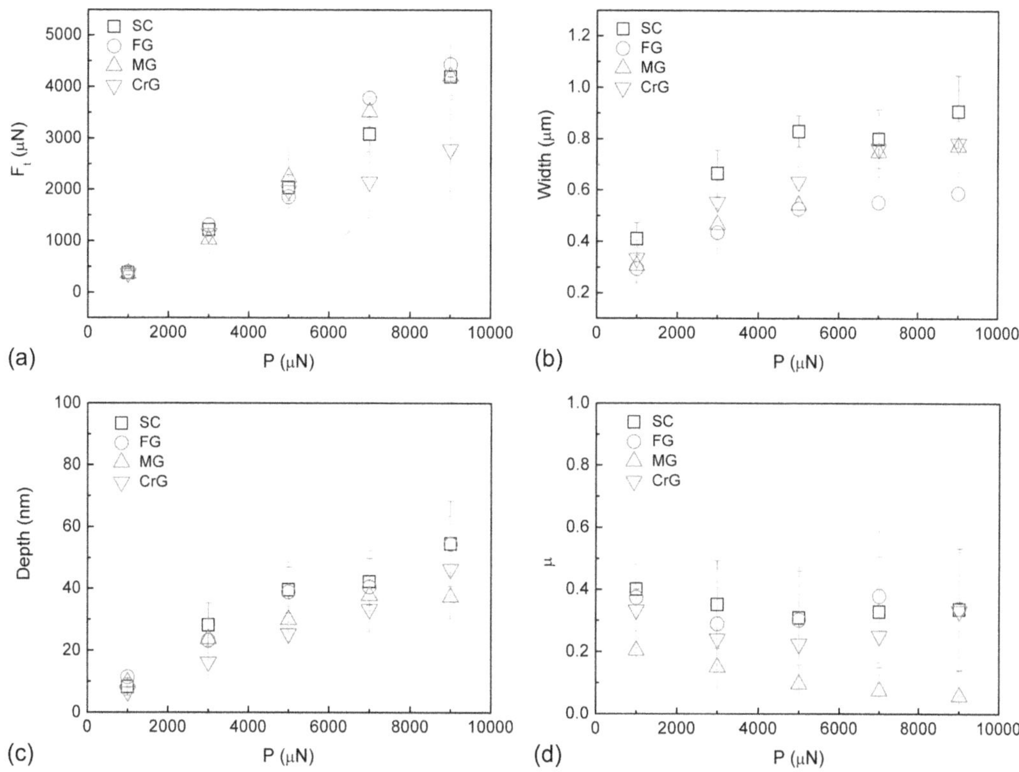

FIGURE 2.33 Variation of (a) lateral force (F_t), (b) width, (c) Depth and (d) friction coefficient as a function of normal load (P) (reprinted with permission from [43].

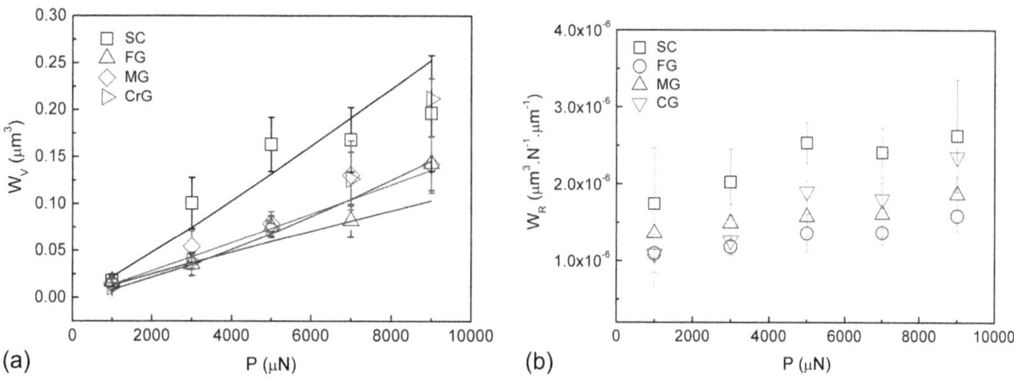

FIGURE 2.34 Variation of (a) wear volume (W_V) and (b) wear rate (W_R) with the applied normal loads (P). The solid lines in panel a indicate the power law fits (reprinted with permission from [43]).

area (A_P) [4]. The value of the nanohardness saturated is in the range of ~28–45 GPa for different applied loads (Figure 2.31).

Further, this significantly high value of nanohardness [43] matches reasonably with what is already reported for other varieties of alumina [44–50] ceramics that do not necessarily possess a microstructural texture. This eventual match also raises the question of whether texturing may provide any additional advantage at the ultralow load-induced scale of static contact. Young's modulus (E) of the present alumina ranged between 380 and 450 GPa, as expected (Figure 2.34).

The interrelationship between a textured alumina microstructure and a relatively high magnitude of Young's modulus is yet to be comprehensively well understood and, hence, may certainly provide an area of future research.

2.7 NANOINDENTATION: EMERGING AREAS

This section focuses on the emerging areas in nanoindentation of especially brittle solids and the related physics of deformation at a small length scale. In general, the areas of emerging importance are the dynamic contact behavior at the nanoscale, the interaction between static contact-induced damages and dynamic contact-induced damages, and the deformation physics of microstructural units (e.g., grains of a polycrystalline microstructure). Further, it needs to be emphasized that analysis of the deformation physics is required by applying artificial intelligence (AI), machine learning (ML), artificial neural network (ANN), generative adversarial network (GAN), finite element modeling (FEM) and molecular dynamic (MD) simulation. There are two general trends. One is to only apply these techniques either individually or in combination to analyze the physics of deformation of various materials at the microstructural length scale. The other is to combine experimental data obtained by conventional nanoindentation or AFM-based nanoindentation with theoretical analysis by AI, ML, ANN, GAN, FEM or MD simulation. The second approach is no doubt challenging but the fruits that it may deliver guarantee that such a challenge is worthy of acceptance. In what follows, an attempt is made to explore some specific recent observations in the emerging areas just as illustrative examples.

2.8 NANOTRIBOLOGICAL BEHAVIOR OF SINGLE-CRYSTAL AND POLYCRYSTALLINE ALUMINA

The experimental observations are mostly reported for static contact behavior [38–42, 44–50]. It generally reveals gradual increments in nanohardness (H) of different varieties of alumina with either load [38] or the loading rates [44–50] or both [38, 39]. Perhaps, the same situation does not really prevail during dynamic contact behavior of different types of alumina. To support this additional viewpoint, the anomalous case of dynamic contact at the microstructural viewpoint is presented. Based on recent work [43], this section therefore discusses the dynamic contact behavior of different varieties of alumina starting from single-crystal alumina to polycrystalline fine-grain alumina to polycrystalline coarse-grain alumina microstructures. Thus, the various types of alumina are reported [43] to be single-crystal alumina (SC), fine-grain alumina (FG) having a grain size of ~1 μm, textured alumina having mixed grains (MG; 2 to 30 μm size) and coarse-grain alumina (CrG) having a grain size of ~20 μm. Further, the nano scratch experiments are reported [43] to be conducted at different loads ranging from 1000 to 9000 μN at a constant scratch speed of 10 $\mu m.s^{-1}$.

The SPM images showing the nanoscratch grooves for these four varieties of alumina at four different loads in the range of 1000 to 9000 μN are shown respectively in Figures 2.29 to 2.32. The lateral force (F_l) in the four different types of alumina increases with the applied normal loads (P) (Figure 2.33a). The scratch width and scratch depth also increase with the applied normal loads for four different types of alumina, as shown respectively in Figure 2.33b and c. The coefficient of friction (μ) gradually decreases with the increase in the applied normal loads (P) (Figure 2.33d). This decreasing trend of μ with the applied loads [43] matches with the data reported in Ref. [50] for microscratch experiments on a coarse grain (~20 μm) alumina at applied normal loads of 2–15 N.

The wear volume (W_V) is calculated assuming a triangular cross section. It is also reported [43] to increase with the applied normal loads (P), as shown in Figure 2.34a. Interestingly, the wear volume for each type of alumina exhibits the empirical power law that fits with the applied normal loads. The empirical power law relationship exhibited by the single-crystal alumina (SC) is given by Equation 2.2:

$$W_V = 10^{-5} P^{1.11} \tag{2.2}$$

The corresponding goodness of fit (R^2) is reported [43] to be ~0.93. The empirical power law relationship exhibited by polycrystalline fine-grain alumina (FG) is given by Equation 2.3 as

$$W_V = 2\times10^{-5} P^{0.94} \tag{2.3}$$

The corresponding goodness of fit (R^2) is reported [43] to be as high as ~0.96. The empirical power law relationship exhibited by polycrystalline textured alumina (MG) is given by Equation 2.4:

$$W_V = 10^{-5} P^{1.05} \tag{2.4}$$

The corresponding goodness of fit (R^2) is reported to be as high as ~0.99. The empirical power law relationship exhibited by polycrystalline coarse-grain alumina (CrG) is given by Equation 2.5 as

$$W_V = 10^{-6} P^{1.3} \tag{2.5}$$

The corresponding goodness of fit (R^2) is reported to be as high as ~0.98. The corresponding wear rate (W_R) also increases with the applied loads (P) (Figure 2.34b).

One interesting fact is noticeable from the aforementioned data: irrespective of the grain size, the nanomechanical properties of the different types of alumina in dynamic contact follow the same trend of variations with the applied normal loads (Figures 2.29 to 2.34). The lateral force (F_l), scratch width, scratch depth, coefficient of friction (μ), wear volume (W_V) and wear rate (W_R) increase with an increase in the applied normal loads (P).

The scratch width is the highest for the single-crystal and lowest for the fine-grain alumina. Similar data for the fine-grain and mixed-grain (textured) alumina are reported [43] to be intermediate between these two limits and are almost coincident with those of one another. Again, the scratch depth is the highest for single-crystal and fine-grain alumina, while the scratch depth is the lowest for mixed-grain (textured) alumina. Contrary to conventional wisdom, the scratch width is the second lowest in the case of the coarse-grain alumina. Classically, one would expect the single crystal to have the least scratch width and depth followed by those of the fine-grain, mixed-grain (textured) and coarse-grain alumina microstructures. This is clearly not happening at the ultralow scale dynamic contact. The reason for such behavior is yet to be known and therefore it is an important area for future research.

Similarly, the lateral force (F_l) is the highest for the fine-grain alumina but the lowest for the coarse-grain alumina. Similar data for the single-crystal and mixed-grain (textured) alumina are reported [43] to be intermediate between these two limits and are almost coincident with those of one another. The reason for such behavior is yet to be known. Hence, it should constitute an important area for future research.

Very interestingly, again, the friction coefficient (μ) is indeed the highest for the single-crystal alumina and the least for the mixed-grain (textured) alumina. Similar data for fine-grain and coarse-grain alumina are reported [43] to be intermediate between these two limits and are almost coincident with those of one another. The scientific reason behind such an unexpected behavior is yet to be known and, hence, may provide an opportunity for future research.

On the other hand, the wear volume (W_v) is the highest for the single-crystal and the lowest for the fine-grain alumina. Data for the mixed-grain (textured alumina) and the coarse-grain alumina are reported [43] to be intermediate between these two limits and are sometimes almost coincident with those of one another. As per conventional ideas, the wear volume should have been the highest for the coarse-grain alumina and the lowest for the single-crystal alumina. The results of the present nanotribology experiments are not totally in accordance with the expectations from the conventional idea. The reason for such behavior is yet to be known and, hence, should definitely constitute an area of future research.

Again, in tune with the case of wear volume data, the wear rate (W_R) is the highest for the single-crystal and the least for the fine-grain alumina. Similar data for the mixed-grain (textured alumina) and the coarse-grain alumina are reported [43] to be intermediate between these two limits and are sometimes almost coincident with those of one another. As per conventional ideas, the wear rate should have been the highest for the coarse-grain alumina and the lowest for the single-crystal alumina. Therefore, it is confirmed that the results of the nanotribology experiments conducted in the recent work [43] are not totally in accordance with the expectations from conventional ideas. The reason for such behavior is yet to be known and hence, should be given priority as an area of future research.

Therefore, it seems plausible to argue that grain size only is not a single factor that determines the dynamic contact resistance of the various aforesaid alumina samples in the present nanotribology experiments. Rather, it may be suggested that the whole-contact dynamics in the nanotribology experiments is a complex yet strongly sensitive function of the grain size, applied load, loading rate, surface roughness, relative density, residual porosity, built-in frozen residual stress and the presence or absence of texture. In addition, the nanohardness and Young's modulus of the different alumina samples might also play major roles in the development of the dynamic contact resistance of the present alumina samples in the current nanotribology experiments [43]. However, further confirmatory experiments as well as detailed postmortem examinations are required to be conducted in the future to verify the validity of the suggestions made in the recent work.

2.9 INTERACTION BETWEEN DYNAMIC CONTACT-INDUCED DAMAGES AND STATIC NANOSCALE CONTACT-INDUCED DAMAGES

It is interesting to note that there is a unique recent attempt [50] about the effect of loading rate variation (1 to 1000 mN s^{-1}) on the nanomechanical response of an intermediate grain size, i.e., 8 μm alumina ceramic. This study uses a Berkovich nanoindenter. What makes this study unique is that it attempts to understand two other important scenarios in the domain of nano- and microscale contact-induced damage interactions. There are hardly any studies reported on these aspects in alumina ceramics.

In the first case, it introduces Vicker's microindents at a load of 100 N to ensure that radial cracks emanate from the four corners of the microindent. Next, it places nanoindents made at 500, 700, and 1000 mN loads at various angles (i.e., 0°, 30°, 45° and 90°) close to the radial cracks and studies how the nanomechanical properties are affected [50]. This study is done [50] with the idea of understanding how the presence of microscale static contact damage affects the nanomechanical properties.

In the second case, it makes conventional microscratch grooves at normal loads of 2, 5, 10 and 15 N. Then, this study introduces linear nanoindent arrays made at 1000 mN and makes them traverse perpendicularly across the different scratch grooves. This study is done [50] to understand how the presence of microscale dynamic contact damage affects the nanomechanical properties.

It is reported [50] further that the magnitude of H increases by ~30% with an increase in the loading rates. This enhancement is suggested [50] to be due to the combined contributions of several factors. The first and the most important one is the nucleation of dislocations [50]. It happens due to very high shear stress generation. This shear stress is generated just underneath the nanoindenter during the nanoindentation process [50]. It is noted that there is a spatial gradient of this shear stress. This gradient is suggested to be active along the walls of the nanoindentation cavity. The next most important factor that contributes to this scenario is shown [50] to be the generation of shear deformation bands within the nanoindentation cavities. Finally, along with these highly plastic and shear deformation-induced damages, the presence of microcracks is noted from the FESEM-based evidence. It is suggested that all these factors may combinedly contribute to the enhancement in (H) of the intermediate grain-sized alumina [50].

As far as the interaction between the static microscale contact damage and the nanoscale damages deliberately created by the nanoindents is concerned, very interesting observations are reported

[50]. It is found [50] that the nanomechanical properties are highly sensitive to the presence of the radial cracks and the angular orientation of the nano indents with respect to the direction along which the radial cracks propagate. Depending on the nanoindentation load (e.g., 500 mN, 700 mN and 1000 mN) and the relative orientation angle (ROA) (e.g., 0°, 30°, 45° and 90°), the magnitude (H) is reported [50] to degrade by as much as 45% to 90%. Similarly, the magnitude (E) is reported [50] to degrade by as much as 15% to 90%.

The degradations in both (H) and (E) are reported to be the highest when the ROA is lowest, that is, 0°, and the lowest when the ROA is the highest, that is, 90°. These degradations are attributed [50] to the relative degrees of severity of interactions between the nanoindents, which act as microstructural flaws, and the radial cracks associated with the microscale contact damage.

As far as the interactions between the dynamic microscale contact damages created at normal loads of 2, 5, 10 and 15 N, and the nanoscale damages deliberately created by the nanoindents made at a fixed load of 1000 mN are concerned, very interesting observations are reported [50]. The results illustrate that there can be very severe effects on the nanomechanical properties of alumina under such a scenario. For instance, due to the interaction between the microcracks underneath the scratch grooves and the nanoscale contact damages, the magnitude of (H) can degrade by as much as 15% to 45% [50]. Further, it is reported that the magnitude of (E) can degrade by about 5% to 45% as well [50]. The maximum degradation of (H) and (E), about 45%, are reported [50] to occur at the center of the dynamically created scratch grooves. It happens because the maximum population of microcracks underneath the scratch grooves most probably resides here. However, as one moves away from the center of the scratch grooves on both sides the center of the scratch grooves, the degradation in (H) becomes about 15% as compared to those measured at the control location away from the scratch groove. These facts provide strong support to the hypothesis about the interaction between the subsurface microcracks and the nanoindents themselves. In a similar fashion, the degradation in (E) becomes about 5% as compared to those measured at the control location away from the scratch groove. Overall, this singular study [50] presents very useful information about how the nanoindentation technique can be utilized to map the degree of interaction between (a) the nanoscale contact damages and the statically created microscale contact damages, and (b) the nanoscale contact damages and the dynamically created microscale contact damages, i.e., the scratch damages.

2.10 NANOINDENTATION RESPONSE OF A SINGLE ALUMINA GRAIN

There is a lot of study reported on the nanoindentation response of single crystalline and polycrystalline alumina ceramics [1, 4, 38–41, 43–48, 50]. However, there is hardly any study reported on the nanoindentation response of a single alumina grain, especially in a coarse-grain microstructure. A recent attempt [49] provides very interesting input in this regard.

This study [49] reports the nanoindentation response of a single alumina grain in a coarse-grain size (e.g., 10 μm) alumina ceramic. The nanoindentation experiments are reported to be conducted at 1000, 5000, 10,000, and 12,000 μN loads with a Berkovich nanoindenter. From the nanoindentation made at 12,000 μN load, the experimental data suggest two very important findings. The first one is that at ultralow depth, ≤ 40 nm, the nanohardness value could be as high as about 50 GPa. This is slightly higher than the nanohardness data, 30–35 GPa, measured for the polycrystalline AMT samples mentioned earlier (Table 2.1) but an order of magnitude similarity exists. Further at relatively higher depths $40 \leq h \leq 180$ nm, the depth-independent nanohardness is about 21 GPa.

Additional nanoindentation experiments conducted at 1000, 5000 and 10,000 μN loads measure nanohardness to be about 30, 26 and 23 GPa, respectively. Thus, these data reveal a surprising nano ISE present in the (H) data of the single alumina grain. Analyzing the data by the Nix and Gao model [5] yields a nanohardness of about 21.2 GPa at infinite depth with a characteristic depth parameter of about 33 nm, which physically signifies that up to such depth the strain gradient will be the most severe. As a result, the nanohardness should be the highest at such a small depth.

Such a picture matches very nicely with the fact [49] that H is reported to be about 50 GPa at ≤40 nm. The presence of nano ISE is explained in terms of two related concepts. The first one is the presence of strain gradient plasticity [5]. The second one is a load-dependent change in relative stiffness [49].

Further, it is noted that the profuse presence of NSP events as reflected in occurrences of micro-pop-ins are related to two important contributors. The first one is the homogeneous nucleation of dislocations. The second one is shear stress. The nucleation process of dislocations is linked to the generation of huge shear stress that is proposed [49] to be active just underneath the nanoindenter.

Furthermore, the experimental results show that the nanoscale contact deformation resistance is uniquely defined by the critical load, and increases as the nanoindentation load (P) is increased. In addition, both critical load and critical depth data are reported [49] to exhibit the power law dependencies on (P) with positive exponents. These unique observations suggest that the grain of this coarse-grain alumina possesses an inherent capability to suitably adjust and, if required, enhance its resistance against nanoscale contact-induced deformation as and when it is required due to an increase in (P).

Thus, it reveals a unique possibility that grains of coarse-grain alumina can be microstructurally tuned to have a built-in defense mechanism against nanoscale contact-induced deformation. The model proposed in this work [49] establishes that both critical load and critical depth should indeed have power law dependencies on (P) as is experimentally observed.

2.11 SUMMARY AND CONCLUSION

This chapter focuses on the physics of deformation at the microstructural or small length scale of different materials, i.e., metals, ceramics, polymers and biomaterials. However, due to the complexity of nanoindentation of brittle solids that cannot deform by plastic deformation, special importance is given to them. The most current work reported on metals, ceramics, polymers, biomaterials, etc. is briefly summarized. Next, the global literature is briefly reviewed. Then the recent works on the nanoindentation response of the MIEC are discussed with an emphasis on the microstructure, load, loading rate and nanomechanical properties of quadrangular interdependence, especially in the cases of magnesia densified, titania densified and magnesia-titania co-densified pressureless sintered alumina ceramics. Further, the recent studies on the nanoindentation response of various varieties of alumina ceramics, i.e., fine grain, intermediate grain size, textured and coarse grain are discussed in light of subtle variations in the physics of deformation at the microstructural length scale. Finally, emerging areas of importance, including the dynamic contact behavior at the nanoscale, the interaction between static contact-induced damages and dynamic contact-induced damages, and the deformation physics of microstructural units (e.g., grains of a polycrystalline microstructure) are adequately discussed.

ACKNOWLEDGEMENTS

The authors acknowledge the kind support and encouragement received from various directors of the Council of Scientific and Industrial Research (CSIR)–Central Glass and Ceramic Research Institute (CGCRI), Kolkata, India, during the tenures of whom parts of the research work carried out by the authors were done at CSIR-CGCRI. They also gratefully acknowledge the various infrastructural supports received from various divisions of CSIR-CGCRI, Kolkata, India. The financial support received from Indian sponsoring agencies, including the CSIR, UGC, DST-INSPIRE, DST(SERB), DAE-BRNS and IPR, of the Government of India (GOI) in terms of various project sponsorships and fellowships is gratefully acknowledged by the authors. Finally, the author AKM acknowledges the kind and continued support received from the authorities of Sharda University, Greater Noida, Uttar Pradesh, India.

REFERENCES

1. Arjun Dey and A. K. Mukhopadhyay, *Nanoindentation of Brittle Solids*, CRC Press, Taylor and Francis Group, USA, 2014, https://www.crcpress.com/Nanoindentation-of-Brittle-Solids/Dey-Mukhopadhyay/p/book/9781138076532.
2. Arjun Dey and A. K. Mukhopadhyay, *Microplasma Sprayed Hydroxyapatite Coatings*, CRC Press, Taylor and Francis Group, USA, 2015, https://www.crcpress.com/Microplasma-Sprayed-Hydroxyapatite-Coatings/Dey-Mukhopadhyay/p/book/9781138748866.
3. Arjun Dey and A. K. Mukhopadhyay, *Nanoindentation of Natural Materials with Hierarchical and Functionally Graded Microstructure*, CRC Press, Taylor and Francis Group, USA, 2018, https://www.crcpress.com/Nanoindentation-of-Natural-Materials-Hierarchical-and-Functionally-Graded/Dey-Mukhopadhyay/p/book/9781498784054.
4. W. C. Oliver and G. M. Pharr, Measurement of hardness and elastic modulus by instrumented indentation: Advances in understanding and refinements to methodology, *Journal of Materials Research* 19 (2004) 3–20.
5. W. D. Nix and H. Gao, Indentation size effects in crystalline materials: A law for strain gradient plasticity, *Journal of the Mechanics and Physics of Solids* 46 (1998) 411–425.
6. Y. Huang, F. Zhang, K. C. Hwang, W. D. Nix, G. M. Pharr and G. Feng, A model of size effects in nanoindentation, *Journal of the Mechanics and Physics of Solids* 54 (2006) 1668–1686.
7. Yufei Sun, Yujia Wang, Enze Wang, Bolun Wang, Hengyi Zhao, Yongpan Zeng, Qinghua Zhang, Yonghuang Wu, Lin Gu, Xiaoyan Li and Kai Liu, Determining the interlayer shearing in twisted bilayer MoS_2 by nanoindentation, *Nature Communications* 13 (2022) 3898. https://doi.org/10.1038/s41467-022-31685-7 | www.nature.com/naturecommunications.
8. Anwar UlHamid, Deposition, microstructure and nanoindentation of multilayer Zr nitride and carbonitride nanostructured coatings, *Scientific Reports* 12 (2022) 5591. https://doi.org/10.1038/s41598-022-09449-6.
9. JiaWen Wang, Kai Yu, Man Li, Jun Wu, Jie Wang, ChangWu Wan, ChaoLun Xiao, Bing Xia and Jiang Huang, Application of nanoindentation technology in testing the mechanical properties of skull materials, *Scientific Reports* 12 (2022) 8717. https://doi.org/10.1038/s41598-022-11216-6.
10. Seyed Ali Delbari, Abbas Sabahi Namini, Seonyong Lee, Sunghoon Jung, Jinghan Wang, SeaHoon Lee, Joo Hwan Cha, Jin Hyuk Cho, Ho Won Jang, Soo Young Kim and Mohammadreza Shokouhimehr, Microstructural and nanoindentation study of TaN incorporated ZrB_2 and ZrB_2–SiC ceramics, *Scientific Reports* 12 (2022) 13765. https://doi.org/10.1038/s41598-022-17797-6.
11. Abhijeet Dhal, Saket Thapliyal, Supreeth Gaddam, Priyanka Agrawal and Rajiv S. Mishra, Multiscale hierarchical and heterogeneous mechanical response of additively manufactured novel Al alloy investigated by high resolution nanoindentation mapping, *Scientific Reports* 12 (2022) 18344. https://doi.org/10.1038/s41598-022-23083-2.
12. Vinay Sharma, Xinfeng Shi, George Yao, George M. Pharr and James Yuliang Wu, Surface characterization of an ultrasoft contact lens material using an atomic force microscopy nanoindentation method, *Scientific Reports* 12 (2022) 20013. https://doi.org/10.1038/s41598-022-24701-9.
13. Diancheng Geng, Hao Yu, Yasuki Okuno, Sosuke Kondo and Ryuta Kasada, Practical method to determine the effective zero point of indentation depth for continuous stiffness measurement nanoindentation test with Berkovich tip, *Scientific Reports* 12 (2022) 6391. https://doi.org/10.1038/s41598-022-10490-8.
14. Xu Long, Ziyi Shen, Qipu Jia, Jiao Li, Ruipeng Dong, Yutai Su, Xin Yang and Kun Zhou, Determine the unique constitutive properties of elastoplastic materials from their plastic zone evolution under nanoindentation, *Mechanics of Materials* 175 (2022) 104485. https://doi.org/10.1016/j.mechmat.2022.104485.
15. Giyeol Han, Karuppasamy Pandian Marimuthu and Hyungyil Lee, Evaluation of thin film material properties using a deep nanoindentation and ANN, *Materials & Design* 221 (2022) 111000. https://doi.org/10.1016/j.matdes.2022.111000.
16. Yunhong Yu, Chenguang Shi, Gang Xu, Zeheng Yao, Tianling Wang, You Wu, Xingyu Yi, Houzhi Wang and Jun Yang, Application of nanoindentation in asphalt material aging and characterization of actual pavement aging, *Construction and Building Materials* 331 (2022) 127348. https://doi.org/10.1016/j.conbuildmat.2022.127348.
17. Lokanath Patra and Ravindra Pandey, Mechanical properties of 2D materials: A review on molecular dynamics based nanoindentation simulations, *Materials Today Communications* 31 (2022) 103623. https://doi.org/10.1016/j.mtcomm.2022.103623.

18. Bowen Si, Zhiqiang Li, Xuexia Yang, Xuefeng Shu and Gesheng Xiao, Characterizations of specific mechanical behaviors for metallic materials with representative crystal structures using nanoindentation, *Vacuum* 207 (2023) 111661. https://doi.org/10.1016/j.vacuum.2022.111661.
19. Norbert Huber and Tobias Beirau, A modeling approach to predict the mechanical response of materials to irradiation damage from external sources: Nanoindentation of Pb-implanted $ZrSiO_4$, *Materialia* 24 (2022) 101506. https://doi.org/10.1016/j.mtla.2022.101506.
20. Zhaopeng Hao, Han Zhang and Yihang Fan, Mechanical response of nanoindentation and material strengthening mechanism of nt-cBN superhard materials based on molecular dynamics, *International Journal of Refractory Metals and Hard Materials* 106 (2022) 105844. https://doi.org/10.1016/j.ijrmhm.2022.105844.
21. Giuseppe Bianco, Tanaji Paul, Ambreen Nisar, Abderrachid Hamrani, Benjamin Boesl and Arvind Agarwal, Nanoindentation mapping defects filtration for heterogeneous materials using generative adversarial networks, *Materials Characterization* 191 (2022) 112107. https://doi.org/10.1016/j.matchar.2022.112107.
22. Soowan Park, João Henrique Fonseca, Karuppasamy Pandian Marimuthu, Chanyoung Jeong, Sihyung Lee and Hyungyil Lee, Determination of material properties of bulk metallic glass using nanoindentation and artificial neural network, *Intermetallics* 144 (2022) 107492. https://doi.org/10.1016/j.intermet.2022.107492.
23. Jianguo Wu, Hongwei Zhao, Yu Wang, Jinbao Lin, Gesheng Xiao, Erqiang Liu, Qiang Shen and Xuexia Yang, Micro mechanical property investigations of Ni-based high-temperature alloy GH4169 based on nanoindentation and CPFE simulation, *International Journal of Solids and Structures* 258 (2022) 111999. https://doi.org/10.1016/j.ijsolstr.2022.111999.
24. Jiaji Xiong, Lusheng Liu, Haozhe Song, Mengrui Wang, Tianwen Hu, Zhaofeng Zhai, Bing Yang, Xin Jiang and Nan Huang, Mechanical properties evaluation of diamond films via nanoindentation, *Diamond and Related Materials* 130 (2022). https://doi.org/10.1016/j.diamond.2022.109403.
25. Andrea Bonicelli, Tabitha Tay, Justin P. Cobb, Oliver R. Boughton, Ulrich Hansen, Richard L. Abel and Peter Zioupos, Association between nanoscale strains and tissue level nanoindentation properties in age-related hip-fractures, *Journal of the Mechanical Behavior of Biomedical Materials* (2022) 105573. https://doi.org/10.1016/j.jmbbm.2022.105573.
26. Yonglong Lai, Jiaxin Yu, Laixi Sun, Fang Wang, Qiuju Zheng and Hongtu He, Nanoindentation creep dependent deformation process of silica and soda lime silicate glass, *Journal of Non-Crystalline Solids* 597 (2022) 121906. https://doi.org/10.1016/j.jnoncrysol.2022.121906.
27. S. Gordon, H. Besharatloo, J.M. Wheeler, T. Rodriguez-Suarez, J.J. Roa, E. Jiménez-Piqué and L. Llanes, Micromechanical mapping of polycrystalline cubic boron nitride composites by means of high-speed nanoindentation: Assessment of microstructural assemblage effects, *Journal of the European Ceramic Society* (2022). https://doi.org/10.1016/j.jeurceramsoc.2022.08.047.
28. Shaopeng Pei, Yilu Zhou, Yihan Li, Tala Azar, Wenzheng Wang, Do-Gyoon Kim and X. Sherry Liu, Instrumented nanoindentation in musculoskeletal research, *Progress in Biophysics and Molecular Biology* 176 (2022) 38–51. https://doi.org/10.1016/j.pbiomolbio.2022.05.010.
29. Dichu Xu, Terence Harvey, Eider Begiristain, Cristina Domínguez, Laura Sánchez-Abella, Martin Browne and Richard B. Cook, Measuring the elastic modulus of soft biomaterials using nanoindentation, *Journal of the Mechanical Behavior of Biomedical Materials* 133 (2022) 105329. https://doi.org/10.1016/j.jmbbm.2022.105329.
30. T. Khvan, L. Noels, D. Terentyev, F. Dencker, D. Stauffer, U.D. Hangen, W. Van Renterghem, C.C. Chang and A. Zinovev, High temperature nanoindentation of iron: Experimental and computational study, *Journal of Nuclear Materials* 567 (2022) 153815. https://doi.org/10.1016/j.jnucmat.2022.153815.
31. W.C. Oliver and G.M. Pharr An improved technique for determining hardness and elastic modulus using load and displacement sensing indentation experiments. *Journal of Materials Research* 7 (1992) 1564–1583. https://doi.org/10.1557/JMR.1992.1564.
32. C. A. Schuh, Nanoindentation studies of materials, *Materials Today* 9 (2006) 32–40.
33. George M. Pharr, Erik G. Herbert and Yanfei Gao, The indentation size effect: A critical examination of experimental observations and mechanistic interpretations, *Annual Review of Materials Research* 40 (2010) 271–292.
34. Farid Alisafaei and Chung-Souk Han, Indentation depth dependent mechanical behavior in polymers, *Advances in Condensed Matter Physics* 2015 (2006) 20. http://dx.doi.org/10.1155/2015/391579.
35. George Z. Voyiadjis and Mohammadreza Yaghoobi, Review of nanoindentation size effect: Experiments and atomistic simulation, *Crystals* 7 (2017) 321. https://doi.org/10.3390/cryst7100321.

36. T. Ohmura and M. Wakeda, Pop-in phenomenon as a fundamental plasticity probed by nanoindentation technique, *Materials* 14 (2021) 1879. https://doi.org/10.3390/ma14081879.
37. Jeena George, Sowjanya Mannepalli and Kiran S. R. N. Mangalampalli, Understanding nanoscale plasticity by quantitative in situ conductive nanoindentation, *Advanced Engineering Materials* (2021) 2001494. https://doi.org/10.1002/adem.202001494.
38. P. Maiti, J. Ghosh and A. K. Mukhopadhyay, Modelling of nanoindentation behaviour in MgO doped alumina, *Ceramics International* 47 (2021) 9090–9110.
39. P. Maiti, J. Ghosh and A. K. Mukhopadhyay, Loading rate effect on nanoscale plasticity in Mgo doped alumina, *Journal of International Academy of Physical Sciences* 25 (2021) 127–135.
40. Payel Maiti, Dhrubajyoti Sadhukhan, Jiten Ghosh and Anoop Kumar Mukhopadhyay, Nanoscale plasticity in titania densified alumina ceramics, *Journal of Applied Physics* 131 (2022) 135107. https://doi.org/10.1063/5.0081872.
41. Payel Maiti, *Studies on Microstructurally Engineered Ceramics and Composites for Structural Applications*, Ph.D. Dissertation, Jadavpur University, 2022.
42. M. D. Barros, H. Jelitto, D. Hotza and R. Janssen, Microstructure and mechanical behaviour of TiO_2-MnO-doped alumina/alumina laminates, *Journal of the American Ceramic Society* 104 (2020) 1047–1057.
43. M. Bhattacharya, *Contact Deformations of Brittle Solids*, Ph.D. Dissertation, Department of Physics, Jadavpur University, Kolkata, 2016.
44. M. Bhattacharya, R. Chakraborty, A. Dey, A. K. Mandal and A. K. Mukhopadhyay, Improvement in nanoscale contact resistance of alumina, *Applied Physics A: Materials Science & Processing* 107 (2012) 783.
45. M. Bhattacharya and A. K. Mukhopadhyay, Contact deformation of alumina, *ISRN Ceramics* (2012) 9.
46. M. Bhattacharya, R. Chakraborty, A. Dey, A. K. Mandal and A. K. Mukhopadhyay, New observations in micro-pop-in issues in nanoindentation of coarse grain alumina, *Ceramics International* 39 (2013) 999.
47. A. Dey, M. Bhattacharya and A. K. Mukhopadhyay, Grain boundary nanohardness of coarse grain alumina, *International Journal of Applied Ceramic Technology* (2014). https://doi.org/10.1111/ijac.12351.
48. Manjima Bhattacharya, Arjun Dey and Anoop Kumar Mukhopadhyay, Influence of loading rate on nanohardness of sapphire, *Ceramics International* 42 (2016) 13378–13386. http://dx.doi.org/10.1016/j.ceramint.2016.05.091.
49. Manjima Bhattacharya, Arjun Dey and Anoop Kumar Mukhopadhyay, Self-adjusting unique nanoscale contact resistance of a single alumina grain, *Materials Research Express* 3 (2016) 045017. https://doi.org/10.1088/2053-1591/3/4/045017.
50. Manjima Bhattacharya and Anoop Kumar Mukhopadhyay, Interaction of nanoscale damages with static and dynamic contact induced damages in alumina: A novel approach using nanoindentation, *Ceramics International* (2019). https://doi.org/10.1016/j.ceramint.2019.04.120.

3 Applications of Machine Learning Techniques to Predict Behaviour of Materials

Charanjeet Singh Tumrate, Sujal Laxmikant Vajire, and Dhaneshwar Mishra

3.1 DATA PREPARATION IN MACHINE LEARNING APPROACH

Machine learning is a sub-branch of artificial intelligence. Arthur Samuel coined the term "machine learning" in 1950s, but it did not receive much fame at that time due to less computational power and the unavailability of datasets required to train the machine. This field began to flourish in the 1980s due to the free availability of big data. Machine learning is different from the hard coding technique in which the programmer writes the instructions and the code behaves similarly. Rather the data along with the output is given to the computer and it learns the trends and behaviour of the data, devises the instructions by itself, and is able to recognize and predict the behaviour of the unknown. The whole basis of machine learning (ML) lies in the dataset. The data used for training the model and the quality of the model depends on the data quantity and quality. If the dataset used for the training of the model is faulty or is biased, then the model trained also gives false predictions. The dataset could be alphanumeric values in the form of tables (Excel file), images (of objects, medical reports, etc.), or structural data (of chemical structure, design of instruments, architectural design, or mechanical equipment). Depending on data nature the ML algorithm is decided. The dataset could be labelled as well as unlabelled for supervised and unsupervised learning respectively. Depending on the nature of the dataset the ML model could perform the predictive task (regression), classify the objects in different categories (classification), recognize the objects (pattern recognition), and decision-making with the help of artificial neural networks. The dataset one selects for training should have the following five properties known as the 5Vs:

1. Volume: The data used for ML should be large enough so that the model can have every aspect of the data. If the dataset is small, the model accuracy will be less.
2. Velocity: The speed by which data is generated. If the velocity of data is slow it means a long time period is required for the training of the model.
3. Veracity: It means how accurate and truthful the data is. The data should be accurate and match with the real world. It should not be imaginary or fairy-tale. As data determine the model quality, it should be reliable and collected from a trustworthy source.
4. Validity: The validity of data is also linked to the quality of the data. The data should be correct and not possess any missing values or unusual entries.
5. Volatility: The data attributes should relate to the problem and not lose their impact as the data grows old. Even after 20–25 years, the same dataset could be used with new data.

A computerized archive used to store and organize data for retrieval through a variety of criteria is termed the database. A variety of databases relevant to each domain of knowledge is available for

 DOI: 10.1201/9781003359364-4

free public access or commercial purposes (which give access to their content after subscription). Components of databases are as follows:

a. Hardware and software.
b. Entry: The record that contains the number of fields to hold actual data.
c. Value: A particular piece of information to retrieve a record.
d. Making a query: The retrieval of data collection against a particular value/keyword.
e. Database Management System (DMS): Sophisticated computer programs for storing, organizing, searching, accessing, modifying, and maintaining data in the database.

There are four major types of databases. A brief description of each is given next.

1. Hierarchical databases: It is a tree-like structure in which different child nodes (subrecords) originate from a single parent node. The parent node is linked to many or a single subentry. The lower data structures or the child node data could be accessed by using pointers. Complex relations are used to retrieve the data from such databases. These types of databases were very common in the 1970s. The information management system by IBM is an example of a hierarchical database.
2. Network databases: This type of database allows uploading the data in the network, with multiple owners of each single record. The data is organized according to the ownership of the records and provides multiple paths to access the data. The CODASYL (Conference on Data Systems Languages) database management system deals with network databases. Another example of a network database is Univac DMS-1100. This type of database is an advancement of hierarchical databases. Instead of having only one path to a specific record, it could be accessed via multiple paths, i.e., having more than one parent node to each node. It is difficult to maintain and implement, yet it has had more flexibility than the hierarchical-based one.
3. Relational databases: The model for this type of database was presented in 1970. At that time, hierarchical and network databases were widely used. Relational databases became very popular because of their flexibility of structure and ease of use. They use a set of tables to organize data in which each table presents a relation (rows and columns). The columns are indexed according to the common feature or attribute. A query is executed by selecting linked data items using a set of operators from different tables to generate one combined report. The Microsoft SQL Server, Oracle Database, MySQL, and IBM DB2 are popular examples of relational databases.
4. Object-oriented databases: Data is stored as objects. An object is a concept or abstraction with meaning for the problem at hand. Objects are linked through pointers that are defined according to predetermined relationships between objects. Searching involves navigation through objects with the aid of pointers. Object-oriented databases are the most advanced type of database and widely used. They are based on programming languages, and easily accessed and managed. GemStone/S, ObjectDB, ObjectDatabase++, and Versant are examples of this type of database.

On the basis of the type of data in the databases, there are two types.

a. Primary databases: Store raw or experimental data submitted by scientists.
b. Secondary databases: Store data that is a refined form of the primary data.

Some specialized databases also exist that contain data about a specific area of interest, such as a database that contains data about breast cancer only. With advancements in the fields of science and technology, there is a great explosion of data as well as databases. There is a huge list of databases about each domain of knowledge. There are databases for all types of data about every aspect of the universe, from minute biological entities like DNA, RNA, and protein sequences and structures to interstellar and space objects like stars, comets, and galaxies. The building of machine learning models stands on the data. The first step is to collect data from the application area. With the availability of databases, it is easy to retrieve data for your area of interest. First, identify your problem, then find the relevant database and eventually the data related to your problem. Then you are free to apply different algorithms to best address and solve your problem. Data visualization is to present the information of the data graphically. It aims to find the trends, tendencies, patterns, and nature of data. The tabular data is a matrix of numbers, but one can never explain the behaviour of data by seeing the data matrix. To infer information about data, we must present it in the form of plots, charts, or graphs so that one can understand the nature and behaviour of data by merely looking at it. Data visualization is like an art that captures the attention of viewers and tells the whole scenario. Once you have searched and retrieved the data related to your problem from the database, the next step is to analyze and visualize the data. This is known as the data preprocessing step. In this process, statistical analysis is done on the data to find the trends and tendencies of the data; useful information is drawn from the huge dataset. The outliers, noise, missing, and null values are treated; data dimensionality is reduced; and the data is cleaned from abnormalities and redundancies to leave only a handful of features or attributes that can best explain the data. These features and attributes are transformed into the format accepted by the algorithm and also relevant to the problem. The data collected from the databases is in a variety of formats such as large datasets with a high number of rows and columns, structured tables, images, audio files, and videos. However, the machine does not understand these formats, as they only interpret the 0s and 1s. Thus, data preprocessing is done in which the data is transformed, or encoded, to bring it to such a state that the machine can easily parse it. Thus, features of the data can be easily interpreted by the algorithm. First, the data is converted to the specific data type required to solve the problem. There are four data types in which the data can be transformed to:

a. Nominal (ID and names of people, colours of cars)
b. Categorical (ranking or categories like small, medium, and large)
c. Interval (numerical values and differences are meaningful)
d. Ratio (numerical values, differences, and ratio are meaningful)

This could be done by normalization, standardization, transformation, or discretization depending on the nature of the problem and the dataset. After type conversion, the data is analyzed and visualized to find the behaviour of the data. The null or missing values are either dropped or replaced by the mean value or the probability value. Then different statistical procedures are used to visualize the trends and behaviour of the data like measures of central tendency (mean, mode, median, etc.), dispersion (range, quartile deviation, mean deviation, standard deviation), symmetry, and kurtosis. Different probability distribution techniques are also used to find how the data is distributed. Dimensionality reduction techniques like principal component analysis (PCA) can be applied to get a smaller subset of attributes or variables to be used for model training and results in a good quality model. The user can perform analysis either on a single variable (univariate analysis), taking two at a time (bivariate analysis), or many simultaneously (multivariate analysis).

3.1.1 Data Visualization

3.1.1.1 Types of Data Visualization for a Single Variable

For a continuous invariable, a histogram is used to find its frequency distribution. A histogram is the presentation of groups with their respective frequencies. It uses the mean as the centre point of the data. Mostly the data is divided into four equal parts via three quartiles each representing 25% of the data. Most of the data lies within the first and third quartiles, known as the interquartile range.

A boxplot is the pictorial representation of the quartiles, where a box depicts the interquartile range, small circles show the outliers of the data, the line at the bottom is the first quartile and the top line is the third quartile. This plot takes the median as the centre point. An amalgam of the boxplot and the probability distribution is given by the Villon plot. It shows the boxplot in the middle and around it the dispersion of the data around the centre point is demonstrated.

3.1.1.2 Types of Data Visualization for Two Variables

Line and bar charts are used to show the relative trends of two variables in a single graph. Each axis represents one variable, and the values are plotted in the graph either as points, in the case of line charts, or solid bars, in the case of bar charts. To show the interdependencies of two variables on each other, a scatterplot is used. Two variables that strongly depend on each other show a high correlation and their plot shows a linear trend. If the linear line is upward, it means with an increase of one variable the other also increases in a uniform manner. For a downward linear trend, it means both variables are inversely related to each other. In machine learning, two variables having strong correlation are not used simultaneously while training, instead any one of these two is used. The variable or the features that have a strong correlation with the target variable are used in model training.

3.1.1.3 Types of Data Visualization for Multiple Variables

For visualizing the trend of multiple variables with each other, a correlation matrix is used. It is the graphical representation of the correlation of variables with each other. In this matrix, each variable is compared with all the other variables including itself to find the correlation. The value of the correlation lies between 0 and 1; 0 means no correlation, above 0.25 means weak correlation, 0.50–0.75 means there is a strong correlation, and 1 means perfect correlation. The variables that have a strong or perfect correlation with the target variable are used in model training. The variables that are strongly correlated to each other than the target variables are not used in the training procedure. The most basic and useful way to extract the important variables for training a model from the dataset is the correlation analysis. Instead of dealing with numbers in the correlation matrix, one can assign colours to the degrees of correlation. The colour gradient shows if there is no, weak, strong, high, or perfect correlation between different variables. Thus, making the task of data visualization and feature selection easy and attractive.

3.2 MACHINE LEARNING ALGORITHMS

The aim of machine learning is to use data from past events and situations to learn the possible outcomes and predict future events. It is the simplified mathematical representation of the training and learning process. Different machine learning algorithms are available that learn the trend of data to make future predictions depending on the types of data and the nature of the problem. These algorithms are classified into two main categories based on the dataset used for training: supervised learning and unsupervised learning.

3.2.1 Supervised Learning

Supervised learning is the type of machine learning in which the dataset is provided to the algorithm along with the output, i.e., labelled data to learn the data behaviour. So that later it could correctly classify the data or make accurate predictions. There is a distinction between the dependent variable (target) and the independent variable (explanatory features). The algorithm learns the trends of explanatory features and the respective outcome. The output could be a continuous value or a class label. Supervised machine learning is used to solve two types of tasks: prediction and classification. For the predictive task, regression analysis is used. In this type of supervised learning, the model gives continuous value as output. There are many different types of regression. Some examples are:

i. Linear and nonlinear regression
ii. Partial least squares
iii. Support vector regression

The regression uses a mathematical equation to define a line, plane, or curve that best fits the data points using one or more than one explanatory variable. For the categorizing data into different classes, classification is used. In this type of supervised learning, class labels are given as output. Many algorithms are used for classification. Some of these are as follows:

a) Decision trees: A tree-like structure where decision-making is done on the internal nodes and the leaves, or the child node represents the final output.
b) K-nearest neighbor: The data points are classified based on the distances from other data points. Two points having the smallest distance between them are assigned to the same class.
c) Bayesian classifier: Based on the probability values, the object is classified using variables that are independent of each other.
d) Artificial neural networks: Based on the principle of the natural brain, each neuron is assigned a specific weight that is adjusted repeatedly to bring the error low and produce the correct output.
e) Support vector machines: Aim to find the best decision boundary in the N-dimensional space that can best classify the data points into different classes.
f) Ensemble techniques: Bagging, boosting, etc. Random forest is the most widely used ensemble method. Multiple decision trees are clustered together in the forest. The final output given by the random forest is generated after majority voting of the outputs given by different decision trees.

These classification algorithms can be used for predicting two classes in data (binary classification task) or multiple classes (multiclassification tasks).

3.2.2 Unsupervised Learning

In the unsupervised learning type of machine learning, the data given to the algorithm does not have any class labels assigned, i.e., no target variable is present in the dataset. The algorithm has to learn the underlying pattern of the data itself and classify similar data points to the same class. Clustering is an example of the unsupervised machine learning approach. A group of data points that have been combined due to some commonalities is referred to as a cluster.

Some of the clustering techniques are:

i. K-means clustering: Data points are clustered into a number (K) of mutually exclusive clusters.
ii. Hierarchical clustering: Data points are clustered into parent and child clusters
iii. Probabilistic clustering: Data points are clustered on a probabilistic scale.

Depending on the nature of the problem and the data in hand, the appropriate algorithm or algorithms are selected. The dataset is either labelled or unlabelled and is then split into training and test data. The splitting is done on some principles. If the dataset is small, then 80% of the data is used for training and 20% for testing. If one has a huge dataset, then 60% of the data is used for training and the remaining 40% is used for testing. To avoid biases in the splitting, random shuffling or y-scrambling can be used before splitting. The training set is given to the algorithm so that it can learn from the data. The choice of the algorithm is made based on the nature of the problem and the data. One could also use all the algorithms, and then find which performs best or by selecting the one appropriate for the problem and the data. Then the performance of the model is checked on the training data. The model should perform best on the training data as it has learned from the same data. If the model does not perform better on the training data it means the algorithm chosen is not appropriate, hyperparameters need to be fine-tuned, or there is an artefact in the data. Thus the model is trained again using a different algorithm, a new set of hyperparameters, or even a new dataset. If the model works well on the training data, then it is tested on the test data. The output given by the model and the actual output of the test data are compared. The difference between the two is the error of the model, which should be as low as possible. Low error term means the predicted outcome of the model and the actual output of the data have the same result. Model validating is done to check whether the model was overfitted to the training data or performed well on the other dataset of a similar nature. The whole data is divided into a number of folds and the model is trained on all the folds one by one, leaving one for testing. The process goes on until the model is tested on each fold of data. If the model performance on each test fold is approximately the same, the model is termed a "generalized model". After training and testing and validating, the model can be used to predict the values or classes of the unknown dataset and real-world problems.

3.3 DISTANCE MATRIX

The similarities and differences between the data points in a dataset are calculated by figuring out the distances between them. The numerical measure of how alike two data objects are is known as similarity. This often falls in the range of 0 to 1. Similarity is higher when objects are more alike. The numerical measure of how different two data objects are is termed dissimilarity. It is lower when objects are more alike, and less when dissimilar. The minimum dissimilarity is 0, but the upper limit varies. Distance matrices are constructed to depict how much the objects in the data are similar or dissimilar to each other. The distance matrix is mostly built based on dissimilarity. Distance matrices are symmetric matrices and the values of the distances in the distance matrix are relative to each other. The distance (dissimilarity) of an object from itself is zero and it lies on the main diagonal of the distance matrix. While performing hierarchical clustering and multidimensional scaling, a distance matrix is used as a data format. In machine learning, the distance matrix improves the performance of the classification and the clustering tasks. The most used distance matrices in machine learning are described next.

3.3.1 Euclidean Distance

The Euclidean distance is the linear distance between two points. The Euclidean distance is computed by taking the square root of the square of the sum of the difference between the same attributes of two data points. It tells the magnitude of the dissimilarity or difference between two data points of the dataset. It is similar to the calculation of Pythagoras' theorem. Data should be normalized before the calculation of Euclidean distance.

$$\sqrt{(y_1 - x_1)^2 + (y_2 - x_2)^2 + \ldots + (y_n - x_n)^2} \tag{3.1}$$

It is the most commonly used for the calculation of distances. It is less informative when dealing with more than three-dimensional data, i.e., when each data point has more than three attributes.

3.3.2 Manhattan Distance

The Manhattan distance, also known as the "city-block method", is the sum of the horizontal and vertical distances between two points. It is useful in calculating the distance of the points that move at a right angle to each other. The Manhattan distance is calculated by taking the sum of the modulus of the distances between values of the same attribute of two data points.

$$|y_1 - x_1| + |y_2 - x_2| + \ldots + |y_n - x_n| \tag{3.2}$$

It also requires normalized data for calculation and is not affected by the higher dimensionality of data.

3.3.3 Minkowski Distance

The Minkowski distance is a more generalized form of the Euclidean and Manhattan distances. It is the *p* root of the sum of the modulus of the differences between the same attributes of two data points raised to power *p*, where *p* could be any number depending on the dimensionality of the data.

$$D = \left(\sum_{i=1}^{n} | p_i - q_i |^p \right)^{1/p} \tag{3.3}$$

For P =1, its value is equal to the Manhattan distance and for P = 2, its value is the same as the Euclidean distance. It is also used for ordinal variables.

3.3.4 Cosine Distance

The cosine distance measure is used to find the difference between the cosine angles of the vectors irrespective of their magnitude. The greater the angle between two vectors, the greater the difference present between them. It is calculated by the dot product of two vectors divided by the product of their lengths.

$$cos\theta = \frac{\vec{a}.\vec{b}}{\|\vec{a}\| \|\vec{b}\|} \tag{3.4}$$

The cosine distance works well with higher dimensional data and is used as an alternative to Euclidean distance for such datasets.

3.3.5 Evaluation Matrix

After the model training, the model is evaluated to check its performance. Without performance evaluation, merely training the model is of no use. The model evaluation parameters tell us about the model quality and its accuracy. If the performance is below the desirable criteria, the model is trained again after fine-tuning the hyperparameters. The model is again evaluated. The process goes on until the desirable accuracy and quality of the model are achieved. There are many performance evaluation matrices available. The use of these metrics depends on the nature of the problem and the output of the model. The performance evaluation matrices for regression problems having a continuous numerical value are different from those of classification problems where the output is the class labels. Some commonly used regression evaluation matrices are given next.

3.3.5.1 Mean Absolute Error (MAE)

Mean absolute error (MAE) is the absolute difference between the actual and the predicted outputs given by the model without considering the direction of the error. It is calculated by taking the mean of the difference between the actual and predicted outputs. MAE is less affected by the outliers in the data. The formula for MAE is:

$$\mathrm{MAE} = \frac{1}{n}\sum_{i=1}^{n} | y_i - y_i \tag{3.5}$$

3.3.5.2 Mean Squared Error (MSE)

Mean squared error (MSE) tells the closeness of the regression line to the data points. It is used to find the average squared error between the predicted and actual values. It is calculated by the following formula:

$$\mathrm{MSE} = \frac{1}{n}\sum_{i=1}^{n} (y_i - y_i)^2 \tag{3.6}$$

3.3.5.3 Root Mean Squared Error (RMSE)

Root mean squared error (RMSE) is a frequently used measure to find the difference between values predicted by a model and the observed values. RMSE provides a higher overall error, and the gap widens as the errors are larger. The calculation of RMSE is quite similar to that of MAE, except that the error is squared rather than taken as the absolute value to eliminate the sign on the individual errors because the square of a negative number is positive. The formula of the RMSE is:

$$\mathrm{RMSE} = \sqrt{}\sum_{i=1}^{n} \frac{\left(y_i - y_i\right)^2}{n} \tag{3.7}$$

3.3.5.4 R Squared (R^2)

R^2 tells us the degree to which the model is able to explain the variance in the data. It tells how much the model learned the trends and the underlying pattern of the data. It also depends on the feature selected for building the model. Its value ranges from 0 to 1. The higher the value of R^2, the better the model quality. If its value is close to 0, it means the set of variables used for model training is not good enough to explain the trends and tendencies of the data. A value above 0.80 is encouraged. The formula for the calculation of R^2 is:

$$R^2 = \frac{\Sigma\left(y - \bar{y}\right)^2 - \Sigma\left(y - \hat{y}\right)^2}{\Sigma\left(y - \bar{y}\right)^2} \tag{3.8}$$

There are many different performance evaluation matrices for classification problems. Some of them are given next:

i. Accuracy: The total number of predictions that are correct.
ii. Precision: The proportion of positive case predictions that are actually correct.
iii. Recall: The proportion of actual positive cases that are correctly identified.
iv. F-measure: Measures the effectiveness of a model with respect to a user. It is the harmonic mean of the precision and recall values.
v. ROC and AUC: The plot between the true positive rate and false positive rate.

3.4 EXTRACTING DATASETS FOR MATERIALS APPLICATION

The development of the dataset can be classified into two categories. The experiment-based dataset consists of negative datasets (datasets that did not give expected results) along with positive datasets.

TABLE 3.1
Development in Computational Material Science Datasets

Source	Title	Dataset Type	Outcome
Schmidt J et al. [1]	Recent advances and applications of machine learning in solid-state materials science	i. Crystal graph convolutional networks [2]; MatErials Graph Networks [3] ii. QM7 [4,5], QM8 [5,6], and QM7b [7,8] dataset	i. For a wide materials space, prediction of various material properties such as formation energies, fermi energies, band gaps, shear and bulk moduli, and Poisson ratios are possible with accuracy. ii. Datasets in chemistry to measure the progress in features and algorithms.
Computational Materials Science [9]	Mendeley datasets	Crystal structure classification in ABO3 perovskites [10], alkali borates structural response at Fe_2O_3 sliding interface [11], etc.	Cloud-based repository of 11 million datasets including materials datasets.
Dösinger C et al. [12]	Applications of data driven methods in computational materials design	Description of segregation and cohesion of grain boundaries in W–25 at % Re alloys [13]	The dataset contains 219 segregation energies for segregation sites in 15 different GBs.
Xu B et al. [14]	Computational materials design: Composition optimization to develop novel Ni-based single crystal superalloys	Creep data of single crystal superalloys [15–17]	Predicting creep strength.
Talirz L et al. [18]	Materials Cloud, a platform for open computational science [18]	Materials Cloud three- and two-dimensional crystals database, hidden spontaneous polarisation in the chalcohalide photovoltaic, and pyrene-based metal organic frameworks etc.	Cloud-based computational material science datasets for various requirements.

Source: *Computational Material Science*

Simulations can be used to generate the datasets in cases where an experimental setup is expensive or hazardous. Special care is to be taken in checking the accuracy and estimating the computational cost to generate simulation datasets. In recent years, machine learning has become one of the most intriguing tools to enter the arsenal of material scientists. A brief review of the recent development of datasets using computational material science is in Table 3.1.

The application of synthetic dataset for predictive maintenance data set [19][20] can reduce the cost and time to collect and train datasets. The ML support vector machine algorithm can accurately predict the concrete material strength [21], whereas the regression model were also used to predict yiel strength of varying steel grades [22]. The Cambridge Structural Database consists of more than 1.1 million 3D structures with data from X-ray and neutron diffraction analyses and additional curation from the Cambridge Crystallographic Data Centre (CCDC). The research divisions in pharmaceutical, agrochemical, and fine chemicals industries access these databases for research.

3.5 TEST CASE

The machine learning classification method decision tree involves modelling, learning, and prediction. The decision tree algorithm architecture breaks the data using the attribute selection measure

(ASM) feature. The attributes developed are considered as decision nodes and thus the dataset is reduced in subsets. The repetition of the steps results in the development of the tree, which is continuous until all tuples have the same attribute values or no remaining attributes are left or no instances exist. A study conducted to classify the surface defect types in stainless steel plates included a dataset with 34 columns consisting of 27 indicators used as input vectors and 7 classes of defects [23]. The aim is to check the accuracy with which the decision tree model can predict the results after being trained. The Jupyter notebook is used to add libraries, and the faults.csv dataset file is uploaded using the Pandas read CSV function. The feature selection process is performed by dividing the available columns into target variables and feature variables. The selection of variables is performed by assigning dependent variables as target variables and independent variables as feature variables. The training and testing datasets are generated with 70% to 30% of the data, respectively, as shown in Figure 3.1. The decision tree classifier is called to check the accuracy of the classifier, which was noted to be 70.5%.

Various other ML algorithms can be used based on the data availability and the output requirement. A comparative study between the FEM-based nanoindentation modelling approach and

Step 1: Division of the training and testing data set.

```
# Split dataset into training set and test set
X_train, X_test, y_train, y_test = train_test_split(X, y, test_size=0.3, random_state=1) # 70% training and 30% test
```

Step 2: Decision Tree classifier

```
# Create Decision Tree classifer object
clf = DecisionTreeClassifier()

# Train Decision Tree Classifer
clf = clf.fit(X_train,y_train)

#Predict the response for test dataset
y_pred = clf.predict(X_test)
```

Step 3: Accuracy

```
# Model Accuracy, how often is the classifier correct?
print("Accuracy:",metrics.accuracy_score(y_test, y_pred))

Accuracy: 0.7049742710120068
```

FIGURE 3.1 Decision tree algorithm.

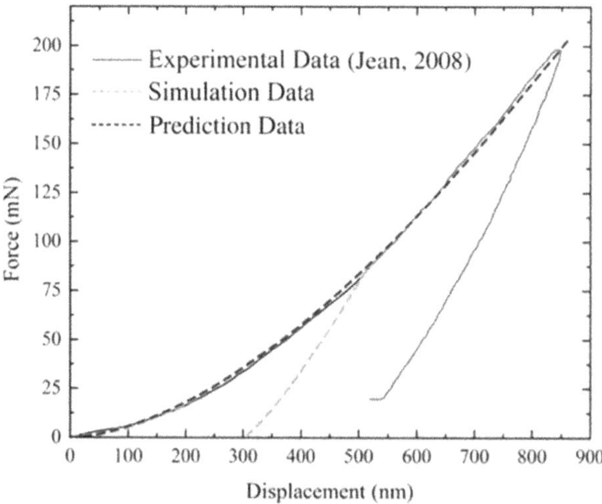

FIGURE 3.2 Load (P)–displacement (h) comparison with experimental, FEM simulations and ML results on nanoindentation deformation [21].

trained ML model application in the prediction of the load-deformation behaviour subjected to larger misfit strain for the thin film on a substrate was conducted [24]. A supervised ML polynomial regression technique was applied to generate a nonlinear relation between dependent and independent variables from the finite element (FE)-based ABAQUS dataset. The experimental, FE, and ML results were generated based on the load (P)–displacement (h) with experimental results on nanoindentation deformation. The cases are shown in Figure 3.2.

3.6 CONCLUSION

The chapter provides a comprehensive explanation of the fundamental ideas underlying machine learning. The nature of the ML method is determined by the datasets, which contain images or alphanumeric data. To understand the foundations on which the ML algorithms are selected, the concept of labelled and unlabelled datasets for supervised and unsupervised learning is addressed. A review of the application of ML algorithms based on computational material science datasets demonstrates the enormous advantages of ML in material science. In addition to the literature of comparative study among experimental, FE, and ML approaches, an example test case is also used to demonstrate how machine learning techniques, such as decision trees, can be applied to predict the required results from a given dataset.

ACKNOWLEDGEMENTS

The authors would like to acknowledge the high-end computational facility of Manipal University Jaipur, and the Multiscale Simulation Research Centre (MSRC) for providing computational support.

REFERENCES

1. Schmidt, J., Marques, M. R., Botti, S. & Marques, M. A. Recent advances and applications of machine learning in solid-state materials science. *NPJ Comput. Mater.* 5(1), 1–36 (2019).
2. Xie, T. & Grossman, J. C. Crystal graph convolutional neural networks for an accurate and interpretable prediction of material properties. *Phys. Rev. Lett.* 120, 145301 (2018).
3. Chen, C., Ye, W., Zuo, Y., Zheng, C. & Ong, S. P. Graph networks as a universal machine learning framework for molecules and crystals. *Chem. Mater.* 31, 3564–3572 (2019).
4. Ramakrishnan, R., Dral, P. O., Rupp, M. & von Lilienfeld, O. A. Quantum chemistry structures and properties of 134 kilo molecules. *Sci. Data* 1, 140022 (2014).
5. Ruddigkeit, L., van Deursen, R., Blum, L. C. & Reymond, J.-L. Enumeration of 166 billion organic small molecules in the chemical universe database GDB-17. *J. Chem. Inf. Model.* 52, 2864–2875 (2012).
6. Ramakrishnan, R., Hartmann, M., Tapavicza, E. & von Lilienfeld, O. A. Electronic spectra from TDDFT and machine learning in chemical space. *J. Chem. Phys.* 143, 084111 (2015).
7. Blum, L. C. & Reymond, J.-L. 970 million druglike small molecules for virtual screening in the chemical universe database GDB-13. *J. Am. Chem. Soc.* 131, 8732–8733 (2009).
8. Montavon, G. Rupp, M., Gobre, V. Vazquez-Mayagoitia, A., Hansen K Tkatchenko A. Müller K. R. & Von Lilienfeld O. A. Machine learning of molecular electronic properties in chemical compound space. *New J. Phys.* 15 (9), 095003 (2013).
9. *Computational Materials Science.* https://www.journals.elsevier.com/computational-materials-science/mendeley-datasets.
10. Thomas, T., Behara, S. & Poonawala, T. Data for: Crystal structure classification in ABO3 perovskites via machine learning. *Mendeley Data* V1 (2020), doi: 10.17632/dfsf6n6g7y.1.
11. Zhu, H. Data for: Structural response of alkali metal borates at Fe2O3 sliding interface: The effect of alkali cations. *Mendeley Data* V1 (2020), doi: 10.17632/5g44b7vn22.1.
12. Dösinger, C., Spitaler, T., Reichmann, A., Scheiber, D. & Romaner, L. Applications of data driven methods in computational materials design. *BHM Berg-und Hüttenmännische Monatshefte* 167(1), 29–35 (2022).
13. Scheiber, D., Razumovskiy, V. I., Puschnig, P., Pippan, R. & Romaner, L. Ab initio description of segregation and cohesion of grain boundaries in W–25 at.% Re alloys. *Acta Mater* 88, 180–189 (2015).

14. Xu, B., Yin, H., Jiang, X., Zhang, C., Zhang, R., Wang, Y. & Qu, X. Computational materials design: Composition optimization to develop novel Ni-based single crystal superalloys. *Comput. Mater. Sci.* 202, 111021 (2022).
15. Yutaka, K., Toshiharu, K., Tadaharu, Y., Makoto, O., Hiroshi, H., Yasuhiro, A. & Mikiya, A. Effects of alloying additions on the creep strength of a fourth generation single crystal superalloy. *J. Japan Inst. Metals.* 68, 206–209 (2004).
16. Liu, Y., Wu, J., Wang, Z., Lu, X.-G., Avdeev, M., Shi, S., Wang, C. & Yu, T. Predicting creep rupture life of Ni-based single crystal superalloys using divide-and-conquer approach-based machine learning *Acta Mater.* 195, 454–467 (2020).
17. Du, J., Lv, X., Dong, J., Sun, W., Bi, Z., Zhao, G., Deng, Q., Cui, C., Ma, H., & Zhang, B. Research progress of wrought superalloys in China. *Acta Metall Sin.* 55(9), 111532. 2019.
18. Talirz, L., Kumbhar, S., Passaro, E., Yakutovich, A. V., Granata, V., Gargiulo, F., Borelli, M., Uhrin, M., Huber, S. P., Zoupanos, S. & Adorf, C. S. Materials Cloud, a platform for open computational science. *Sci. Data.* 7(1), 1–2 (2020).
19. Matzka, S. Explainable artificial intelligence for predictive maintenance applications. 2020 Third International Conference on Artificial Intelligence for Industries (AI4I), 2020, pp. 69–74, doi: 10.1109/AI4I49448.2020.00023.
20. Tumrate, C. S., Saini, D. K., Gupta, P., & Mishra, D. Evolutionary computation modelling for structural health monitoring of critical infrastructure. *Archives of Computational Methods in Engineering.* 30(3), 1479–93. (2023).
21. Malhotra, K., Mishra, D., & Tumrate, C. S. Prediction of concrete compressive strength employing machine learning techniques. *Materials Today: Proceedings.* (2023).
22. Tumrate, C. S., Chowdhury, S. R., & Mishra, D. Development of regression model to predicting yield strength for different steel grades. *InIOP Conference Series: Earth and Environmental Science.*796, 1, p. 012033). IOP Publishing, (2021).
23. Semeion Centro Ricerche, Semeion: research center of sciences of communication. *Via Sersale 117*Rome. Italy, 2021. www.semeion.it.datasets/sureshmecad/steel-plates-faults/
24. Vajire, S. L., Singh, A. P., Saini, D. K., Mukhopadhyay, A. K., Singh, K. & Mishra, D. Novel machine learning-based prediction approach for nanoindentation load-deformation in a thin film: Applications to electronic industries. *Comput. Ind. Eng.* 174, 108824 (2022).

Part 2

Experimental Methods to Study Mechanics and Physics of Deformation and Fracture in Materials

4 Deformation and Fracture of Metals

Experimental Methods and Challenges

Mohit Jain, Anamitra Datta, and Amitava Basu Mallick

4.1 INTRODUCTION

An array of engineering materials is used in structural applications. A structure's resistance to deformation (alteration of form or shape) during mechanical loading is the primary factor to be concerned about. Therefore, when designing structural components, besides considering the acceptable level of performance and economy of the fabrication process, safety and durability issues are of prime importance. In service, it is essential that twisting, bending, stretching, etc., causing failure should be resisted. Thus, a priori knowledge of the deformation and fracture properties or evaluation of the material's mechanical properties is essential for designing structural components. Mechanical properties of the material are assessed by conducting a wide variety of mechanical tests. The type of test to be performed and the test conditions depend mainly on the type of deformation and mechanical loading expected to happen in service.

4.2 MATERIALS

Among the different classes of engineering materials, viz. ceramics and glasses, metals and their alloys, polymers and composites, in most cases, metals and alloys are chosen for fabricating structural components. Metals refer to substances that have a crystalline nature in solid form and exhibit properties like ductility, electrical and thermal conductivity, high density, malleability and high reflectivity of light [1]. Although metals exhibit high strength and ductility, they are seldom used in their pure form for load-bearing applications. Alloys regarded as combinations of metals or metal and nonmetal, e.g., steel, brass and bronze, provide improved strength properties and desired qualities compared to pure metals and are recommended for structural applications.

The mechanical properties of metals and alloys depend on their crystal structure, the packing efficiency of the constituent atoms and imperfections in the arrangement of atoms. Dislocations are line imperfections in the arrangement of atoms in a crystal. They are generated during processing or introduced when the material undergoes permanent deformation. The deformation and fracture in metals and materials are intimately related to the ease of dislocation motion along the span of the crystal lattice. Therefore, the strength and ductility depend on the resistance to dislocation motion at the microscopic length scale.

4.3 BASIC ISSUES IN DEFORMATION AND FRACTURE

Deformation can be defined as the change in an object's original dimensions after applying a certain amount of load. Deformations in metals can be broadly divided into two categories: elastic deformation and plastic deformation.

DOI: 10.1201/9781003359364-6

Elastic deformation is when the object regains its original size and dimensions after removing the applied load and is usually the desirable response of structural components in service. The limiting load to which the material exhibits elastic behaviour is known as the elastic limit for that material. If the elastic limit exceeds, permanent deformation occurs [2].

Plastic deformation is when the object is permanently deformed and does not regain its original dimensions even after removing the applied load. It begins after exceeding the elastic limit [2].

It should be noted that plasticity and fracture are different events. However, a permanent shape change occurs under the applied load in both cases. Still, they differ from each other in terms of cracking. The former's change in shape is independent of cracking. In contrast, the latter completely depends on creating and propagating a crack, forming fragmented individual components.

Hooke's law, involving the relationship between deformation and applied load, is valid for most materials. More commonly, this is stated as stress (σ) being directly proportional to strain (*e*). According to the law, the load–deformation relationship remains linear, however, it is not true for all materials. For example, rubber shows a nonlinear stress–strain relationship that follows the definition of elastic materials [2].

$$\frac{\sigma}{e} = E = constant \tag{4.1}$$

where E refers to Young's modulus.

When a tensile force is applied in, say, x direction, it causes the contraction in the lateral y and z directions. The ratio of transverse strain to longitudinal strain is called Poisson's ratio, which is a constant.

$$\varepsilon_z = \varepsilon_y = -v\varepsilon_x = -\frac{v\sigma_x}{E} \tag{4.2}$$

where v is Poisson's ratio; ε_z, ε_y and ε_x are the normal strain in the x, y and z directions; and σ_x is longitudinal stress in the x direction.

Stress (σ) is defined as the force applied per unit cross-sectional area on a material. Assuming a cylindrical bar, subjected to an axial-tensile load, F, which has an initial and instantaneous cross-sectional area of A_o and A, if the stress is computed using the original area A_O, then the stress is referred to as engineering stress. In contrast, if the stress is computed using instantaneous cross-sectional area A, it is known as true stress [3]. Thus, the mathematical definitions of engineering stress and true stress are given as follows:

$$\text{Engineering stress}: \sigma = \frac{F}{A_0} \tag{4.3}$$

$$\text{True stress}: \sigma_t = \frac{F}{A} \tag{4.4}$$

Stresses can be classified into two ways depending on the direction of their application:

i. Normal stress: The applied load and the material's surface are perpendicular.
ii. Shear stress: The applied load and the material's surface are parallel.

It is also well known that the strain (*e*) is defined as the ratio of change in the length of an object to its initial length. Suppose we assume that a cylindrical bar, having an initial length, L_O, is acted upon by an axial-tensile load, *F*, causing its length to become *L*. In that case, the engineering strain is the ratio between the object's length and the original length change. On the other hand, true strain (ε) can be defined as the change in linear dimension divided by the instantaneous value of length. Thus, the mathematical definitions of engineering strain and true strain are given as follows:

$$\text{Engineering strain} : e = \frac{\Delta L}{L_0} \tag{4.5}$$

$$\text{True strain} : \varepsilon = ln\frac{L}{L_0} \tag{4.6}$$

Strain can be again classified as normal and shear strain.

i. Normal strain: In the case of normal strain, the change in length of the material and the surface of the material are perpendicular to each other.
ii. Shear strain: In the case of shear strain, the change in length of the material and the surface of the material are parallel to each other.

When the continuous load is applied to a material, in the initial stage, a region in the specimen undergoes an extensive increase in length and gradual decrease in cross-sectional area; the feature is known as necking, as shown in Figure 4.1. Cracks develop at a later stage of loading, resulting in fracture. In metals and alloys, the cracks are typically formed by the aggregation of dislocations, e.g., at grain boundaries, second phase precipitate, inclusion-matrix interface and free surfaces [5]. The process of fracture development includes void nucleation, void growth and void linking (coalescence).

Depending on the material's ability to endure the extent of plastic deformation before fracture, the fracture type can be differentiated as ductile or brittle. In ductile fracture, the crack is preceded by extensive plastic deformation. It has stable cracks that resist further crack propagation. As ductile materials are pulled in tension, necking occurs, and voids are generated at the grain boundaries, inclusions, inclusion/matrix interfaces and at the centre of the neck. As the material (say, a bar) is deformed, a 45° angle shear lip develops in the material, which causes the cup-and-cone-shaped crack to appear, which results in a cup and cone fracture as shown in Figure 4.2.

The brittle fracture in a mild steel tensile specimen and the fractured surface are shown in Figure 4.3a and b, respectively. In brittle fracture, the plastic deformation is insignificant, which causes the cracks to become unstable and spread swiftly without any increment in the applied stress. Here, the crack propagation and applied stress are orthogonal. Multiplication of cracks happens via the breaking of atomic bonds in specific cleavage planes.

Figure 4.4 shows a mixed mode fractured surface. A mixed mode fracture is a type of fracture where both cleavage and microvoid coalescence occur, and features of both ductile and brittle fractures are seen [7].

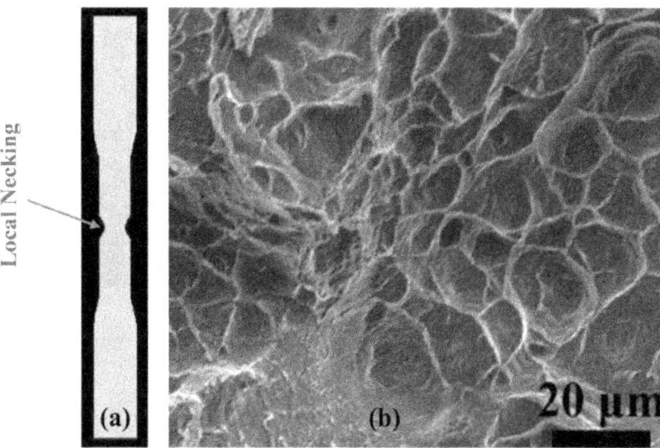

FIGURE 4.1 (a) Necking in the tensile specimen and (b) fractography of necking fracture surface (reprinted with permission from [4]).

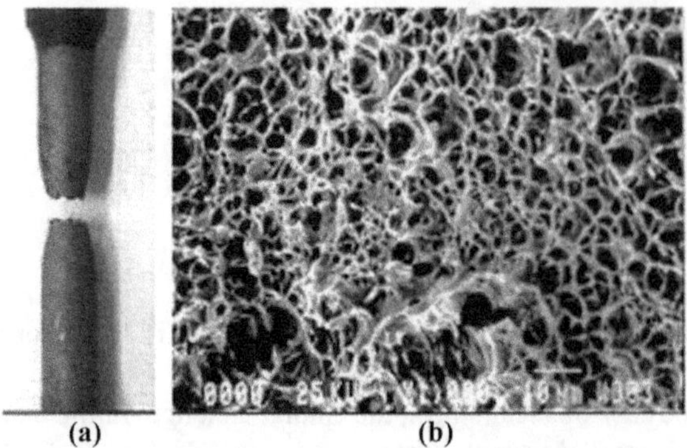

FIGURE 4.2 (a) Cup and cone ductile fracture in a tensile specimen and (b) fractography of a ductile fracture (reprinted with permission from [6]).

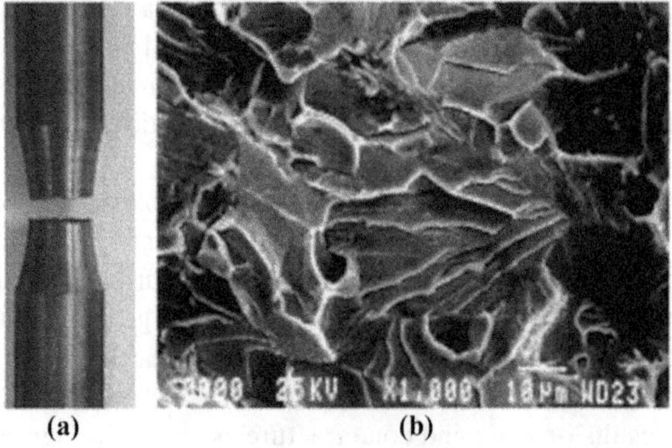

FIGURE 4.3 (a) Brittle fracture in mild steel and (b) fractography of the brittle fracture (reprinted with permission from [6]).

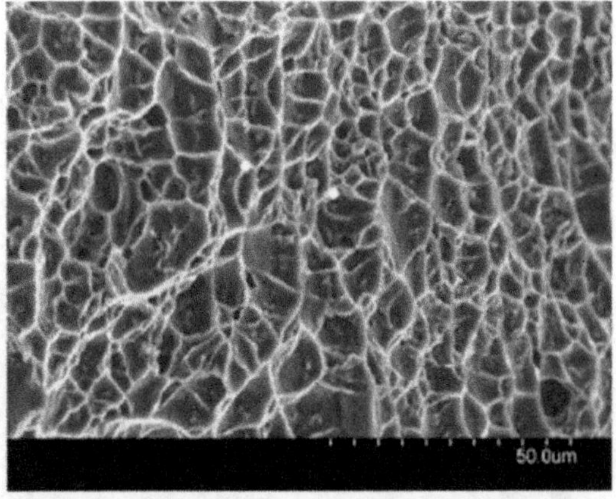

FIGURE 4.4 Fractography of a mixed fracture (reprinted with permission from [8]).

In material science, DBTT refers to the temperature at which a material undergoes ductile to brittle transition. It is a temperature range of around 10°C. It is one of the important properties to consider during the selection of materials. It is important to understand the failure processes (fatigue, overloading, stress cracking, etc.). The standard to measure DBTT is ISO 6603-2 (multiaxial instrumented impact).

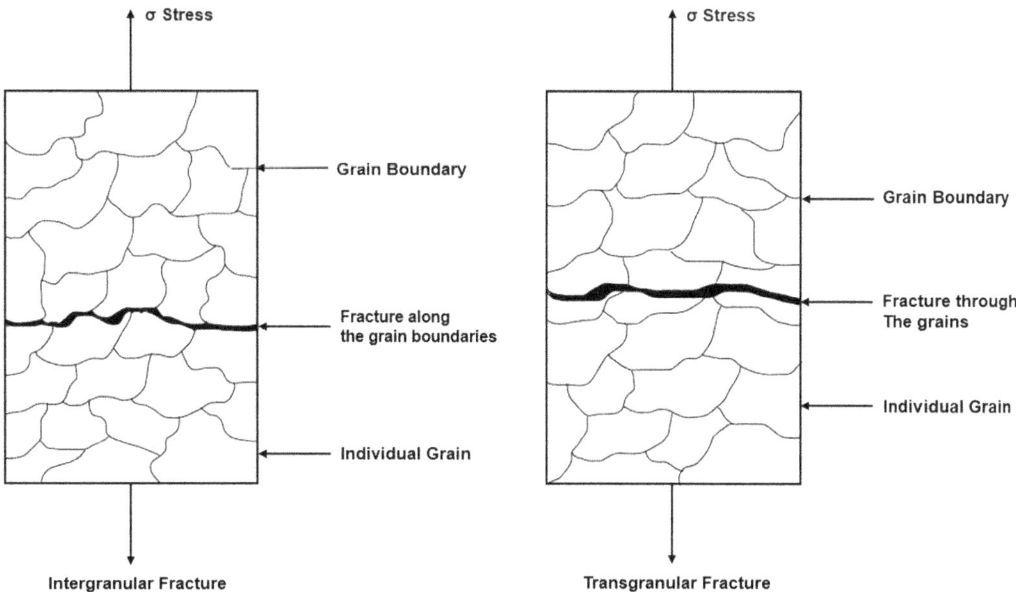

FIGURE 4.5 Diagram showing transgranular and intergranular fracture.

The fracture can be further classified as transgranular and intergranular based on the crack propagation path. In transgranular, characteristically, the cracks travel directly through the grain along certain crystallographic planes of the material, as shown in Figure 4.5. These cracks are sometimes referred to as cleavage cracks because they arise from the repeated breaking of atomic bonds. The type of fracture can be both ductile and brittle. The crack propagation in intergranular fracture occurs along the grain boundaries, where the grain boundaries are weakened by impurity segregation, element depletion or preferential chemical attack. Generally, the types of fractures in these cases are brittle.

4.4 TESTING AND STRENGTH EVALUATION OF METALS AND ALLOYS

Strength is the supreme mechanical property of a material that determines its application. Strength refers to the material's ability to withstand plastic deformation. The strength of metal or alloy can be further classified into three types based on the application of load: tensile, compressive and impact strength.

4.4.1 Tensile Strength

Tensile strength can be defined as the ability of a metal to restrict deformation and fracture under tension. It measures the utmost tensile stress a material can withstand unaccompanied by failure. A material's tensile strength is usually measured by a universal testing machine (UTM), which applies uniaxial tension until the breaking point is reached. The cross-section of the test specimen is either circular or rectangular, with enlarged ends to provide extra area for greater support. During the test, the specimen is slowly pulled at a constant speed by controlling the movement between the fixed and moving crossheads. The machine is connected to an extensometer for measuring the change in the gauge length of the sample during the tensile test. To maintain the constant speed of the crosshead, the applied axial force (P) varies as the test proceeds. The engineering stress (σ) at any time is determined by dividing P by the initial cross-sectional area A_o. The engineering strain is estimated by dividing the change in length by the original length.

A stress versus strain curve may be drawn from which it is possible to determine yield strength, an indication of the amount of stress the material can support before it encounters plastic deformation (as shown in Figure 4.6). To eliminate other factors such as temperature and impurities from playing a role, the test is conducted in a controlled environment.

4.4.2 Compressive Strength

In some cases, materials are used to resist deformation by compressive stresses; such materials behave differently in compression than in tension. The compressive strength of a metal or alloy specimen is the maximum stress or pressure that it can withstand before it ruptures or deforms by a certain percentage, e.g., 1%–10% deformation. The specimen is usually cylindrical with a length-to-diameter (L/d) ratio between 1 and 3. In some cases, the L/d ratio up to 10 is used to determine the elastic modulus accurately. Square or rectangular cross-sectional specimens can also be used for the test. Like the tension tests, in the compression test, the specimen is slowly pressed between two plates at a constant speed between the crosshead, and deformation versus the applied load is recorded. Figure 4.7 shows the schematic diagram of the compression testing machine.

4.4.3 Impact Strength

Impact strength is defined as the capacity of a metal or an alloy to absorb the energy imparted by a shock or sudden loading. Impact strength is also known as toughness. Therefore, the evaluation of impact strength is concerned with measuring the amount of energy a material can withstand while resisting cracking or fracturing. The impact strength is expressed in joules per metre. An impact strength measurement involves the application of force within a millisecond or less. When a load is applied nearly instantaneously, the material will crack, tear or suffer damage whenever it cannot withstand the energy being applied to it. A range of factors, such as temperature, thickness of the specimen and notch tip radius, influences impact strength. The material's impact strength can be measured by the Charpy impact test and the Izod impact test.

4.4.4 Charpy Impact Test

In the Charpy impact test, a heavy pendulum weight is swung from a set height and struck against a notched specimen, producing a high strain rate at the point of impact. The length of the Charpy test

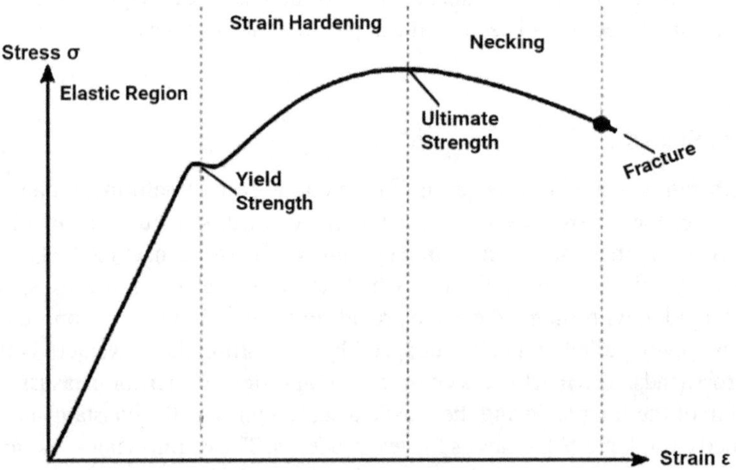

FIGURE 4.6 Stress–strain curve observed during tensile testing.

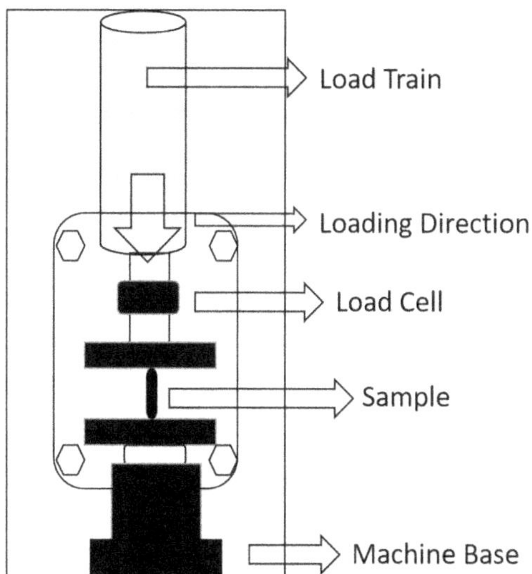

FIGURE 4.7 Schematic of compression testing machine (adapted following [9]).

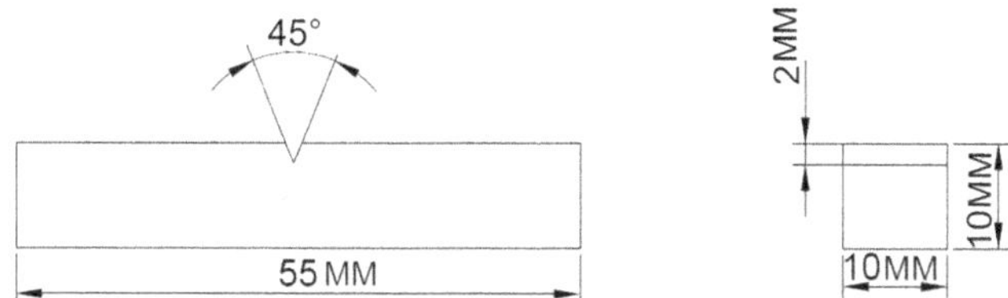

FIGURE 4.8 Dimensions of Charpy impact test sample.

specimen is usually 55 mm, 10 mm square, with a depth of 2 mm on one face of 0.25 mm radius [10] (as shown in Figure 4.8). This test allows one to measure the energy a specimen can absorb during the occurrence of fracture.

The pendulum strikes the specimen on the opposite face to the notch while it is supported on two sides by an anvil. The notch toughness of the test material is calculated by estimating the amount of energy absorbed in fracturing the test material. The height of the swing is one of the factors that determines how much energy is absorbed during the fracturing process. A typical test involves testing three specimens at a given temperature and averaging the results. In addition, the test material's ductile to brittle transition curves may also be obtained by testing the specimens at a specified range of temperatures.

A material's brittle or ductile nature can be found using the Charpy test by examining the microstructural features of the fracture. It can be particularly useful in ferritic steels because it becomes increasingly brittle with a gradual fall in temperature. Suppose a metal/alloy is tested for impact resistance. In that case, the amount of absorbed energy is less or more dependent on the brittle or ductile nature of the metal.

4.4.5 Izod Impact Test

Like the Charpy impact test, the Izod impact test is also a single-point test, where a heavy pendulum weight is swung from a set height and made to strike a notched specimen at a particular point for the fracture to occur at the point of impact. The Izod impact test generally uses an ASTM D256 standardized specimen of 64 × 13 × 13 mm dimension. The preferred depth under the notch of the sample is 10.2 mm, as shown in Figure 4.9 [11].

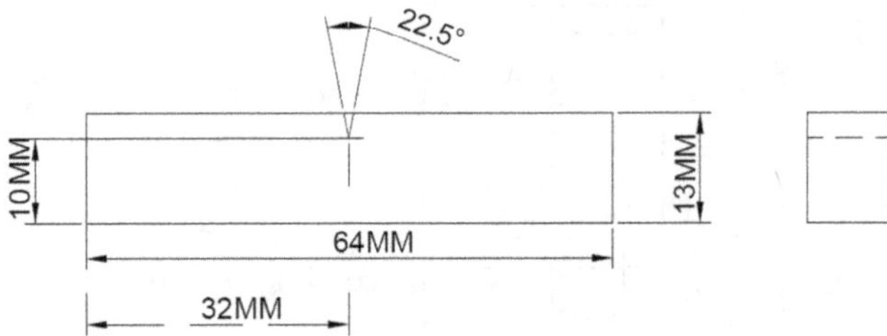

FIGURE 4.9 Dimensions of Izod impact test sample.

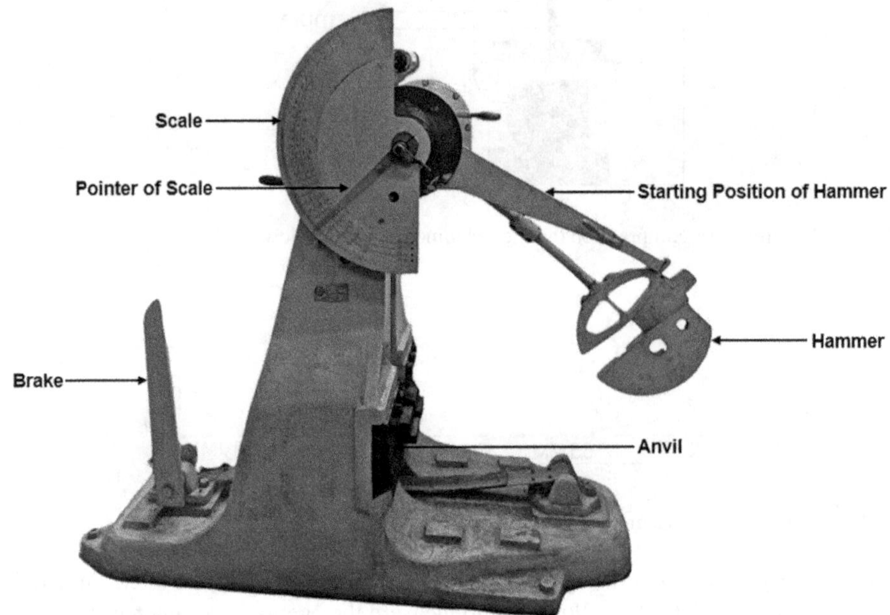

FIGURE 4.10 Experimental setup for Charpy and Izod impact tests.

One of the major differences between the Charpy and Izod tests is that during the Izod impact test, the test specimen is placed in a vertically cantilevered position. It is clamped onto an anvil at one end. After the pendulum is allowed to swing (as shown in Figure 4.10), it strikes the specimen on the notched face above the V-notch.

It has been observed that different test materials exhibit lower impact strength as the temperature decreases. For this reason, it is necessary to perform impact tests at temperatures for which a particular material is intended to be used. Another major difference between the Charpy and Izod tests is that the Charpy impact test can measure the impact strength for only metals. In contrast, the Izod impact test can measure the impact strength of all materials, including metals.

4.5 FRACTURE TOUGHNESS OF METALS

Fracture toughness indicates a material's capability to withstand further cracks in the presence of already existing ones. Numerically, the resistance of a material to brittle fracture can be expressed using fracture toughness. It also indicates the maximum energy absorbed by a material before failure. The fracture toughness values also determine the material performance evaluation, and quality assurance of conventional engineering structures such as nuclear pressure vessels, petrochemical tanks, oil and gas pipelines, aircraft structures, automobiles, ships and other vehicles. Therefore, testing and assessing fracture toughness are important [12]. Materials fracture toughness is a highly popular parameter due to the ease of preparation of small specimens and the requirement for very small specimens. The fracture toughness of materials is measured using several types of tests, namely the indentation test and the plane strain test. The indentation test typically has a notched specimen that can be configured in several ways. Charpy impact testing, which applies a sudden load to samples with V- or U-notches from behind, has become one of the most widely used standardized indentation test methods. Crack displacement tests, such as three-point beam bending tests, are also widely used [13].

To measure fracture toughness, thin cracks are introduced into the specimens before loading them. Fracture toughness can be measured with three types of coupons specified in the ASTM standard E1820: single edge bending coupons [SE(B)], compact tension coupons [C(T)] and disk-shaped compact tension coupons [DC(T)]. There are three dimensions to every specimen configuration: length of the crack (*A*), thickness (*B*) and width (*W*). Depending on the test type performed, these dimension values vary. Flexural tests are usually conducted on compact or three-point configurations (as shown in Figure 4.11).

Two types of fracture toughness specimens are mostly used in testing plane strain fracture toughness: (a) the single edge notch bend (SENB) and (b) the compact tension (CT) specimens. Generally, conditions for plane strain hold good when [14]

$$B, a \geq 2.5\left(\frac{K_{IC}}{\sigma_{YS}}\right)^2 \tag{4.7}$$

where minimum thickness corresponds to B and fracture toughness corresponds to the K_{IC} (stress intensity factor) of the samples. Finally, the yield strength of the material is indicated by σ_{YS} [15]. In this test, the load is steadily increased until K_{IC} reaches 2.75. Testing continues until the maximum load is attained, and the load and crack mouth opening displacement (CMOD) are recorded [14]. There are various ASTM standardized testing methods for measuring the fracture toughness of a material. Some of them are [15]:

i. ASTM C1161 Flexural Strength Test Method for Advanced Ceramics at Ambient Temperature
ii. ASTM C1421 for Determining the Fracture Toughness of Advanced Ceramics at Ambient Temperature
iii. ASTM E399 Standard Test Method for Linear-Elastic Plane-Strain Fracture Toughness of Metallic Materials
iv. ASTM E740 Standard Practice for Fracture Testing with Surface-Crack Tension Specimens
v. ASTM E1820 Standard Test Method for Measurement of Fracture Toughness
vi. ASTM E1823 Standard Terminology Relating to Fatigue and Fracture Testing

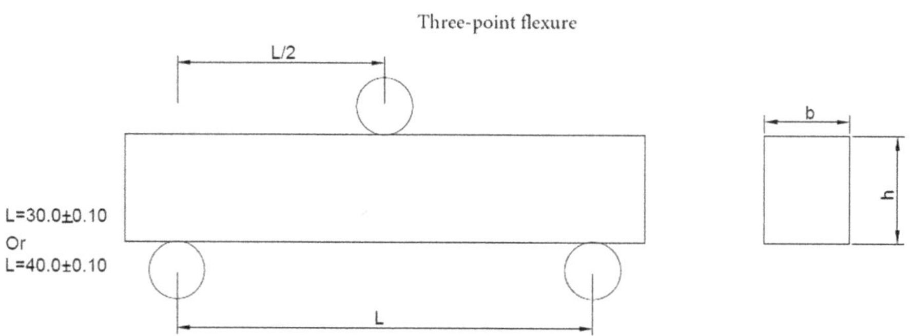

FIGURE 4.11 Schematic of three-point flexure for flexure tests.

4.6 CREEP STRENGTH OF METALS: DIFFERENT STAGES OF CREEP CURVE

The stress at a particular temperature (*T*) is called creep strength, whose steady-state values fall in the range of 10^{-11} to 10^{-8} s^{-1} [16]. Creep occurs when a material is permanently deformed (plastically deformed) with time under constant load (or constant stress), usually at high temperatures. Creep is usually appreciable above $T > 0.4\ T_m$, where T_m corresponds to the absolute melting temperature. In engineering, creep design is important in high-temperature applications, e.g., steam turbines in power plants, jets, rocket engines and nuclear reactors. The non-performance of light bulb filaments is also an example of creep.

In metals and alloys, creep is large, therefore importance should be given to structures used in high-temperature conditions. There are various tests designed to measure the dimensional changes and load time-bearing characteristics like creep and rupture. The creep strength is usually measured by creep testing; the schematic of the creep testing machine is shown in Figure 4.12.

In the creep test, load and temperature are kept constant, and values of time and elongation are measured concerning each other. Correspondingly, the rupture timing is noted when it happens during the test. The creep curve is obtained by plotting the strain variation at the constant load and temperature as a function of time (as shown in Figure 4.13).

The creep curve obtained can be divided into three distinct stages: primary, secondary and tertiary [17].

Stage 1 is known as primary or transient creep. In this stage, the effect of work hardening is more than the recovery. Here, the creep rate decreases with time.

Stage 2 is called secondary creep or steady-stage creep. Here, the creep rate is the lowest and nearly constant. In this stage, work hardening is balanced by recovery.

Stage 3 is known as tertiary creep. In this stage, the creep rate increases until failure. As a result, grain boundary separation and internal cracks, cavities and voids form, causing failure.

4.7 HIGH STRAIN RATE STRENGTH OF METALS IN SHPB AND GAS GUN TECHNIQUES

Materials used for various industrial applications like impact-resistant pressure vessels, the body of the car, bulletproof armour and shipping casks to transport nuclear materials are expected to

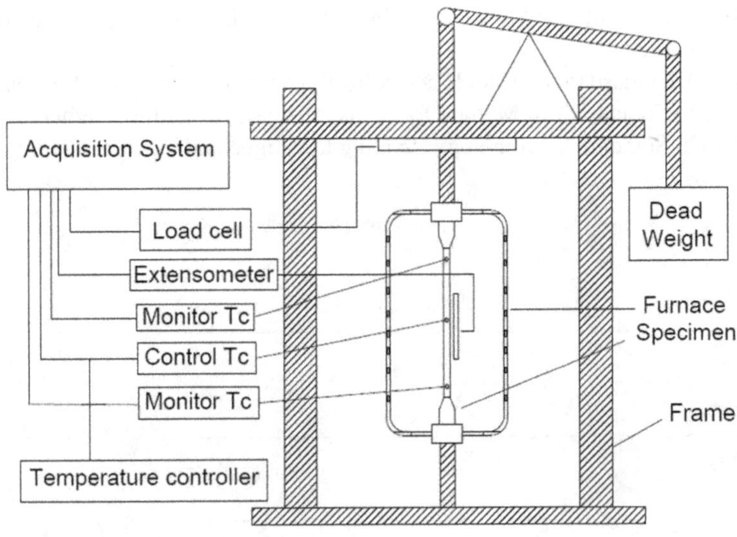

FIGURE 4.12 Schematic of creep testing machine.

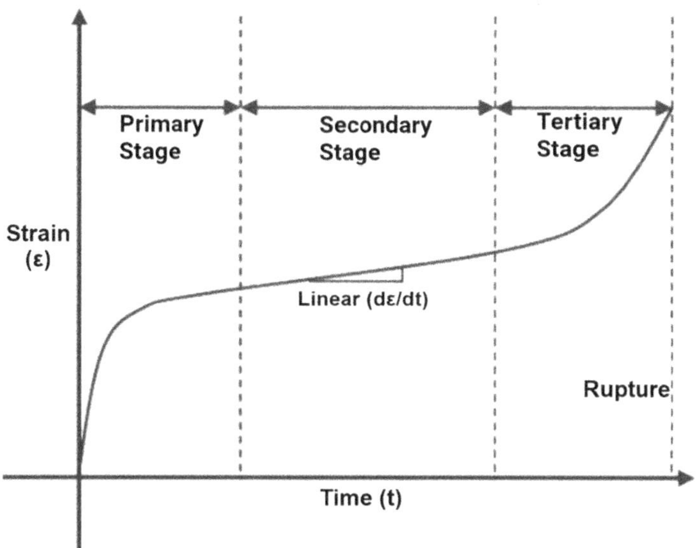

FIGURE 4.13 Three stages of creep.

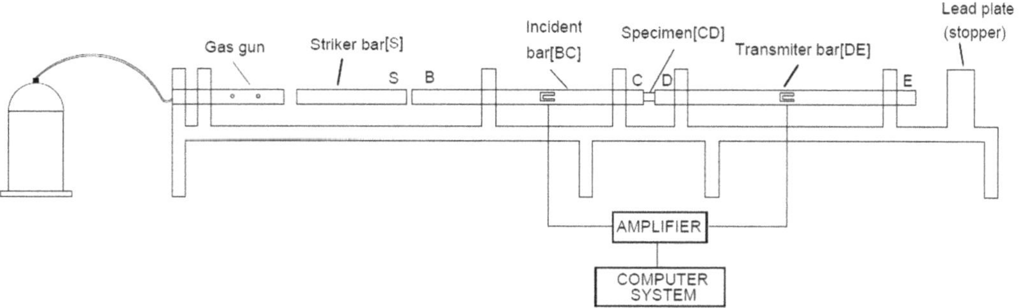

FIGURE 4.14 Experimental setup of split-Hopkinson pressure bar.

withstand deformation at a very high strain rate. Various forming methods like extrusion, rolling and machining put materials under high strain. So, components subject to high strain rates of loading must be designed and analyzed based on the constitutive behaviour of materials under high strains. However, a strain rate greater than $10s^{-1}$ can rarely be obtained using conventional tensile testing machines. So, we need to use modern methods like the split-Hopkinson pressure bar (SHPB) method and gas gun techniques for testing the high strain strength of materials [18].

SHPB operates on the theory of unidimensional propagation of the wave. The principal parts of the SHPB setup are a gas gun, a striker bar and two straight bars called the incident bar and transmission bar (shown in Figure 4.14). Gas gun chambers have striker bars in their barrels. All three bar types use identical materials and have the same cross-sectional area. The striker, incident bar and transmission bar must be elastic throughout the test. The sample is positioned in the centre of the incident and transmission bar. By applying gas pressure, the striker bar moves towards the incident bar.

The incident bar transmits an elastic wave toward the sample on impact. The incident wave splits into two when it reaches the specimen. The transmitted wave, among the two, passes through the specimen and reaches the transmitted bar, resulting in plastic deformation. On the other hand, the reflected wave is reflected away from the specimen and returns to the incident bar [19].

Strain gauges are attached to the bars to record strains generated by the elastic waves. If the specimen is uniformly deformed, the incident, transmitted and reflected wave amplitudes can be used to calculate stress and strain [20]. The impact load pulse on a specimen can be determined by measuring the impact strain pulses recorded by the strain gauges connected to the metallic bars as a function of time.

Let's consider $\varepsilon_I(t)$ and $\varepsilon_R(t)$ as the incident strain and reflected strains measured by the strain gauge attached to the incident bar. These values can be used to calculate the strain rate of deformation $\varepsilon^{\cdot}(t)$ and the plastic flow stress of the specimen $\sigma(t)$ [21].

$$\sigma(t) = \mathrm{E}(\mathrm{A}_o / \mathrm{A})(\varepsilon_I + \varepsilon_R) \tag{4.8}$$

$$\varepsilon^{\cdot}(t) = 2\mathrm{C}_o\varepsilon_R(t) / \mathrm{L} \tag{4.9}$$

$$\varepsilon = ln\frac{L}{L_0} \tag{4.10}$$

where E is Young's modulus of the incident bar, C_o is the speed at which the elastic wave propagates, A_o is the cross-sectional area of the incident bar, A is the cross-sectional area of the test material, and L and L_0 are the final and initial length of the test material.

The gas gun technique comprises an instrumented target support attached to a gas gun's muzzle. It measures the force-time pulses of the impact on the specimen. During impact, the specimen is supported by two metallic bars. Strain gauges are connected to the bars to measure the test specimen's strain during impact. Assuming that the wave propagates longitudinally, the impact load pulse on a specimen can be determined by measuring the impact strain pulses recorded by the strain gauges connected to the metallic bar [22]. This technique is designed for ballistic and impact studies of specimens at temperatures of 100–1000 K and can test materials undergoing uniform radial expansion at very high strain rates (10^3–10^5 s^{-1}) or even more [23].

4.8 CURRENT OBSERVATIONS: LITERATURE REVIEW

Metallic materials possess complex mechanical properties, and improving their mechanical properties requires in-depth knowledge of their deformation and fracture behaviours. It is now known that deformation and fracture mechanisms are closely associated with the initial microstructure of materials. Therefore, the evolution of microstructural changes occurring during deformation is considered critical to understanding the deformation stages and fracture mechanisms.

During the past few years, research has been conducted to develop relationships between microstructural evolution during deformation and underlying deformation and fracture mechanisms. An overview of some of the research carried out in this direction is provided next.

Deformation behaviour and microstructures of deformed metals have been investigated since the invention of light microscopy. To date, many researchers have investigated the microstructural features in deformed metals. They used automatic or semiautomatic techniques for determining parameters which would be useful in understanding and correlating the deformation behaviour with the microstructure [24]. An assessment of the deformation and fracture behaviour of WC-Co tool material based on the microstructural features and its effect is available in the open source literature [24].

Research related to finding a relationship between plastic deformation, microstructure and crystallography of metals was also reported, where it was claimed that during plastic deformation, the microstructures showed a pattern of grain subdivision up to nanometer length scale. Physical models were used to predict the formation and evolution of dislocation boundaries and to develop relationships between the processing conditions, structure and properties, which can be useful for future applications [25].

Studies were also conducted to identify the mechanisms responsible for forming fine grains in materials subjected to severe plastic deformation (SPD). The effect of SPD and subsequent annealing on strength, ductility and superplastic properties was also reported [26].

However, new or modified deformation monitoring techniques are needed to investigate metallic materials' deformation and fracture behaviour of advanced technologies and engineering. Among the various deformation monitoring techniques, a new method for material testing was devised, which was claimed to be effective under the dynamic deformation condition. While developing this method, tests on various materials of different groups were conducted, and correlations between "strain rate and strained structure" were found. The correctness of the results was verified by using the extensively available data. It was claimed that the results can be used for designing the structure of vehicles and aircraft [27].

In another method, various uniaxial tensile tests were performed on Cu, Fe and Ti specimens under constant, direct current levels and identical temperatures. The test data show constant strain, current density and temperature conditions along the gauge length of the specimen, which is essential for achieving a clear understanding of electrically assisted deformation of metals interpretation [28].

Recently, electromagnetic radiation detection has attracted considerable attention from researchers. Electromagnetic emissions from materials under external loading, treated as signals, are analyzed for predicting fracture and failure of the material; investigations have shown that electromagnetic emission detection techniques can be effective for monitoring deformation [29].

4.9 EMERGING AREAS AND CHALLENGES

The demand for improved communication, transportation, healthcare, infrastructure developments and people's living has caused technology to change rapidly. The emergence of advanced technology is deeply connected with developing new materials with improved properties or manufacturing processes and design improvements. Therefore, the material testing capabilities should be raised to a higher standard or suitably modified to support the advancement. Some areas where technology development needs special attention are mentioned next.

In natural disasters like earthquakes and cyclones, engineering structures have to withstand impulsive loads due to the propagation of the shock waves caused by the earthquake or collisions with objects during cyclones. Therefore, material tests for understanding the deformation and fracture behaviour must be performed under simulated conditions [30].

Another concern is unpredictable fire and explosions in lithium-ion batteries used in laptops, smartphones, electronic gadgets and electric vehicles. In most cases, such failures arise due to their misuse of devices. The failure may be caused by mechanical, thermal or electrical abuse or the combined effect. To ensure battery safety, mechanical, thermal and electrical abuse tests were conducted; however, shear and strain-rate-dependent failure phenomena are yet to be considered in such tests. The deformation mechanism associated with the failure is yet to be fully understood [31].

Biodegradable magnesium alloys have recently emerged as an important material because of their excellent biocompatibility and desired strength for implants. However, the sudden decrease in strength due to inhomogeneous dissolution is still unsettled. Therefore, static and dynamic deformation tests under simulated corrosive environments are needed for their acceptance as implant material [32].

Introducing nanoelectromechanical systems (NEMS) and microelectromechanical systems (MEMS) in mechanical and electronic devices necessitates an assessment of their mechanical properties and reliability. Although NEMS and MEMS have sizes well below the standard test specimens, performance assessment based on the conventional macroscopic test data needs to be more accurate due to the dominance of size effect in the systems mentioned earlier. Therefore, specific nano- and micromechanical tests must be formulated to determine the mechanical properties, reliability and performance [33].

REFERENCES

1. Stephan Handschuh-Wang, Tomasz Gancarz, Sergey Uporov, Tao Wang, Eryuan Gao, and Florian J. Stadler, "A short history of fusible metals and alloys – Towards room temperature liquid metals," Eur J Inorg Chem, vol. 2022, pp. 1–27, 2022.
2. Hyon-Jee Lee, "Chapter 4 Deformation behaviours in metals and alloys 4.1," pp. 1–45, 2003. https://thesis.library.caltech.edu/2237/5/chapter_4.pdf
3. S. Tu, X. Ren, J. He, and Z. Zhang, "Stress–strain curves of metallic materials and post-necking strain hardening characterization: A review," Fatigue Fracture Eng Mater Struct, vol. 43, pp. 3–19, 2020.
4. X. H. Sun, J. W. Qiao, Z. M. Jiao, Z. H. Wang, H. J. Yang, and B. S. Xu, "An improved tensile deformation model for in-situ dendrite/metallic glass matrix composites," Sci Rep, vol. 5, pp. 1–11, 2015.
5. M. T. Albdiry and M. F. Almensory, "Failure analysis of drill string in petroleum industry: A review," Eng Fail Anal, vol. 65, pp. 74–85, 2016.
6. Y. Enomoto, "Advances in Steam Turbines for Modern Power Plants", Elsevier, pp. 397–436, 2017.
7. E. S. Dzidowski, "Physical concept of shear fracture mesomechanism and its applications," Cent Eur J Eng, vol. 1, pp. 217–233, 2011.
8. S. Chatterjee, A. Sinha, D. Das, S. Ghosh, and A. Basu Mallick, "Microstructure and mechanical properties of Al/Fe-aluminide in-situ composite prepared by reactive stir casting route," Mater Sci Eng A, vol. 578, pp. 6–13, 2013.
9. A. Nazir and J. Y. Jeng, "Buckling behaviour of additively manufactured cellular columns: Experimental and simulation validation," Mater Des, vol. 186, p. 108349, 2020.
10. A. Q. Barbosa, L. F. M. da Silva, A. Öchsner, J. Abenojar, and J. C. del Real, "Influence of the size and amount of cork particles on the impact toughness of a structural adhesive," J Adhes, vol. 88, no. 4–6, pp. 452–470, 2012.
11. J. R. Wagner, E. M. Mount, and H. F. Giles, "Testing properties," In: *Extrusion*, Elsevier, pp. 241–253, 2014.
12. F. Test and F. M. Testing, "ASTM E1820-13," ASTM Standards, vol. I, no. C, pp. 1–56, 2014.
13. A. Fujishima and J. L. Ferracane, "Comparison of four modes of fracture toughness testing for dental composites," Dent Mater, vol. 12, no. 1, pp. 38–43, 1996.
14. X. K. Zhu and J. A. Joyce, "Review of fracture toughness (G, K, J, CTOD, CTOA) testing and standardization," Eng Fracture Mech, vol. 85, pp. 1–46, 2012.
15. ASTM, "Test Method for Plane Strain Fracture Toughness of Metallic Materials ASTM-E399.pdf." pp. 413–433, 1997. https://doi.org/10.1520/E0399-90R97
16. F. M. Shuaeib, K. Y. Benyounis, and M. S. J. Hashmi, "Material behaviour and performance in environments of extreme pressure and temperatures," In: Reference Module in Materials Science and Materials Engineering, Elsevier, pp. 2–13, 2017. https://www.sciencedirect.com/science/article/pii/B9780128035818041709
17. W. S. Loewenthal and D. L. Ellis, "Sources of variation in creep testing," NASA/TM, no. 215493, pp. 1–47, 2011.
18. S. Sharma, V. M. Chavan, R. G. Agrawal, R. J. Patel, R. Kapoor, and J. K. Chakravarty, "Split-Hopkinson pressure bar: An experimental technique for high strain rate tests." BARC External, no. BARC/2011/E/013, p. 4, 2011.
19. A. Gilat and Y. H. Pao, "High-rate decremental-strain-rate test," Exp Mech, vol. 28, no. 3, pp. 322–325, 1988.
20. A. Bagher Shemirani, R. Naghdabadi, and M. J. Ashrafi, "Experimental and numerical study on choosing proper pulse shapers for testing concrete specimens by split Hopkinson pressure bar apparatus," Constr Build Mater, vol. 125, pp. 326–336, 2016.
21. P. Rama Rao, S. V. Kamat, and G. Sundararajan, "Material deformation and fracture under impulsive loading conditions," Bull Mater Sci, vol. 15, no. 1, pp. 3–25, 1992.
22. S. A. Ritt and A. F. Johnson, "Load pulse determination in gas gun impact tests," Appl Mech Mater, vol. 7–8, pp. 259–264, 2007.
23. D. R. Jones, D. J. Chapman, and D. E. Eakins, "A gas gun-based technique for studying the role of temperature in dynamic fracture and fragmentation," J Appl Phys, vol. 114, no. 17, pp. 1–12, 2013.
24. B. Roebuck and E. A. Almond, "Deformation and fracture processes and the physical metallurgy of WC–Co hard metals," Int Mater Rev, vol. 33, no. 1, pp. 90–112, 1988.
25. N. Hansen and R. F. Mehl, "New discoveries in deformed metals," Metall Mater Trans A Phys Metall Mater Sci, vol. 32, no. 12, pp. 2917–2935, 2001.

26. B. Verlinden, "Severe plastic deformation of metals," Metalurgia J Metal, vol. 11, no. 3, pp. 165–182, 2005.
27. J. Pawlicki, Z. Stanik, A. Płachta, and A. Kubik, "A new method of testing the dynamic deformation of metals," Materials, vol. 14, no. 12, p. 3317, 2021.
28. C. Rudolf, R. Goswami, W. Kang, and J. Thomas, "Effects of electric current on the plastic deformation behaviour of pure copper, iron, and titanium," Acta Mater, vol. 209, p. 116776, 2021.
29. S. K. Sharma, V. S. Chauhan, and M. Sinapius, "A review on deformation-induced electromagnetic radiation detection: History and current status of the technique," J Mater Sci, vol. 56, no. 7, pp. 4500–4551, 2021.
30. P. R. Rao, V. Kamat, and G. Sundararajan, "Material deformation and fracture under impulsive loading conditions," Bull Mater Sci, vol. 15, no. 1, pp. 3–25, 1992.
31. B. Liu, Y. Jia, C. Yuan, L. Wang, X. Gao, S. Yin, J. Xu "Safety issues and mechanisms of lithium-ion battery cell upon mechanical abusive loading: A review," Energy Storage Mater, Elsevier BV, vol. 24, pp. 85–112, 2020.
32. F. Kiani, C. Wen, and Y. Li, "Prospects and strategies for magnesium alloys as biodegradable implants from crystalline to bulk metallic glasses and composites—A review," Acta Biomaterialia, vol. 103, pp. 1–23, 2020.
33. T. Connolly, P. E. Mchugh and M. Bruzzi, "A review of deformation and fatigue of metals at small size scales," Fatigue Fracture Eng, Mater Struct, vol. 28, no. 12, pp. 1119–1152, 2005.

5 Deformation and Fracture of Ceramic Materials *Experimental Methods and Challenges*

Payel Maiti, Manjima Bhattacharya, and Anoop Kumar Mukhopadhyay

5.1 INTRODUCTION

Strong, hard, and characteristically brittle ceramics are almost as old as human civilization. Amenable to a wide variety of processing techniques, ceramic materials are evolving today for a wide range of applications [1–10]. Such applications include high-temperature electronics [1], various structural and functional applications [2], refractory applications [3], biomedical applications [4], high-temperature sensing applications [5], microwave applications [6], machining applications [7], structural armour applications [8], ultrahigh entropy material development applications [9], and transparent armour application [10]. All such applications involve contact. That is why the deformation and fracture of ceramics assume the highest importance as they are characteristically brittle. Speaking of the macroscale they almost inevitably fail in a brittle manner without showing any appreciable plastic deformation at all. This is what makes the appearance of micro-/nanoscale plasticity even at a localized scale a very special area of research. Deformation and especially fracture of ceramics are strongly sensitive to processing history, sintering aid, sintering temperature, relative density, volume percent open porosity, pore size, pore size distribution, grain size, grain size distribution, average flaw size, flaw size distribution, critical flaw size, characteristics of the grain boundary phase, and so on. However, the main objective of this chapter is to emphasize currently available experimental techniques to study deformation and especially the fracture of ceramics. This chapter therefore focuses briefly on the techniques used to study the deformation and fracture of ceramic materials at room temperature at the macro-, micro-, and -nanoscale. This approach is followed by a brief overview of current experimental data and knowledge before developing a short final hint on the challenges involved for future developments. High-temperature deformation and fracture of ceramics including creep deformation is a very important area that deserves dedicated attention that is beyond the scope of the present work and, hence, is excluded from the current discussions. The most important mechanical characterization techniques for ceramics are macroscale techniques, e.g., tensile strength (σ_t), flexural strength (σ_f), compressive strength (σ_c), modulus of elasticity (E), fracture toughness (K_{IC}), and high strain rate strength; microscale techniques such as microhardness evaluation; and nanoscale techniques such as nanoindentation techniques. These are briefly discussed and details elaborated in the corresponding ASTM standards [11].

5.2 UNIAXIAL TENSILE TEST

Ceramics are most weak in tension as they possess hardly any ability to undergo plastic deformation. These tests are done on dog-bone-shaped samples or cylindrical samples using very specially

 DOI: 10.1201/9781003359364-7

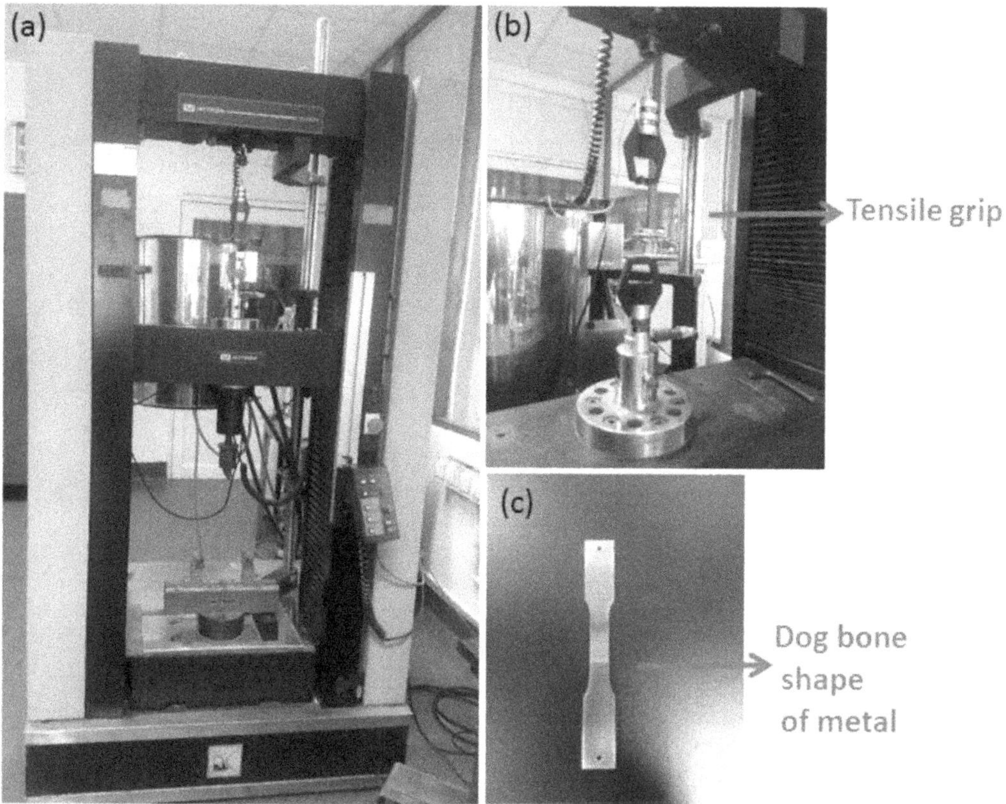

FIGURE 5.1 Tensile test setup.

designed dedicated grips in a universal testing machine (UTM) [11]. The tensile strain is measured through a strain gauge attached over the gauge length portion. The tensile strain (ε_t) of the material is evaluated as $\varepsilon_t = \frac{L - L_0}{L_0}$, where L_0 is the initial gauge length and L is the final length. Similarly, the tensile strength(σ_t) of the material is evaluated as:

$$\sigma_t = \left(\frac{P}{A}\right) \tag{5.1}$$

where P is the applied load measured by the load cell and A is the nominal cross-section of the sample coupon. The details of the experimental techniques are given elsewhere [12] and a schematic representation of the setup is given in Figure 5.1.

5.3 THREE-POINT AND FOUR-POINT FLEXURAL STRENGTH TEST

The three-point and four-point bending tests are the most commonly utilized techniques for evaluating flexural strength (σ_f) of ceramic samples evaluated using a UTM as elaborated in the ASTM standard [11]. This test done by a three-point bend fixture (Figure 5.2) is very sensitive to the sample size, strain rate, and very crucially loading geometry [13]. The flexural strength (σ_f) measured by the three-point bend test is evaluated as:

$$\sigma_f = \frac{3P_fL}{2BW^2} \tag{5.2}$$

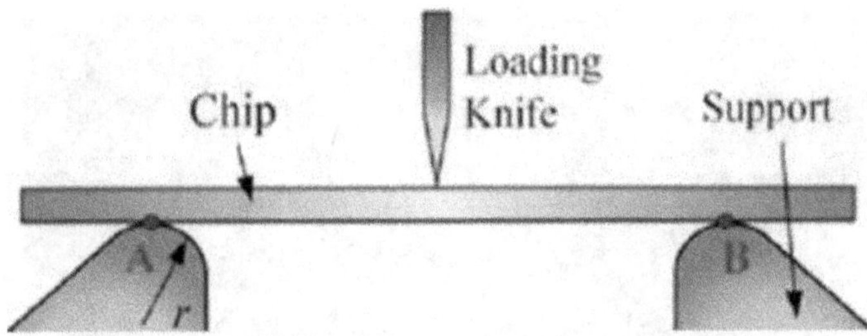

FIGURE 5.2 Schematic representation of three-point bending test (reprinted with permission from [13]).

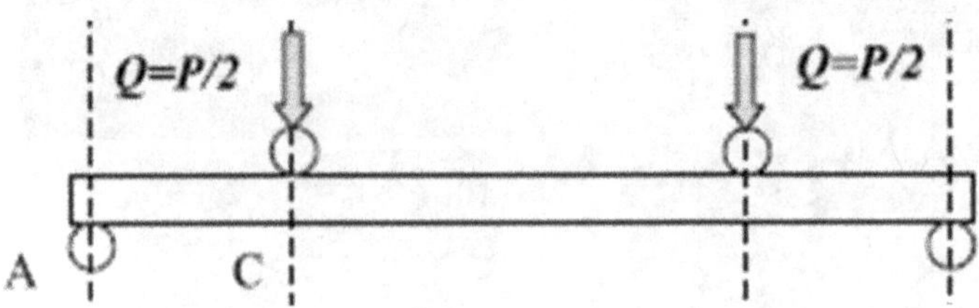

FIGURE 5.3 Schematic representation of four-point bending test (reprinted with permission from [14]).

where P_f, L, B and W represent load, the span length between the two bottom support rollers, the width of the rectangular sample, and the thickness of the rectangular sample, respectively.

Similarly, the flexural strength (σ_f) measured by the four-point bend test is evaluated as:

$$\sigma_f = \frac{3P_f d}{BW^2} \tag{5.3}$$

where d is $(L - l)/2$, L is the span length between the two bottom support rollers, and l is the distance between the two top-loading rollers (Figure 5.3), while the other terms are as defined earlier [14]. These two techniques are also utilized for the evaluation of fracture toughness (K_{1C}) and the modulus of elasticity (E) of ceramics at room and high temperatures.

5.4 UNIAXIAL COMPRESSIVE TEST

Uniaxial compressive tests are conducted in uniaxial compression using hardened stainless steel platens with a tungsten carbide top coating [15] to avoid denting by the ceramics on the platens themselves (Figure 5.4). The compressive strength (σ_c) is evaluated as:

$$\sigma_c = \frac{P}{A} \tag{5.4}$$

where P is the applied compressive load and A is the area of the ceramic disc under the compressive load.

5.5 HARDNESS MEASUREMENT TESTS

Hardness measurements are done to evaluate the microscale deformation behaviour of ceramics. The two most common measurement techniques are the Vickers indentation technique and Knoop indentation technique. These two microscale tests illustrate how easy or how difficult it is

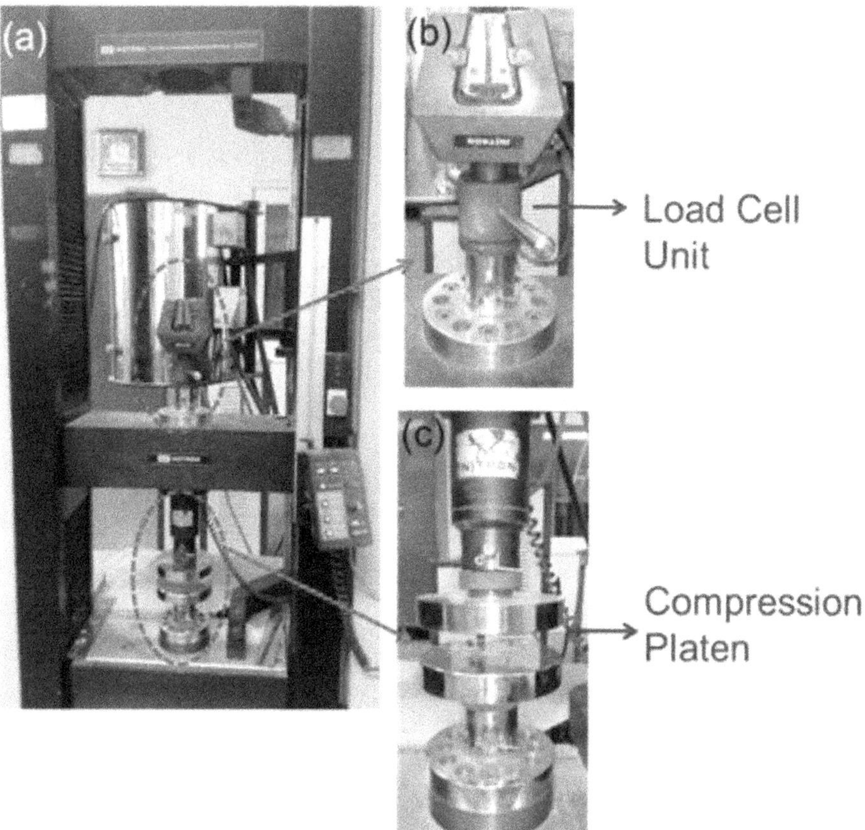

FIGURE 5.4 (a) The universal testing machine and close-up views showing the (b) load cell unit and (c) compression platen (reprinted with permission from [15]).

to plastically deform a given ceramic at a local scale with or without causing localized microcrack generation. The Vickers indentation gives the Vickers microhardness H_v (Figure 5.5) as [16]:

$$H_v = (0.1899P)/d^2 \quad (5.5)$$

where d is the mean diagonal evaluated as $[d = (d_1 + d_2)/2]$, where d_1 and d_2 are the lengths of the two mutually perpendicular diagonals of the pyramidal indent impression caused by the Vickers diamond pyramidal indenter at the given applied load P on the polished ceramic surface under consideration.

The smaller the value of (d), the higher the Vickers microhardness of the ceramic and vice versa. This test is conducted using a Vickers hardness tester. The tester is equipped with a diamond pyramidal indenter with a square base at an angle of about 136° between opposite faces. Here, P is the applied load (N), d is the average diagonal length (in μm), and d_1 and d_2 are the lengths of the two mutually perpendicular diagonals in μm. Thus, the Vickers microhardness is evaluated in GPa.

In a similar manner, the Knoop hardness (H_k) is given as [17]:

$$H_k = \frac{14.22P}{D_l^2} \quad (5.6)$$

where P is the applied load in (N) and D_l is the length of indentation along its long axis (Figure 5.6). The Knoop hardness tester is equipped with a diamond indenter in the shape of an extended pyramid with the length-to-width ratio being 7:1. The face angle is 172° for the long edge and 130°

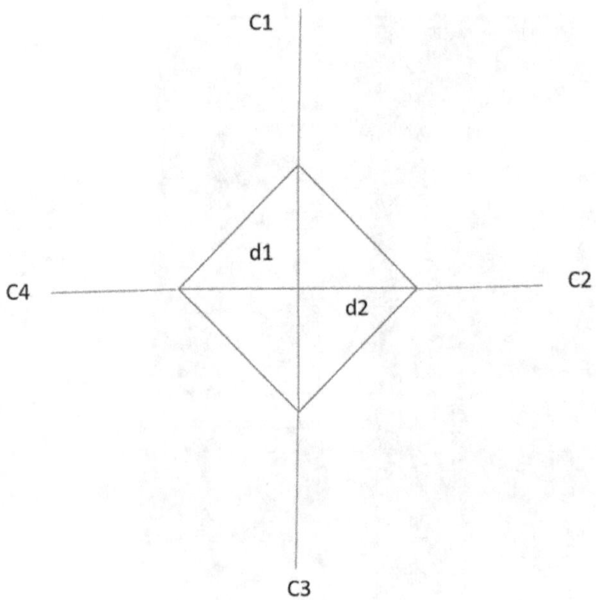

FIGURE 5.5 Schematic representation of Vickers indentation.

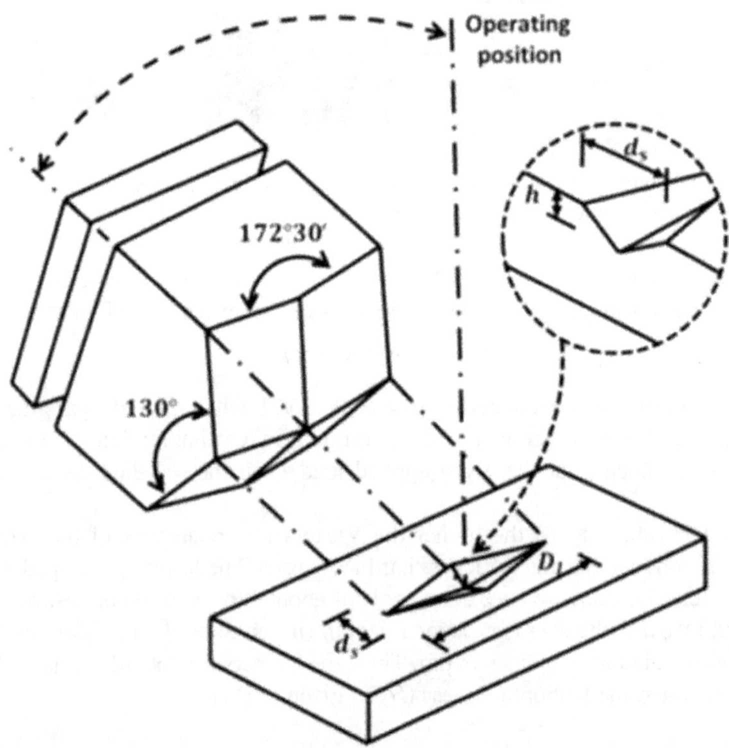

FIGURE 5.6 Schematic representation of Knoop indentation (reprinted with permission from [17]).

for the short edge. In this case, the depth of the indentation is usually approximated as 1/30 of (D_l). Further details of the experimental technique are given in [17].

5.6 FRACTURE TOUGHNESS MEASUREMENTS

5.6.1 Single-Edge Notched Beam (SENB) Bend Test

One of the most popular and user-friendly methods for measuring the fracture toughness of a ceramic material is the SENB bend test conducted in either three-point or four-point loading in flexure. In this case, a sharp notch is created on the ceramic specimen to determine its fracture toughness. The details of the experimental technique are given elsewhere [18]. The mode I critical stress intensity factor in plane strain condition or fracture toughness (K_{1c}) in four-point bending (Figure 5.7) for the initial notch-crack length a is evaluated as [18]:

$$K_{Ic} = \frac{3P_f d\sqrt{a}}{BW^2} f\left(\frac{a}{W}\right) \tag{5.7}$$

$$f\left(\frac{a}{W}\right) = 1.99 - 2.24\left(\frac{a}{w}\right) + 12.97\left(\frac{a}{w}\right)^2 - 23.17\left(\frac{a}{w}\right)^3 + 24.8\left(\frac{a}{w}\right)^4 \tag{5.8}$$

Here, $f\left(\frac{a}{W}\right)$ is the geometry factor, and B and W are respectively, the width and thickness of the sample.

However, the bending moment distribution in three-point bending is more acute and centred in the middle of the loading span. This makes the stress distribution a bit more nonuniform. On the other hand, in four-point bending the bending moment distribution remains more uniform. Hence, the stress distribution is more uniform. Thus, more volume of the experimental sample is exposed to this uniform stress distribution applied during the experiment. Therefore, more flaw volume is exposed to this uniform stress distribution in four-point bending rather than in three-point bending. It should be noted that for the three-point bend test, the corresponding expression for evaluation of fracture toughness is given as [19]:

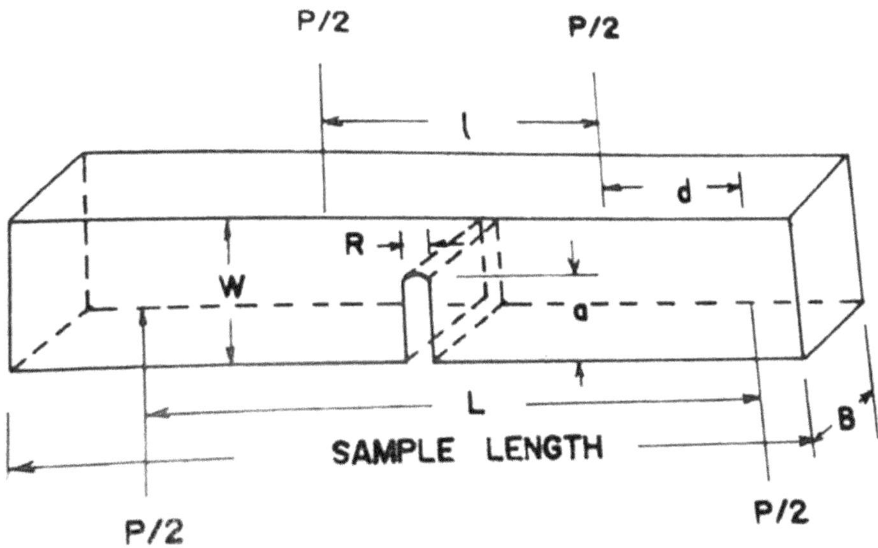

FIGURE 5.7 Schematic representation of SENB test (reprinted with permission from [18]).

$$K_{Ic} = \frac{3P_f L\sqrt{a}}{2BW^2} f\left(\frac{a}{W}\right) \tag{5.9}$$

$$f\left(\frac{a}{W}\right) = \frac{1.99 - \left\{\left(\frac{a}{w}\right)\left(1-\frac{a}{w}\right)\right\}\left\{2.15 - 3.93\left(\frac{a}{W}\right) + 2.7\left(\frac{a}{W}\right)^2\right\}}{\{(1+2\left(\frac{a}{W}\right)\}\left\{\left(1-\left(\frac{a}{W}\right)\right)^{1.5}\right\}} \tag{5.10}$$

As a result, more volume of characteristic flaw is exposed to this more uniform stress distribution. Hence, both flexural strength and fracture toughness data evaluated by the four-point bend test provide a more characteristic, realistic picture in comparison to those provided by the three-point bend test. Therefore, wherever possible, four-point bend tests should be conducted for the evaluation of flexural strength and fracture toughness of ceramics.

5.6.2 Single-Edged V-Notched Beam (SEVNB)

In this technique, a V-notch is prepared [20] in the middle of the already cut sawn notch in a specimen to ensure that failure is initiated in a given ceramic from as sharp a crack as possible (Figure 5.8). Here, K_{Ic} is evaluated through four-point loading as [20]:

$$K_{Ic} = \frac{P_f}{B\sqrt{W}} \cdot \frac{(L-l)}{W} \cdot \frac{3\sqrt{\alpha}}{2(1-\alpha)^{1.5}} \cdot Y^* \tag{5.11}$$

$$Y^* = \left[(1.9887) - (1.326\alpha) - \left\{\frac{\left(3.48 - 0.68\alpha + 1.35\alpha^2\right)\alpha(1-\alpha)}{(1+\alpha)^2}\right\}\right] \tag{5.12}$$

where P_f is the fracture load, B is the width and W is the height of the rectangular parallelopiped coupon specimen, and $(L - l)$ is the difference between the length of the lower support span (L) and upper support span l (mm). Further, α((=(a/W) is the normalized crack length. Furthermore, B is the thickness of the sample and W is the width of the sample.

5.6.3 Chevron Notch Beam (CNB) Method

The CNB method is unique as it has three great advantages. The first advantage is that it does not need a sharp pre-crack introduction, because the sharp crack already forms from the chevron notch

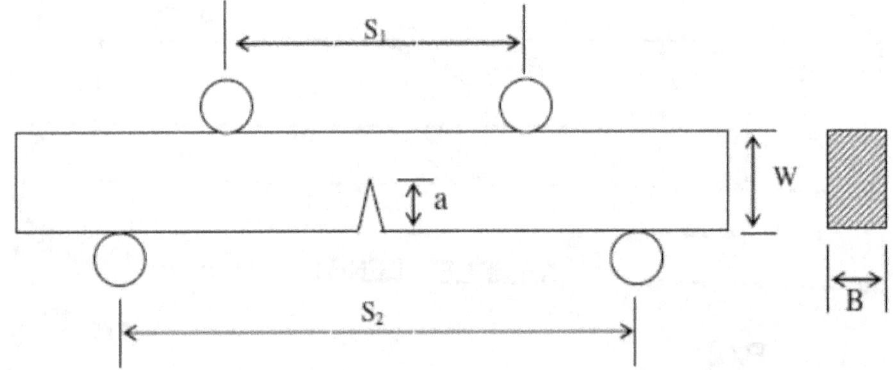

FIGURE 5.8 Schematic representation of SEVNB method (reprinted with permission from [19]).

as the load is enhanced. The second advantage is that there is no specific requirement for crack length measurement. The reason is that here only a change in compliance of the specimen coupon sample is required to be noted. The third and most important advantage is that this notch provides a stable growth of the crack during the failure process. Here, K_{Ic} is evaluated as [21]:

$$K_{Ic} = P_f/(W\sqrt{B})\ Y^* \tag{5.13}$$

where P_f, W, B, and Y^* represent the fracture load, thickness of specimen, width of specimen, and geometry function, respectively. The schematic representation of the CNB method is presented in Figure 5.9. The details of how to evaluate Y^* are given in Ref. [21].

5.6.4 Double Cantilever Beam (DCB) Method

The DCB method is mainly used to evaluate slow crack growth. The same method can also be used to determine the K_{Ic} of ceramics at both room and high temperatures (Figure 5.10) [22]. In this method, while one end of the test specimen is fixed, the other end is separated by the wedges. Hence, by following this technique, the translational movement gets restricted [22]. This method has the unique advantage that the energy release rate is independent of crack length. This method uses the torsional moments of the double cantilever beam. The strain energy release rate G is given as [23]:

$$G = \frac{Eny^2}{2Wa}\left(\frac{Af}{Ey}\right)^{\{(n+1)/(n)\}} \tag{5.14}$$

where E is Young's modulus, W is the gap in the effective thickness between two centre grooves; and A, f, and n are constants. The strain energy release rate in mode I, i.e., G_I becomes G_{Ic} at the critical condition. Under such cases, G_{Ic} is related to K_{Ic} by the relationship $K_{Ic} = [(G_{Ic}E)/(1-\nu^2)]^{0.5}$.

5.6.5 Compact Tension (CT) Technique

Compact tension is a standard technique to determine the fracture toughness of a ceramic material. In this case, a notched sample (Figure 5.11) is used to create a fatigue crack that initiates at the point of the notch and extends through the sample. The fracture toughness is determined as [24]:

$$K_{1c} = \frac{P_f}{B}\sqrt{\frac{\pi}{W}}\left[16.7\left(\frac{a}{W}\right)^{1/2} - 104.7\left(\frac{a}{W}\right)^{3/2} + 369.9\left(\frac{a}{W}\right)^{5/2} - 573.8\left(\frac{a}{W}\right)^{7/2} + 360.5\left(\frac{a}{W}\right)^{9/2}\right] \tag{5.15}$$

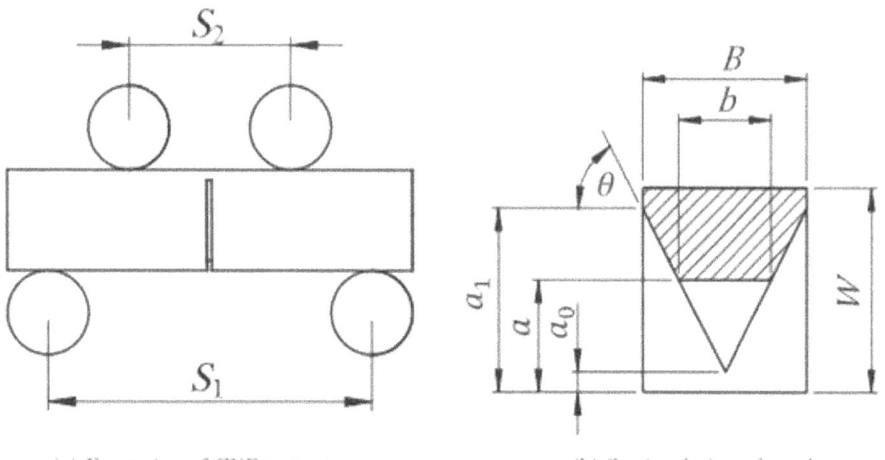

FIGURE 5.9 Schematic representation of CNB method (reprinted with permission from [20]).

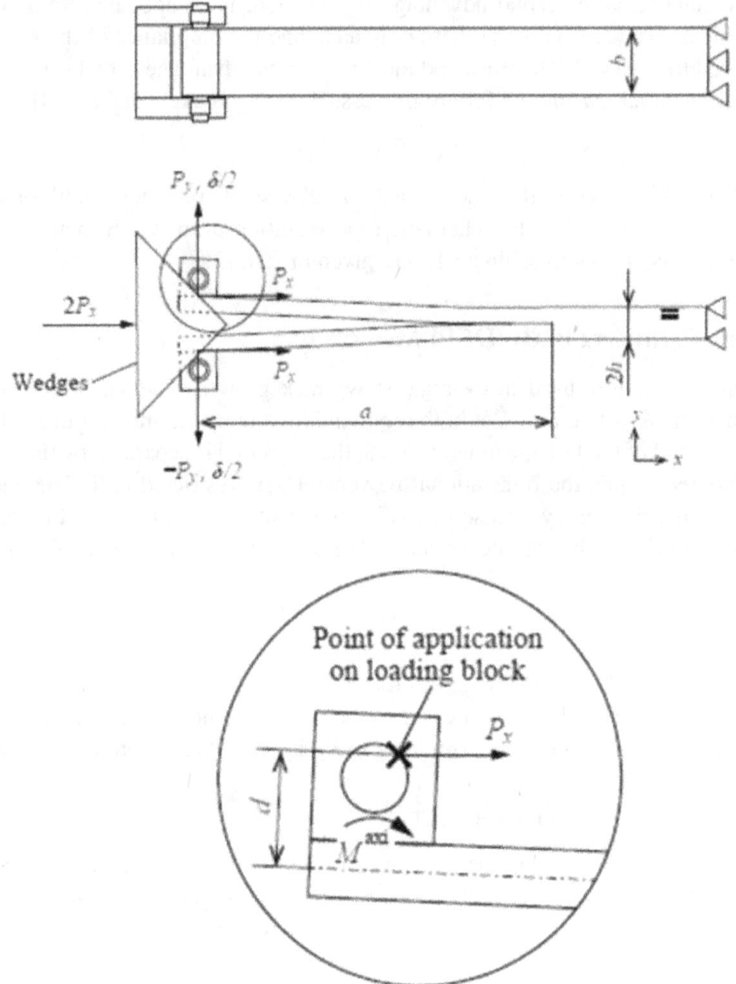

FIGURE 5.10 Schematic representation of DCB method (reprinted with permission from [21]).

where P_f is the applied load, B is the width of the specimen, a is the crack length, and W is the effective thickness of the specimen in the direction of the crack growth. It may be noted from Figure 5.11 that it is the distance between the centreline of the holes and the backface of the coupon.

5.6.6 Indentation Fracture

The indentation fracture toughness (K_{IC}) is calculated (Figure 5.5) as [25]:

$$K_{Ic} = 0.016\left(\frac{E}{H_V}\right)\left(\frac{P}{D^{1.5}}\right) \tag{5.16}$$

where 0.016 is a constant factor that is related to the geometry of the indenter and calibration data. In Equation 5.16, E is Young's modulus and H_v is the Vickers microhardness of the ceramic sample. Here, D is the characteristic crack length that comprises half the length of the average indentation diagonal length plus the average crack length measured from the tip of the indentation impression up to the crack tip created at the externally applied load (P).

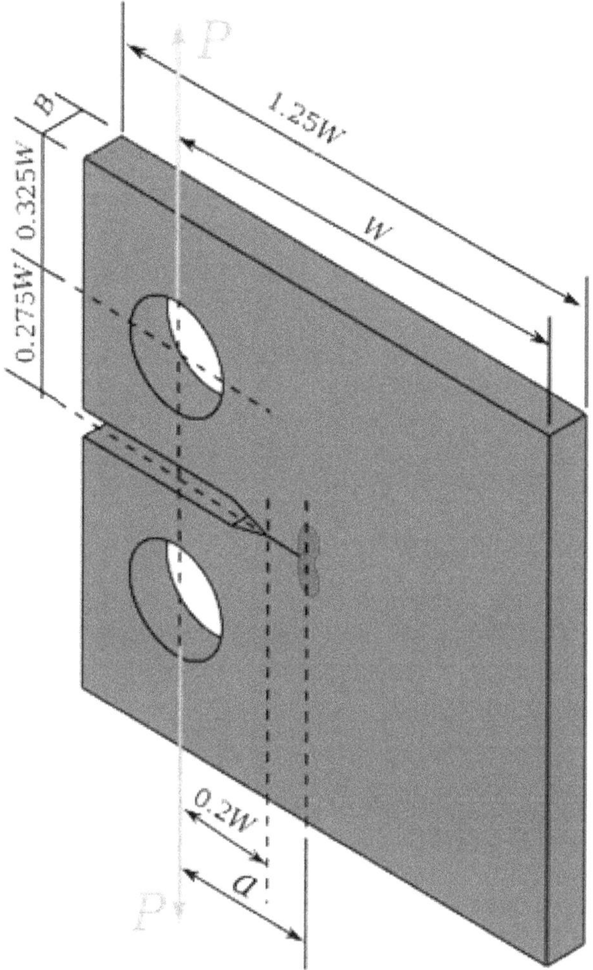

FIGURE 5.11 Schematic representation of CT specimen (from Wikipedia).

5.7 NANOINDENTATION TECHNIQUE

The aforementioned experimental techniques involve the determination of the mechanical properties of ceramic materials at micro and macro length scales. However, any kind of deformation in a ceramic material actually initiates at the nanoscale level. Although ceramics, unlike metals, usually undergo deformation to a much lesser extent, it becomes essentially vital to determine the mechanical properties of the ceramic material at the very nanoscale level. The most common technique that is used to determine mechanical properties at the nanoscale level is nanoindentation.

During nanoindentation, a high-resolution instrument continuously monitors the load, P, and depth of penetration, h, of an indenter. The nanoindenter used in most of the cases is a Berkovich tip, which is in the form of a triangular pyramid. The load (P) and depth (h) data are utilized to get the load versus depth of penetration (P–h) plot as shown schematically in Figure 5.12.

The important physical quantities obtained from the load versus depth of penetration plot are the peak load, P_{max}; maximum depth of penetration, h_{max}; final depth of penetration, h_f; and the contact stiffness, S. From the Oliver and Pharr model [26], the nanohardness (H) of a material is:

$$H = \frac{P_{max}}{A_{cr}} \tag{5.17}$$

where P_{max} is the maximum applied load and A_{cr} is the real contact area between the nanoindenter and the ceramic material.

According to Oliver and Pharr [], the polynomial form of A_{cr} can be expressed as Equation 5.18:

$$A_{cr} = 24.56h_c^2 + C_1h_c + C_2h_c^{1/2} + C_3h_c^{1/4} + ... + C_8h_c^{1/128} \tag{5.18}$$

where C_1 to C_8 are constants to be determined by standard calibration method and h_c is the penetration depth determined from the following Equation 5.19 [24]:

$$h_c = h_{max} - \kappa \frac{P_{max}}{S} \tag{5.19}$$

where $\kappa \approx 0.75$ for a Berkovich indenter [24].

Again, the contact stiffness (S), which is the slope of the first one-third of the linear part recorded during the unloading cycle of the load versus depth (Figure 5.12) of the penetration plot, can be expressed by Equation 5.20 [24]:

$$S = \frac{dP}{dh}\bigg| h = h_{max} = \alpha C_A E_r \sqrt{A_{cr}} \tag{5.20}$$

where $\alpha = 1.034$ and $CA = 2/\sqrt{\pi}$ for a Berkovich indenter [24] and E_r is the reduced Young's modulus. Following the Oliver and Pharr model, E_r can be expressed by Equation 5.21 [24]:

$$\frac{1}{E_r} = \frac{1 - v_i^2}{E_i} + \frac{1 - v_s^2}{E_s} \tag{5.21}$$

where E_i and ν_i are the Young's modulus and Poisson's ratio, respectively; and the subscripts i and s denote the indenter and the sample, respectively. For the Berkovich diamond indenter, the values of E_i and ν_i are taken as 1140 GPa and 0.07, respectively, following Ref. [26].

However, according to Oliver and Pharr, the unloading curve simply obeys the following power law relationship given in Equation 5.22 [24]:

$$P = \alpha\left(h - h_f\right)^m \tag{5.22}$$

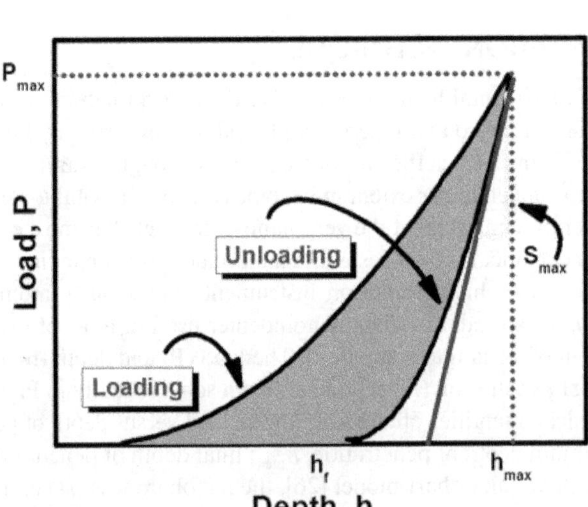

FIGURE 5.12 A typical load (P)–depth (h) plot obtained from the nanoindentation experiment.

where α and m are empirical constants that can be determined by fitting the experimentally measured data from the load versus depth of penetration data plot to Equation 5.22. Thus, the contact stiffness can also be determined using the following Equation 5.23 [24]:

$$S = \frac{dP}{dh}\Big|h - h_{\max} = \alpha m\left(h \quad h_f\right)^{m-1} \tag{5.23}$$

Therefore, substituting the values of S, α, C_A, and A_{cr} in Equation 5.20, the value of the reduced modulus E_r is calculated. The Young's modulus value of the sample, E_s, can be easily obtained then from Equation 5.22, using the known values of E_r and E_i.

5.8 HIGH STRAIN RATE STRENGTH TESTING OF CERAMICS

5.8.1 Split-Hopkinson Pressure Bar (SHPB) Technique

SHPB is a well-known technique widely used to study the compressive strength of brittle ceramics under high-impact dynamic loading at high strain rates (Figure 5.13). This works mainly in the strain rate range of 10^{-1} to 10^{1} s^{-1}. This experiment provides data on the compressive strength of ceramics at strain rates in the range of 10^{-1} to 10^{1} s^{-1}.

The setup comprises of an incident, transmitter, and striker bars. The ceramic specimen to be tested is placed as inserted in between the incident and the transmitter bars. In order to prevent damage at the end of the bars from the hard fragments of the ceramic sample, impedance-matched tungsten carbide platens are used. These platens are covered with shrunk fit maraging steel collars so that their failure during the test can be prevented. In addition, a soft copper disc is also used to change the rise time of the pulse. Moreover, to reduce the frictional effects during high-impact loading, a grease, preferably molybdenum disulfide, is used between the bars and platens and also between the platens and the specimen. A high-speed camera is used to capture high-speed real-time video images. The details of the experimental method are given in Refs. [27–29] and hence are not included once again here for the sake of brevity.

5.8.2 Gas Gun Technique

The gas gun is a superior technique to generate shock waves. This works mainly in the strain rate range of 10^{1} to 10^{3} s^{-1} (Figure 5.14). In this method, precise tuning of the projectile velocity, impactor acceleration without alteration of its initial thermodynamic state, and generation of a shock wave in a target with a flat top followed by a well-defined unloading wave can be easily achieved.

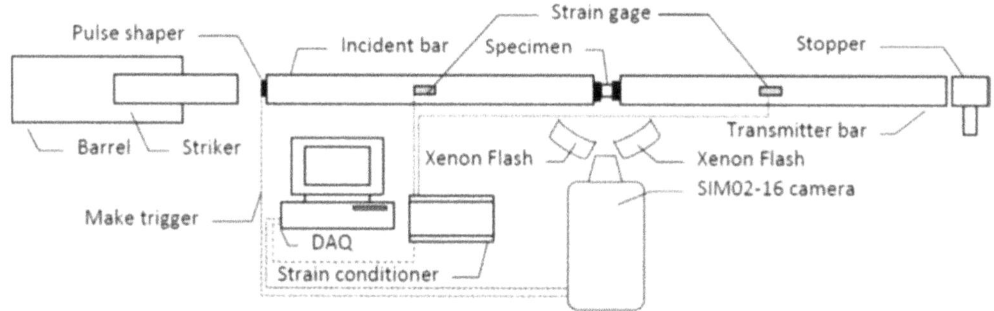

FIGURE 5.13 Schematic diagram of SHPB setup with high-speed camera (reprinted with permission from [26]).

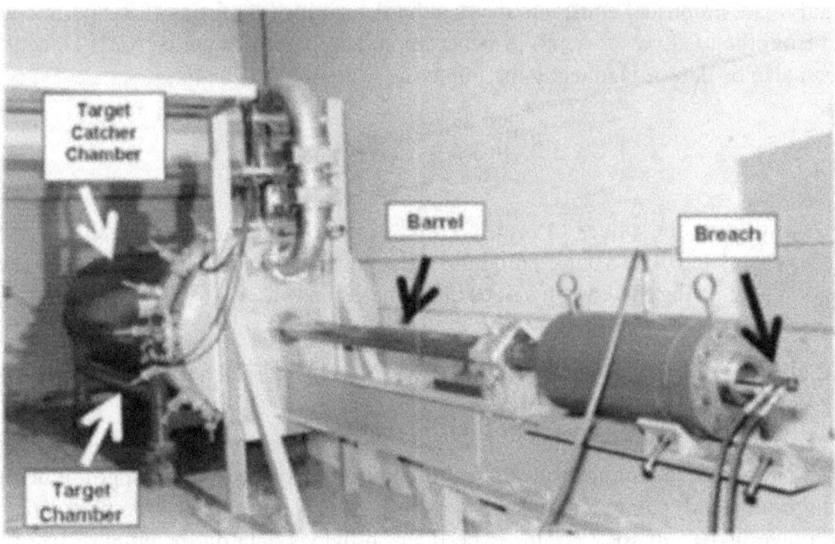

FIGURE 5.14 Photograph of the laboratory setup of gas gun shock experiment (reprinted with permission from [27]).

The single-stage gun consists of a breech, a barrel, and a target catcher system. The description and function of each part of the system and the mechanism of operation of this technique are given in detail elsewhere [27]. Here, the generated shock wave, whose speed depends up on the stress amplitude, propagates at a supersonic speed in the undisturbed ceramic sample. The strain rates, in this case, are as high as 10^3 s^{-1} and the rise time of the shock wave front is of the order of a few micro- to nanoseconds. However, there may be discontinuous changes in the physical quantities, e.g., density, temperature, energy density, and particle velocity across the shock wave front. Across the shock wave front, the following Equations 5.24–5.26 are obeyed [27, 29]:

$$\rho_0 U_S = \rho_1 \left(U_S - U_P\right) \tag{5.24}$$

$$\sigma_1 - \sigma_0 = \rho_0 U_S U_P \tag{5.25}$$

$$E_1 - E_0 = \frac{1}{2\left(\sigma_1 + \sigma_0\right)\left(V_1 - V_0\right)} \tag{5.26}$$

where U_s is the shock front velocity; U_p is the particle velocity in the compressed region; and σ, E, ρ, and V are the longitudinal stress, specific internal energy, density, and specific volume, respectively. The physical quantities with a subscript 0 refer to the unshocked state. Similarly, the physical quantities with a subscript 1 refer to the shocked state. These equations can be used to obtain the final compressive failure strength of the test sample as a function of the strain rate.

5.9 CURRENT OBSERVATIONS

Table 5.1 gives a brief overview of the current global scenario where the so far discussed experimental techniques are applied to determine the mechanical properties of a wide variety of ceramic materials [30–55]. There are 26 references. Out of these, only about 15% are from a period before 2021–2022, while the remaining 85% are from the most recent period, i.e., 2021–2022.

The data presented in Table 5.1 offer very interesting insight into the variety of methods that are currently employed. It also identifies the materials that are being considered from various practical viewpoints. In at least eight cases, the compressive strength is measured for a wide variety of

TABLE 5.1

Current Literature Survey of the Application of Different Experimental Techniques on Ceramics

Material	Experimental Method	Compressive Strength (MPa)	Tensile Strength (MPa)	Flexural Strength (MPa)	Vickers Hardness (GPa)	Fracture Toughness (MPa. $m^{1/2}$)	Reference
Porous silica ceramics (Silica = 5–20 wt%)	UTM (crosshead speed = 0.2 mm/min)	0.084–0.337	—	—	—	—	[30]
Mullite ceramics	UTM (loading rate = 1.5 MPa/s)	2.1–187.5	—	—	—	—	[31]
Ice-templated alumina	UTM (crosshead speed = 0.5 mm/min)	118.7	—	—	—	—	[32]
Monolithic zirconia crowns	UTM (crosshead speed = 0.5 mm/min)	1514–1988(N)	—	—	—	—	[33]
Transparent LiF ceramics	UTM	382	—	—	1.36	—	[34]
3D-printed porous alumina	UTM	280					[35]
3D-printed porous HA	UTM	0.6–1.1	—	—	—	—	[36]
3D printed porous zirconia	UTM (crosshead speed = 0.5 mm/min)	0.14–9.43	—	—	—	—	[36]
Porous yttria-stabilized zirconia	UTM (strain rate = 0.0 5 s^{-1})	124–1455	—	—	—	—	[37]
12Ce-TZP + (10–20 wt% Al_2O_3)	UTM, 3-point bending test	—	—	620–760	—	—	[38]
12Ce-TZP + (10–20 wt% Al_2O_3) + MnO	UTM, 3-point bending test	—	—	273–648	—	—	[39]
12Ce-TZP + 30 vol% SrO	UTM, 4-point bending test	—	—	726	—	—	[40]
10Ce-TZP + 30 vol% Al_2O_3 + 0.05 mol% TiO_2	UTM, 4-point bending test	—	—	941	—	—	[41]
Single-phase (Hf, Zr, Ti, Ta, Nb) C high-entropy carbide (HEC) ceramics	UTM, 4-point bending test	—	—	421	—	—	[42]
(Ti, Nb) $3SiC_2/Al_2O_3$ ceramics	UTM, 3-point bending test	—	—	672	—	—	[43]
Hollow-grained $La_2Zr_2O_7$ ceramics	UTM, 4-point bending test	—	—	55–105	—	—	[44]
Al_2O_3 reinforced by kyanite coating	UTM, 3-point bending test	—	—	584	—	—	[45]
Mini-SiC/SiC ceramics	UTM, (strain rates of 10^{-5}/s,10^{-4}/s, 10^{-3}/s, and 10^{-2}/s)	—	144.59–168.83	—	—	—	[46]
Hollow-grained $La_2Zr_2O_7$ ceramics	UTM	251	—	—	—	—	[44]
Oxide-oxide CMCs (AS-N610)	UTM (crosshead speed 1 mm/min)	—	396–428	—	—	—	[47]
Single-phase (Hf, Zr, Ti, Ta, Nb) C high-entropy carbide (HEC) ceramics	CNB	—	—	—	—	3.5	[42]
Y-TZP	Indentation	—	—	—	1.2–1.6	3.43	[48]
3Y-TZP	Indentation and (SEVNB)	—	—	—	—	5.5 and 9.5	[49]
Nano-ATZ ceramics	(SEVNB)	—	—	—	16.32	5.68	[50]
$(Hf_{0.25}Zr_{0.25}Ta_{0.25}Ti_{0.25})C$ high-entropy carbide ceramics	Indentation method	—	—	—	25.7	4.3	[51]
α/β-Si_3N_4 composite ceramics	Indentation method	—	—	—	17–18	8.25–8.53	[52]
Zirconia ceramics	Indentation method	—	—	—		2.2–4.2	[53]
$Na_{0.4}K_{0.1}Bi_{0.5}TiO_3$ Ferroelectric ceramics	Indentation method	—	—	—	7.5	1.45	[54]
5Y-TZP, Si_3N_4, SiC and Al_2O_3	(SEVNB)	—	—	—	—	5.37, 5.20, 2.75 and 2.93	[55]
12Ce-TZP + (10–20 wt% Al_2O_3) + MnO	Short-rod (SEVNB)	—	—	—	—	6.5–8.5	[39]
12Ce-TZP + 30 vol% SrO	DCB	—	—	—	—	15.1	[40]

materials [30–37]. The preferred crosshead speed range is 0.2 to 0.5 mm/min. In a single occasion, the loading stress rate, e.g., 1.5 MPa/min is used for mullite ceramics [31]. The porous silica ceramics [30] exhibit a very low strength of 0.08 to 0.34 MPa, while ice-templated alumina exhibits a compressive strength of about 120 MPa [32]. Transparent LiF ceramics exhibit a compressive strength of about 390 MPa [34], while 3D-printed alumina ceramics exhibit a compressive strength of about 300 MPa [35]. On the other hand, 3D-printed porous HA and ZrO_2 ceramics exhibit very low compressive strength in the range of about 0.6 to 10 MPa [35, 36]. In a singular occasion, a strain rate of $0.05s^{-1}$ is used in the case of compressive strength evaluation of porous YSZ ceramics [37]. Depending on the porosity, the compressive strength could vary from as low as 124 MPa to 1.455 GPa [37]. Super-insulating $La_2Zr_2O_7$ ceramics exhibit a compressive strength of about 250 MPa. The microstructure plays a very important role in defining the compressive strength of these advanced ceramics.

Similarly, in eight instances the method of choice is the evaluation of flexural strength by either the three-point [38, 39, 43, 45] or four-point bending technique [40–42, 44]. Recent and past research on various Ce-TZP ceramics (12Ce-TZP + (10–20 wt% Al_2O_3) + MnO)) confirm that depending on various microstructural parameters and the variety of the type and amount of reinforcements, the flexural strength could vary from about 270 to 950 MPa [38–41]. That is why these materials exhibit a K_{1c} of about 6 to 9 $MPa.m^{0.5}$ when evaluated by the short-rod SEVNB technique [39], while the DCB method evaluates a K_{1c} of about 15 $MPa.m^{0.5}$ for the 12Ce-TZP + 30 vol% SrO ceramics. There are only two recent efforts [46, 47] that evaluate the tensile strength of SiC/SiC ceramics and oxide/oxide ceramic matrix composites (CMCs). These values span the range of about 150 to 430 MPa.

However, most researchers adopted the indentation method for evaluation of K_{1c} [48, 49, 51–54]. On a few occasions, the SEVNB method is also used [49, 50, 55]. The CNB method is used only on a singular occasion [42], wherein the high-entropy carbide (HEC) materials register a K_{1c} of about 4.3 $MPa.m^{0.5}$ along with a relatively higher microhardness of about 26 GPa. The nano-ATZ ceramics also register a relatively higher microhardness of about 16 GPa along with a K_{1c} of about 5.68 $MPa.m^{0.5}$ [50]. Generally, 5Y-TZP and Si_3N_4 ceramics exhibit a relatively higher K_{1c} of about 5 $MPa.m^{0.5}$, while the SiC and Al_2O_3 ceramics exhibit a relatively lower K_{1c} of about 2 to 3 $MPa.m^{0.5}$ [55]. However, the aforesaid literature survey is only on the macroscale fracture aspect of advanced ceramics. It is imperative to discuss the deformation and/or fracture scenario at the nanoscale of the microstructure. This is done through the usage of the nanoindentation technique. The same is presented next.

5.10 DEFORMATION AND FRACTURE AT THE NANOSCALE AND ACROSS LENGTH SCALES: A CASE STUDY WITH ALUMINA CERAMICS

Since alumina is one of the most well-researched ceramics in academics and industry, it is decided to present the global scenario of deformation and fracture at the nanoscale and across length scales on alumina ceramics as a case study. We start with the nanoscale mechanical properties evaluated by the nanoindentation technique. A typical survey of the relevant literature is given in Table 5.2.

It may be noted from the studies conducted in the relevant literature (Table 5.2) that in some cases, there are abrupt discontinuities in the load–depth (P–h) plots obtained during the nanoindentation technique experiments. They are universally accepted as multiple micro-pop-ins that initiate at an ultralow load called the critical load (P_c) corresponding to a critical depth of penetration (h_c) [56]. The P_c values range from 360 μN to 10 mN. The corresponding h_c values span a range as wide, from 2.5 nm to 120 nm. It is also sensitive to the plane of specific crystallographic orientation. For instance, h_c (~50 nm) for (1012) plane is less than that (e.g., ~80–100nm) for (0001) plane of sapphire [56]. It also happens in both load- and depth-controlled nanoindentations. The multiple micro-pop-in behaviour signifies the transition from classically elastic to elastoplastic deformation of sapphire under nanoindentation [56–65].

TABLE 5.2
Nanoindentation Behaviour of Single Crystal (SC) and Polycrystalline (PC) Alumina Ceramics

Plane	Experimental Technique	Indenter Type	H (GPa)	Type of Deformation	Reference
				SC#	
(0001), $(10\bar{1}2)$	Nanoindentation	Berkovich	20	Dislocation loops	[56]
(0001)		Berkovich	24	Pyramidal, basal and prismatic slip and rhombohedral twinning.	[57]
(0001)		Berkovich	40	Cracks at the Berkovich corners. Fracture effects more pronounced for the cube-corner indenter than for the Berkovich tip.	[58]
(0001), $(01\bar{1}0)$ $(2\bar{1}\bar{1}0)$		Berkovich	37–47	Dislocations (prism, basal, pyramidal) along with stacking faults in (0001), while for the other two planes, only dislocations (prism for $(01\bar{1}0)$ and pyramidal for $(2\bar{1}\bar{1}0)$) by MD simulation.	[59]
(0001)		Berkovich*	39	Dislocation	[60]
—		Berkovich+	29	—	[61]
(0001)		Berkovich	29.1	Dislocations	[62]
(0001)		Berkovich	46.7–27.5	Slip planes and dislocations	[63]
$(10\bar{1}2)$		Berkovich	27.5	Slip system	[64]
(0001)		Berkovich	27	Dislocations	[65]
$(11\bar{2}0)$ (0001) $(1\bar{1}00)$ $(1\bar{1}02)$		Spherical		Twin formation at low stress and twin formation together with cracks propagation at high stress. The applied load, the tip radius, and the tip angle influence the crack length and depth and thus the degree of chipping. The higher the indenting load, the longer the created cracks. The sharper the tip, the higher the stress and as a consequence, more severe chipping is created.	[66]
PC§					

(Continued)

TABLE 5.2 (CONTINUED)
Nanoindentation Behaviour of Single Crystal (SC) and Polycrystalline (PC) Alumina Ceramics

Plane	Experimental Technique	Indenter Type	H (GPa)	Type of Deformation	Reference
			SC[#]		
Grain Size (μm)	**Experimental Technique**	**Indenter**	**H (GPa)**	**Type of Deformation**	**Reference**
0.1	Nanoindentation	Berkovich	42–29	Dislocation	[67]
2		(a) Berkovich (b) Cube-corner	25–30	(a) Radial cracking (b) No radial cracking but severe chipping	(68)
(a) 0.31 (residual porosity: 4.8%) (b) 0.47 (residual porosity: 1.2%) (c) 0.63 (residual porosity: 0.1%)		Vickers	(a) 25.9 (b) 27 (c) 30.4 at 10 mN load	Cracking beyond a threshold load	[69]
0.9		Berkovich	20	—	[70]
1.65		Berkovich	24	—	[71]
0.290– 1.3		Berkovich	23–25	Slip	[72]
10		Berkovich	19		[73]
20		Berkovich	11.5	Dislocation	[74]

Why does it happen? A majority of researchers suggest that it happens due to the homogeneous nucleation of dislocations [56, 59, 60, 62–65]. Others opine that formation of slip bands [57] is the cause. Another view suggests twin formation along with microcrack generation [66] in the corresponding sapphire crystals under nanoindentation as the cause. Experimental data obtained by transmission electron microscopy (TEM) studies confirm that the pop-in events correspond to the nucleation of 100–200 nm diameter dislocation loops in the otherwise perfect sapphire crystals [56]. The driving force for loop nucleation stems from the loop relaxation of elastic energy associated with the localized flexture of the sapphire surface [56].

The process and its propensity are also sensitive to the particular plane, e.g., the C-, M-, or R-planes under nanoindentation [63, 64]. The homogeneous nucleation of dislocation initiates when the maximum shear stress (Φ_{max}) below the nanoindenter becomes of the order of or even greater than the theoretical shear strength of alumina, 26–30 GPa [63, 64]. Depending again on the particular plane under nanoindentation (e.g., the C-, M-, or R-planes) and the experimental conditions (e.g., load, loading rate, presence or absence of ion implantation), the nanohardness of sapphire varies from 20 to 28 GPa [56–58, 61–64] barring a few instances wherein data as high as even ~40 GPa is reported [59, 60].

Theoretical considerations [60] also reflect that the magnitude of maximum shear stress (τ_{max}) for initiation of rhombohedral twinning in the C-plane is ~30% higher than that required for initiation of rhombohedral twinning in the M-plane while the same for initiation of basal twinning in the C-plane was ~45% higher than that required for initiation of basal twinning in the M-plane. The corresponding number of possible geometrically necessary dislocations depends on the particular sapphire crystal structure and the orientation of the evaluated surface. Molecular dynamic simulations [60] on nanoindentation suggest further that nanohardness could be as high as ~50 GPa at an ultralow depth of 2 nm for (0001), ($0\bar{1}\bar{1}0$), and ($2\bar{1}\bar{1}0$) planes of sapphire. The most striking prediction that this work made is that even at such nanoscale contact the number of disordered atoms drastically reduces with depth decreasing below about 0.2 nm and thus could produce an indentation size effect.

Irrespective of the grain size, the multiple occurrences of abrupt discontinuities i.e., multiple micro-pop-ins in the load–displacement plots of polycrystalline alumina, are also reported to be present [67–74]. The P_c values range from 400 μN to 350 mN. The corresponding h_c values span a range as wide, 4.5 nm to 1 μm. For polycrystalline alumina, the majority of researchers suggest that it happens due to the homogeneous nucleation of dislocations and the formation of slip bands [67–74] in the alumina microstructure under nanoindentation. Recent experimental evidence (Figure 5.15) suggests a similar scenario to be valid for polycrystalline alumina [15].

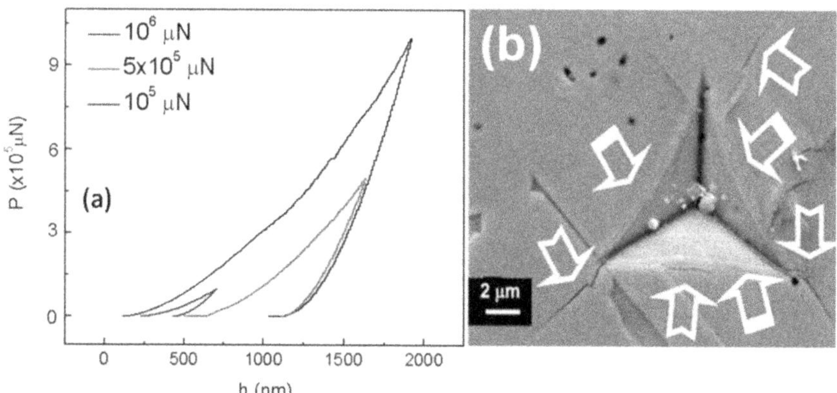

FIGURE 5.15 A typical representation of P–h plot along with the impression of nanoindentation cavity on alumina ceramic (reprinted with permission from [15])

TABLE 5.3
Microhardness and Indentation Fracture Toughness of Alumina Ceramics

GS+ (μm)	ρ# (%)	Experimental Technique	Indenter	P$ (N)	IH (GPa)	IIE (GPa)	IIIK$_{IC}$ (MPa.m$^{0.5}$)	Type of Deformation	Reference
1 –400	99	Indentation	(a) Vickers (b) Knoop	(i) 5 (ii) 1	(i) (a) 23 (b) 26 (ii) (a) 18 (b) 19	—	—	Cracking	[75]
0.4–0.6	>99.9		Vickers	100	20–21	—	—	—	[76]
0.35	100		Vickers	25	20	—	3.30	—	[77]
1.1–1.6	>99.2		Vickers	100	18–21	—	3.25–3.45	—	[78]
0.5	99.1		Vickers	100	25	—	4.5 (improvement due to addition of small amount of MgO)	—	[78]
0.5	<99.95		Vickers	—	20.5–21.5	400	3.5	—	[79]
3.05	>99%		Vickers	—	20.29	—	—	—	[80]
5	98		Vickers	10	15	416	3.5	—	[81]
(a) 3 (b) 5			Vickers		(a) 20.8 (b) 17.2	—	(a) 2.5 (b) 4.6	Cracking	[82]
8	98.7		Vickers	9.8–200	14	371	4.1	Transgranular to intergranular fracture with change in temperature	[83]
(a) 3 (b) 4 (c) 6 (d) 20 (e) 25 (f) 35	99.9 90 99 99.9 99.9 99.7		Vickers		20.1 13.1 19 19.1 16 17		(a) 3.9 (b) 2.9 (c) 4.6	Cracking	[84]
(i) 0.9 (ii) 1.9 (iii) 4.5	— — —		Vickers	3, 5, 10, 20 150, 300, 450	—	—	(i) 3.5–4.5 (ii) 3.3–4.3 (iii) 3.2–5.3	Lateral cracking and chipping increased with the grain size at higher loads but not detected at low loads.	[85]

5.10.1 Mechanical Properties of Alumina from Micro- to Macroscale: Microhardness and Indentation Fracture Toughness

The microhardness of alumina is known to be a strong function of grain size according to the Hall–Petch relationship [75]. However, from the extensive literature survey (Table 5.3) [75–85], the microhardness of alumina is found to be dependent on not only the grain size of alumina but also the density of alumina.

The microhardness is noted to be as high as 25 GPa for a dense (e.g., ~99% of theoretical value, 3.98 g.cc^{-1}), fine-grained (~0.5 μm) alumina [78]. Again, for still finer (~0.3 μm) grain size alumina, the microhardness of the alumina is ~18 GPa. This low value is due to the relatively lower density (~68% of theoretical value) of the alumina [80]. Again, for a grain size of ~1.48 μm, the microhardness enhances to ~20.3 GPa due to the relatively higher density (~98% of theoretical) of the alumina [80]. However, the problem is that the nearly similar value of microhardness (~19 GPa) is also achieved for a much coarser grained (~25 μm) alumina [82].

The question that comes to mind is what actually controls the hardness of alumina. Is it only grain size? Is it only relative density? Is it the optimum combination of these two factors [78, 80]? Is there any other factor that is beyond the conventional observations, e.g., made in the work reported in Ref. [82]. The relative density matters because the density of the alumina sample in this case is ~99.9% of the theoretical [82]. But then it possesses a very coarse grain size of ~25 μm. *These anomalies suggest that further work is needed since an unequivocal picture is yet to emerge.*

Again, the microhardness is reported to be ~13 GPa for alumina having an intermediate grain size of ~4 μm. This lower value of microhardness is attributed to the relatively lower density (~90%) of the alumina as compared to that reported for a coarse grain size (~25 μm) alumina [82]. Again, for denser (~98% of theoretical) alumina of nearly similar grain size (~6 μm) the microhardness increases to ~19 GPa [83]. The controversy is substantial when we note that the microhardness of fine grain size (~0.9 μm) alumina is reported to be ~18 GPa [84]. The density of the alumina in this case is comparatively a little bit lower (~95% of theoretical) [84].

Again, the fracture toughness of alumina is found to span a wide range, irrespective of the grain size. It could be as low as 2.5 MPa.m$^{0.5}$ [82] or as high as 4.6 MPa.m$^{0.5}$ [84], depending on the density of alumina (Table 5.3). On the contrary, the indentation fracture toughness as well as the crack resistance curve behaviour of alumina significantly improves with grain size [86]. *Therefore, it can be said that the grain size of alumina can have a complex influence on the evaluated magnitudes of the indentation fracture toughness as well as microhardness.* However, the presence of microindentation radial cracks in the alumina can generate a huge amount of residual stress in the material [75]. Nanoindentation is used in a singular instance to quantify the extent of residual stress by placing nanoindents near the microindent [75].

5.10.2 Macroscale Mechanical Properties: Compressive Strength Behaviour of Alumina Ceramics

The compressive strength of hot-pressed and subsequently hot-isostatically pressed (HPed plus HIPed) alumina [87, 88], vacuum hot-pressed alumina [88, 89], Lucalox alumina pressureless sintered without [90] or with [93] MgO doping is reported to be a strongly sensitive function of the relative density, residual porosity, presence or absence of a brittle or tough grain boundary phase, processing conditions, and grain size. The compressive strength of these alumina samples [93] generally remains strain rate insensitive in the typical low to intermediate strain rate range of 10^{-5} to $10^{-1}s^{-1}$ but increases to different extents with a strain rate in the typical intermediate to high strain rate regime of 10^{-1} to 10^{4} s^{-1} (Figure 5.16).

It is the subcritical growth of axial microcracks that controls the strain rate sensitivity of compressive strength in the low to intermediate strain rate regime of about 10^{-6} s^{-1} $\leq \dot{\varepsilon} \leq 10^{-1}$ s^{-1} [93].

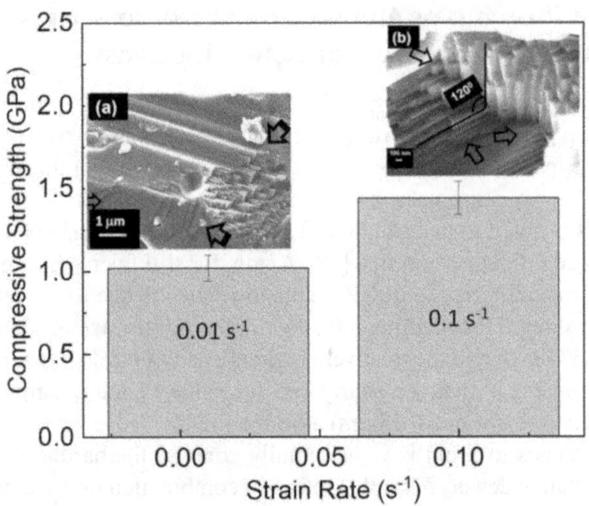

FIGURE 5.16 Compressive strengths and typical fractographic images of alumina ceramic fractured at (a) 0.01 s^{-1} and (b) 0.1 s^{-1} strain rates (reprinted with permission from [15]).

However, for the intermediate to high strain rate regime of about 10^{-1} $s^{-1} \leq \dot{\varepsilon} \leq 10^5$ s^{-1}, the control of the compressive strength in alumina is suggested to be exercised by the crack inertia [93].

Thus, based on the literature data [87–95] presented in Table 5.4, the present knowledge on compressive strength (σ_c) of alumina could best be summarized by stating that it is a strongly sensitive yet complex function of (a) strain rate, (b) processing condition, (c) grain size, and (d) strain rate sensitivity. Depending on the particular high strain rate regime concerned, the typical literature scenario (Table 5.4) on compressive failure mechanisms of alumina involves a multitude of possibilities [87–95]. These are, e.g., the generation of dislocations on single- or multiple-slip planes and their pile-up at the grain boundaries [87, 88] and nucleation of microcracks ahead of such dislocation pile-up [88]. Similarly, generation of microcracks, especially in coarse grain alumina, at stress levels much lower than that required to cause plasticity is also proposed as an important mechanism of compressive failure in alumina [88]. It is also proposed that the presence or absence of porosity and/or intrinsic flaw population may play an important role in the compressive failure mechanism of alumina [88]. Another important factor of compressive failure is the presence or absence of a glassy grain boundary phase [89], generation of microcracks, thermally activated subcritical, transgranular cracks, nucleated athermally by twins formed over a characteristic stress range [90–93], grain localized micro-cleavages, plasticity, intragranular microcracking, as well as sub-grain formation and tensile microcracks and/or wing cracks [26, 27]. *Thus, these numerous possibilities indicate that further dedicated research is very much needed to understand the complex dependencies of compressive strength on a multitude of microstructural variables.*

5.11 EMERGING HORIZONS: CHALLENGES AHEAD

In spite of the wealth of literature, many aspects of ceramic fracture [94] as well as nanoscale plasticity are still far from well understood [95]. Recent in situ experiments conducted on MgO-doped Al_2O_3 indicate the complex interrelation between the atomic coordination geometry at the core of the grain boundary and ensuing fracture processes [94]. Recent nanoindentation studies and extensive characterization by field-emission scanning electron microscopy (FESEM) prove that in MgO- and TiO_2-doped microstructurally engineered Al_2O_3 as well as ZTA ceramics, nanoscale plasticity due to homogeneous nucleation of dislocation, shear band formation, and microcracking can coexist

TABLE 5.4
Compressive Strength Behaviour of Alumina

Processing Details and Alumina Type	GS (μm)	Experimental Technique	$\dot{\varepsilon}$. (s^{-1})	σ_s (GPa)	Deformation Mechanism	Reference
(1)AD995: VHP (2)JSI: VHP + HIP (3)JSII: VHP	(1) 17 (2) 1.45 (3) 3.9	(a) Servo controlled hydraulic test (b) SHPB	(a) 10^{-4}–10^{0} (b) 500–2500	(1) 3.8–9 (2) 6–9 (3) 1.2–5	(a) Dislocation pileup (b) Nucleation of MC ahead of dislocation pileup (c) Generation of MC, especially in coarse grain alumina, at stress levels much lower than that required to cause plasticity	[87]
HP + HIP	1.5	(a) Four-point bend tests, (b) servohydraulic test, (c) SHPB	(b) 10^{-4} – 10^{1} (c) 10^{2} – 10^{3}	(a) 805, 916 and 961 MPa; (b) and (c) 5.5–8.3 GPa	(b) Dislocation bands and pile-ups in the form of a narrow microshear band with subsequent microcrack nucleation	[88]
Lucalox	25	(a) electrohydraulic, servocontrolled testing machine, (b) Hopkinson pressure bar	(a) 6×10^{-5} – 2×10^{3}, (b) 2×10^{3}	(a) 3–3.5, (b) 4	MC generation due to twin GB interaction	[89]
Lucalox	—	(a) Electrohydraulic, servocontrolled testing machine, (b) Hopkinson pressure bar	10^{-5}–10^{3}	3–3.5	Formation of axial MC	[90]
Lucalox with MgO doped	25	(a) Electrohydraulic, servocontrolled testing machine, (b) Hopkinson pressure bar	7×10^{-5}–1×10^{4}	2.5–5.8	Athermal nucleation of MC at intrinsic flaws	[91]
PS	20	Servocontrolled hydraulic testing machine	10^{-5}–10^{0}		Dislocation nucleation	[92]
PS	5	SHPB	0.9×10^{3}	3	Grain localized microcleavages, plasticity, intragranular MC, and SG formation	[93]
CIP	2	Gas gun			Subsurface cracking and surface radial cracks	[94]
PS	10	Gas gun	4×10^{4}		Microcleavages, grain boundary microcracks, stacking faults, dislocations, intragranular shear deformation bands	[95]

[95]. How to use this knowledge further for the development of damage-tolerant ceramics and their mechanical characterization is a challenge that needs to be addressed. Recent work shows the possibility that intelligent defect engineering can be a suitable way to have the simultaneous achievement of high strength and toughness in biogenic ceramics [96]. As such, these are contradictory requirements. The challenge is how to characterize these defects and their effect on mechanical characteristics across length scales involving a polymeric interface.

ACKNOWLEDGEMENTS

The authors acknowledge the kind support and encouragement received from various directors of the Council of Scientific and Industrial Research (CSIR)–Central Glass and Ceramic Research Institute (CGCRI), Kolkata, India, during the tenures of whom parts of the research work were carried out. They also gratefully acknowledge the various infrastructural support received from various divisions of CSIR-CGCRI, Kolkata, India. The financial support received from Indian sponsoring agencies, including the CSIR, UGC, DST(SERB), DAE-BRNS, and IPR, of the Government of India (GOI) in terms of various project sponsorships and fellowships is gratefully acknowledged by the authors. Finally, the author AKM acknowledges the kind and continued support received from the authorities of Sharda University, Greater Noida, Uttar Pradesh, India.

REFERENCES

1. K. Lina, G. Nie, P. Sheng, S. Zhao, S. Wu, Effects of doping Al-metal powder on thermal, mechanical and dielectric properties of AlN ceramics, Ceram. Int. 2022. https://doi.org/10.1016/j.ceramint.2022.08.178.
2. S. R. N. Kiran Mangalampalli, P. Dobbidi, L. N. Ramasubramanian, E. P. Korimilli, S. Perumal, S. R. Bakshi, Advances in functional and structural ceramics: Development, characterization, and applications, Ceram. Int. 48 2022 28763–28765.
3. A. Harrati, Y. Arkame, A. Manni, A. E. Haddar, B. Achiou, A. E. Bouari, I. E. A. E. Hassani, A. Sdiri, C. Sadik, Cordierite-based refractory ceramics from natural halloysite and peridotite: Insights on technological properties, J. Indian Chem. Soc. 99 2022 100496.
4. C. Wang, Y. Wang, Z. Bao, J. Dong, Y. Geng, S. Liu, C. Wang, P. Nie, Characterization of microstructure and mechanical properties of titanium-based bioactive ceramics laser-deposited on titanium alloy, Ceram. Int. 48 2022 28678–28691.
5. R. Xia, J. Chen, R. Liang, Z. Zhou, The positive effect of A-site nonstoichiometry on the electrical properties of $Sr_2Nb_2O_7$ ceramics for high temperature piezoelectric sensor application, Ceram. Int. 48 2022 22500–22508.
6. Y. Zheng, L. Jia, F. Xu, G. Wang, X. Shi, H. Zhang, Microstructures and magnetic properties of low temperature sintering NiCuZn ferrite ceramics for microwave applications, Ceram. Int. 45 2019 22163–22168.
7. S. K. Sahu, J. Thrinadh, S. Datta, Parametric studies on SiC-abrasive jet assisted machining of alumina ceramics, Mater. Today: Proc. 44 2021 1643–1652.
8. A. B. Dresch, J. Venturini, S. Arcaro, O. R. K. Montedo, C. P. Bergmann, Ballistic ceramics and analysis of their mechanical properties for armour applications: A review, Ceram. Int. 47 2021 8743–8761.
9. S. Akrami, P. Edalati, M. Fuji, K. Edalati, High-entropy ceramics: Review of principles, production and applications, Mater. Sci. Eng. R Rep. 146 2021 100644.
10. O. J. Akinribide, G. N. Mekgwe, S. O. Akinwamide, F. Gamaoun, C. Abeykoon, O. T. Johnson, P. A. Olubambi, A review on optical properties and application of transparent ceramics, J. Mater. Res. Technol. 21 2022 712–738.
11. A. C. Fischer-Cripps, Introduction to contact mechanics, Springer, New York, Vol. 101, 2007.
12. E. Martin, D. Leguillon, O. Sevecek, R. Bermejo, Understanding the tensile strength of ceramics in the presence of small critical flaws, Eng. Frac. Mech. 201 2018 167–175.
13. Z. Liu, Y. A. Huang, L. Xiao, P. Tang, Z. Yin, Nonlinear characteristics in fracture strength test of ultrathin silicon die, Semicond. Sci. Technol. 30 2015 045005.
14. F. Mujika, On the difference between flexural moduli obtained by three-point and four-point bending tests, Polym. Test. 25 2006 214–220.

15. M. Bhattacharya, Contact deformations of brittle solids, Ph.D. Dissertation, Department of Physics, Jadavpur University, Kolkata, India, 2016.
16. Standard Test Method for Microhardness of Materials, Designation No. E 384. 1995 Annual Book of ASTM Standards, Vol. 03.01. American Society for Testing and Materials, Philadelphia, PA, 1995.
17. F. Knoop, C. C. Peters, W. B. Emerson, A sensitive pyramidal diamond tool for indentation measurements, J. Res. Nat. Bur. Stand. 23 1939 39.
18. A. K. Mukhopadhyay, Investigations on parameters affecting fracture toughness of silicon nitride and its composites, PhD. Dissertation, Department of Physics, Jadavpur University, Kolkata, India, 1988.
19. T. G. T. Nindhia, J. Schlacher, T. Lube, Fracture Toughness (KIC) of lithography based manufactured alumina ceramic, IOP Conf. Ser.: Mater. Sci. Eng. 348 2018 012022.
20. H. Zielke, M. Abendroth, M. Kuna, B. Kiefer, Determining the fracture toughness of ceramic filter materials using the miniaturized chevron-notched beam method at high temperature, Ceram. Int. 44 2018 13986–13993.
21. S. Oshima, A. Yoshimura, Y. Hirano, T. Ogasawara, Experimental method for mode I fracture toughness of composite laminates using wedge loaded double cantilever beam specimens, Composites Part A Appl. Sci. 112 2018 119–125.
22. B. N. Rao, A. R. Acharya, Evaluation of fracture toughness through Jic testing with standard compact tension specimen. Exp. Tech. 16 1992 37–39.
23. G. R. Anstis, P. Chantikul, B. R. Lawn, D. B. Marshall, A critical evaluation of indentation techniques for measuring fracture toughness: II, strength method, J. Am. Ceram. Soc. 64 1981 539–543.
24. W. C. Oliver, G. M. Pharr, An improved technique for determining hardness and elastic modulus using load and displacement sensing indentation experiments, J. Mater. Res. 7 1992 1564–1583.
25. J. E Field, S. M. Walley, W. G. Proud, H. T. Goldrein, C. R. Siviour, Review of experimental techniques for high-rate deformation and shock studies. Int. J. Impact Eng. 30 2004 725–775.
26. S. Acharya, S. Bysakh, V. Parameswaran, A. K. Mukhopadhyay, Deformation and failure of alumina under high strain rate compressive loading. Ceram. Int. 41 2015 6793–6801.
27. A. K. Mukhopadhyay, K. D. Joshi, A. Dey, R. Chakraborty, A. Rav, S. K. Biswas, S. C. Gupta, Shock deformation of coarse grain alumina above Hugoniot elastic limit, J. Mater. Sci. 45 2010 3635–3651.
28. D. F. Carroll, S. M. Widerhorn, High temperature creep testing of ceramics, Int. J. High Technol. Ceram. 4 1988 227–241.
29. X. Li, L. Yan, Y. Zhang, X. Yang, A. Guo, H. Du, F. Hou, J. Liu, Lightweight porous silica ceramics with ultra-low thermal conductivity and enhanced compressive strength, Ceram. Int. 48 2022 9788–9796.
30. B. Xia, Z. Wang, L. Gou, M. Zhang, M. Guo, Porous mullite ceramics with enhanced compressive strength from fly ash-based ceramic microspheres: Facile synthesis, structure, and performance, Ceram. Int. 48 2022 10472–10479.
31. S. Akurati, S. Qian, D. Ghosh, AC electric field-assisted fabrication of ice-templated alumina materials and remarkable enhancement of compressive strength, Scr. Mater. 206 2022 114264.
32. M. Mohaghegh, M. Mohaghegh, S. Baseri, M. H. Kalantari, R. Giti, Influence of sintering temperature on the marginal fit and compressive strength of Monolithic Zirconia Crowns, J. Dent. Shiraz. Univ. Med. Sci. 23 2022 307–313.
33. B. Liu, F. L. Lin, C. C. Hu, K. X. Song, J. H. Zhang, C. Lu, Y. H. Huang, Novel transparent LiF ceramics enabled by cold sintering at 150 °C, Scr. Mater. 220 2022 114917.
34. F. Zhang, Z. Li, M. Xu, S. Wang, N. Li, A review of 3D printed porous ceramics, J. Yang, J. Eur. Ceram. Soc. 42 2022 3351–3373.
35. S. Lee, C. Y. Lee, J. H. Ha, J. Lee, I. H. Song, S. H. Kwon, The effects of process conditions on improvement of the compressive strengths of reticulated porous zirconia, Appl. Sci. 12 2022 1591.
36. A. Abaza, J. Laurencin, A. Nakajo, M. Hubert, T. David, F. Monaco, C. Lenser, S. Meille, Fracture properties of porous yttria-stabilized zirconia under micro-compression testing, J. Eur. Ceram. Soc. 42 2022 1656–1669.
37. S. Ray, M. Haque, M. M. Rahman, M. N. Sakib, K. Al Rakib, Experimental investigation and SVM-based prediction of compressive and splitting tensile strength of ceramic waste aggregate concrete, J. King Saud Univ. - Eng. Sci. 36 2021 112–121.
38. T. Sato, T. Endo, M. Shimada, Postsintering hot isostatic pressing of Ceria-Doped Tetragonal Zirconia/Alumina composites in an Argon–Oxygen Gas atmosphere, J. Am. Ceram. Soc. 72 1989 761–764.
39. J. F. Tsai, U. Chon, N. Ramachandran, D. K. Shetty, Transformation plasticity and toughening in CeO_2-partially-stabilized zirconia-alumina (Ce-TZP/Al_2O_3) composites doped with MnO, J. Eur. Ceram. Soc. 75 1992 1229–1238.

40. R. A. Cutler, R. J. Mayhew, K. M. Prettyman, A. V Virkar, High-toughness Ce-TZP/Al_2O_3 ceramics with improved hardness and strength, J. Am. Ceram. Soc. 74 1991 179–186.
41. K. Tanaka, J. Tamura, K. Kawanabe, M. Nawa, M. Oka, M. Uchida, T. Kokubo, T. Nakamura, Ce-TZP/Al_2O_3 nanocomposite as a bearing material in total joint replacement, J. Biomed. Mater. Res. 63 2002 262–270.
42. L. Feng, W. T. Chen, W. G. Fahrenholtz, G. E. Hilmas, Strength of single-phase high-entropy carbide ceramics up to 2300°C, J. Am. Ceram. Soc. 104 2021 419–427.
43. J. Ji, H. Yang, Z. Zhang, Y. Chen, S. Chen, Q. Li, M. Sun, D. Li, G. Shi, Z. Wang, Synthesis and strengthening mechanism for $(Ti,Nb)_3SiC_2/Al_2O_3$ ceramics: A combined experimental and first-principles investigation, Ceram. Int. 48 2022 21299–21308.
44. S. Li, C. A. Wang, F. Yang, L. An, K. So, J. Li, Hollow-grained "Voronoi foam" ceramics with high strength and thermal superinsulation up to 1400 °C, Mater. Today. 46 2021 35–43.
45. H. Y. Li, H. Hao, Y. Tian, X. G. Liu, D. Wan, Y. Bao, Temperature dependence of flexural strength and residual stress of Al_2O_3 reinforced by kyanite coating, Ceram. Int. 48 2022 28745–28750.
46. B. Xu, D. Chen, H. Yang, R. Luo, L. Wang, Z. Chen, M. Li, Q. Yuan, Y. Hua, J. Zhou, Y. He, Y. Huo, T. Liu, Effect of strain rate on the tensile properties of mini-SiC/SiC composites, Ceram. Int. 48 2022 2092–2096.
47. K. Ramachandran, S. Leelavinodhan, C. Antao, A. Copti, C. Mauricio, Y. L. Jyothi, D. D. Jayaseelan, Analysis of failure mechanisms of Oxide-Oxide ceramic matrix composites, J. Eur. Ceram. Soc. 42 2022 1626–1634.
48. Z. Mei, Y. Lu, Y. Lou, P. Yu, M. Sun, X. Tan, J. Zhang, L. Yue, H. Yu, Determination of hardness and fracture toughness of Y-TZP manufactured by digital light processing through the indentation technique, Biomed. Res. Int. 2021 2021 6612840.
49. M. F. R. P. Alves, C. D. Santos, C. N. Elias, J. E. V. Amarante, S. Ribeiro, Comparison between different fracture toughness techniques in zirconia dental ceramics, J. Biomed. Mater. Res. - Part B Appl. Biomater. 111 2022 103–116.
50. Y. Yu, X. Liu, Y. Yuan, W. Yu, H. Yin, Z. Yin, Y. Zheng, X. He, Microstructural evolution and mechanical properties of nano-ATZ ceramics by solid solution precipitation, J. Mater. Res. Technol. 16 2022 1293–1304.
51. M. Ma, Y. Sun, Y. Wu, Z. Zhao, L. Ye, Y. Chu, Nanocrystalline high-entropy carbide ceramics with improved mechanical properties, J. Am. Ceram. Soc. 105 2022 606–613.
52. J. Huang, X. Lv, X. Dong, M. I. Hussain, C. Ge, Microstructure and mechanical properties of α/β-Si_3N_4 composite ceramics with novel ternary additives prepared via spark plasma sintering, Ceram. Int. 48 2022 30376–30383.
53. S. M. Čokić, M. Cóndor, J. Vleugels, B. Van Meerbeek, H. Van Oosterwyck, M. Inokoshi, F. Zhang, Mechanical properties–translucency–microstructure relationships in commercial monolayer and multilayer monolithic zirconia ceramics, Dent. Mater. 38 2022 797–810.
54. P. Varade, N. Shara Sowmya, N. Venkataramani, A. R. Kulkarni, Microstructural and mechanical behavior of $Na_{0.4}K_{0.1}Bi_{0.5}TiO_3$ ferroelectric ceramics, Ceram. Int. 48 2022 26546–26552.
55. A. Wang, X. Zhao, M. Huang, Y. Cheng, D. Zhang, A systematic study on the quality improving of fracture toughness measurement in structural ceramics by laser notching method, Theor. Appl. Fract. Mech. 114 2021 102981.
56. T. F. Page, W. C. Oliver, C. J. Mchargue, The deformation-behavior of ceramic crystals subjected to very low load (nano)indentations, J. Mater. Res. 7 1992 450–473.
57. S. J. Lloyd, J. M. Molina-Aldareguia, W. J. Clegg, Deformation under nanoindents in sapphire, spinel and magnesia examined using transmission electron microscopy, Philos. Mag. A, 82 2002 1963–1969.
58. T. Chudoba, P. Schwaller, R. Rabe, J.-M. Breguet, J. Michler, Comparison of nanoindentation results obtained with Berkovich and cube corner indenters, Philos. Mag. 86 2006 5265–5283.
59. K. Nishimura, R. K. Kalia, A. Nakano, P. Vashishta, Nanoindentation hardness anisotropy of alumina crystal: A molecular dynamics study, Appl. Phys. Lett. 92 2008 161904.
60. C. Lu, Y. W. Mai, P. L. Tam, Y. G. Shen, Nanoindentation-induced elastic–plastic transition and size effect in α Al_2O_3 (0001), Philos. Mag. Lett. 87 2007 409–415.
61. K. H. Zhang, D. H. Wen, T. Hong, and J. L. Yuan, Study on the mechanical properties of sapphire by numerical simulation and nanoindentation technology, Adv. Mater. Res. 69–70 2009 103–107.
62. S. N. Dub, V. V. Brazhkin, N. V. Novikov, G. N. Tolmachova, P. M. Litvin, L. M. Lityagina, T. I. Dyuzhev, Comparative studies of mechanical properties of Stishovite and Sapphire single crystals by nanoindentation, J. Superhard Mater. 32 2010 406–414.

63. W. G. Mao, Y. G. Shen, C. Lu, Deformation behavior and mechanical properties of polycrystalline and single crystal alumina during nanoindentation, Scripta Mater. 65 2011 127–130.
64. W. Mao, Y. Shen, Nanoindentation study of pop-in phenomenon characteristics and mechanical properties of sapphire (1012) crystal, J. Am. Ceram. Soc. 95 2012 3605–3612.
65. M. Bhattacharya, A. Dey, A. K. Mukhopadhyay, Influence of loading rate on nanohardness of sapphire, Ceram. Int. 42 2016 13378–13386. https://doi.org/10.1016/j.ceramint.2016.05.091.
66. V. Trabadelo, S. Pathak, F. Saeidi, M. Parlinska-Wojtan, K. Wasmer, Nanoindentation deformation and cracking in sapphire, Ceram. Int. 2019. https://doi.org/10.1016/j.ceramint.2019.02.022.
67. E. J. Piqué, L. Ceseracciu, Y. Gaillard, M. Barch, G. D. Portu, M. Anglada, Instrumented indentation on alumina-alumina/zirconia multilayered composites with residual stresses, Phil. Mag. 86 2006 5371–5382.
68. A. Krell, S. Schadlich, Nanoindentation hardness of submicrometer alumina ceramics, Mater. Sci. Eng. A 307 2001 172–181.
69. J. Gong, Z. Peng, H. Miao, Analysis of the nanoindentation load–displacement curves measured on high-purity fine-grained alumina, J. Euro. Ceram. Soc. 25 2005 649–654.
70. J. Meng, N. H. Loh, B. Y. Tay, S. B. Tor, G. Fu, K. A. Khor, L. Yu, Pressureless spark plasma sintering of alumina micro-channel part produced by micro powder injection molding, Scripta Mater. 64 2011 237–240.
71. L. Huang, W. Yao, A. K. Mukherjee, J. M. Schoenung, Improved mechanical behavior and plastic deformation capability of ultrafine grain alumina ceramics, J. Am. Ceram. Soc. 95 2012 379–385.
72. R. Chakraborty, A. K. Mukhopadhyay, A. Dey, K. D. Joshi, A. Rav, S. K. Biswas, S. C. Gupta, Nanomechanical properties of coarse grain alumina ceramic, Jour. Inst. Eng. (India), Part MM, Metall. Mater. Sci. Div. 91 2010 9.
73. M. Bhattacharya, R. Chakraborty, A. Dey, A. K. Mandal, A. K. Mukhopadhyay, New observations in micro-pop-in issues in nanoindentation of coarse grain alumina, Ceram. Int. 39 2013 999–1009.
74. B. R. Lawn, Fracture and deformation in brittle solids: A perspective on the issue of scale, J. Mater. Res. 19 2004 22–29.
75. R. W. Rice, C. C. Wu, F. Borchel, Hardness-grain-size relations in ceramics, J. Am. Ceram. Soc. 77 1994 3539–3553.
76. A. Krell, P. Blank, H. Ma, T. Hutzler, M. Bruggen and R. Apetz, Transparent sintered corundum with high hardness and strength, J. Am. Ceram. Soc. 86 2003 12–18.
77. G. D. Zhan, J. D. Kuntz, R. G. Duan, A. K. Mukherjee, Spark-plasma sintering of silicon carbide whiskers (SiC_w) reinforced nanocrystalline alumina, J. Am. Ceram. Soc. 87 2004 2297–2300.
78. D. Chakravarty, S. Bysakh, K. Muraleedharan, T. N. Rao, R. Sundaresan, Spark plasma sintering of magnesia-doped alumina with high hardness and fracture toughness, J. Am. Ceram. Soc. 91 2008 203–208.
79. A. Krell, J. Klimke, T. Hutzler, Advanced spinel and sub-mm Al_2O_3 for transparent armour applications, J. Euro. Ceram. Soc. 29 2009 275–281.
80. H. Erkalfa, Z. Mlslrh, M. Demirci, C. Toy, T. Baykara, The densification and microstructural development of $A1_2O_3$ with manganese oxide addition, J. Euro. Ceram. Soc. 15 1995 165–71.
81. R. G. Munro, Evaluated material properties for a Sintered a-Alumina, J. Am. Ceram. Soc. 80 1997 1919–28.
82. M. Li, M. J. Reece, Influence of grain size on the indentation-fatigue behavior of Alumina, J. Am. Ceram. Soc. 83 2000 967–970.
83. V. M. Sglavo, E. Trentini, Fracture toughness of high-purity alumina at room and elevated temperature, J. Mater. Sci. Lett. 18 1999 1127–1130.
84. R. F. Cook, B. R. Lawn, C. J. Fairbanks, Microstructure-strength properties in ceramics: I, effect of crack size on toughness, J. Am. Ceram. Soc. 68 1985 604–615.
85. L. C. Lim, A. Muchtar, Micro- and macro-indentation fracture toughness of alumina, J. Mater. Sci. Lett. 21 2002 1145–1147.
86. P. Chantikul, S. J. Bennison, B. R. Lawn, Role of grain size in the strength and R-curve properties of Alumina, J. Am. Cerom. Soc. 73 1990 2419–27.
87. J. Lankford, W. W. Predebon, J. M. Staehler, G. Subhash, B. J. Pletka, C. E. Anderson, The role of plasticity as a limiting factor in the compressive failure of high strength ceramics, Mech. Mat. 29 1998 205–218.
88. J. M. Staehler, W. W. Predebon, B. J. Pletka, G. Subhash, Micromechanisms of deformation in high-purity hot-pressed alumina, Mater. Sci. Eng. A 291 2000 37–45.

89. J. Lankford, Compressive strength and microplasticity in polycrystalline alumina, J. Mater. Sci. 12 1977 791–796.
90. J. Lankford, Temperature-strain rate dependence of compressive strength and damage mechanisms in aluminium oxide, J. Mater. Sci. 16 1981 1567–1578.
91. J. Lankford, Mechanisms responsible for strain-rate-dependent compressive strength in ceramic materials, Comm. Am. Ceram. Soc. 64 1981 C33–C34.
92. M. Bhattacharya, S. Dalui, N. Dey, S. Bysakh, J. Ghosh, A. K. Mukhopadhyay, Low strain rate compressive failure mechanism of coarse grain alumina, Ceram. Int. 42 2016 9875–9886.
93. C. E. J. Dancer, H. M. Curtis, S. M. Bennett, N. Petrinic, R. I. Todd, High strain rate indentation-induced deformation in alumina ceramics measured by Cr^{3+} fluorescence mapping, J. Euro. Ceram. Soc. 31 2011 2177–2187.
94. S. Kondo, A. Ishihara, E. Tochigi, N. Shibata, Y. Ikuhara, Direct observation of atomic-scale fracture path within ceramic grain boundary core, Nat. Commun. 10 2019 2112. https://doi.org/10.1038/s41467-019–10183-3.
95. P. Maiti, Studies on microstructurally engineered ceramics and composites for structural applications, Ph. D. Dissertation, Jadavpur University, 2022.
96. Z. Deng, H. Chen, T. Yang, Z. Jia, J. C. Weaver, P. D. Shevchenko, F. De. Carlo, R. Mirzaeifar and L. Li, Strategies for simultaneous strengthening and toughening via nanoscopic intracrystalline defects in a biogenic ceramic, Nat. Commun. 11 2020 5678. https://doi.org/10.1038/s41467-020–19416-2.

6 Deformation and Fracture of Polymeric Materials

Experimental Methods and Challenges

Sangeeta Rawal, Monika Srivastava, and Anoop Kumar Mukhopadhyay

6.1 INTRODUCTION

It is needed to first understand what a polymer is. *Poly* means "many" in Greek. Similarly, *mer* means "part". Therefore, a polymer is a material which has many repeats of similar parts. A polymer consists of very large macromolecules. The macromolecules are mostly contently bonded. The macromolecules are comprised of many subunits. These subunits keep repeating. Polymers can be synthetic (e.g., polyurethane) as well as natural (e.g., various biopolymers). These macromolecules have a large molecular mass that provides toughness to the structure. It also provides the ability to deform without breaking to a large extent. It also supports the formation of both amorphous and semicrystalline structures in polymers.

Because of such interesting properties, polymers and their nanocomposites find use in a wide range of applications. Today, polymeric materials are considered useful for water remediation [1], matrix materials for dispersion of liquid crystals for optoelectronic applications [2], and pipeline applications [3]. In thin film form, polymeric materials are proposed to be useful for a host of biomedical applications [4]. Such applications include the controlled delivery of drugs as well as the controlled delivery of gene. Other possibilities include tissue engineering. Designed polymeric thin film brushes are useful for imaging purposes as well. Advanced research also suggests that it is possible to design polymeric thin film brushes with the characteristic feature of self-actuating behaviour, which means that they can respond to various stimuli [4].

An area of emerging importance today is the 3D printing of complex-shaped conductive polymers which form the backbone of a myriad of electronic systems and devices [5]. Further, zwitterionic polymers are emerging as designable materials for anti-fouling coatings and sensor applications [6]. These materials have both positive and negative charges in the same repeat unit. As a result, they have tuneable large dipole moments along with the presence of many charged groups. These help them to achieve major applications in the areas of drug delivery and reduction of water pollution [6].

Recent work [7] reveals that reusable, degradable, and self-healing amine epoxy polymers can be developed for shape memory applications. Recent works also emphasize the huge roles played by polymers for a variety of armour applications [8]. Further, the usage of various advanced polymeric and elastomeric fibres for armour development is also under active focus [9]. The application of weak hydrogen bonds and strong coordination bonds resulted in the recent development of high-strength artificial polymeric muscles, which can self-heal up to 95% [10]. Thus, this chapter focuses on providing a brief overview of the deformation and fracture of polymeric materials. First, the different aspects of polymers as a material are briefly discussed. Next, the various types of polymers

DOI: 10.1201/9781003359364-8

and their applications are touched upon. Further, the various important mechanical properties of relevance to polymer deformation and fracture are discussed. Furthermore, a detailed survey of the literature on the mechanical properties of polymers and polymer composites is presented. Finally, the future direction of research needs is identified.

6.2 BASIC ASPECTS OF POLYMER: TYPES AND THEIR APPLICATION PROSPECTS

A polymer is a chain of monomers, and the properties of the polymers are greatly influenced by the extent of tight and/or loose bonding between these monomers. There are various kinds of polymers, including elastomers or rubbers, plastics, coatings, foams, films, and fibres. Depending on their properties, polymers find applications in various fields. Polymers are categorized in different categories (Figure 6.1) as described next.

6.2.1 Based on Source

According to the origin of the source, polymers can be of three types: (a) natural (exist in nature), (b) synthetic (man-made), and (c) semisynthetic (naturally occurring but modified by humans).

6.2.2 Based on Molecular Forces

The primary interaction force (forces within the monomer chain) in the polymer materials is a covalent bond, whereas the secondary interaction (interaction between two monomer chains) can be Van der Waal's forces, dipole–dipole interactions (hydrogen bonds), or polar–dipole interactions. Depending on the secondary interaction, polymers can be categorized as (a) thermoplastic (weak Van der Waal's forces), (b) thermosetting (having irreversible cross-linking between polymer chains) such that these polymers cannot reshape after heating, and (c) elastomers (weakest Van der Waal's forces).

6.2.3 Based on the Structure of the Chains

Polymers can also be differentiated based on the structure of the chains, i.e., the kind of functional groups present in the polymer chain. Accordingly, they are classified as (a) linear polymers (only two functional monomer groups without any interaction are present), (b) branched polymers

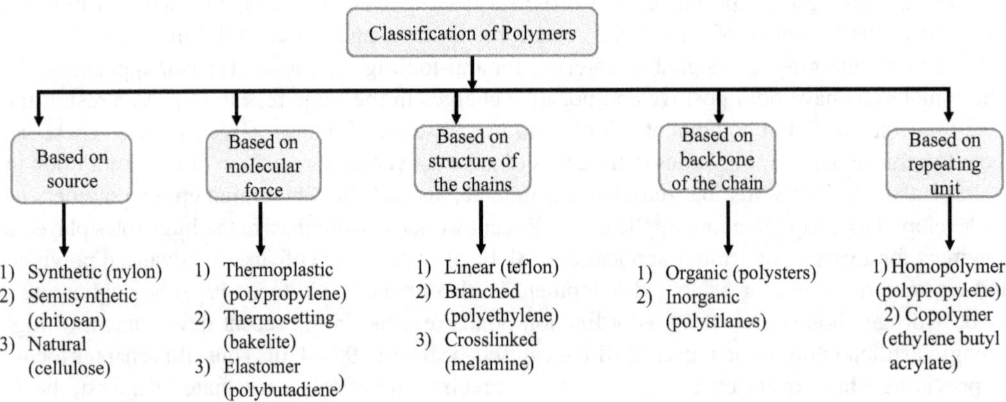

FIGURE 6.1 Classification of polymers.

(branching starts and two or three functional monomer groups are present), and (c) cross-linked polymers (three-dimensional network of monomer chains).

6.2.4 Based on the Backbone of the Polymer Chains

According to the elements present in the monomer units, polymers are mainly of two types: (a) organic polymers (only C-C bonds are present) and (b) inorganic polymers (elements other than carbon are also present).

6.2.5 Based on the Repeating Units

Based on whether the repeating unit is of the same type or a different type the polymers can be classified as (a) homopolymer (only one type of monomer is present in the structure) or (b) copolymer (more than one type of polymer is present in the structure).

Polymers can also be classified according to their property and application, as follows.

6.2.6 Plastics

Plastic polymers are those polymers that show structural rigidity when a load is applied. Thus, polypropylene, polyvinylchloride, fluorocarbon, polyethylene, phenolics, epoxies, and polyesters are some examples of plastic materials. Certain plastic polymers are brittle and rigid, whereas others are flexible. The flexible polymers show both plastic and elastic deformation. Generally, plastic polymers are used in certain general-purpose applications, e.g., ice trays, toys, tumblers, and film wrapping material.

6.2.7 Elastomers

Polymers which exhibit the property of viscoelasticity are known as elastomers. Viscoelasticity is the property where materials show the properties of both a viscous material and an elastic material. Elastomers generally possess high strain to failure, high yield strength, and low Young's modulus as molecules in an elastomer are interconnected by weak intermolecular forces. Due to their properties, elastomers regain their original size and shape once the deforming forces are withdrawn, e.g., a simple rubber band. Natural rubber, polybutadiene, silicone, neoprene, and polyurethanes are some examples of elastomers.

6.2.8 Fibres

Fibre polymers are those polymers that can be drawn into long filaments (100:1 length-to-diameter ratio). Due to this property, fibre polymers find applications in the textile industry. Fibre polymers also possess high abrasion resistance, high modulus of elasticity, and high tensile strength over a wide range of temperatures. The reason is that while manufacturing, fibres are subjected to a large number of deformation steps like shearing, twisting, abrasion, and stretching. The tensile strength of fibre polymers is enhanced in proportion to the degree of crystallinity. The molecular weight of fibre polymers is found to be high. Fibre molecules also show chemical stability in different environments, such as bases, acids, sunlight, and bleaches. Further, these materials are amenable to drying and non-flammable.

Crystallization (a process in which an ordered solid structure is obtained from a liquid material after cooling) of polymer material affects its thermal and mechanical properties. Liquid polymers crystallize by the nucleation and growth process. Recrystallizing plays a very important role in the polymer crystallization process. Melting of a solid polymer crystal leads to the transformation of the ordered aligned polymer chains to highly random viscous liquid at the melting temperature (T_m).

Melting of solid polymers occurs over a range of temperatures and it depends on its crystallization temperature (the temperature at which the material crystallizes) and rate of heating [11]. Molecular structure and chemistry influence the ability of rearrangement of the polymer chain.

Amorphous and semicrystalline polymer crystals have another property. This very important property is called the glass transition temperature (T_g). The glass transition temperature is defined as the temperature at which a polymer transforms into a hard brittle material from a ductile or rubbery material. At the glass transition temperature carbon structure starts moving. This process leads to the enlargement of the gap between two molecular chains. Other properties such as heat capacity, stiffness, and coefficient of thermal expansion also change at the glass transition temperature.

Glass transition temperature and melting temperature play important roles in deciding whether the material will be suited for rigid or flexible application. Further, T_m and T_g are determined from a plot of specific volume versus temperature. Viscoelasticity is another property which is represented by the fibre polymers and it can be conceptually defined as a property which measures both elasticity and viscosity. Viscosity in a polymer fibre material exists due to connections between the particles of fibres, although these connections are temporary.

For a viscoelastic material, the rate of strain defines whether the deformation is viscous or elastic. Silicone, being a viscoelastic material when rolled into a ball, shows elastic behaviour when it is dropped horizontally and the rate of deformation is found to be very rapid during the bounce. If tangential stress is applied to the silicone polymer, it elongates just like a highly viscous material.

6.2.9 Application Prospects

Polymer nanofibers show promising characteristics, e.g., wettability, easy preparation, elasticity, surface-to-volume ratio, and tailored surface and interface properties. They can be suitably utilized to develop new products in various fields. Polymer materials are also used for the coating of surfaces. This is usually done to protect surfaces from the environment, to provide electrical insulation, and to improve the appearance [12].

Nowadays, synthetic polymers such as polysiloxanes, polyurethane acrylics, and rubber materials are also used as adhesives. Polymer material is used as a pressure-sensitive adhesive. Polymers have applications in the automotive industry, aerospace industry, packaging industry, electronics industry, and medical devices industry [13]. Generally, such applications include polymer compounds, raw materials, foams, composites, structural adhesives, fillers, films, fibres, membranes, emulsions, sealing materials, solvents, adhesive resins, inks, and pigments [14–18]. Polymers are used in a range of engineering applications, from avionics to tissue engineering. Further, polymers are used for biomedical applications, biosensor devices, drug delivery systems, and cosmetics. For polymers and polymer matrix composites (PMCs), some typical applications of current importance are:

- 3D-printed parts
- Sports equipment, aircraft, and aerospace
- Biopolymers for molecular recognition and biosensing
- Organic polymers for water purification
- Polymers for holography
- Substrates of printed circuit boards
- Polymeric biomolecules
- Bulletproof vests and fire-resistant jackets
- Renewable biomass operations

6.3 BASIC IDEAS ABOUT THE MECHANICAL PROPERTIES OF POLYMERS

Discussing the mechanical properties of a polymer before its use in an application is very crucial because it gives a suitable macroscopic description of the polymer behaviour under consideration.

Both physical structure and chemical composition affect the mechanical properties of polymers [11, 19]. The mechanical properties of anisotropic polymers are controlled by their elastic parameters (Poisson's ratio and moduli values, e.g., Young's modulus, bulk modulus, and shear modulus). If the material is homogeneously isotropic and elastic, then only the Poisson ratio and Young's modulus become important for determining the mechanical behaviour.

In this section, we will touch upon once again the basic mechanical properties of importance. Then we discuss the literature scenario. Thus, the strength properties are the basic properties to start with. The tensile strength data are often reported. It reflects how much stress the polymer can endure before fracture when pulled longitudinally along its longitudinal axis. The corresponding strain is called the fracture strain or, equivalently speaking, elongation at break. The corresponding load is termed the tensile fracture load. Even when fractured, the polymer can still carry some load. However, beyond a critical deformation strain, it cannot sustain further load. This is called the yield strain and the corresponding strength is called the yield strength, as mentioned earlier.

6.3.1 Stress–Strain Behaviour

Typical stress–strain curves for different types of polymeric materials, i.e., brittle, ductile, rubber-like, polymer gels, and semicrystalline polymers, are represented in Figure 6.2.

Brittle polymers such as polystyrene and polymethyl methacrylate don't support much deformation. Brittle polymers fracture while deforming elastically as is evident from the stress–strain curve of Figure 6.2a. The load elongation relationship for plastic polymer is displayed in Figure 6.2b. The schematic plot depicts that, initially, the behaviour is elastic. The elastic deformation regime is followed by a region of plastic deformation. On the other hand, elastomers (i.e., rubber-like polymers) produce large strain or elongation even for low stress values as displayed in Figure 6.2d. Semicrystalline polymers show an elastic behaviour initially and then a rubber-like behaviour as shown by the curve of Figure 6.2c.

6.3.2 Factors Affecting the Strength of Polymers

6.3.2.1 Crystallinity

The crystalline polymer material shows more strength in comparison to an amorphous material as crystalline material has good intermolecular bonding. Generally, strength increases with an increase in the crystallinity. The relatively higher strength is linked to the orientation of the chains [20].

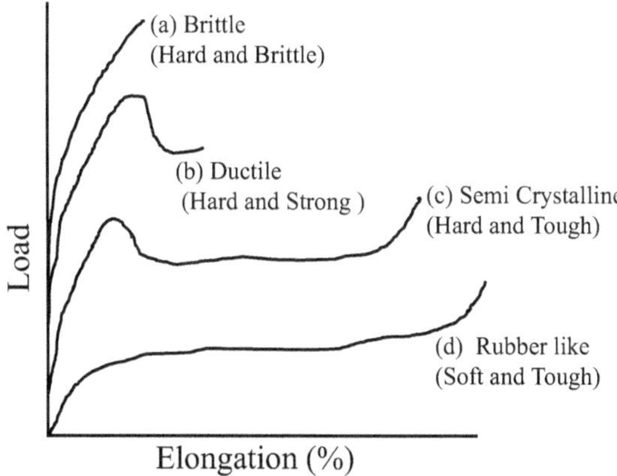

FIGURE 6.2 Stress–strain behaviour of different types of polymers.

6.3.2.2 Molecular Weight

The molecular weight of a material affects its tensile strength. So, the tensile strength of a polymer material increases with an increase in its molecular weight. In low molecular weight polymers, the tensile strength is low although the crystallinity is present. This happens due to the weak Van der Waals forces between the polymer chains [21]. In high molecular weight polymers, e.g., ultra-high molecular weight polyethylene (UHMWPE), the long chain of polymers gets entangled. This ultimately gives rise to high tensile strength in the polymer.

The molecular weight of the polymer material also affects its melting temperature. For low molecular weight polymers, an increase in polymer length leads to an increase in melting temperature. Polymers have a range of melting temperatures because polymers are made up of molecular chains of varying molecular weight. The melting temperature of a polymer material can be altered by varying the (a) degree of cross-linking in the polymer as it limits the movement of the chains and hence enhances the melting temperature, and (b) by enhancing Young's modulus. The stress–strain curve for a typical polymer is represented in Figure 6.3. The yield point is taken as the point in the stress–strain curve (Figure 6.3) corresponding to the maximum value (beyond the linear elastic region). Thus, the applied stress is maximum at the yield point (σ_y). The tensile strength of a polymer may be greater than or lesser than its yield strength.

In plastic polymers, the maxima point (beyond the linear elastic region) in the stress–strain curve corresponds to the yield point and the stress at this point is known as the yield strength (σ_y) of the material. Hence, in the stress–strain curve, yield strength is the maximum stress value point, whereas tensile strength (σ_t) corresponds to the stress value at which the material gets fractured (Figure 6.3). The typical stress–strain curve of a semicrystalline polymer material (Figure 6.3) displays a lower yield point followed by a near horizontal region and then an upper yield point. There is a small neck formation at the upper yield point in the gauge section. The polymer chain becomes oriented within this neck as the chain axis aligns parallel to its elongation direction. This alignment of the polymer chain strengthens the material and the material becomes resistive to continued deformation, and the material shows further elongation along the gauge length due to propagation of this neck region. The tensile strength value for these polymer materials (plastic polymers) can be greater than or lesser than the yield strength, as mentioned earlier. The strength of plastic polymers is their tensile strength value only. The maximum tensile strength for certain polymers can be as high as 100 MPa.

Next comes the role of compressive strength. These data are also reported. What it reflects is how much stress the polymer can endure before fracturing when it is compressed longitudinally. The corresponding fracture load is termed the compressive fracture load. Here the corresponding strain is termed the compressive strain.

Many applications of polymers and polymer composites involve load application in a direction perpendicular to the longitudinal axis of the sample or structure. Therefore, this brings flexural deformation into the picture. These strength data are called the flexural strength data, or the

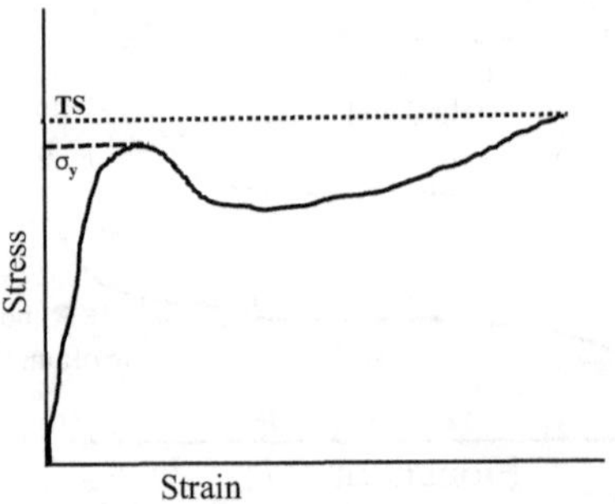

FIGURE 6.3 Schematic representation of the stress–strain curve for a typical polymer.

bending strength or transverse strength of polymers. The stress at which the polymer cannot sustain further the flexural load applied perpendicular to its longitudinal axis is called the flexural strength. The corresponding load at which the flexural fracture happens is called the flexural failure load or the transverse failure load. The corresponding strain is termed the flexural or transverse strain at failure.

Further, many applications of polymers require their structural integrity to be preserved against the applied stress that wants to twist the structure. Thus, there can be in-plane shear forces on the polymer trying to twist it. Also, there may be out-of-plane shear forces acting on the polymer to twist it out-of-plane. This is where the shear deformation ability of the polymer without getting fractured comes into the picture. Thus, the maximum stress up to which the polymer can sustain its structural integrity before suffering fracture due to shear is called the shear strength of the polymer. The corresponding stress is termed the maximum shear stress. Furthermore, the strain suffered by the polymer under shear deformation at fracture is called the shear strain at failure. Many possibilities of mixed-mode fracture also happen when polymers are actually applied in complex stress-bearing scenarios. For the sake of simplicity and convenience, it is desired to keep the discussion limited to simple modes of loading.

The separation or breaking of a material in pieces under applied stress is defined as fracture. A fracture may be brittle or ductile. The deforming ability of the material determines whether the fracture will be brittle or ductile. The fracture strengths for polymeric materials are relatively low in comparison to those of ceramics and metals. Thermosetting polymers show brittle fracture, as these materials as heavily cross-linked polymers. Thermoplastic polymers exhibit mostly ductile fracture. Fracture mechanics of polymers is a subject by itself and it is beyond the scope of this chapter to consider it in detail. Polymers exhibit behaviour that is different from those of ceramics and metals. The metals contain grain and grain boundaries. They can easily deform plastically. So, they generally exhibit ductile fracture. On the other hand, ceramics are strong, hard, and, hence, brittle by nature. So, they exhibit brittle fracture. The fracture of polymers is thus guided by the deformation of the chains that form the structural subunits of a polymer.

The property that comes as the second most important factor to be considered is Young's modulus. Simply, it is defined as the stress-to-strain ratio. The stress and strain to be considered are from the linear response part of the total stress–strain response curve for a given polymer. However, Young's modulus of polymers follows the case of the simple loading direction-dependent strength properties. Deformation produced in the material after the application of stress defines the modulus. The ratio of longitudinally applied stress and the resulting longitudinal strain due to the longitudinally applied stress gives the value of Young's modulus. The reciprocal of modulus is known as compliance. Elastic deformation in a material is generally instantaneous, i.e., strain is developed in the material until the stress is applied, and this deformation is totally recovered once the stress is released. Elastic deformations in a polymeric material start at low stress levels. Elastic deformation in semicrystalline materials occurs due to the elongation of chain molecules in the applied stress direction in the amorphous region. Along with the elongation, stretching and bending of the material are responsible for the elongation. Figure 6.4 represents the deformations in a typical stress–stain curve.

Thus, based on the ratio of linear stress to linear strain obtained during tensile strength measurements, the tensile Young's modulus is measured. In a similar manner, based on the ratio of linear stress to linear strain obtained during compressive strength measurements, the compressive Young's modulus is measured. In an analogous manner, based on the ratio of linear stress to linear strain obtained during flexural strength measurements, the flexural Young's modulus is measured. This is also termed as the bending Young's modulus or transverse Young's modulus. In a similar manner, based on the ratio of linear stress to linear strain obtained during shear strength measurements, the shear modulus is measured. This modulus is also termed the torsional modulus or modulus of rigidity. The modulus of rigidity value for very stiff polymers is generally found to be as high as about 4 GPa, whereas elastic polymers show comparatively low value (about 7MPa).

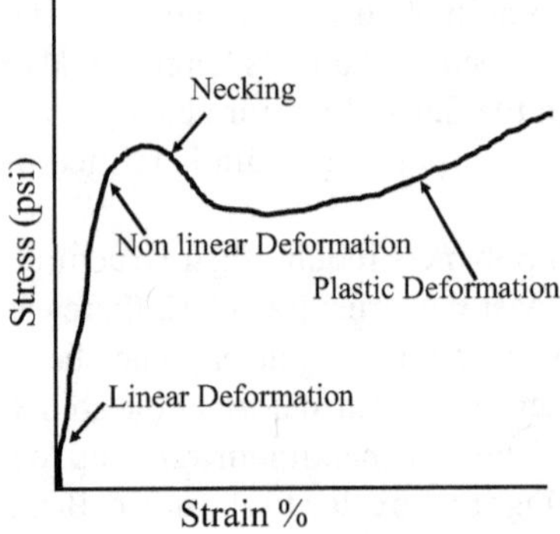

FIGURE 6.4 Schematic illustration of stress–strain behaviour of a polymer displaying regions of linear, nonlinear, and plastic deformations.

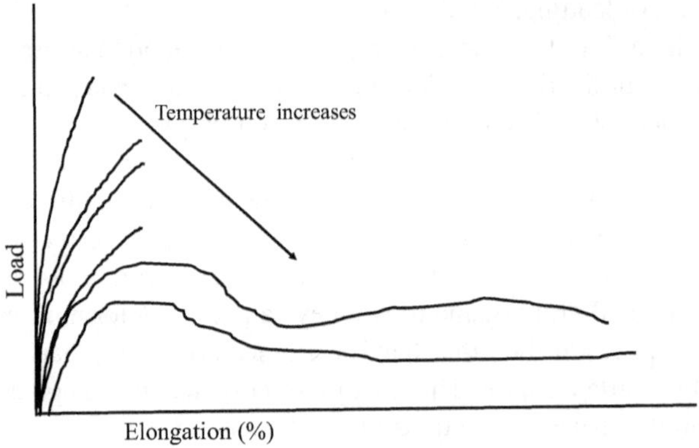

FIGURE 6.5 Schematic demonstration of stress–strain behaviour of a polymer at different temperatures.

In addition, there can be the bulk modulus, which refers to the bulk deformation test. In this case, the volumetric stress and volumetric strain response parts are taken into consideration. It is also important to note that Young's modulus, shear modulus, and bulk modulus are related to each other. The ratio of lateral strain to longitudinal strain is called Poisson's ratio. This is also considered an important mechanical property as far as the deformation and fracture of polymers are concerned.

Another important property of huge importance is hardness. It is related to localized plastic deformation. This property basically reflects the ability of a given polymer to resist localized permanent plastic deformation caused by an externally applied load on an indenter that tries to intrude into the material. Many different types of hardness measurements are in practice. Some typical well-known examples are Barcol hardness, Shore hardness, and Brinell hardness. All such hardness values are evaluated as force per unit area, i.e., stress. Thus, hardness reflects the maximum stress sustainable by a given polymer just prior to suffering permanent localized plastic deformation area. This area is measured. Then the ratio of applied load to this contact deformation area is expressed as hardness.

6.3.3 Effect of Temperature on Mechanical Properties

Temperature is an important parameter when studying mechanical properties, as the change in temperature affects the mechanical properties of polymer materials. Figure 6.5 represents the temperature effect on the stress–strain behaviour of a Plexiglas polymer in the temperature range of 4°C to 60°C. It can be inferred from the data plotted in Figure 6.5 that with an increase in temperature, the ductility of the material increases, whereas elastic modulus and tensile strength decrease. The material shows brittle behaviour at low temperatures and plastic behaviour at higher temperatures.

As far as fracture of polymers is concerned, the most important property is fracture toughness. This reflects the intrinsic mechanical resistance of the given polymer against catastrophic crack propagation. The maximum energy absorbed by a material before it breaks defines its toughness. The toughness of a polymer material is measured by the area under the curve in a stress–strain curve (Figure 6.6) and is given by the formula

$$\text{Toughness} = \int \sigma \, d\varepsilon \tag{6.1}$$

The brittle polymers possess high Young's modulus. However, ductile polymers have higher fracture toughness. Both brittle and ductile polymers have a similar elastic modulus. However, the elastomers possess generally lower values of Young's modulus.

The nature of crack propagation is dependent on the load application mode. Thus, there are three modes: mode I, mode II, and mode III. Mode I involves in-plane load application. This load tries to tear apart the sample. Mode II involves in-plane shear loading, which tries to break the sample through in-plane shear. However, mode III involves out-of-plane shear loading. This load tries to break the sample by out-of-plane shear.

The resistances to mode I, mode II, and mode III loading-induced failure are represented by the corresponding critical stress intensity factors, i.e., K_{Ic}, K_{IIc}, and K_{IIIc}. These are also equivalently termed mode I fracture toughness, mode II Fracture toughness, and mode III fracture toughness. However, for the sake of convenience and simplicity, the discussions in this chapter will be mostly limited to mode I fracture toughness (K_{Ic}). The corresponding strain energy release rate is defined as G_{Ic}. These two quantities are related to each other as:

For plane stress:

$$G_{Ic} = \left((K_{Ic}^2\right)/\left(\text{E}\right) \tag{6.2}$$

For plane strain:

$$G_{Ic} = \left[\left(K_{Ic}^2\right)\right]/\left[\left(\text{E}\right)/\left(1-\mu^2\right)\right] \tag{6.3}$$

Further discussions on these aspects will follow later.

However, fractures can also happen due to impact. Thus, impact fracture strength is also important for polymers considered for load-bearing applications involving impulsive force. These impact fracture strength values of polymers are also referred to as impact toughness or simply toughness. These reflect the ability of the polymer to absorb impact energy transferred by the hammer during the Charpy impact strength or Izod impact strength measurements.

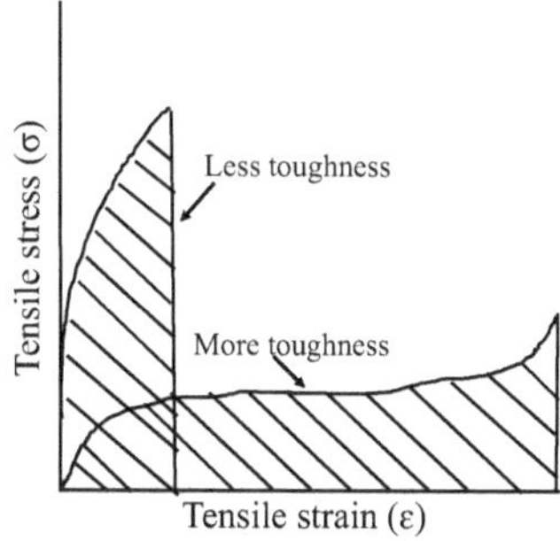

FIGURE 6.6 Tensile stress versus tensile strain curve giving schematic demonstration of toughness in a polymer material.

In both tests, the impacts are done by a calibrated "hammer" whose energy at impact is known. The sample may be notched or unnotched. The machine measures how much energy is absorbed by the sample when it undergoes the impact strike made by the hammer. This energy is measured as the area of the load-deformation curve divided by the cross-sectional area of the sample which faced the impact hammer.

As polymers are viscoelastic in nature, they can also have creep at room temperature and at high temperature. Creep is the amount of permanent deformation that happens due to load application that is deliberately sustained over a predefined period of time. This property reflects how much the material will deform when a load is applied over a period of time that is typically much more than the time applied for rapid loading to fracture. If the polymer is being used for structural application with known load values, then how much deformation is going to happen over a period of time can be estimated. At temperatures higher than room temperature, the extent of creep deformation will be much greater.

Initially, when the strain is linearly proportional to stress, the deformation in the material is linear deformation followed by nonlinear deformation. Then necking occurs and after that plastic deformation occurs. An amorphous polymer behaves differently at different temperatures. At low temperatures, it behaves like a glass. It becomes a rubbery solid at intermediate temperatures. Finally, it becomes a viscous liquid with a further increase in temperature. So, at intermediate temperature ranges when the polymer is rubbery solid it exhibits the mechanical properties of both materials, i.e., an elastic material and a viscous liquid. This property is known as viscoelasticity, as mentioned earlier.

In viscous polymers, the deformation is neither instantaneous nor reversible (Figure 6.7b). Whereas in the intermediate temperature range when the behaviour of the material is viscoelastic, initially there is instantaneous strain. This part is followed by a time-dependent strain (Figure 6.7c). For totally viscous behaviour of the materials, strain or deformation is not instantaneous. In response to an applied stress, deformation is dependent on time or delayed, as represented in Figure 6.7d. Furthermore, once this stress is released, the deformation is not reversible (or completely recovered).

Thus, generally, creep is measured in terms of steady-state creep strain rate as a function of applied stress and temperature. The activation energy of creep deformation is calculated when stress is kept constant and temperature is varied. By fitting the data to an Arrhenius-type equation, the activation energy required to initiate creep deformations is calculated. The creep exponent gives an idea about the mechanism that is involved at the atomic scale to drive creep deformation.

Similarly, the activation energy of creep deformation is also calculated when stress is varied and temperature is kept constant. Again by fitting the data to an Arrhenius-type equation, the activation energy required to initiate creep deformation is calculated. Here also, the creep exponent gives an idea about the mechanism that is involved at the atomic scale to drive creep deformation. For all kinds of practical applications, the lower the steady-state creep strain rate, the better the

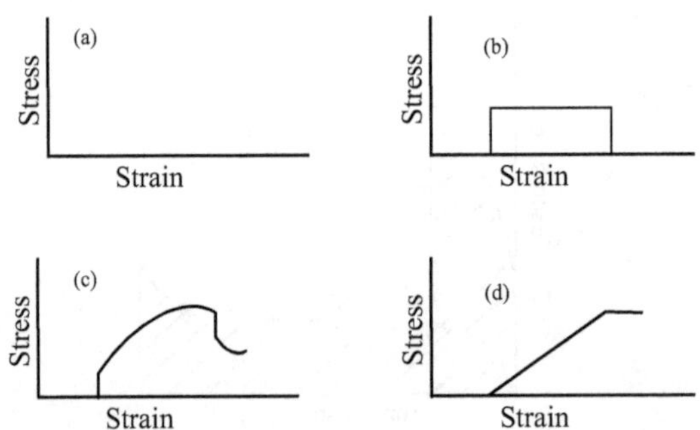

FIGURE 6.7 Schematic illustration of the strain versus time responses for viscous polymers: (a) totally elastic, (b) non-instantaneous, (c) viscoelastic, and (d) viscous behaviours.

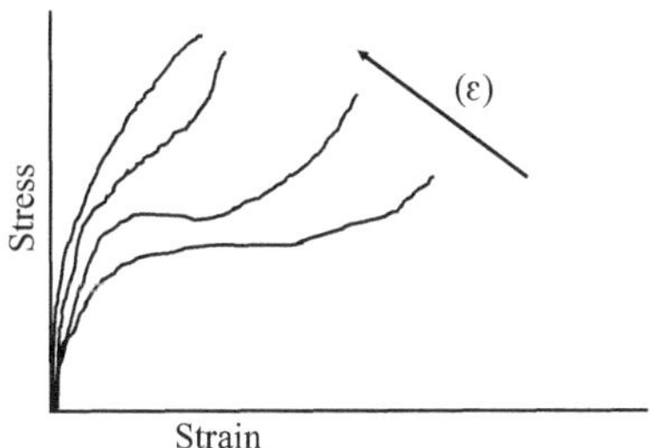

FIGURE 6.8 Schematic representation of strain rate effect in polymers.

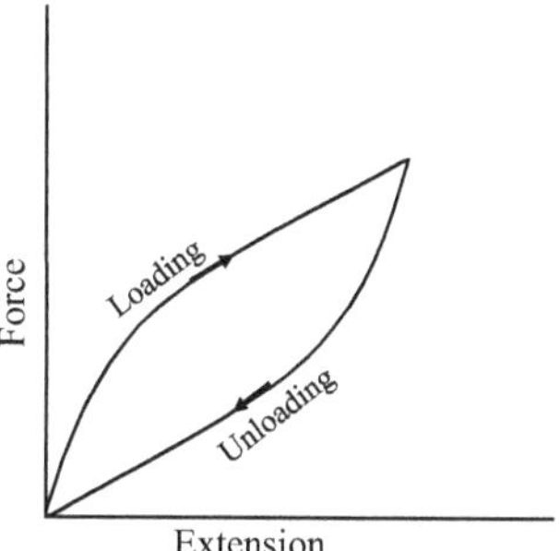

FIGURE 6.9 Schematic representation of hysteresis stress–strain curve.

acceptability of the given polymer. The creep of metals, ceramics, and polymers is by itself a vast topic. Therefore, further discussion on this topic is limited.

The strength of polymers is also very sensitive to the strain rate applied during fracture experiments. Thus, high strain rate experiments are conducted by using a split-Hopkinson pressure bar impact test, gas gun impact test, and hypervelocity gas gun test as per the application of strain rate ranging from 10 to 10^2, or from 10^2 to 10^4, or from 10^4 to 10^6, respectively. Generally, the strength increases with an increase in strain rate. Polymer becomes more ductile and softer with a decrease in strain rate. A schematic representation of the effect of strain rate is displayed in Figure 6.8. Hence, it can be concluded that a decrease in strain rate has a similar effect as the increase in temperature.

Generally speaking, the mechanical properties of polymers are also sensitive to the presence of oxygen, water, and organic solvents. Polymers may be ductile, brittle, or elastic depending on the conditions of loading such as temperature, strain rate, size of the specimen, and mode of loading. Generally, both amorphous and crystalline polymers possess low-impact strength and are found to be brittle at low temperatures [11, 19–21].

On many occasions, the polymeric parts may be exposed to cyclic loading and unloading. Hence, the cyclic fatigue life data as expressed by the plot of fracture strength as a function of the number of cycles to failure, e.g., the S-N curve, is also a very important aspect of the mechanical properties of polymers [22–24]. The fatigue behaviour analysis of polymers is gaining more attention due to the increasing applications of polymers in the engineering field. Polymer fatigue life is sensitive to oxidation, temperature, and crystallization. Polymer fatigue can be measured by different experimental tests including a simple extension test, pure shear test, tearing test, edge crack test and cyclic tension test. The cyclic tension test is most widely used as polymers are sensitive to cyclic loading. Polymers don't show cyclic hardening like metals; instead, they exhibit cyclic softening. Figure 6.9 displays the schematic representation of hysteresis in the stress–strain curve in polymer material exposed to cyclic fatigue loading. In certain cases, polymers also exhibit cyclic stability. Cracks in polymers generally occur at the polymer fibre and matrix interface of PMCs.

The cyclic loading process in polymers for the measurement of fatigue performance generally starts with nucleation or crack initiation followed by crack growth. So, initially, a crack is formed under the applied stress. A good fibre in a PMC is expected to absorb a large amount of energy before its failure or fracture.

6.4 ADVANCED POLYMERS AND POLYMER COMPOSITES FOR VARIOUS IMPORTANT APPLICATIONS

Polymers undergo different modifications to make conjugates [22–24]. Polymer–protein and polymer–drug conjugates are widely used these days as polymeric carriers in biological applications. They offer resistance to drugs from the biological environment and prolong the drug's release for better stability [25–28]. Further, fibre-reinforced PMCs are being used these days for several structural and non-structural applications [29]. The distinctive properties of electrically conducting polymers and their composites ensure their application in shape memory alloys, cost-effective surface coatings, flexible electronics, and actuators [30].

6.5 LITERATURE REVIEW OF MECHANICAL PROPERTIES OF POLYMERS AND COMPOSITES

When we talk of polymers, it needs to be understood that there are two major types of polymers. These are called thermosetting polymers and thermoplastic polymers. Thermosetting polymers are those polymers which are taken through a temperature-driven curing process to achieve a particular shape. However, once they cool down they become rigid. It is not possible to further thermally reprocess them. Examples of thermosetting polymers are epoxy and polyurethane. Thermoplastic polymers are those which are taken through a temperature-driven curing process to achieve a particular given shape. However, they are not rigid. As a result, they can be again reformed using the temperature-driven process if required. Polycarbonate, PEEK, UHMWPE, and nylon 6-6 are typical examples of thermoplastic polymers. To enhance the mechanical properties of polymer properties, particulate reinforced polymer matrix composites (PRPMCs), fibre-reinforced polymer matrix composites (FRPMCs), and both particulate and fibre, i.e., hybrid reinforced polymer matrix composites (HRPMCs) are developed. These are discussed in detail in Chapter 7.

What follows next is a very brief overview of relevant literature [31–64]. However, some data on PMCs based on both thermosetting and thermoplastic composites are also included, as most of the significant applications and developments are happening in PMC materials (e.g., PRPMCs, FRPMCs, and HRPMCs).

6.5.1 Brief Literature Overview of Mechanical Properties of Polymers and Polymer Matrix Composites

The data presented in Table 6.1 gives a brief overview of various experimental techniques applied to determine the mechanical properties of the polymeric materials and PMCs. As such, most of the recent work involves a wide variety of PMCs which cannot be avoided. Hence, data from the literature on various PMCs are included.

Recent work [31] investigated the mechanical properties of PLA (polylactic acid) polymers for various infill structures, e.g., gyroid, rhombic, and hexagonal structures. Several properties such as tensile strength, flexural strength, and hardness are measured. For instance, a specimen is reported to have a tensile strength of about 32 MPa. The hexagonal infill structure is reported to possess good flexural strength due to a good bonding area [31]. Optimization or improvisation is the key property of structural properties of polymeric materials in applications like aerospace, marine, and civil engineering which often require a lighter and improved version of materials. Another study

TABLE 6.1
Brief Literature Overview of Mechanical Properties of Polymers and Polymer Matrix Composites

Materials	Experimental Method	Factors Affecting	Flexural Strength (MPa)	Elastic Modulus (GPa)	Tensile Strength (MPa)	References
Polyurethane core glass fibre-based sandwiched PMCs	Vacuum bag moulding technique	Core thickness and face sheet thickness	192.5	31.72	—	32
Polyester AKAVINA and titanium oxide	Hounsfield H50K-S tester	Composite structure	—	1.43	—	33
Polyvinyl chloride (PVC)	Test force 20 KN	Heating and cooling environments	—	—	63.67	36
UHMWPE and basalt fibre	Mechanical co-activation	Increased adhesion of the fibre	—	Increases by 19%	Increases by 14%	37
UHMWPE + PEG + HDPE	Kokubo method	Variation in HDPE concentration	Increases by 90%	—	—	38
UHMWPE/HA/BO	Parakeet Mill	Addition of BO	—	370 MPa	Decreases with addition of BO	39
UHMWPE + Graphene + HA	Solvent mixing	Addition of graphene	Increases by 24%	Increases by 114%	—	40
UHMWPE/ATZ composite	Compression moulding	(80–20 wt%) (ATZ)		Increases by 13%		41
(UHMWPE)/hydroxyapatite (HAP) + titanium dioxide	Solvent dispersing technique	Additives in polymer matrix	Increases by 95%	Increases by 71%	—	42
UHMWPE + nanodispersed TiO_2	Hot moulding	Vacuum plasma arc modification	Increases	Increases	—	43
UHMWPE–TiO2 composites	Homogeneous dispersion	TiO_2 as nucleating agent	—	800 MPa	—	44
UHMWPE + nanoparticles of boron carbide	Twin-screw extrusion processing,	Varying concentration of NPs		Increase in viscous and elastic components		47
(UHMWPE) + carbon fibres	Lubricated techniques	Varying concentrations in carbon fibres	Increases by 105%	Increases by 208%	—	51
UHMWPE + carbon fillers	MWCNT Fluorination	Nanofibrils volume fraction	300	—	—	52

[32] focuses on the effect of face sheet and bond strength on the fracture strength of sandwiched PMCs. The various experiments conducted with different core sheets indicate that the elastic modulus of the composite polymer material increases with the core thickness. However, the fracture strength decreases with the increase in core thickness [32]. Another work [33] illustrates the comparative study between the theoretical and experimental analysis of the elastic modulus of three-phase polymer composites of polyester, glass fibre, and titanium dioxide (TiO_2) [33]. These PMCs have numerous uses in the fields of waterproof, corrosion-proof, and crack-proof shipbuilding and fishing vessels.

In some theoretical research, matrix and particle interactions are taken into account. In this approach, effective matrix properties are calculated. This is followed by estimating the elastic moduli for PMCs, which are comprised of the effective matrix and unidirectionally reinforced fibres [34, 35].

The heating and cooling effects are also reported to be observed in PVC (polyvinyl chloride) [36]. This study shows that, as expected, with an increase in the heating temperature, the strength of the PVC decreases, i.e., the material becomes softer. However, a reduction in temperature increases the strength of the sample. Because of the huge importance of UHMWPE, a large number of studies concentrate on various aspects of UHMWPE and UHMWPE-based composites [37–43].

For instance, a recent study indicates that as the wt. % of basalt fibre is increased in the UHMWPE composite, ~~and~~its tensile strength and elastic modulus increase~~s~~ by about 14% and 7%, respectively [37]. Further, the usage of different additives like polyethylene glycol (PEG) and HAP [38], organophilic bentonite (BO) and HAP [39], graphene nanoplatelets (GNPs) and HAP [40], alumina-toughened zirconia (80–20 wt.%) (ATZ) [41], and single-walled carbon nanotubes (SWCNTs) with UHMWPE modify its mechanical properties. In other words, these UHMWPE composites find many avenues of applications, e.g., biomedical prosthesis, biomedical devices, and antibacterial usage.

Furthermore, the inorganic fillers are also used to improve the mechanical properties of UHMWPE [42]. The use of TiO_2 as filler in UHMWPE prevents the growth of bacterial infections. Such materials are proposed to have huge prospects as biomaterials [43–45]. The hybrid composite of UHMWPE, HAP, and TiO_2 are fabricated and used as artificial cartilage in joint prostheses. This hybrid approach drastically improves Young's modulus, hardness, and flexural strength, respectively, by e.g., about 70%, 30%, and 100%, as compared to those of the only UHMWPE [46]. In a similar fashion, the incorporation of B_4C in UHMWPE leads to the lowest bullet penetration depth. This attribute makes it more suitable for use in armour applications [47].

UHMWPE 65 wt.% biomass-activated carbon composites exhibit an increase in tensile strength by as much as 300% [48]. Similarly, UHMWPE–nanoclay composites exhibit about 110% improvement in impact resistance [49]. It is further reported [50] that cellulose nanocrystal (CNC)–silica (SiO_2) incorporated UHMWPE hybrid nanocomposites exhibit a significant increase in tensile strength, flexural strength, flexural modulus, and softening temperature. In addition, 30 wt.% recycled carbon fibre–UHMWPE composites exhibit 280%, 105%, and 146% increases in Young's modulus, ultimate tensile strength, and hardness, respectively [51]. Moreover, it causes about 62% and 32% reductions in the coefficient of friction and wear rate, respectively [51]. Further, the fluoridated carbon nanotube–UHMWPE composites exhibit huge enhancement in tensile strength [52].

Work on the compressive strength of Kevlar 49-reinforced PMCs shows promising results [53]. The compressive strength of crumpled polymer thin films is also reported [54]. A good deal of work has been reported on the mechanical properties of geopolymers with and without reinforcement by fibres, e.g., carbon fibre, polypropylene fibre, and PVA fibre [55–63]. Carbon fibre reinforcement leads to poor toughness and inadequate impact resistance [60, 61], while polypropylene fibre reinforcement gives lower tensile strength and Young's modulus [62]. However, the use of PVA fibres leads to a 34% increase in the compressive strength of geopolymers [63]. The Young's modulus of softer polymers is reported to be higher than those of relatively harder polymers [64].

Another polymer which comes under the category of thermoplastics is polycarbonate. It has exceptional mechanical properties and a superior weight-to-strength ratio. This is what makes these materials suitable for many engineering applications. The data presented in Table 6.2 illustrate the mechanical properties of some thermoplastic polymeric materials [1–30].

Polycarbonate (PC) has high ductility and yield strain with considerable strain hardening because it has impressive impact and perforation resistance [65]. The recent computational prediction of the ballistic performance of thick PC targets shows that there can be five types of deformation mechanisms under ballistic impact. These are reported to be dishing, petalling, deep penetration, cone cracking, and plugging [65]. Further studies reported [66] that a thin plate of PC under ballistic impact exhibits elastic dishing deformation followed by deep penetration, while thick PC plates display deep penetration and yawning penetration. The PC under study is reported to be capable of deflecting the path of flight of the projectile from its initial angle of penetration. Furthermore, for projectile penetration obliquely on PC with a speed of 720 ms^{-1} the projectile follows an S-shaped trajectory with 30° impact, and the maximum stress of 2.29 GPa is reported to be sustainable [66]. Figure 6.10 shows the effect of projectile velocity on the depth of penetration and on the target's trajectory.

The effect of temperature on the mechanical properties of polycarbonate resin is also reported [67]. The results show that the yield stress increases under the annealing condition, however, the impact strength decreases at all annealing temperatures. There is also a clear indication of the permanent deformation under the annealing effect. Transparent polymer materials such as PCs have now replaced glasses, eye lenses, and compact disks, but they are prone to scratches and abrasion, thus limiting their application. Therefore, a scratch- and abrasion-proof coating is required to overcome the limitations.

Recent work [68] reports about 25% and 270% enhancements in hardness and Young's modulus of sol–gel silica-coated PCs. Uniaxial tensile and compressive tests confirm that the material behaviour of PC does not depend on the loading and unloading stages, and therefore any non-linearity in its behaviour may be due to the hypoelastic behaviour of the PC material [69]. Earlier research compendium [70] reports that thermosetting polymers like epoxy and polyester have average densities of about 1300 and 1400 $Kg.m^{-3}$, respectively. Their average Young's moduli are about 4.5 and 3.5 GPa, respectively. Further, the average tensile strength is about 68 and 50 MPa, respectively. However, depending on the density, the ranges are about 35–100 MPa and 40–60 MPa for epoxy and polyester resins, respectively. Thus, the average elongation at break is about 4% for epoxy and 2% for

TABLE 6.2
Mechanical Properties of Thermoplastic Polymeric Materials

Polymers	Specific Gravity	Elasticity Modulus (GPa)	Tensile Strength (MPa)	Fracture Strength (MPa)	Elongation to Break (%)
Polyethylene (low density)	0.917–0.932	0.17–0.28	8.3–31.4	9.0–14.5	100–650
Polyethylene (high density)	0.952–0.965	1.06–1.09	22.1–31.0	26.2–33.1	10–1200
Polyvinyl chloride (PVC)	1.30–1.58	2.4–4.1	40.7–51.7	40.7–44.8	40–80
Polytetrafluoroethylene (PTFE)	2.14–2.20	0.40–0.55	20.7–34.5	—	200–400
Polypropylene (PP)	0.90–0.91	1.14–1.55	31–41.4	31.0–37.2	100–600
Polystyrene (PS)	1.04–1.05	2.28–3.38	35.9–51.7	—	1.2–2.5
Polymethyl methacrylate (PMMA)	1.17–1.20	2.24–3.24	48.3–72.4	53.8–73.1	2.0–5.5
Phenol-formaldehyde	1.24–1.32	2.76–4.83	34.5–62.1	—	1.5–2.0
Nylon 6-6	1.13–1.15	1.58–3.80	75.9–94.5	44.8–82.8	15–300
Polyester	1.29–1.40	2.8–4.1	48.3–72.4	59.3	30–300
Polycarbonate (PC)	1.20	2.38	62.8–72.4	62.1	110–150

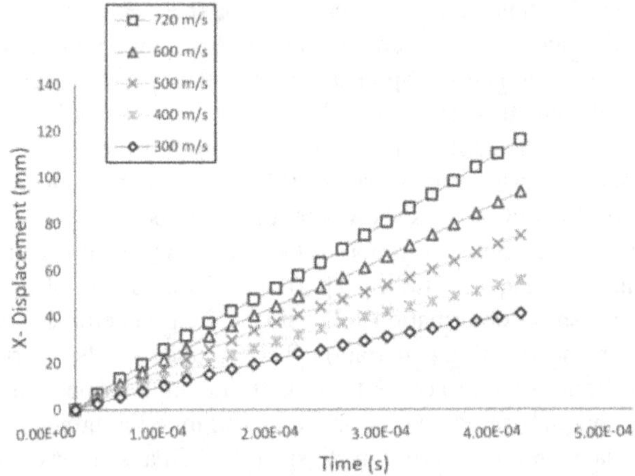

FIGURE 6.10 Projectile's depth of penetration at various velocities and an angle of 30° [65].

polyester resins. Similarly, the mechanical properties of thermoplastic polymers are also sensitive to densities as well as the presence or absence of a large number of repetitive complex chain structures.

For instance, PCs with an average density of 1130 $Kg.m^{-3}$ have an average Young's modulus of about 2.3 GPa, tensile strength of about 58 MPa, and elongation at break of about 75%; whereas nylon 6-6 with an average density of 1140 $Kg.m^{-3}$ has an average Young's modulus of about 2.1 GPa, tensile strength of about 68 MPa, and elongation at break of about 60% [70]. These data establish that both density and molecular structural repeat unit matter in defining the overall mechanical properties of thermoplastic polymers. Other thermoplastic polymers, e.g., polysulphone and polyetherimide, are reported [70] to have average values of 1250 and 1270 $Kg.m^{-3}$ for density (ρ), 2.2 and 3.3 GPa for Young's modulus (E), 76 and 110 MPa for tensile strength (σ_t), and 75% and 60% for elongation at break (e_b), respectively.

In a similar manner, polyamide-imide and polyphenylene sulphide are reported [70] to have average values of 1270 and 1400 $Kg.m^{-3}$ for density (ρ), 4.8 and 3.8 GPa for Young's modulus (E), 190 and 65 MPa for tensile strength (σ_t), and 4% and 50% for elongation at break (e_b), respectively. These data reflect the complex interdependence of density and molecular chain structure on one hand and mechanical properties on the other. Further, the PEEK polymers are reported [70] to have an average value of 1290 $Kg.m^{-3}$ for density (ρ), 3.6 GPa for Young's modulus (E), 93 MPa for tensile strength (σ_t) and 50% for elongation at break (e_b). However, the polypropylene polymers are reported [70] to have average values of 900 $Kg.m^{-3}$ for density (ρ), 2.5 GPa for (E), 32 MPa for tensile strength (σ_t), and greater than 300% for elongation at break (e_b).

An extensive curing temperature range helps the thermosetting epoxy polymers to have specific chemical properties needed to have anti-corrosive properties as well as toughness, flexibility, adhesion, and chemical resistance [71, 72]. Recent work [72] on the mechanical properties of epoxy and its composites with glass fibre, carbon fibre, and a hybrid of both glass carbon fibres show that the glass fibre incorporation increases the fracture strength by about 68% as compared to that of the carbon fibre-reinforced epoxy. However, the hybrid composite is even more effective in increasing the fracture strength of the epoxy polymer. On the other hand, the toughness of carbon fibre-reinforced epoxy is about 53% higher than that of glass fibre-reinforced epoxy. This happens because the carbon fibres have high modulus and strength as compared to the glass fibres. As a result, they add up more to the fracture toughness [72]. The data presented in Table 6.3 illustrate the variations in the mechanical properties of the epoxy matrix composites with different fibres as well as hybrid reinforcements.

Incorporation of waste containing B [73] leads to significant enhancement of compressive strength in epoxy composites. Thus, there are interesting variations in the compressive strength

TABLE 6.3
Effect of Fibre Reinforcement on the Mechanical Properties of Epoxy [72]

Fibre Epoxy Composites	Epoxy Glass Fibre Composites	Epoxy Carbon Fibre Composites	Hybrid Epoxy Glass and Carbon Fibre Composites
Strength (MPa)	28	30	39
Fracture toughness (MPa$\sqrt{m}$)	10.96	16.34	18.36

TABLE 6.4
Mechanical Properties of Natural Fibre-Reinforced Polymer Composites [76]

Matrix	Fibre	Improvement in Mechanical Properties
Epoxy	Glass fibre	Flexural modulus increased by 19% and flexural strength by 20% as compared to neat composite.
Epoxy	Carbon fibre	Flexural strength and modulus of the epoxy composite increased by 25% and 28% as compared to neat epoxy by adding GO filler.
Polypropylene	Rice husk fibre	Stiffness increased from 2100 MPa to more than 3600 MPa.
Polymer polyester resin	Kenaf fibre	Tensile strength, modulus, and bending resistance composite improved. Flexural strength increased by 121% and tensile strength by 38%.
Epoxy	Glass fibre with ZnO	ZnO harms flexural strength and tensile strength while having a positive impact on impact strength.
Polyester	Glass fibre with ZnO	Flexural strength of 3 wt.% ZnO-filled GFRP composite was substantially improved up to 62.12% compared to unfilled composite. Hardness impact strength also improved by increasing ZnO loading.

and fracture strength of neat epoxy and epoxy with waste containing B [73]. Earlier work [74] also shows that both the critical strain energy release approach and critical stress intensity factor approach can successfully explain the fracture of epoxy polymers in spite of the slight inelastic deformation present at the crack tip. Recent work done on epoxy with increasing density from 180 to 500 kg.m^{-3} [75] shows that the stress intensity factor (K_{Ic}) increases with an increase in density, from 0.1 to 0.79 MPa$\sqrt{m}$.

From the viewpoint of sustainable development, the use of natural fibre-reinforced polymer composites filled with inorganic nanoparticles is claimed to be more beneficial (Table 6.4) [76]. This approach is successful because natural fibres have low weight, low cost, easy availability, high flexibility during processing, excellent biodegradability, minimal health hazards, and better mechanical properties.

Polymeric materials and their composites have a wide range of applications. It happens due to their extremely favourable properties like low cost, high chemical resistance, tuneable thermal and electrical insulation, and light weight. But they possess serious limitations of fatigue resistance, low fracture toughness, microcracks, and other structural defects. If the defects like microcracks appear in the polymer composites they degenerate the reinforcement fillers and decrease the service life of the structure formed from the composite. This puts the safety of the product and people at stake [77, 78]. To tackle this problem, self-healing polymers are reported to be developed. Generally, the addition of a reinforcing material to a polymer matrix improves the mechanical properties as well as the service life of the composite material. Thus, the short carbon fibre (SCF) reinforced nylon (i.e., nylon SCF) is reported [80] to have an average tensile strength of ~53.8 MPa, compressive strength of ~229.8 MPa, tensile modulus of ~1.97 GPa, tensile toughness of ~5.15 MJm^{-3}, and compressive toughness of ~68.11 MJ m^{-3}.

6.6 FUTURE RESEARCH NEEDS

There are practically innumerable emerging areas in the research and development of polymers and polymer nanocomposites. So, what follows is just a brief overview mainly in line with the theme of the chapter. There is a lot of focus on the future development of self-healing polymers so that the worry about their deformation and subsequent fracture can be significantly minimized [79].

A completely new approach is emerging to enhance the mechanical characteristics of biopolymers, e.g., PLA, by introducing PVA and PET fibres with a commendable degree of success [81]. However, the most recent focus is on understanding as well as enhancing the mechanical characteristics of 3D polymers as the new advent of additive manufacturing (AM) has opened a completely new vista for multifunctional applications of polymers and a variety of PMCs [82] as well as their fracture resistance behaviour [31].

The challenges are very complex and the understandings to date are very basic, but the challenges are worth taking and pursuing in terms of future application prospects and commercial exploitation of AM technology. For instance, there is a significant effort dedicated to understanding the role of ballistic impact in affecting the mechanical integrity of additively manufactured polymers and composites [80].

It is interesting to note that carboxyl-terminated butadiene-acrylonitrile (CTBN) rubber is an important material for enhancing the failure resistance of bulk epoxy polymers and foams. Thus, the 12 wt.% CTBN addition enhances the failure energy of syntactic epoxy foam from about 200 to 300 J.m^{-2} and of bulk epoxy polymer from about 100 to 1100 J.m^{-2} [83]. Considering the importance, this area will be worth pursuing in the future.

The importance of AM is also amply established further in a study [84] that focuses on fatigue and fracture of two additively manufactured polymers. The first one is polyethylene terephthalate glycol (PET-G). The second one is acrylonitrile butadiene styrene (ABS). The low cycle fatigue tests, 36,050 cycles at 0.40% strain, illustrate that the low cycle fatigue resistance of the PET-G samples is better than that of ABS samples [84]. However, such studies are very rare to find and this fact indicates this to be an emerging area of extraordinary importance in the future.

Another emerging area is fracture of foams [75]. When epoxy foams with densities ranging from 0.18 to 0.50 g.cc^{-1} are tested in compression as well as tension, and in single-edge notched beam (SENB) experiments, it is noted that K_{Ic} increases from about 0.1 to 0.8 MPa.m$^{0.5}$. Further, there is reported to be a significant transition in fracture behaviour at a critical density between 0.23 and 0.25 g.cc^{-1}. This is an important area where further optimization may need to be done [75].

Governed by the need for the development of complex parts for multifunctional applications, one major area of emerging importance is the application of AM techniques to join different polymers and to understand how their structural and mechanical behaviours are interconnected to each other [85]. This is an area which will be very much worthy of pursuit for future research. Similarly important will be to the need to understand the interrelation between the degree of crystallinity and the fracture resistance (K_{Ic}) behaviours of the interface between a metal (e.g., Ti) and polymer (e.g., PEKK) [86] as such joints have huge practical application prospects in aerospace and spacecraft industries.

From a sustainable development point of view, it is important to pursue how the conversion of waste plant-based fibres into advanced plant-based fibres can be exploited in the development of advanced PMCs for load-bearing applications [87]. Such efforts are very rare indeed [76, 87] and hence it constitutes a major area of study for future research and development.

In addition, a new concept, graded interface enhanced phase field modelling [88], is now also emerging as a vibrant area of study to predict the complex 3D fracture behaviour of polymeric nanocomposites. Such efforts are definitely to be enhanced in the future to develop a better understanding of crack growth and fracture mechanics in polymeric nanocomposites [88].

A comprehensive understanding of failure at the interface of polymer coating and substrate is very much needed to examine their durability. The same is provided by a microscale digital image correlation (DIC) technique applied to the pure epoxy coating as well as to two types of nanoparticle-reinforced epoxy coating for the same given substrate [89]. The DIC technique is able to measure crack tip opening displacement (CTOD) with an accuracy of about 80 nm and hence it is suggested to be exploitable for evaluation of both mode I and mode II stress intensity factors at the interface. This is proposed to reflect the strength of the adhesion working at the interface [89]. This is poised to be a major area of future research.

PMMA and GPPS are important transparent materials for shatter-resistant applications [90]. Thus, further emphasizing the importance of mixed-mode I/II fracture, the corresponding stress intensity factors are evaluated by the SCB and BD techniques in recent work. The same is also obtained by theoretical analysis [90]. This recent work identifies both mixed-mode stress intensity factor evaluation and transparent thermoplastic polymers (e.g., PMMA and GPPS) as an area of huge importance.

It is therefore not surprising to note that detailed reviews are presented on factors affecting the fracture toughness of various additively manufactured polymers and PMCs [91]. Thus, the current state-of-the-art knowledge and future research directions are provided in recent work on the development and challenges in fracture characterization of polymers and PMCs by various established standardized techniques as well as some new standardization techniques proposed to be fruitful [92].

6.7 SUMMARY AND CONCLUSION

This chapter presents the basics of polymers and their mechanical properties with a particular emphasis on their deformation and fracture. Various important mechanical properties and their measurement techniques are discussed in simple terms, although it is appreciated that the deformation and fracture of polymers are in reality far more complex. Further, a brief overview of the literature scenario is presented. Finally, the emerging areas of importance for future research and development in the deformation and fracture of polymers are briefly discussed.

ACKNOWLEDGEMENTS

The authors acknowledge the kind support and encouragement received from various directors of the Council of Scientific and Industrial Research (CSIR)–Central Glass and Ceramic Research Institute (CGCRI), Kolkata, India, during the tenures of whom parts of the research work were carried out. They also gratefully acknowledge the various infrastructural support received from various divisions of CSIR-CGCRI, Kolkata, India. The financial support received from Indian sponsoring agencies, including CSIR, UGC, DST(SERB), DAE-BRNS, and IPR, of the Government of India (GOI) in terms of various project sponsorships and fellowships is gratefully acknowledged by the authors. Finally, the author AKM acknowledges the kind and continued support received from the authorities of Sharda University, Greater Noida, Uttar Pradesh, India.

REFERENCES

1. Herbert Musarurwa, Nikita Tawanda Tavengwa, Recyclable polysaccharide/stimuli-responsive polymer composites and their applications in water remediation, *Carbohydrate Polymer*, Volume 298, 15 December 2022, Page 120083. https://doi.org/10.1016/j.carbpol.2022.120083.
2. Huimin Zhang, Zongcheng Miao, Wenbo Shen, Development of polymer-dispersed liquid crystals: From mode innovation to applications, *Composites Part A: Applied Science and Manufacturing*, 163, December 2022, Page 107234. https://doi.org/10.1016/j.compositesa.2022.107234.
3. Jie Zhang, Chao Wei, Jingyu Ran, Yang Li, Jiajun Chen Properties of polymer composite with large dosage of phosphogypsum and its application in pipeline, *Polymer Testing*, 116, December 2022, Page 107742. https://doi.org/10.1016/j.polymertesting.2022.107742.
4. Adnan Murad Bhayo, Yang Yang, Xiangming He, Polymer brushes: Synthesis, characterization, properties and applications, *Progress in Materials Science*, Volume 130, October 2022, Page 101000. https://doi.org/10.1016/j.pmatsci.2022.101000.

5. Kirstie R. Ryan, Michael P. Down, Nicholas J. Hurst, Edmund M. Keefe, Craig E. Banks, Additive manufacturing (3D printing) of electrically conductive polymers and polymer nanocomposites and their applications, *eScience*, Volume 2, Issue 4, July 2022, Pages 365–381. https://doi.org/10.1016/j.esci.2022.07.003.
6. Keyu Qu, Zhiang Yuan, Anyan Wang, Zhaohui Song, Xuyang Gong, Yi Zhao, Qiyu Mu, Qinghong Zhan, Wenlong Xu, Linlin Wang, Structures, properties, and applications of zwitterionic polymers, *ChemPhysMater*, Volume 1, Issue 4, October 2022, Pages 294–309. https://doi.org/10.1016/j.chphma.2022.04.003.
7. Xiaohong Liu, Ending Zhang, Jiaming Liu, Jingjing Qin, Mengqin Wu, Chaolong Yang, Liyan Liang, Self-healing, reprocessable, degradable, thermadapt shape memory multifunctional polymers based on dynamic imine bonds and their application in nondestructively recyclable carbon fiber composites, *Chemical Engineering Journal*, Volume 454, Part 1, 15 February 2023, Page 139992. https://doi.org/10.1016/j.cej.2022.139992.
8. Aayush Bhat, J. Naveen, M. Jawaid, M. N. F. Norrrahim, Ahmad Rashedid, A. Khane, Advancement in fiber reinforced polymer, metal alloys and multi-layered armour systems for ballistic applications – A review, *Journal of Materials Research and Technology*, Volume 15, November–December 2021, Pages 1300–1317. https://doi.org/10.1016/j.jmrt.2021.08.150.
9. Dakshitha Weerasinghe, M. R. Bambach, Damith Mohotti, Hongxu Wang, Sheng Jiang, Paul J. Hazell, Development of a coated fabric armour system of aramid fibre and rubber, *Thin-Walled Structures*, Volume 179, October 2022, Page 109679. https://doi.org/10.1016/j.tws.2022.109679.
10. Wenpeng Zhao, Yuan Li, Jian Hu, Xianqi Feng, Hao Zhang, Jun Xu, Shouke Yan, Mechanically robust, instant self-healing polymers towards elastic entropy driven artificial muscles, *Chemical Engineering Journal*, Volume 454, Part 1, 15 February 2023, Page 140100. https://doi.org/10.1016/j.cej.2022.140100.
11. D.W. Van Krevelen, K. te Nijenhuis, *Properties of Polymers*, Elsevier, 2008, Pages 1–1031.
12. A William, R Ashcroft, *Industrial Polymer Applications: Essential Chemistry and Technology*, Royal Society of Chemistry, 2019, Pages 1–221.
13. A Muhammad, S Zafar, Saad Liaqat, Saria Najeeb, Zohaib Khurshid, Motasem Alrahabi, Sana Zohaib, *Polymer Science: Research Advances, Practical Applications and Educational Aspects*, ISBN 13: 9788494213489, Publisher Formatex Research Center S.L, 2016.
14. Wenjing Hou, Yaoming Xiao, Gaoyi Han, Jeng-Yu Lin, The applications of polymers in solar cells: A review, *Polymers*, Volume 11, Issue 1, 15 January 2019, Page 143. https://doi.org/10.3390/polym11010143.
15. S Ramakrishna, J Mayer, E Wintermantel, Kam W Leong, Biomedical applications of polymer-composite materials: A review, *Computer Science and Technology*, Volume 61, Issue 9, July 2001, Pages 1189–1224. https://doi.org/10.1016/S0266-3538(00)00241-4.
16. Sindhu Doppalapudi, Anjali Jain, Wahid Khan, Abraham J. Domb, Biodegrable polymers – an overview, *Polymers for Advanced Technology*, Volume 25, Issue 5, April 2014, Pages 427–435. https://doi.org/10.1002/pat.3305.
17. Debasish Sahoo, Sarmila Sahoo, Priyanka Mohanty, S. Sasmal, P.L. Nayak, Chitosan: A new versatile biopolymer for various applications, *Designed Monomers and Polymers*, Volume 12, April 2012, Pages 377–404. https://doi.org/10.1163/138577209X12486896623418.
18. G. H. Michler, F. J. Balta-Calleja, Eds, *Mechanical Properties of Polymers Based on Nanostructure and Morphology*, ISBN: 9780429118609, Pages 1–784. https://doi.org/10.1201/9781420027136.
19. J. T. Seitz, The estimation of mechanical properties of polymers from molecular structure, *Journal of Applied Polymer Science*, Volume 49, Issue 8, August 1993, Pages 131-1351. https://doi.org/10.1002/app.1993.070490802.
20. J. R. Martin, Julian F. Johnson, A. R. Cooper, Mechanical properties of polymers: The influence of molecular weight and molecular weight distribution, *Journal of Macromolecular Science*, Volume 8, Issue 1, December 2006, Pages 57–199. https://doi.org/10.1080/15321797208068169.
21. R. F. Boyer, Dependence of mechanical properties on molecular motion in polymers, *Polymer Engineering & Science*, Volume 8, Issue 3, July 1968, Pages 161–185. https://doi.org/10.1002/pen.760080302.
22. John Sweeney, I. M. Ward, *Mechanical Properties of Solid Polymers*, John Wiley & Sons, 2012, ISBN: 1119967112, 9781119967118.
23. Francesca Luzi, Luigi Torre, Debora Puglia, Chapter 10- *Polymer composites and nanocomposites containing lignin: Structure and applications, Micro and Nanolignin in Aqueous Dispersion and Polymers, Interactions, Properties, and Application*, 2022, Pages 293–324. https://doi.org/10.1016/B978-0-12-823702-1.00007-4.

24. Rohan A. Hule, Darrin J. Pochan, Polymer nanocomposites for biomedical applications, *MRS Bulletin*, Volume 32, January 2011, Pages 354–358. https://doi.org/10.1557/MRS2007.235.
25. P Venkateshwar Reddy, Saikumar Reddy, Lakshmana Rao Jeeru, D. Mohan Krishnudu, An overview on natural fiber reinforced composites for structural and non-structural applications, *Materials Today: Proceedings*, Volume 45, Part 7, 2021, Pages 6210–6215. https://doi.org/10.1016/j.matpr.2020.10.523.
26. Wei Zhao, Liwu Liu, Fenghua Zhang, Jinsong Leng, Yanju Liu, Shape memory polymers and their composites in biomedical applications, *Material science and Engineering: C*, Volume 97, April 2019, Pages 864–883. https://doi.org/10.1016/j.msec.2018.12.054.
27. Anwesha Mukherjee, Sangita Panda, Peerzada Gh. Jeelani, Abdel-Tawab Mossa, Ramalingam Chidambaram, Chapter 20 - Biodegradable polymers/silica nanocomposites: Applications in food packaging, *Nanotechnology Applications for Food Safety and Quality Monitoring*, 2023, Pages 395–414. https://doi.org/10.1016/B978-0-323-85791-8.00001-X.
28. Shaily Shabnam Khan, Anujit Ghosal, Shahnawaz Ahmad Bhat, Fahmina Zafar, Mudsser Azam, Manawwer Alam, M. Shahid, Qazi Mohd Rizwanul Haq, Nahid Nishat, Phenolic lipid derived coordination polymer nanocomposites: Synthesis, characterization and surface protective coating applications, *Applied Surface Science Advances*, Volume 11, October 2022, Page 100290. https://doi.org/10.1016/j.apsadv.2022.100290.
29. Shubham Sharma, P Sudhakara, Abdoulhdi A Borhana Omran, Jujhar Singh, R A Ilyas, Recent trends and developments in conducting polymer nanocomposites for multifunctional applications, *Polymers (Basel)*, Volume 13, Issue 17, 2021. https://doi.org/10.3390/polym13172898.
30. Pawan Kaur, Rita Choudhary, Anamika Pal, Chanchal Mony, Alok Adholeya, Polymer - metal nanocomplexes based delivery system: A boon for agriculture revolution, *Current Topics in Medicinal Chemistry*, Volume 20, Issue 11, 2020, Pages 1009–1028. https://doi.org/10.2174/1568026620666200330160810.
31. S Ganeshkumar, SD Kumar, U Magarajan, S Rajkumar, B Arulmurugan, S Sharma, C Li, RA Ilyas, MF Badran, Investigation of tensile properties of different infill pattern structures of 3D-printed PLA polymers: Analysis and validation using finite element analysis in ANSYS, *Materials*, Volume 15, Part 15, 25 July 2022, Page 5142. https://doi.org/10.3390/ma15155142.
32. W Ashraf, MR Ishak, MY Zuhri, N Yidris, AM Ya'acob. Experimental investigation on the mechanical properties of a sandwich structure made of flax/glass hybrid composite facesheet and honeycomb core, *International Journal of Polymer Science*, 11 March 2021. https://doi.org/10.1155/2021/8855952.
33. ND Duc ND, DK Minh. Experimental study on mechanical properties for a three-phase polymer composite reinforced by glass fibers and titanium oxide particles, *Vietnam Journal of Mechanics*, Volume 105, Part 10, 10 June 2011. https://doi.org/10.15625/0866-7136/33/2/42.
34. G. A. Vanin, Micro - Mechanics of composite materials, *Nauka Dumka*, Kiev, 1985.
35. G. A. Vanin, Nguyen Dinh Duc, The theory of spherofiberous composite.1: The input relations, hypothesis and models, *Mechanics of Composite Materials*, Volume 32, Part 3, 1996, Pages 291–305.
36. S Rostam, AK Ali, FH Abdal Muhammad, Experimental investigation of mechanical properties of PVC polymer under different heating and cooling conditions, *Journal of Engineering*, 1 January 2016. https://doi.org/10.1155/2016/3791417.
37. SN Danilova, AA Dyakonov, AP Vasilev, AA Okhlopkova, Examination of the influence of the mechanical co-activation time on the properties of UHMWPE filled with basalt fibers, *Procedia Structural Integrity*, Volume 40, 1 January 2022, Pages 118–123. https://doi.org/10.1016/j.prostr.2022.04.015.
38. M Ahmad, MU Wahit, MR Abdul Kadir, KZ Mohd Dahlan, Mechanical, rheological, and bioactivity properties of ultra high-molecular-weight polyethylene bioactive composites containing polyethylene glycol and hydroxyapatite, *The Scientific World Journal*, 1 January 2012. https://doi.org/10.1100/2012/474851.
39. DL Macuvele, G Colla, K Cesca, LF Ribeiro, CE da Costa, J Nones, ER Breitenbach, LM Porto, C Soares, MA Fiori, HG Riella, UHMWPE/HA biocomposite compatibilized by organophilic montmorillonite: An evaluation of the mechanical-tribological properties and its hemocompatibility and performance in simulated blood fluid, *Materials Science and Engineering: C*, Volume 100, 1 July 2019, Pages 411–23. https://doi.org/10.1016/j.msec.2019.02.102.
40. SM Taromsari, M Salari, R Bagheri, MA Sani, Optimizing tribological, tensile & in-vitro biofunctional properties of UHMWPE based nanocomposites with simultaneous incorporation of graphene nanoplatelets (GNP) & hydroxyapatite (HAp) via a facile approach for biomedical applications. *Composites Part B: Engineering*, Volume 175, 15 October 2019, Page 107181. https://doi.org/10.1016/j.compositesb.2019.107181.

41. D Duraccio, V Strongone, G Malucelli, F Auriemma, C De Rosa, FD Mussano, T Genova, MG Faga, The role of alumina-zirconia loading on the mechanical and biological properties of UHMWPE for biomedical applications, *Composites Part B: Engineering*, Volume 164, 1 May 2019, Pages 800–808. https://doi.org/10.1016/j.compositesb.2019.01.097.
42. A Salama, BM Kamel, TA Osman, RM Rashad. Investigation of mechanical properties of UHMWPE composites reinforced with HAP+ TiO2 fabricated by solvent dispersing technique, *Journal of Materials Research and Technology*, Volume 21, 11 November 2022, Pages 4330–4343.
43. AV Ushakov, IV Karpov, LY Fedorov, AA Lepeshev, AA Shaikhadinov, VG Demin VG. Nanocomposite material based on ultra-high-molecular-weight polyethylene and titanium dioxide electroarc nanopowder, *Theoretical Foundations of Chemical Engineering*, Volume 49, Part 5, September 2015. https://doi.org/10.1134/S0040579515050176.
44. GC Efe, C Bindal, AH Ucisik. Characterization of UHWPE-TiO2 composites produced by gelation/crystallization method, *Acta Physica Polonica A*, Volume 132, September 2017, Pages 767–769. https://doi.org/10.12693/APhysPolA.132.767.
45. JF Vega, S Rastogi, GWM Peters, HEH Meijer. Rheology and reputation of linear polymers, Ultrahigh molecular weight chain dynamics in the melt, *Journal of Rheology*, Volume 48, Part 3, May 2004, Pages 663–678. http://doi.org/10.1122/1.1718367.
46. L Fang, Y Leng, P Gao. Processing and mechanical properties of HA/UHMWPE nanocomposites, *Biomaterials*, Volume 20, Part 20, 1 July 2006, Pages 3701–3707. http://doi.org/10.1016/j.biomaterials.2006.02.023.
47. NP da Silva Chagas, V de Oliveira Aguiar, F da Costa Garcia Filho, AB da Silva Figueiredo, SN Monteiro, NR Huaman, MD Marques, Ballistic performance of boron carbide nanoparticles reinforced ultra-high molecular weight polyethylene (UHMWPE), *Journal of Materials Research and Technology*, Volume 17, 1 March 2022, Pages 1799–1811. https://doi.org/10.1016/j.jmrt.2022.01.104.
48. R Wang, T Meng, B Zhang, C Chen, D Li. Preparation and characterization of activated carbon/ultra-high molecular weight polyethylene composites, *Polymer Composites*, Volume 42, June 2021, Pages 2728–2736. https://doi.org/10.1002/pc.26008.
49. RR Dias, A Lavoratti, D Piazza, CR da Silva, AJ Zattera, RM Lago, PS de Oliveira Patricio, IM Pereira. Effect of molecular structures on static and dynamic compression properties of clay and amphiphilic clay/carbon nanofibers used as fillers in UHMWPE/composites for high-energy-impact loading, *Journal of Applied Polymer Science*, Volume 136, Part 8, 20 February 2019, Page 47094. https://doi.org/10.1002/app.47094.
50. Y Li, H He, B Huang, L Zhou, P Yu, Z Lv, In situ fabrication of cellulose nanocrystal-silica hybrids and its application in UHMWPE: Rheological, thermal, and wear resistance properties, *Polymer Composites*, Volume 39, Part S3, June 2018, Pages 1701–1713. https://doi.org/10.1002/pc.24690.
51. AM Ventura, LM Kneissl, S Nunes, N Emami. Recycled carbon fibers as an alternative reinforcement in UHMWPE composite. Circular economy within polymer tribology, *Sustainable Materials and Technologies*, Volume 34, 1 December 2022, Page e00510. https://doi.org/10.1016/j.rinp.2017.02.024.
52. AV Maksimkin, KS Mostovaya, FS Senatov, DI Chukov, SG Nematulloev, LK Olifirov, The influence of fluorinated MWCNT distribution quality on the mechanical properties of the bulk oriented UHMWPE-based composites, *Results in Physics*, Volume 7, 1 January 2017, Pages 1044–1049. https://doi.org/10.3390/polym9110629.
53. SR Allen, Tensile recoil measurement of compressive strength for polymeric high performance fibres, *Journal of Materials Science*, Volume 22, Part 3, March 1987, Pages 853–859. https://doi.org/10.1007/bf01103520.
54. AB Croll, T Twohig, T Elder, The compressive strength of crumpled matter, *Nature Communications*, Volume 10 Part 1, 2019, Pages 1–8. https://doi.org/10.1038/s41467-019-09546-7.
55. SY Oderji, B Chen, MR Ahmad, SF Shah, Fresh and hardened properties of one-part fly ash-based geopolymer binders cured at room temperature: Effect of slag and alkali activators, *Journal of Cleaner Production*, Volume 225, 10 July 2019, Pages 1–10. https://doi.org/10.1016/j.jclepro.2019.03.290.
56. R Xiao, Y Ma, X Jiang, M Zhang, Y Zhang, Y Wang, B Huang, Q He. Strength, microstructure, efflorescence behavior and environmental impacts of waste glass geopolymers cured at ambient temperature, *Journal of Cleaner Production*, Volume 252, 10 April 2020, Page 119610. https://doi.org/10.1016/j.jclepro.2019.119610.
57. R Bajpai, K Choudhary, A Srivastava, KS Sangwan, M Singh, Environmental impact assessment of fly ash and silica fume based geopolymer concrete, *Journal of Cleaner Production*, Volume 254, 1 May 2020, Page 120147. https://doi.org/10.1016/j.jclepro.2020.120147.

58. L Wang, G Li, X Li, F Guo, S Tang, X Lu, A Hanif, Influence of reactivity and dosage of MgO expansive agent on shrinkage and crack resistance of face slab concrete, *Cement and Concrete Composites*, Volume 26, 1 February 2022, Page 104333. https://doi.org/10.1016/j.cemconcomp.2021.104333.
59. L Wang, Z Yu, B Liu, F Zhao, S Tang, M Jin, Effects of fly ash dosage on shrinkage, crack resistance and fractal characteristics of face slab concrete, *Fractal and Fractional*, Volume 6, Part 6, 16 June 2022, Page 335. https://doi.org/10.3390/fractalfract6060335.
60. SH Chu, H Ye, L Huang, LG Li, Carbon fiber reinforced geopolymer (FRG) mix design based on liquid film thickness, *Construction and Building Materials*, Volume 269, 1 February 2021, Page 121278. https://doi.org/10.1016/j.conbuildmat.2020.121278.
61. S Ma, H Yang, S Zhao, P He, Z Zhang, X Duan, Z Yang, D Jia, Y Zhou. 3D-printing of architectured short carbon fiber-geopolymer composite. *Composites Part B: Engineering*, Volume 226, 1 December 2021, Page 109348. https://doi.org/10.1016/j.compositesb.2021.109348.
62. G Humur, A Çevik. Effects of hybrid fibers and nanosilica on mechanical and durability properties of lightweight engineered geopolymer composites subjected to cyclic loading and heating–cooling cycles, *Construction and Building Materials*, Volume 326, 4 April 2022, Page 126846. https://doi.org/10.1016/j.conbuildmat.2022.126846.
63. P Zhang, S Wei, Y Zheng, F Wang, S Hu. Effect of single and synergistic reinforcement of PVA fiber and nano-SiO2 on workability and compressive strength of geopolymer composites, *Polymers*, Volume 14, Part 18, 8 September 2022, Page 3765. https://doi.org/10.3390/polym14183765.
64. S Kim, Y Lee, M Lee, S An, SJ Cho. Quantitative visualization of the nanomechanical Young's modulus of soft materials by atomic force microscopy, *Nanomaterials*, Volume 11, Part 16, 17 June 2021, Page 1593. https://doi.org/10.3390/nano11061593.
65. QM Li, EA Flores-Johnson. Hard projectile penetration and trajectory stability, *International Journal of Impact Engineering*, Volume 38, Part 10, 1 October 2011, Pages 815–823. http://doi.org/10.1016/j.ijimpeng.2011.05.005.
66. SS Esfahlani, Ballistic performance of Polycarbonate and Polymethyl methacrylate under normal and inclined dynamic impacts, *Heliyon*, Volume 7, Part 4, 1 April 2021. http://doi.org/10.1016/j.heliyon.2021.e06856.
67. GA Adam, A Cross, RN Haward. The effect of thermal pretreatment on the mechanical properties of polycarbonate, *Journal of Materials Science*, Volume 10, Part 9, September 1975, Pages 1582–1590. https://doi.org/10.1007/bf01031859.
68. LY Wu, E Chwa, Z Chen, XT Zeng. A study towards improving mechanical properties of sol–gel coatings for polycarbonate, *Thin Solid Films*, Volume 516, Part 6, 30 January 2008, Pages 1056–62. https://doi.org/10.1016/j.tsf.2007.06.149.
69. E Parsons, MC Boyce, DM Parks, An experimental investigation of the large-strain tensile behavior of neat and rubber-toughened polycarbonate, *Polymer*, Volume 45, Part 8, 1 April 2004, Pages 2665–2684. https://doi.org/10.1016/j.polymer.2004.01.068.
70. M Kutz, editor. Mechanical engineers' handbook, Volume 1: *Materials and Engineering Mechanics*, John Wiley & Sons, 2 March 2015. https://doi.org/10.1002/9781118985960.meh110.
71. S Fakhreddini-Najafabadi, M Torabi, F Taheri-Behrooz, An investigation on the effects of synthesis on the mechanical properties of nanoclay/epoxy, *Journal of Materials Research and Technology*, Volume 15, 1 November 2021, Pages 5375–5395. https://doi.org/10.1016/j.jmrt.2021.10.129.
72. KC Shekar, B Singaravel, SD Prasad, N Venkateshwarlu, B Srikanth, Mode-I fracture toughness of glass/carbon fiber reinforced epoxy matrix polymer composite, *Materials Today: Proceedings*, Volume 41, 1 January 2021, Pages 833–837. https://doi.org/10.1016/j.matpr.2020.09.160.
73. T Uygunoglu, I Gunes, W Brostow, Physical and mechanical properties of polymer composites with high content of wastes including boron, *Materials Research*, Volume 18, 4 December 2015, Pages 1188–1196. https://doi.org/10.1590/1516-1439.009815.
74. WJ Cantwell, HH Kausch, Fracture behaviour of epoxy resins, In *Chemistry and Technology of Epoxy Resins*, Springer, Pages 144–174, 1993. https://doi.org/10.1007/978-94-011-2932-9_5.
75. G Irven, D Carolan, A Fergusson, JP Dear, Fracture performance of epoxy foam: Low density to bulk polymer, *Polymer*, Volume 261, 18 November 2022, Page 125420. https://doi.org/10.1016/j.polymer.2022.125420.
76. T Mishra, P Mandal, AK Rout, D Sahoo. A state-of-the-art review on potential applications of natural fiber-reinforced polymer composite filled with inorganic nanoparticle, *Composites Part C: Open Access*, 2 July 2022, Page 100298. https://doi.org/10.1016/j.jcomc.2022.100298.
77. KL Reifsnider, K Schulte, JC Duke, Long-term fatigue behavior of composite materials. *Long-term Behavior of Composites*, Volume 813, January 1983, Pages 136–159. https://doi.org/10.1520/STP31820S.

78. K Mphahlele, SS Roy, A Kolesnikov, Self healing polymeric composite material design, failure design, and future outlook: A review, *Polymers*, Volume 9, Page 537. https://doi.org/10.3390/polym9100535.
79. CR Ratwani, A Abdelkader. Self-healing by Diels-Alder cycloaddition in advanced functional polymers: A review, *Progress in Materials Science*, Volume 131, Page 101001, 2023. https://doi.org/10.1016/j.pmatsci.2022.101001.
80. MN Islam, KP Baxevanakis, VV Silberschmidt, Dynamic fracture behaviour of additively manufactured polymers and composites under ballistic impact, *Procedia Structural Integrity*, Volume 37, 1 January 2022, Pages 217–224. https://doi.org/10.1016/j.prostr.2022.02.050.
81. Milán Ferdinánd, Róbert Várdai, János Móczó, Béla Pukánszky, Poly(lactic acid) reinforced with synthetic polymer fibers: Interactions, structure and properties, *Composites Part A: Applied Science and Manufacturing*, Volume 164, 2023, Page 107318. https://doi.org/10.1016/j.compositesa.2022.107318.
82. Gonghe Zhang, Qinglin Wang, Yinxu Ni, Pei Liu, Fenghua Liu, Dominique Leguillon, Luoyu Roy Xu. A systematic investigation on the minimum tensile strengths and size effects of 3D printing polymers, *Polymer Testing*, Volume 117, 2023, Page 107845. https://doi.org/10.1016/j.polymertesting.2022.107845.
83. Sammy He, Declan Carolan, Alexander Fergusson, Ambrose C. Taylor, Investigating the transfer of toughness from rubber modified bulk epoxy polymers to syntactic foams, Composites Part B 245, 2022, Page 110209. https://doi.org/10.1016/j.compositesb.2022.110209.
84. Janusz Kluczynski, Ireneuszszachog łuchowicz, Janusz Torzewski, Lucjan Sniezek, Krzysztof Grzelak, Grzegorz Budzik, Łukasz Przeszłowski, Marcin Małek, Jakub Łuszczek. Fatigue and fracture of additively manufactured polyethylene terephthalate glycol and acrylonitrile butadiene styrene polymers, *International Journal of Fatigue*, Volume 165, 2022, Page 107212. https://doi.org/10.1016/j.ijfatigue.2022.107212.
85. A. García-Collado, J.M. Blanco, Munish Kumar Gupta, R. Dorado-Vicente, Advances in polymers based multi-material additive-manufacturing techniques: State-of-art review on properties and applications, *Additive Manufacturing*, Volume 50, 2022, Page 102577. https://orcid.org/0000-0002-0777-1559.
86. V.M. Marinosci, N.G.J. Helthuis, L. Chu, W. J. B. Grouve, M.B. de Rooij, S. Wijskamp, R. Akkerman. The role of process induced polymer morphology on the fracture toughness of titanium–PEKK interfaces, *Engineering Fracture Mechanics*, Volume 268, 2022, Page 108475. https://doi.org/10.1016/j.engfracmech.2022.108475.
87. Jyotishkumar Parameswaranpillai, Jineesh Ayippadath Gopi, Sabarish Radoor, CD Midhun Dominic, Senthilkumar Krishnasamy, Kalim Deshmukh, Nishar Hameed, Nisa V. Salim, Natalia Sienkiewicz. Turning waste plant fibers into advanced plant fiber reinforced polymer composites: A comprehensive review, *Composites Part C: Open Access*, 2022. https://doi.org/10.1016/j.jcomc.2022.100333.
88. Paras Kumar, Paul Steinmann, Julia Mergheim, A graded interphase enhanced phase-field approach for modeling fracture in polymer composites, *Forces in Mechanics*, Volume 9, 2022, Page 100135. https://doi.org/10.1016/j.finmec.2022.100135.
89. Wenjie Qian, Huiying Zhang, Jianguo Zhu, Jian Li, Jian Zhang, Muyu Zhang. Determination of fracture toughness of polymer coating using micro-scale digital image correlation technique, *Polymer Testing*, Volume 93, 2021, Page 106896. https://doi.org/10.1016/j.polymertesting.2020.106896.
90. A. R. Torabia, M. Jabbaria, J. Akbardoost. Mixed mode notch fracture toughness assessment of quasi-brittle polymeric specimens at different scales, *Theoretical and Applied Fracture Mechanics*, Volume 109, 2020, Page 102682. https://doi.org/10.1016/j.tafmec.2020.102682.
91. S. Sharafi, M. H. Santare, J. Gerdes, S.G. Advani, A review of factors that influence the fracture toughness of extrusion-based additively manufactured polymer and polymer composites, *Additive Manufacturing*, Volume 38, 2021, Page 101830. https://doi.org/10.1016/j.addma.2020.101830.
92. Andreas J. Brunnera, Laurent Warnet, Bamber R.K. Blackman. 35 years of standardization and research on fracture of polymers, polymer composites, and adhesives in ECIS TC4: Past achievements and future directions, *Procedia Structural Integrity*, Volume 33, 2021, Pages 443–455. https://creativecommons.org/licenses/by-nc-nd/4.0.

7 Deformation and Fracture of Composite Materials
Experimental Methods and Challenges

Payel Maiti, Manjima Bhattacharya, Aniruddha Samanta, and Anoop Kumar Mukhopadhyay

7.1 INTRODUCTION

A composite material comprises of a matrix phase and at least one or more reinforcing phase(s). These phases exist separately in the composite in a way that is distinctly different from a mixture or a solid solution. Further, the properties of the composite are different from those of the matrix and the reinforcing phases. The history of composites is almost as old as human civilization. The basic types of materials are metals (M), polymers (P) and ceramics (C). Therefore, the basic types of composites are metal matrix composites (MMCs), polymer matrix composites (PMCs) and ceramic matrix composites (CMCs). Thus, the particulate reinforced MMCs are known as (PRMMCs). Similarly, the fibre reinforced MMCs are known as (FRMMCs). Similarly, both particulate and fibre reinforced MMCs are known as hybrid reinforced MMCs (HRMMCs).

It is worth noting that the fibres could be unidirectional, i.e., parallel to the length of the composite. It could be perpendicular to the length of the sample. Thus the fibre orientation angles are 0°, 90° and 45° in general. Also, short fibres randomly distributed in the matrix are commonly used for reinforcement purposes. Similar to the fibres, whiskers are also utilized for reinforcing purposes. In an analogous manner, carbon nanotubes (CNTs), e.g., multi-walled CNTs (MWCNTs) or single-walled (SWCNTs) could be also used for reinforcement purposes in a matrix. Such a scenario also holds good for fibre reinforced PMCs and CMCs. Further, particulate reinforced PMCs are known as (PRPMCs). Similarly, the fibre reinforced PMCs are known as (FRPMCs). Both particulate and fibre reinforced PMCs are known as hybrid reinforced PMCs (HRPMCs).

Furthermore, a similar situation pertains in the case of CMCs. Thus, the particulate reinforced CMCs are known as (PRCMCs), fibre reinforced CMCs are known as (FRCMCs) and both particulate and fibre reinforced CMCs are known as hybrid reinforced CMCs (HRCMCs). *Apart from these conventional composites, there are also advanced structural composites which may be multilayer metal–metal micro-/nanolaminates, multilayer metal–ceramic micro-/nanolaminates and multilayer ceramic–ceramic micro-/nanolaminates.* Figures 7.1–7.4 present the schematic pictures of the aforementioned composite designs.

All such composites, i.e., MMCs [1–3], PMCs [4–6] and CMCs [7–9], provide a huge range of applications. For instance, Cu-based MMCs are developed for precision electronics and aerospace applications [1], Mg-based MMCs are developed for stent applications [2], and W-Ag-based MMCs are developed for electrical switching applications [3]. Various PMCs are being developed for antibacterial [4], hydrogen storage [5], and pipeline applications [6]. Similarly, the CMCs are being developed for high-temperature stealth applications [7], aerospace applications [8] and thermal conductive applications [9].

DOI: 10.1201/9781003359364-9

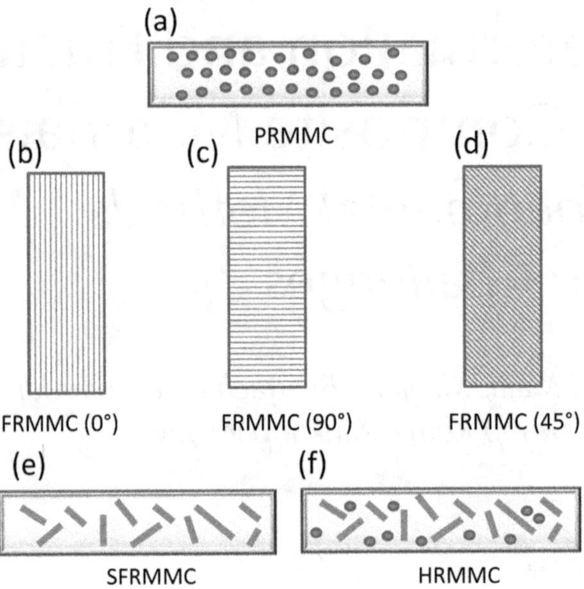

FIGURE 7.1 Various designs of MMCs. (a) Particulate reinforced MMC (PRMMC). Fibre reinforced MMC (FRMMC) with fibres oriented at (b) 0 degrees, (c) 90 degrees, (d) 45 degrees. (e) Short Fibre reinforced MMC (SRMMC). (f) Short fibre and particulate hybrid reinforced MMC (HRMMC).

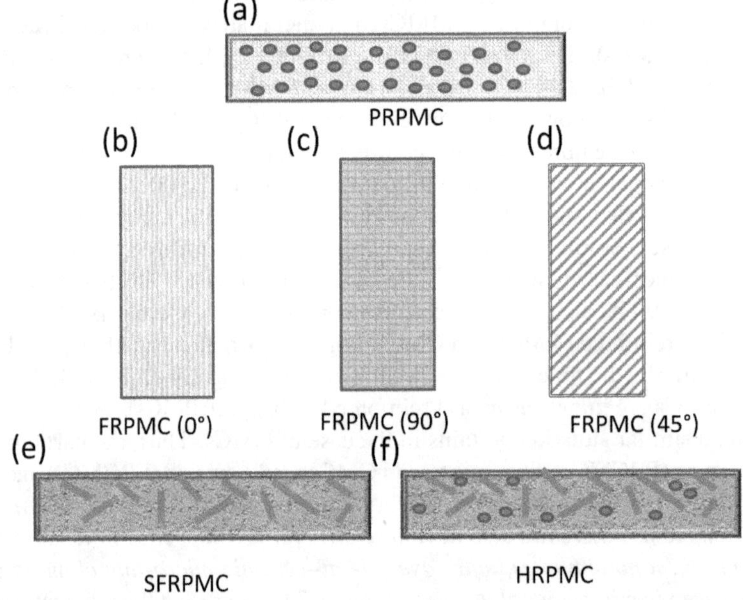

FIGURE 7.2 Various designs of PMCs. (a) Particulate reinforced PMC (PRPMC). Fibre reinforced PMC (FRPMC) with fibres oriented at (b) 0 degrees, (c) 90 degrees, (d) 45 degrees. (e) Short fibre reinforced PMC (SRPMC). (f) Short fibre and particulate hybrid reinforced PMC (HRPMC).

The roles of the particulates, fibres and their hybrid reinforcements in all these cases are to share the applied load and, hence, the applied stress so that the damage to the matrix phase can be minimized. Further, as far as fibre reinforcement is concerned, it could be short fibre, long or continuous. Furthermore, the short as well as the long fibres can be unidirectionally aligned. In addition, the

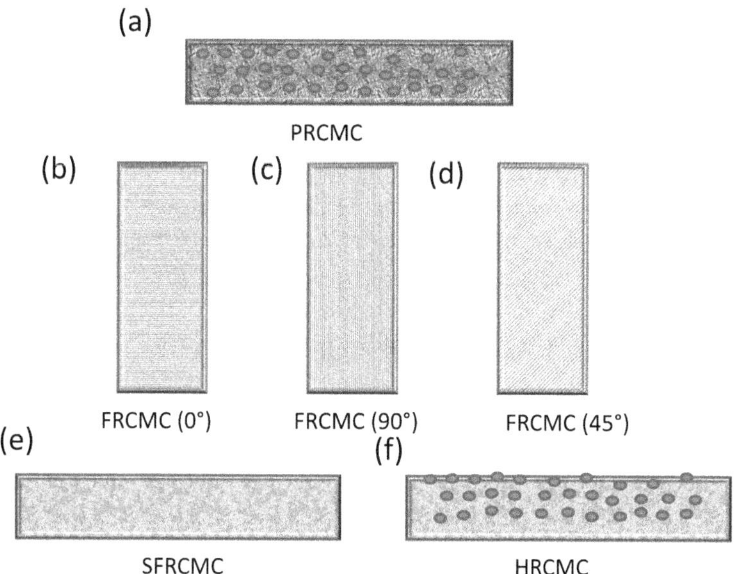

FIGURE 7.3 Various designs of CMCs. (a) Particulate reinforced CMC (PRCMC). Fibre reinforced CMC (FRCMC) with fibres oriented at (b) 0 degrees, (c) 90 degrees, (d) 45 degrees. (e) Short fibre reinforced CMC (SRCMC). (f) Short fibre and particulate hybrid reinforced CMC (HRCMC).

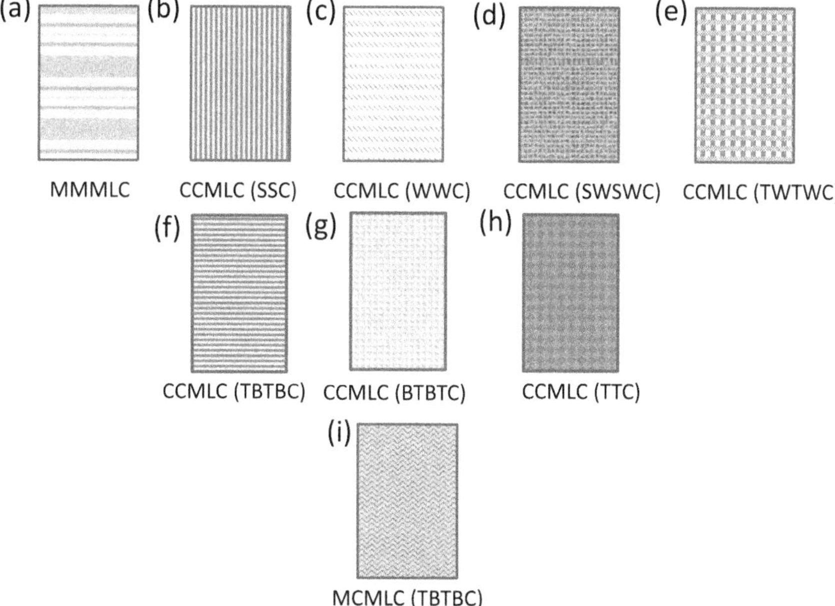

FIGURE 7.4 Various designs of multilayer laminated composites (MLCs). (a) Metal–metal MMC (MMMLC) and ceramic–ceramic MLC (CCMLC) with different combinations of the interface: (b) strong–strong (SS), (c) weak–weak (WW), (d) strong–weak–strong-–weak (SWSW), (e) tough–weak–tough–weak, (f) tough–brittle–tough–brittle (TBTB), (g) brittle–tough–brittle–tough (BTBT), (h) tough–tough (TT), and (i) tough–brittle–tough–brittle (TBTB) with zig-zag interface.

short fibres can be randomly distributed. The fibres could be glass fibres (GFs), carbon fibres (CFs) and metal fibres (MFs). Carbon nanotubes (MWCNTs, SWCNTs) are also used as reinforcement in MMC, PMC and CMCs, as mentioned earlier. Moreover, the most well-known PMCs are glass fibre reinforced polymers (GFRPs) and carbon fibre reinforced polymers (CFRPs). The aforementioned applications involve contact-induced deformation and fracture. Therefore, it is important to understand the techniques used to study the deformation and fracture in composites [10]. Hence, the objective of this chapter is to briefly address the theoretical aspects and experimental techniques involved in the evaluation of deformation and fracture of composites prior to identification of future challenges in this domain.

7.2 BASIC THEORETICAL ASPECTS OF COMPOSITES

7.2.1 Deformation

There can be two conditions of reinforcement, i.e., fibre orientation. These are called the isostrain condition and isostress condition. If the applied stress is parallel in direction to both the matrix (m) and reinforcing (r) phases, the strain in the matrix phases (ε_m) is equal to that of the reinforcing phase (ε_r) and they are also equal to the strain in the composite (ε_c). Let this be denoted as ε. This is called the isostrain condition. It is valid only under the simplifying assumption that there is no damage at the interface between the reinforcement and the matrix. Let the Young's moduli be E_m, E_r and E_c for the matrix, reinforcing and composite phases, in turn. If V_m and V_r are the volume fractions of the matrix phase and the reinforcing phase, respectively, then the stress (σ_c) in the composite is ($\sigma_c = V_m\sigma_m + V_r\sigma_r$). Since ($\sigma_m = E_m\varepsilon$), ($\sigma_r = E_r\varepsilon$) and ($\sigma_c = E_c\varepsilon$), it is evident that $\sigma_c = (E_c)\varepsilon = [(V_m\sigma_m) + (V_r\sigma_r)] = [(V_mE_m + V_rE_r)]\varepsilon$. Therefore, we have:

$$(E_c) = (V_mE_m + V_rE_r) \tag{7.1}$$

This is the isostrain modulus of a simple hypothetical composite. This gives the upper bound of the Young's modulus. This is very useful for particulate reinforced composites (PRMMCs, PRPMCs and PRCMCs). Further, assuming the reinforcing phase as fibres, this is also true for FRMMCs, FRPMCs and FRCMCs.

On the other hand, if the applied stress is perpendicular in direction to both the matrix (m) and reinforcing (r) phases, the stress in the matrix phase (σ_m) is equal to that of the reinforcing phase (σ_r) and they are also equal to the stress in the composite (σ_c). Let this stress be denoted by σ. It is valid only under the simplifying assumption that there is no damage at the interface between the reinforcement and the matrix. Thus, the strain in the composite (ε_c) is given by (ε_c) = $[(V_m\varepsilon_m) + (V_r\varepsilon_r) = \sigma[(V_m/E_m) + (V_r/E_r)] = [\sigma/E_c]$. Therefore, we have:

$$(1/E_c) = [(V_m/E_m) + (V_r/E_r)] \tag{7.2}$$

Or,

$$E_c = \frac{E_mE_r}{V_mE_r + V_rE_m} \tag{7.3}$$

This is the isostress modulus of a simple hypothetical composite. This gives the lower bound of Young's modulus. Assuming the reinforcing phase as fibres, this is true for FRMMCs, FRPMCs and FRCMCs as mentioned earlier.

7.2.2 Fracture Considerations

For a given composite (FRMMC, FRPMC, FRCMC), the stiffness is maximum when all the fibres are parallel to the loading direction. Under such conditions, tensile fibre fracture can happen only when the localized stress exceeds the tensile strength of the matrix phase. In this condition, the composite fracture strength is given by (σ_{pl}) = (P/A), where P is the applied tensile load and A is the cross-sectional fibre area. However, such a condition is somewhat idealized.

More often than not the fibres are misoriented at a small angle θ ($<<\pi/2$). In such a situation, the cross-sectional fibre area is no longer A but (A Cosθ). So, in this situation, the composite fracture strength is given by [(σ_{pl})/(Cos$^2\theta$)]. If all the fibres are at 90° to the loading axis, the composite fracture strength is given by σ_{per} (=P/A). The fibres can no longer carry the applied load. So, transverse matrix failure is most likely to occur. However, if the fibres are not exactly at 90° but still oriented at a large value of θ ($\sim\pi/2$). The area is given by (A/Sinθ), while the load is given by (P/Sinθ). Therefore, under such conditions, the composite strength is given by (σ_{per} / Sin$^2\theta$).

However, for intermediate situations when the fibre misorientation angle the loading axis is ($0 < \theta < \pi/2$). This makes the effective load P/Cosθ and the effective area A/Sinθ. Thus, the composite tensile strength in such a situation is governed by matrix shear strength as it is given by (τm/ ASinθCosθ), where τ_m is the shear strength of the matrix.

7.3 MECHANICAL PROPERTY EVALUATION OF MMCS

The techniques for evaluation of the bulk or macroscale mechanical properties of MMCs are nearly similar to those of metals and also very well described in relevant literature [11, 12]. These techniques are also partially discussed in Chapter 4. Therefore, these will not be repeated for the sake of brevity. Thus, the mechanical properties of major importance for MMCs are the tensile strength (σ_t), percent tensile elongation ($\varepsilon_{t\%}$), yield strength (σ_y), hardness (Brinell/Rockwell/Vickers), compressive strength (σ_c), impact strength (σ_I), fracture toughness (K_{IQ}) and fatigue strength (σ_{fa}) as a function of the number of cycles (N) to failure [11]. The techniques to evaluate these properties are also very well documented in the literature [11]. However, it is worth mentioning that several methods are employed to evaluate the fracture toughness (K_{IQ}) of MMCs [12]. These include single edge notched bar (SENB) with a fatigue loaded pre-crack loaded in bending under three-point or four-point loading, compact tension (CT), double edge notch (DEN), centre notch (CN), chevron notched short rod (CNSR) and Charpy impact (CI). Nevertheless, the most common technique employed is the CT technique [11]. All such tests are conducted as per the corresponding ASTM standards. The literature reported on mechanical properties of MMCs is huge and hence to cover all aspects of the same is beyond the scope of the present work. But it should be noted that the survey will be far from very exhaustive and extensively complete. Rather, the attempt will be here to convey the major findings in a nutshell.

7.4 LITERATURE REVIEW

Since there are innumerable very good reviews as well as original research papers available in the literature [11–20], in the best interest of the reader an attempt is made here to portray only a general scheme of knowledge derived so far. For a given metal (e.g., Al) or its alloys (e.g., Al 6061 or Al 7075-T6) reinforcements with unidirectional alumina and boron fibre enhance the axial modulus, tensile modulus, specific axial modulus, axial tensile strength (Figure 7.5a–d), axial compressive strength, specific axial tensile strength and specific compressive strength (Figure 7.6a–d).

A similar picture holds for SiC fibre reinforced Ti6Al4V alloys [11]. Further, for Al 7075-T6, Ti6Al4V and SS 4340 steel reinforced with SiC particulate reinforcement, all the relevant mechanical properties, e.g., modulus, tensile strength, ultimate tensile strength, and specific modulus, improve significantly (Figure 7.7a–d) [11].

Moreover, all these properties increase significantly with an increase in the volume fraction of SiC particulates from 25 to 70 [11]. A recent review confirms that a similar scenario holds good for metallic Al [13]. Further, for both Al and Al 6061, the tensile strength and hardness (BHN) improve with an increase in the amount of SiC reinforcements [13]. However, generally, there is a reduction

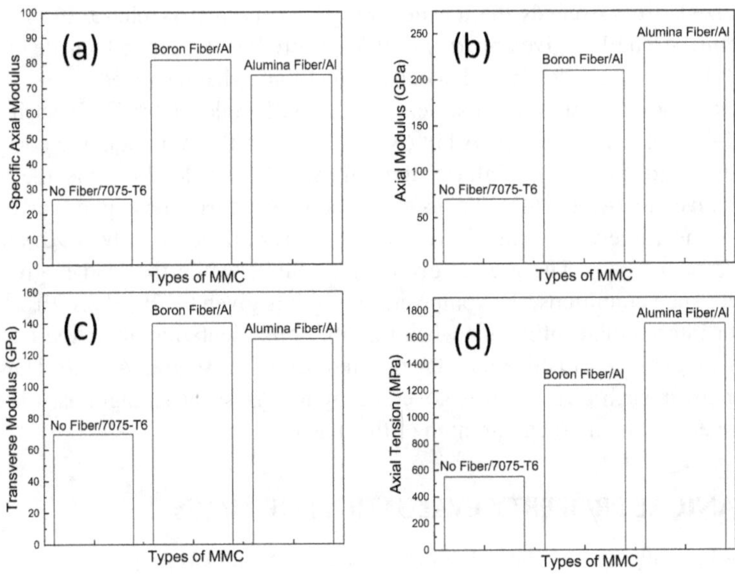

FIGURE 7.5 Variation of mechanical properties of FRMMCs based on Al 7075-T6 alloy: (a) specific axial modulus, (b) axial modulus, (c) transverse modulus and (d) axial modulus as a function of fibre type used for reinforcement [11].

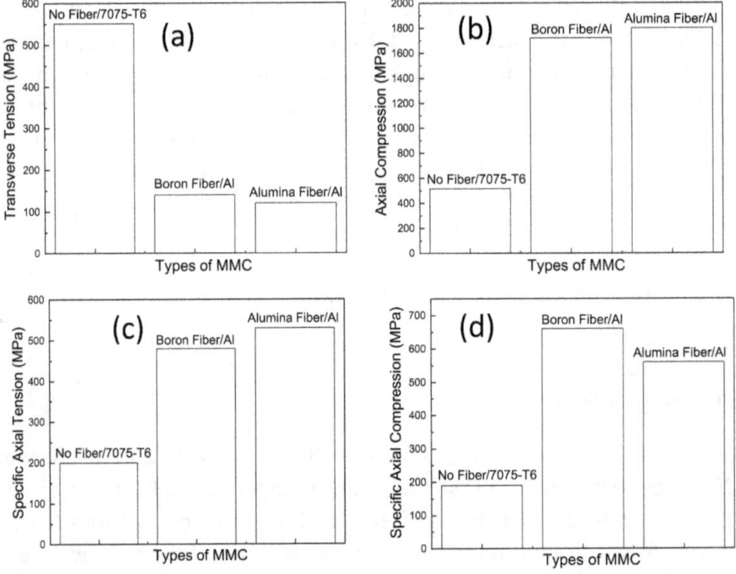

FIGURE 7.6 Variation of mechanical properties of FRMMCs based on Al 7075-T6 alloy: (a) transverse tension, (b) axial compression, (c) specific axial tension and (d) specific axial compression as a function of fibre type used for reinforcement [11].

in elongation at break and impact toughness because of the incorporation of the brittle ceramic phase [11, 13]. Reinforcement of Al, Al 2024 and Al 6061 with the TiC and B_4C particulates also follows a similar trend as achieved with SiC particulate incorporation. [13]. Similarly, the wear rate of soft metallic Cu is significantly reduced with the incorporation of SiC, Al_2O_3, TiC and B_4C particulates, the last one giving the highest reduction in wear rate [13]. On the other hand, for Mg

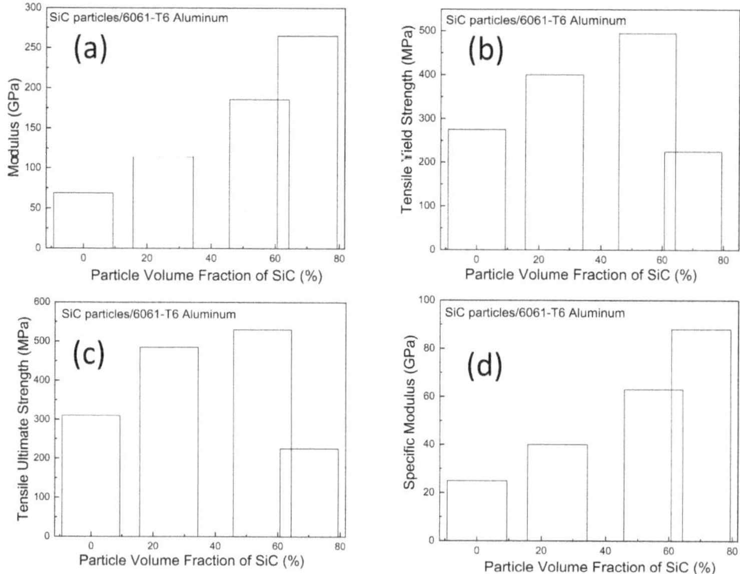

FIGURE 7.7 Variation of mechanical properties of PRMMCs based on Al 7075-T6 alloy: (a) modulus, (b) tensile yield strength, (c) tensile ultimate strength and (d) specific modulus as a function of SiC particle volume in (%) [11].

alloy (e.g., AZ91D), the incorporation of (10%–15%) SiC particulates enhances the ultimate tensile strength significantly from about 250 to 300 MPa [13]. Incorporation of finer-size SiC particulates in Al 6061 [14] and specific heat treatment e.g., T6 of A359 alloy (31 wt.% SiC) [15], also enhances the crack initiation toughness. Moreover, 20 wt% SiC particulate reinforced Al-Mg composites exhibit high cyclic fatigue resistance as compared to that of the Al-Mg alloy and 5 wt.% SiC particulate reinforced Al-Mg alloy [16]. This improvement happens due to two factors. The first one is of course the transfer of local load from the easily deformable alloy phase to the strong SiC phase. The second factor is the high stiffness of the SiC phase [16]. For 1, 5, 10 and 20 μm Al_2O_3 particulate reinforced Al 6061, both Young's modulus and yield strength improve with enhancement in reinforcing phase amount from 10 vol.% to 30 vol.% [17]. Similarly, for 1, 10 and 20 μm SiC particulate reinforced Al 6061 both Young's modulus and yield strength improve with enhancement in reinforcing phase amount from 10 vol.% to 30 vol.% [17]. A very recent review [18] points out that at 10^7 cycles, the fatigue strength of Al 2024 alloy reinforced with SiC particulates decreases with an increase in SiC particulate size from 2.5 to 35 μm. On the other hand, for the same number of cycles, the nanosize SiC particulate reinforced AA7075-T651 alloy is reported [18] to exhibit much better cyclic fatigue resistance. This is attributed to the very high surface area of the SiC nanoparticles, which offers much better interfacial bonding and, hence, enhanced smoothness of the load transfer between the host matrix and the reinforcements [18]. In a very recent review, high amounts of Orwan strengthening and dislocation strengthening are suggested to occur when the interface between B_4C reinforcement and various Al matrices reaches a dimension of less than about 100 nm [19]. It is interesting to note from a recent review that powder metallurgy route processed Al-3.82 vol.% GNP (graphene nanoplatelet) reinforced composites exhibit about 67% increase in hardness, while hot extrusion processed 0.13 vol.% GNP reinforced Al suffers a 15% reduction in hardness [20]. However, the incorporation of about 1.28 to 6.34 vol% of GNP leads to huge enhancements in yield strength, tensile strength, flexural strength, compressive strength and failure strain. On the contrary, if carbide formation occurs, both tensile and compressive strength could be significantly reduced [20]. A recent review [21] concludes that Ti6Al4V reinforced with in situ TiB_2 whiskers is capable of exhibiting high strength, stiffness, creep and fatigue resistance. This happens as the in situ reaction induces better thermodynamic stability between the matrix and reinforcement phases. Further, 10 wt.% TiC particulate reinforced Ti6Al4V alloy exhibits much higher yield strength and tensile strength as compared to those of the basic Ti6Al4V alloy.

7.5 FUTURE CHALLENGES IN MMCS

As far as future challenges in this highly evolved technological domain are concerned, a major focus is now directed [22] to developing Mg-based MMCs as biodegradable implants and their thorough mechanical characterization. Another very prospective challenge is the development of carbon fibre reinforced MMCs and characterization of their interfacial mechanical properties as well as macroscale mechanical properties [23]. A grand challenge today is how to do proper mechanical and tribological characterization of cold-sprayed MMC coatings because they offer significant technological application potential [24]. However, the most crucial future challenge is to be able to sufficiently and comprehensively characterize the local mechanical properties at the interface in MMC materials [25]. The newly developed nanoindentation technique [26] may find a major role here in developing microstructurally designed, advanced MMC materials.

7.6 POLYMER MATRIX COMPOSITES

7.6.1 Brief Introduction

PMCs find the widest industrial, engineering, biomedical and high-end technological applications [11, 27–36]. The concerned literature is already so rich that only the crux of the information deserves any special mention, if at all. It has been most comprehensively reviewed in the literature. Very recent reviews identify a major potential of biodegradable PMCs for biomedical applications [27–29]. Major applications of huge technological importance are reviewed in recent works [30, 31]. Even microwave processing has been reviewed for PMCs with a huge fallout on energy benign scenarios [32]. Numerical modelling has been very recently applied to hole drilling problems in PMCs [33]. Further, PMCs are also emerging as a fantastic material for wave transparent applications [34] as well as for high-end electronics applications [35]. However, the interface characterization of the PMCs is yet to receive the importance that is due to be given [36].

7.7 LITERATURE REVIEW OF PMCS

The basic characterization techniques utilized for PMCs are similar to those already given for polymers in Chapter 6. Although there are numerous possible combinations of polymer matrices and reinforcements such as fibres and particles, the major categories are glass fibre reinforced polymer (GFRP) composites and carbon fibre reinforced polymers (CFRP) composites. However, for the benefit of readers, the basic standards utilized for mechanical properties evaluation of GFRP composites (rods, bars and laminates, e.g., unidirectional GF reinforced, unidirectional short fibre reinforced, random short fibre reinforced) are given in Table 7.1.

7.7.1 Reinforcing Fibres

The typical reinforcing glass fibres include E-glass and HS (high strength) glass fibres. The tensile modulus for E-glass is about 70 GPa, while tensile strength is about 2 GPa. However, the HS fibre has an axial modulus of about 80 GPa and tensile strength of about 4.2 GPa [11]. Polyacrylonitrile (PAN)-based CFs offer very interesting combinations of properties. Thus, the standard modulus (SM) PAN CF, ultrahigh strength (UHS) PAN CF and ultrahigh modulus (UHM) PAN CF provide tensile moduli of about 230, 590 and 290 GPa, respectively. Similarly, they have tensile strength of about 3.2, 3.8 and 7 GPa, respectively [11]. On the other hand, UHM pitch-based CFs exhibit a tensile modulus of

TABLE 7.1
Mechanical Properties Evaluation Standards for GFRP and CFRP Composites

Test Type	ASTM Standard
Ultimate tensile strength	ASTM:D638
Flexural strength	ASTM:D790
Compressive strength	ASTM:D695
Izod impact	ASTM:D256
Charpy impact	ASTM:D256
Izod impact	ASTM:D2344
	ASTM:D3846
Interlaminar shear strength	ASTM:D2344
	ASTM:D3846
Barcol hardness test	ASTM:D2583

about 895 MPa and tensile strength of about 2.2 GPa. Polymeric aramid fibres have a relatively lower axial modulus of about 120 GPa as compared to those of the CF varieties discussed earlier. However, their tensile strength of about 3.2 GPa is comparable to those of some of the CF varieties. The high density polyethylene (HDPE) fibre provides a slightly higher tensile modulus of about 170 GPa and similar tensile strength of about 3 GPa. Thus, the choice of a particular fibre is to be guided by the particular application concerned and the applied stress scenario that the developed PMC is going to encounter. The ceramics, e.g., alumina and SiC fibres [11], provide slightly higher tensile modulus of about 370 and 400 GPa as compared to the aramid and HDPE as well as GF varieties, the SM PAN CF and the UHS PAN CF. The tensile strength of alumina fibre is about 1.9 GPa. This is comparable to that of HS GF but lower than those of the CF varieties. On the other hand, the SiC fibre has tensile strength comparable to those of the SM and UHS PAN carbon fibres. In contrast, nonmetallic boron (B) fibres have a tensile modulus of about 400 GPa and tensile strength of about 3.6 GPa [11]. Thus, there is a wide variety of choices in terms of axial modulus and axial strength of reinforcing fibres.

7.7.2 Matrix Materials

In general, the polymer matrices are relatively weaker than their metallic and ceramic counterparts. They have inherently low stiffness. Therefore, they are more prone to viscoelastic deformation. The two polymer types of most practical utility are thermosetting polymers and thermoplastic polymers. The key types of thermosetting resins used in such thermosetting polymers are epoxies, bismaleimides, phenolics, vinyl esters, cyanate esters, polyesters and polyimides. The thermosetting polymers require heat curing and once cured are not further deformable as they become rigid. The most common matrix in PMCs is the thermosetting epoxy, which has a very small tensile modulus of about 3.5 GPa and a low tensile strength of about 70 MPa [11]. On the other hand, examples of crystalline thermoplastic polymers include nylon, polyethylene, polypropylene and polyether ether ketone (PEEK). Their tensile moduli vary from about 2 to 4 GPa, while tensile strength varies from about 30 to 200 GPa [11]. Thermosetting polymers are polymers that are mouldable at a certain high temperature and solidify when cooled.

7.8 MECHANICAL PROPERTIES OF PMCs

The unidirectional PMCs with E-glass fibres as reinforcement and polyester resin as matrix exhibit the highest tensile strength of 750 MPa with a tensile modulus of about 40 GPa. For woven roving reinforcement, these values reduce to about 250 MPa and 16 GPa. In the case of chopped strand mat reinforcement, these values reduce further to about 100 MPa and 8 GPa. Sheet moulded PMC has an even further reduced tensile strength of about 85 MPa, while the bulk moulded PMC has a tensile strength of about 50 MPa. Thus, the fracture-relevant mechanical properties of the PMCs are highly sensitive to the processing condition for a given type of reinforcement and the physical form of the reinforcement (Figures 7.8–7.11) [11].

Further, measured in the transverse direction of reinforcement, the tensile strength registers a low value of about 50 MPa, while the modulus is about 12 GPa. This means that these PMCs are highly anisotropic as far as their mechanical properties are concerned. It happens because the properties measured in the wrap direction are slightly lower than those in the fill direction. The wrap direction means that the properties are measured along the longitudinal direction of the reinforcement. However, the fill direction means that the properties are measured in a direction perpendicular to the longitudinal direction of the reinforcement. On the other hand, PMCs with CF reinforcement

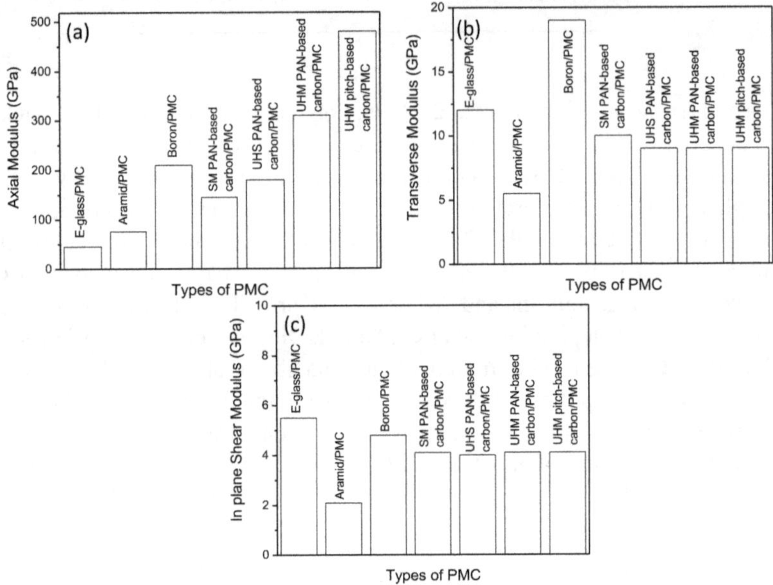

FIGURE 7.8 Variation of mechanical properties of FRPMCs: (a) axial modulus, (b) tensile modulus and (c) in-plane shear modulus as a function of reinforcing fibre type [11].

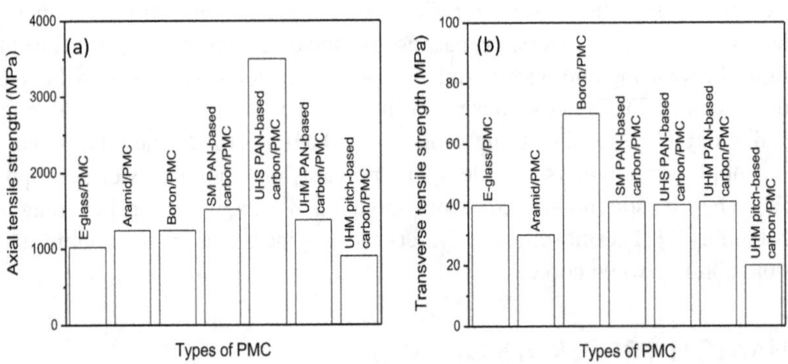

FIGURE 7.9 Variation of mechanical properties of FRPMCs: (a) axial tensile strength and (b) transverse tensile strength as a function of reinforcing fibre type [11].

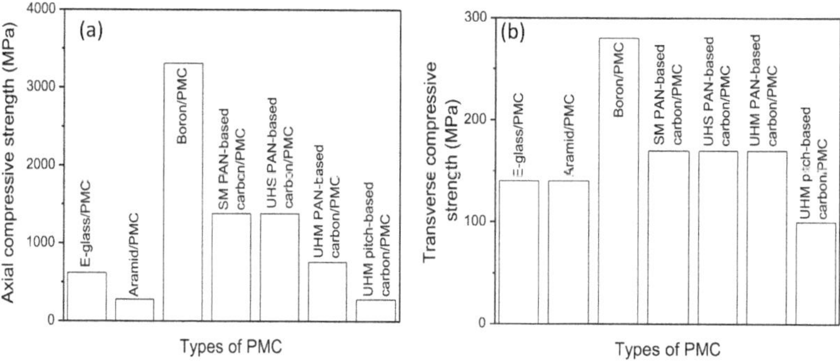

FIGURE 7.10 Variation of mechanical properties of FRPMCs: (a) axial compressive strength and (b) transverse compressive strength as a function of reinforcing fibre type [11].

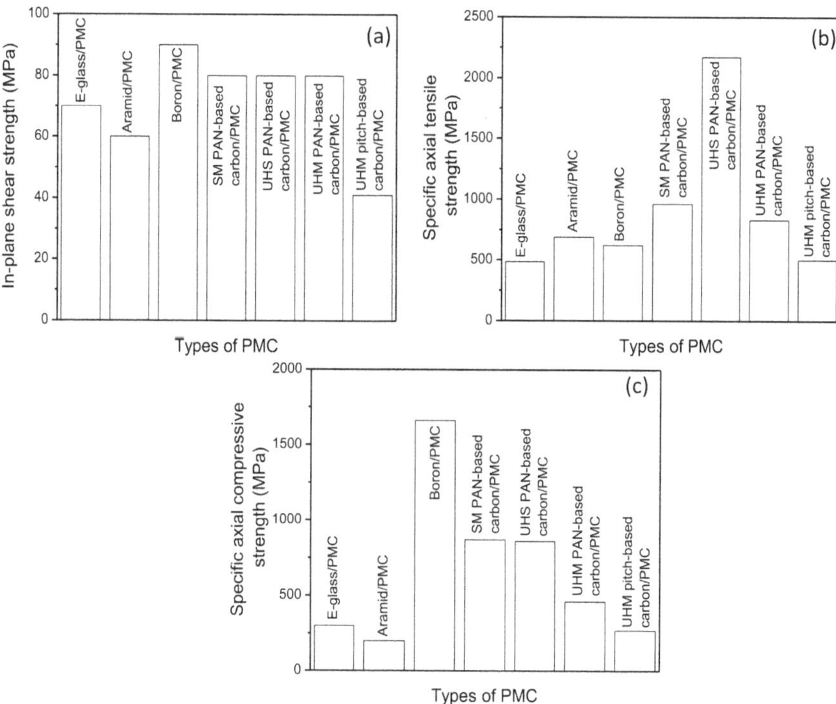

FIGURE 7.11 Variation of mechanical properties of FRPMCs as a function of reinforcing fibre type: (a) in-plane shear strength, (b) specific axial tensile strength and (c) specific axial compressive strength [11].

have much higher properties, as expected. For instance, the tensile moduli of epoxy-based PMCs reinforced with UHS PAN CF and UHM PAN CF are about 200 GPa and 500 GPa, respectively [11]. Their tensile strengths are also very high as compared to that of the matrix, about 3.5 GPa for the UHS PAN CF reinforcement and about 1.4 GPa for the UHM PAN CF reinforcement. This is linked to the better tensile modulus and tensile strength of the different CF varieties, as mentioned earlier. Similarly, the corresponding compressive strengths of these CFRP-based PMCs are about 1.4 GPa and 0.8 GPa, respectively.

However, the methods to measure the delamination resistance of PMCs are yet to be standardized, as the existing standard ISO 4585 has some genuine limitations [37, 38]. The modified split cantilever beam has been used to evaluate the mode II and mode III fracture toughness of woven glass fibre reinforced PMCs [39]. The most recent efforts also focus on human factors as far as the repeatability and reproducibility of the mechanical properties of PMC [40] are concerned. A very recent important effort highlights the need for the usage of digital technology to improve the accuracy of experimental determination of relevant fracture mechanics parameters, e.g., critical strain energy release rate [41]. A successful attempt is reported [42] for mode II delamination fracture toughness determination of woven roving-based CFRP composites by using both numerical modelling and experimental data from similar data of unidirectional composites measured using the standard ASTM D7905.

7.9 FUTURE RESEARCH NEEDS OF PMCS

Recent work identifies that nanoindentation could offer very good insight into the mechanical properties of various constituent phases in PMC and especially at the interface [36, 43, 44]. However, the total quantum of work especially in PMCs is far from significant and hence it deserves special attention for further futuristic development of the PMCs.

7.10 CERAMIC MATRIX COMPOSITES

The ceramic matrix composites can be particulate reinforced (Figure 7.3a), continuous fibre reinforced (Figure 7.3b, c, d), short fibre reinforced (Figure 7.3e) and hybrid reinforced (Figure 7.3f) as schematically shown earlier. Due to their much better mechanical properties at both room temperature and high temperature, CMCs are in high demand for application in aerospace, defence, space, aviation, automotive and other industries [11]. In addition to superior high-temperature mechanical properties, CMCs offer low density. As a result, the specific strength of the CMCs is high. Similarly, the CMCs possess a high specific modulus as well. Moreover, they exhibit good resistance against high-temperature oxidation. Further, their resistance against ablation is much better than those exhibited by their conventional brittle ceramic counterparts. This unique combination of properties makes CMCs one of the best candidates for usage in harsh environments, e.g., hot sections in aerospace vehicular engines. It is opined [11] that out of various varieties, as mentioned earlier, the woven or braided CMC varieties will be in high demand as far as manufacturing complex shapes with exceptionally good mechanical properties are concerned. The general scenario is that within the threshold limit, there can be modest improvements in strength and fracture toughness by the inclusion of relatively stronger nanoparticles, such as SiC in a ceramic matrix (e.g., $Al_2O_{3)}$. However, the greatest challenge is to densify such a PRCMC by conventional pressureless sintering. More often than not costly techniques such as hot pressing (HP) and hot isostatic pressing (HIP) are required for better densification and further reduction of the residual porosity. If and when this is done, it results in some further improvement in strength and fracture toughness. However, ceramics are characteristically brittle, low toughness (1–3 MPa.m$^{0.5}$) materials. Therefore, even a slight improvement in fracture toughness matters a lot. This is where the reinforcements by ceramic fibres (Figure 7.12) and CNTs come into the picture, although both of these reinforcements are costly to produce. In principle, numerous combinations of materials are possible for PRCMCs, FRCMCs and HRCMCs; therefore, only typical instances are to be discussed because a very detailed discussion is beyond the scope of the present chapter.

7.11 LITERATURE REVIEW OF CMCS

As such, the fabrication of FRCMCs is fairly complex. There are three stages involved. The very first stage is the forming of material. This is usually done by well-known processes such as gel casting and fibre lay-up. The next stage is the most critical one. This stage requires the formed FRCMC to be densified. A multitude of techniques and their combinations are used. The most popular method so far adopted by industry is the chemical vapour infiltration technique. This is termed

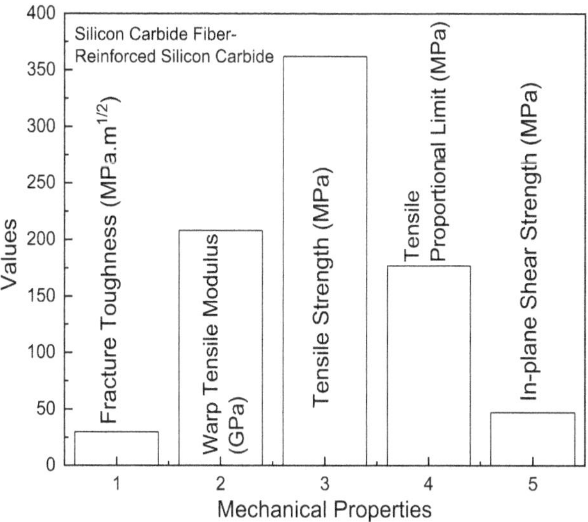

FIGURE 7.12 Variation of mechanical properties of FRCMCs (SiC fibre reinforced SiC) [11].

as the CVI process. Another method is precursor infiltration and pyrolysis. This is called the PIP process. Further, some processes use the liquid silicon infiltration technique. This process is termed the LSI process. Once the densification is done, comes the most challenging part of the job. This is generally known as the post-machining stage. More often than not, it is essential to do it because otherwise the densified FRCMCs cannot be utilized for any useful engineering application. Since, the machining of ceramics in general and the FRCMCs in particular is by itself a very challenging task, a lot of current focus is diverted to the utilization of newly developed additive manufacturing (AM) techniques [45, 46]. The AM techniques are also known as 3-D printing technology. These newly developed AM techniques include many varieties. The first one and most popular technique is stereolithographic addition, in short SLA. Another very useful as well as flexible technique for 3-D product development is selective laser sintering (SLS). In case the product is required in a laminate form, the laminated object manufacturing (LOM) technique is adopted. If the ceramic is to be deposited onto a particular form of already built substrate, then the inkjet printing (IJP) technique is utilized. It also provides a scope to fabricate a functionally graded CMC structure through proper optimization of the operational parameters. Through layer-by-layer deposition, as pre-planned, the laser engineered net shaping (LENS) technique is also getting attention for near net shape development of FRCMCs. On the other hand, from an already-made ceramic paste layer-by-layer casting through extrusion is another evolving technique known as direct ink writing (DIW).

Another very important emerging area is the development of biodegradable CMCs [47]. For instance, cryomilling and uniaxial pressing are exploited to develop CMCs from nanocrystalline hydroxyapatite and silk fibres. Further, the damage resistance of CMCs is tried to be improved by using alumina-based prepeg reinforced alumina CMCs [48] with important fall out in both body armour developments and biomedical prosthesis development. Furthermore, the new concept of atomically laminated ceramics is utilized in improving the toughness and, hence, the damage tolerance of CMCs [49]. This work also highlights the importance of predesigning the microstructurally engineered CMCs as well as ceramics for multifunctional applications. Moreover, it must be kept in mind that the interphase plays a very vital role in the sharing of applied stress in ordinary CMCs and especially in 3-D-printed CMCs [50].

Recent work indicates that high processing temperature is necessary for reactive melt infiltration processing as it is a crucial requirement for the successful development of multiphase CMCs [51]. A very recent development identifies how UO_2 can be reinforced with the useful dispersion of graphene nanosheets to render huge developments in mechanical properties [52]. The ongoing techniques adopted and future challenges expected in the development of whisker reinforced CMCs have been very comprehensively discussed in recent literature [53, 54]. Very novel ideas for developing mini CMCs are put forward to examine the transverse tensile strength of CMCs [55]. This is an area which often misses the importance that it should have gotten. Short fibre reinforced CNTs containing CMCs emerged as an important material for electromechanical applications. Recent attempts have been made to predict their electromechanical properties through a multiscale modelling approach.

A very big challenge is to be able to machine the CMCs as per the requirement of the intended design applications [56–58]. Therefore, it is not surprising to see that a lot of current research effort is directed at conventional machining of CFRCMCs [56], unconventional machining of various other CMCs [57] and especially the ultrasonic atomization-assisted micromilling of CMCs [58]. Recent works [56–58] as well as earlier work [11] confirm the fantastic combination of beneficial properties of CMCs, e.g., CMCs offer excellent properties such as high specific strength and modulus. Further, they have high resistance to ablation as mentioned earlier. Furthermore, they exhibit the capability to withstand high temperatures without compromising their mechanical properties. This is why they are being considered as ultrahigh temperature (UHT) ceramic matrix composites. A lot of current attention is therefore directed to understanding the strategies to further improve their properties [59, 60] as well as processing [61].

7.12 FUTURE RESEARCH NEEDS OF CMCS

The most important area is identified as studying the interfacial mechanical properties at the fibre matrix interface of both PRCMCs and FRCMCs at both room temperature and high temperature by using well-known techniques such as nanoindentation and high-temperature nanoindentation. Innovative experimental techniques are also needed to experimentally measure the properties of various interphases in CMCs. *Development of microstructurally designed PRCMCs and multilayer CMCs (MLCMCs) as well as microstructurally designed advanced multilayer metal–ceramic hybrid composites (MLMCHCs) are also of paramount importance as described next.*

7.13 FUTURE DEVELOPMENTS IN ADVANCED COMPOSITES

Here it is intended to describe the recent developments in microstructurally engineered PRCMCs, MLCMCs and MCMLCs.

7.13.1 Microstructurally Engineered Particulate Ceramic Matrix Composites

The well-known nanoindentation technique with a Berkovich nanoindenter of tip radius ~150 nm has been used in a number of recent efforts [62–65] to determine the nanomechanical properties of 10, 20 and 40 vol% of 3 mol% Yttria partially stabilized zirconia toughened alumina (ZTA) nanocomposite ceramics (Figures 7.13–7.19). In all these experiments the loading and unloading times are kept fixed at 30 s. Their microstructures are shown in Figure 7.13.

The applied load range varied from 10 to 1000 mN. The presence of indentation size effect (ISE) is explained by the concept of strain gradient plasticity and the geometrically necessary dislocations (GNDs) inside the nanoindent cavity of 10, 20 and 40 ZTA ceramics [62]. The change in the strain gradient becomes much higher at depths of about 20 to 50 nm than beyond it. The variations of the ratio of plastic to total energies with the slightly load-dependent variations in (H/E) are linked to the corresponding ratios ($0.5\ h_f^3/0.5h_{max}^3$) of the present ZTA nanocomposite ceramics. Here, h_f refers to the final nanoindentation depth measured after complete unloading. Further, h_{max} refers to the maximum nanoindentation depth measured at the peak load. Both nanohardness (H) and Young's modulus (E) are reported [62] to decrease with the increase in vol.% of zirconia in the ZTA nanocomposite ceramics.

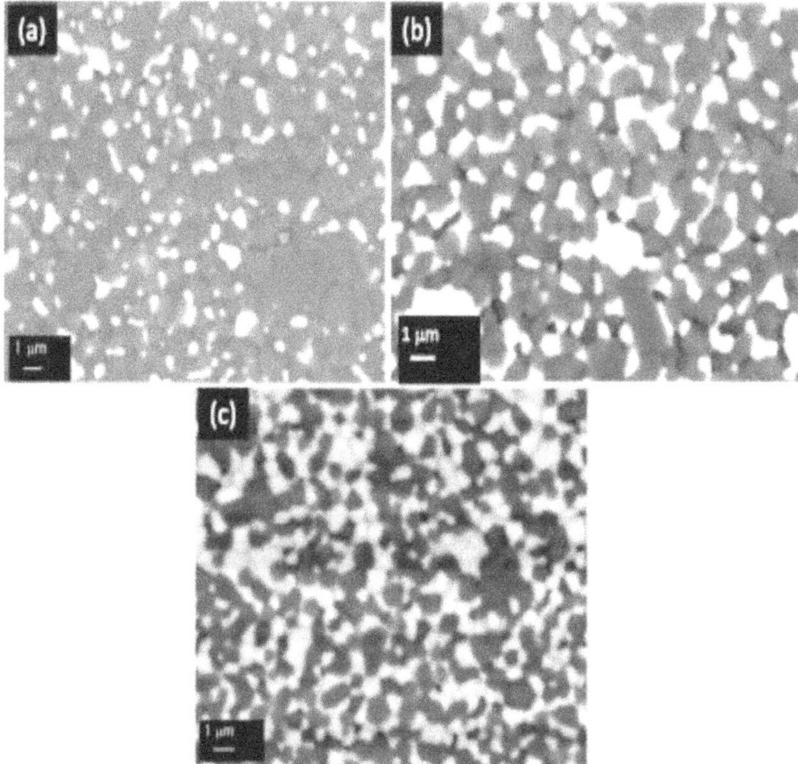

FIGURE 7.13 Microstructures of PRCMCs, e.g., zirconia toughened alumina (ZTA) ceramics: (a) 10 volume percent ZTA, (b) 20 volume percent ZTA and (c) 40 volume percent ZTA ceramics (reprinted with permission from [62]).

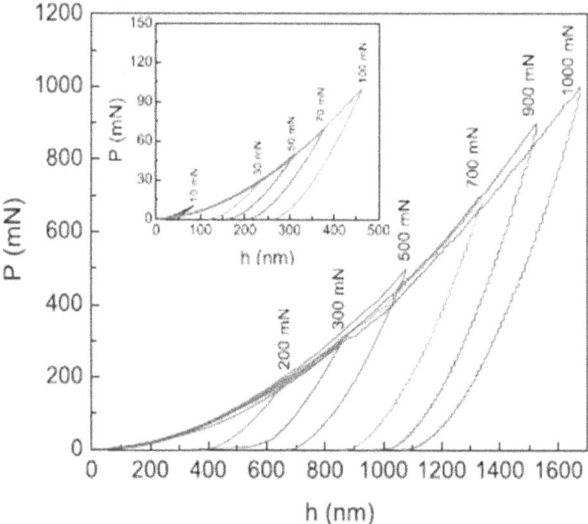

FIGURE 7.14 Nanoindentation-based P–h plots of PRCMCs, e.g., zirconia toughened alumina (ZTA) ceramics 10 volume percent ZTA (reprinted with permission from [63]). Inset: P–h plot at various lower loads up to 100 mN.

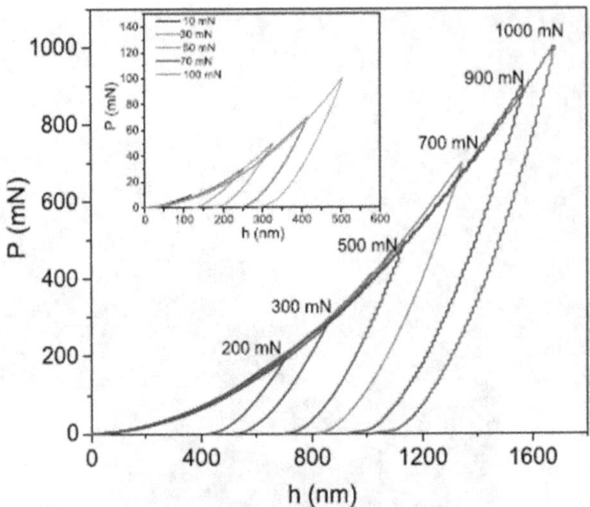

FIGURE 7.15 Nanoindentation-based P–h plots of PRCMCs, e.g., zirconia toughened alumina (ZTA) ceramics 20 volume percent ZTA (reprinted with permission from [64]). Inset: P–h plot at various lower loads up to 100 mN.

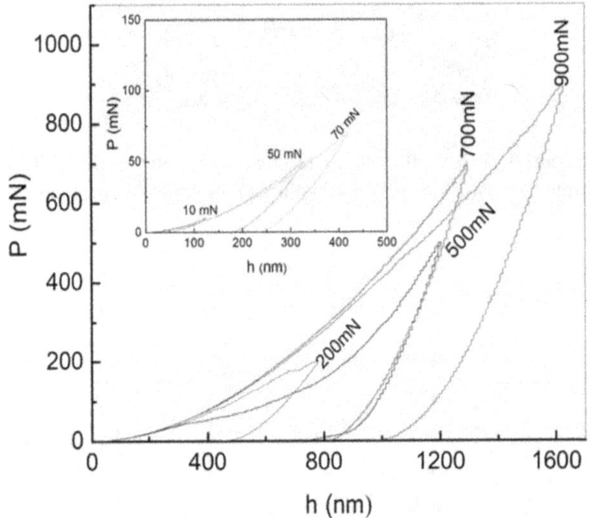

FIGURE 7.16 Nanoindentation-based P–h plots of PRCMCs, e.g., zirconia toughened alumina (ZTA) ceramics 40 volume percent ZTA (reprinted with permission from [65]). Inset: P–h plot at various lower loads up to 100 mN.

The intrinsic contact deformation resistance or the load at which the nanoscale plasticity events/micro-pop-in events start is called the critical load (P_c). The magnitude of P_c increases with the applied loads from 10 to 1000 mN in the case of the 10 vol% of 3 mol% Yttria partially stabilized ZTA (i.e., 10 ZTA) nanocomposite ceramics (Figure 7.14).

The values of maximum shear stress underneath the nanoindentation cavity are found to be much higher than the theoretical shear strength of these 10 ZTA nanocomposite ceramics. The formation of shear deformation bands inside the nanoindentation cavity is confirmed by the field-emission scanning electron microscopy (FESEM) photomicrographs reported elsewhere [63]. The critical resolved shear stress at which localized microcracking occurs is also evaluated to be much smaller

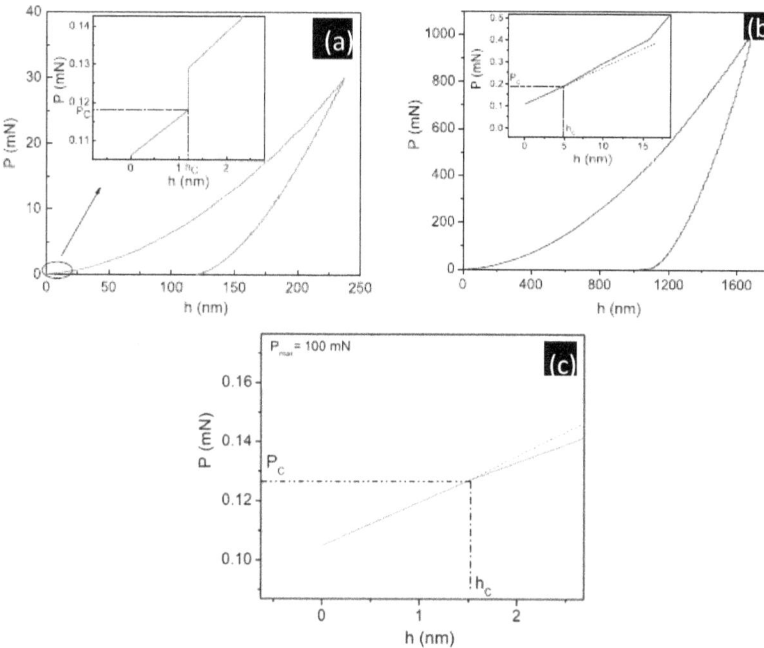

FIGURE 7.17 Identification of critical load (P_c) and critical depth (h_c) values in the nanoindentation-based P-h plots of PRCMCs, e.g., zirconia toughened alumina (ZTA) ceramics: (a) 10, (b) 20 and (c) 40 volume percent ZTA (reprinted with permission from [63, 64, 65]).

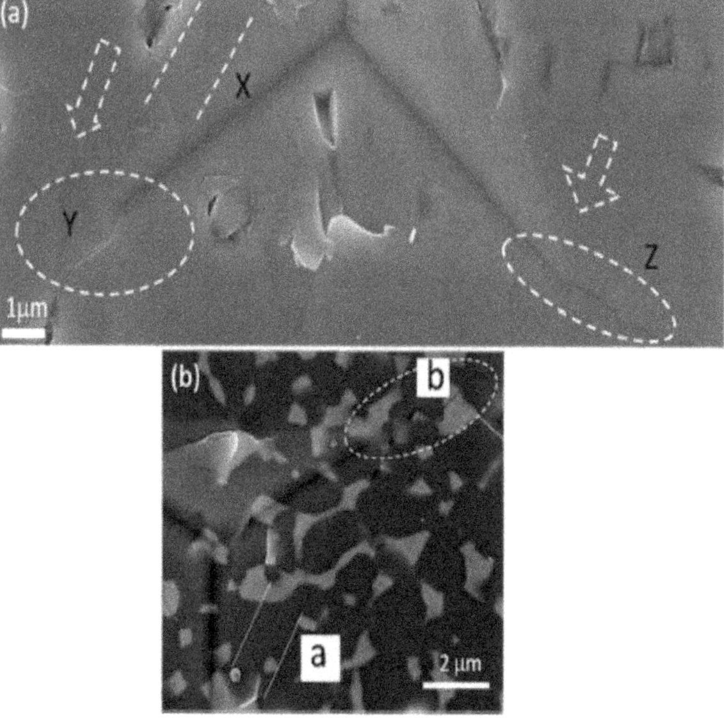

FIGURE 7.18 Shear deformation and microcracking in nanoindentation of PRCMCs, e.g., 10 and 20 volume percent ZTA (reprinted with permission from [65]).

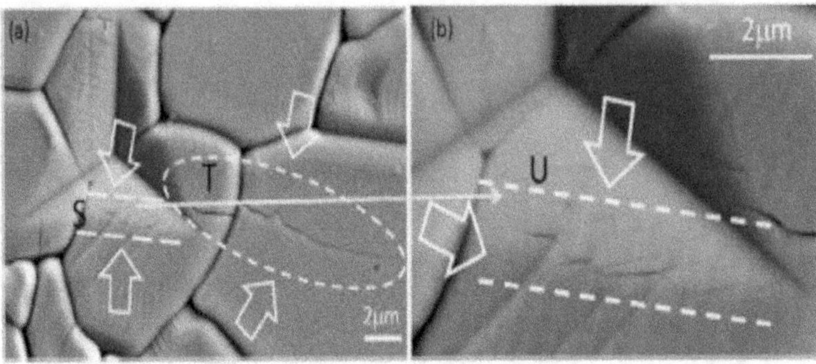

FIGURE 7.19 Shear deformation and microcracking in nanoindentation of PRCMCs, e.g., 40 volume percent ZTA (reprinted with permission from [65]).

than the maximum shear stress [63]. As a result, the microcracking happens around the nanoindentation cavity. This is also reported to be confirmed by FESEM-based evidence reported elsewhere [63]. The genesis of micro-pop-in issues in 10 ZTA nanocomposite ceramics is suggested [63] to be linked to both the dislocation nucleation and the initiation of microcracks (Figures 7.17 and 7.18) during the nanoindentation process.

Further, in the case of the 20 vol% of 3 mol% Yttria partially stabilized zirconia toughened alumina (i.e., 20 ZTA) nanocomposite ceramics, it is found [64] that critical load or intrinsic contact deformation resistance at which the nanoscale plasticity events occur, increases with the applied nanoindentation loads from 10 to 1000 mN following positive power law dependency. In the lower load range (10–100 mN), there is a slight indentation size effect, whereas at a higher load range (500–1000 mN), the magnitudes of both H and E are saturated. The load averaged values of H and E of 20 ZTA nanocomposite ceramics are evaluated to be ~13.74 ± 1.5 GPa and 464 ± 31.6 GPa, respectively [64]. However, the true hardness of 20 ZTA ceramics is predicted as 40.67 GPa by using the well-known Sakai model. This value is much higher than the experimentally measured nanohardness value. More and more dislocation nucleation occurs underneath the nanoindentation cavity, especially at higher loads. The simultaneous occurrence of the maximum shear stress generated inside the nanoindentation cavity, nucleation of dislocation, formation and propagation of microcracks are associated with the micro-pop-in or nanoscale plasticity events in the 20 ZTA nanocomposite ceramics (Figures 7.15, 7.17 and 7.18).

Similarly, in the case of the 40 vol% of 3 mol% Yttria partially stabilized zirconia toughened alumina (40 ZTA) nanocomposite ceramics, incipient plastic events (IPE) or micro-pop-in related parameters like critical load (P_c), critical depth (h_c), critical load difference (ΔP_c) and corresponding depth difference (Δh_c), dislocation density (ρ_G), maximum shear stress (τ_{max}) and critical resolved shear stress (τ_{CRSS}), all follow power law dependencies with the applied loads from 10 to 1000 mN [65]. All such power law dependencies are reported to occur [65] with a positive exponent. However, the dislocation loop radius (R_d) shows negative power law dependency with the applied loads. The magnitude of bi-axial flexural strength (σ_f) and indentation fracture toughness (K_{IC}) both increase with the enhancement of zirconia content in these ZTA nanocomposite ceramics. The values of (σ_f) are measured to be about 250 ± 31, 380 ± 53 and 490 ± 71 MPa, respectively, for the 10, 20 and 40 ZTA nanocomposite ceramics. In addition, the values of (K_{IC}) are measured to be ~4.1 ± 0.39, 4.3 ± 0.67 and 4.5 ± 0.78 MPa.m$^{0.5}$, respectively. All these (K_{IC}) values are measured at a load of 100 N for the 10, 20 and 40 ZTA nanocomposite ceramics. These observations may be explained by the fact that the amount of the tetragonal phases enhances with the increase in zirconia content in the present ZTA nanocomposite ceramics. However, measured at an applied load of 100 N, the values of Vickers microhardness (Hv) are reported [65] to decrease with the increase in zirconia content in the ZTA nanocomposite ceramics (Figures 7.16, 7.17 and 7.19).

7.13.2 Microstructurally Engineered Tape Cast Multilayer Ceramic Matrix Composites

One problem with the PRCMCs is ensuring that the reinforcements are homogeneously dispersed. Similarly, a major problem with FRCMCs is maintaining the distance between the fibres at a constant and appropriate level which will allow them to jumble together. Another major crisis is to make the interface appropriately tuned for seamless transfer of the applied load and, hence, stress from the matrix to the fibre. A similar problem persists with CFRCMCs, CNTRCMCs and HRCMCs. One way to address this is to use the multilayer structure like in metal–metal (Figure 7.4a) and/or ceramic–ceramic micro and nano multilayer composites (Figure 7.4 b, c, d, e, f, g, h, j and i) with a predesigned interface [66–71]. These results obtained earlier confirm the success of the concept [66–71]. It has been demonstrated [66–71] that this can be achieved with a variety of ceramic–ceramic multilayer composites (CCMLCs) with various interface designs, e.g., strong–strong combination (SSC; Figure 7.4b), weak–weak combination (WWC; Figure 7.4c), strong–weak–strong–weak interface combination (SWSWC; Figure 7.4d), tough–weak–tough–weak combination (TWTWC; Figure 7.4e), tough–brittle–tough–brittle combination (TBTBC; Figure 7.4f), brittle–tough–brittle–tough combination (BTBTC; Figure 7.4g) and tough–tough combination (TTC; Figure 7.4h). Another variety is metal–ceramic multilayer composites (MCMLCs) with interface design of, e.g., metal (tough)–ceramic (brittle)–metal (tough)–ceramic (brittle) combination (TBTB, Figure 7.4i).

As an example it may be mentioned [66–71] that nano alumina (nA) single tape shows the lowest failure energy of 14. 6±1.8 kJ.m^{-3}. The highest failure energy of 65.16±16.57 kJ.m^{-3} is observed for nano alumina/nano alumina 20-layer multilayer composites (MLCs). For 10-layer MLC samples, the failure energy was 41.18±7.42 kJ.m^{-3}. Further, the bi-axial strength values of nZ/nZ (nano zirconia/nano zirconia) MLC samples are reasonably good (400–630 MPa), although they may not be as high as that of the single tapes (789.77 MPa). The highest bi-axial flexural strength data (632.60 MPa) are exhibited by the 20-layer samples and it is only ~ 25% lower than that of its monolithic counterpart [66–71]. The most commonly observed trend in MLC material is that toughness is considerably improved. However, the variation in the bi-axial strength of these MLC samples is not so systematic in terms of the variation in the number of layers. Thus, the highest failure energy data (694 kJ.m^{-3}) is obtained [66, 67] for the 30-layer nZ/nZ MLC samples. This is more than six times as high as that of the monolithic nano zirconia tape (96.48 kJ.m^{-3}). Among the porous lanthanum phosphate (LaP) coating-based MLC samples, the failure energy of nZ/LaP is 282.38 kJ.m^{-3}, nearly thrice that of monolithic nZ tape (96.48 kJ.m^{-3}). On the other hand, the failure energy is 262.15 kJ.m^{-3} for the 20-layer nA/5ZTA MLC. Finally, the failure energy of nZ/5ZTA is found [66–71] to be the highest at about 370 kJ.m^{-3}, which is more than three times that of the monolithic nano zirconia tape (96.48 kJ.m^{-3}). It is also identified [66, 67] that the nature and magnitude of residual stress can play a very significant role in affecting the mechanical behaviour of the multilayer composites. The relevant mechanical properties of various CCMLCs are presented in Figure 7.20a–d.

It is easily observable from the data plotted in Figure 7.20a that for the nA/nA combination, which is a brittle–brittle combination, the fracture toughness increases with increases in the number of layers, from about 1.4 MPa.m$^{0.5}$ for a single layer (1L) to about 7 MPa.m$^{0.5}$ for 20-layer (2o L) MLCs. Further, it is easily notable from the data plotted in Figure 7.20b that for the nA/ZTA combination, which is a brittle–tough combination, the fracture toughness increases with increases in the number of layers (from about 1.4 MPa.m$^{0.5}$ for single layer (1L) tape cast nano alumina tapes to about 23.5 MPa.m$^{0.5}$ for 10-layer (10L) MLCs. This happens presumably due to the phase transformation-induced toughening in the ZTA tapes of the MLC architecture. Further, it is interesting to note from the data plotted in Figure 7.20c that for the nZ/ZTA combination, which is a tough–tough combination, the fracture toughness increases with increases in the number of layers, from about 2 MPa.m$^{0.5}$ for single layer (1L) tape cast nZ tapes to about 24 MPa.m$^{0.5}$ for 3- layer (30L) nZ/nZ MLCs. This also happens presumably due to the phase transformation-induced toughening in the ZTA tapes of the MLC architecture. However, there is a slight decrease in fracture toughness at numbers of layers higher than 30 in the MLC. This happens possibly due to the increase in thickness

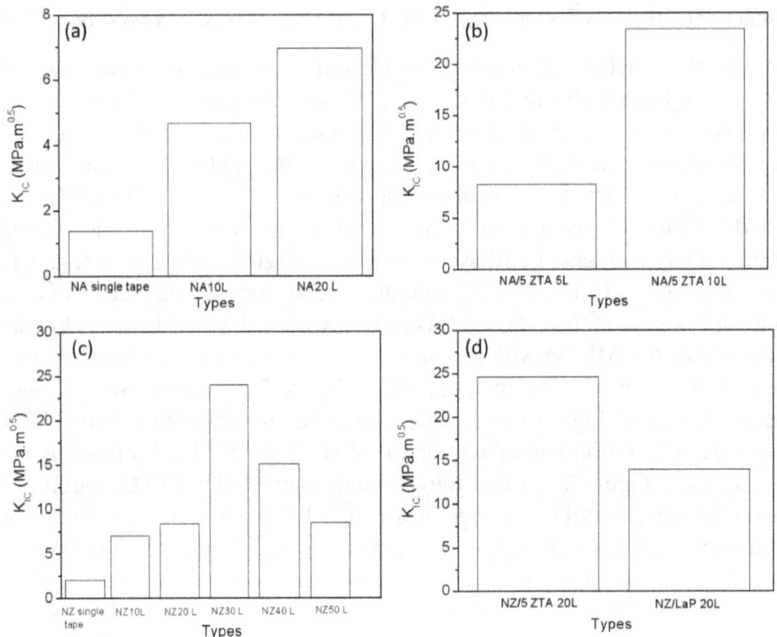

FIGURE 7.20 Variation of fracture toughness (K_{IC}) as a function of microstructural interface architecture design in various multilayer ceramic composites (MLCs).

and hence the interfacial defects in the laminated structure. There is further research necessary to optimize the layer number and layer thickness and to have better control of the interface fracture aspects. Furthermore, it needs to be noted from the data plotted in Figure 7.20d that for the same 20-layer MLC architectures, the nZ/5ZTA combination, which is a tough–tough combination, provides the fracture toughness, i.e., about 25 MPa.m$^{0.5}$, which is a much higher (about 15 MPa.m$^{0.5}$) toughness than that of the nZ/LaP combination, which is tough–weak layer combination design. However, both the MLCs have fracture toughness much higher (about 2 MPa.m$^{0.5}$) than that of the single tape cast nZ material. The corresponding failure energy variations in the MLCs are depicted in Figure 7.21a–d. Thus, there is still further research needed to optimize the tough–weak multilayer combination of MLC architecture.

7.13.3 Microstructurally Designed Advanced Multilayer Metal–Ceramic Hybrid Composites (MLMCHCs)

Microstructurally designed advanced multilayer metal–ceramic hybrid composites (MLMCHCs) are one of the most important latest additions [72–96] in the global scenario of advanced composites for multifunctional applications. In such multilayer structures, the metallic thin layer provides the intrinsic toughness, while the ceramic thin layer provides the intrinsic hardness. The toughness provides in turn the damage tolerance. The hardness in turn provides the wear resistance.

7.13.3.1 Literature Review of MLMCHCs

MLMCHCs provide a wide range of applications including ultrahigh-strength materials; optical devices; high-performance capacitors; thermoelectric material; high wear resistance and low friction coatings for gears, bearings and cutting tools; and thermal protective layers in aircraft and automobile engines [72–96]. In particular, these composite structures consisting of alternating layers of metal and ceramic can offer higher strength-to-weight ratios, less friction and wear, higher

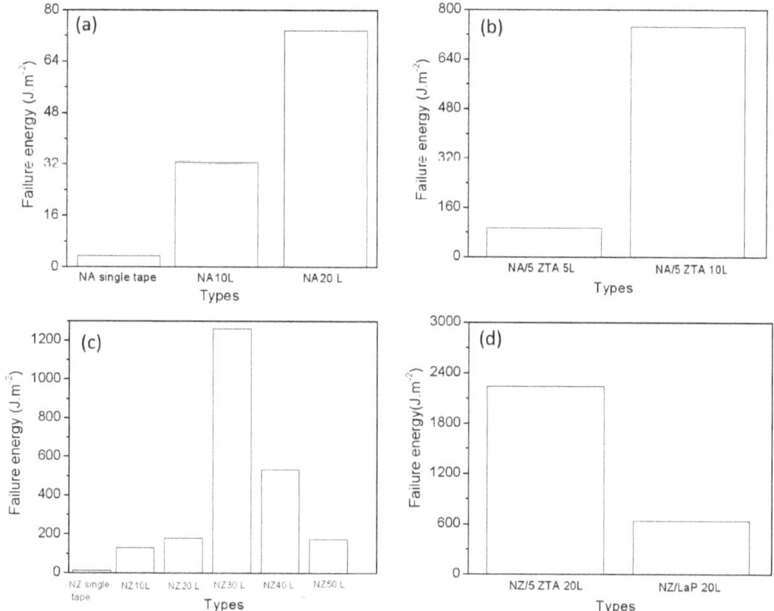

FIGURE 7.21 Variation of failure energy as a function of microstructural interface architecture design in various multilayer ceramic composites (MLCs).

operational temperatures, corrosion resistance, and fracture toughness compared to homogeneous metallic or ceramic structures. These aspects are discussed in the following sections.

Therefore, designing and manufacturing metal–ceramic multilayers at the micro- and nanoscales are attractive strategies for developing a new generation of advanced composite materials [72–96]. Further, the combination of the material properties of metals (ductility, electrical conductivity, load-bearing capacity) and ceramics (high strength even at elevated temperatures, chemical resistance, oxygen–ion conductivity) is able to fulfil the respective application requirements. However, the main problem with metal–ceramic advanced composites is that the manufacturing is often complex if not cost intensive. Since the MLMCHCs have ever-increasing applications in the field of wear resistance, it is very crucial to develop an understanding of their deformation and fracture.

7.13.3.2 Design Aspect, Synthesis Issues, and Application Prospects

The thickness of individual layers during their design has a significant impact on the deformation behaviour under variable normal load. So, this aspect has to be taken care of in the designing of MLMCHCs. These metal and ceramic coatings can be deposited through various processes like chemical vapour deposition and physical vapour deposition (e.g., magnetron spattering, plasma spraying). Wear in multilayer coatings is reported to be less by a factor of almost two in comparison to those of the monolayer ones. MLMCHCs not only cater to the protection requirement against wear and oxidation but also the thermal properties of the coatings play an important role in their performance [72, 73]. For instance, materials with high thermal conductivity are required for electronic devices in order to dissipate heat and act as a heat sink. Materials with low thermal conductivity are required for hard coatings and barrier coated elements in order to reduce the heat transport from the hot to the cold sides and act as heat insulation. In high-speed machining operations like hot forging and casting, the temperature at which the tool–chip interface moves up depends upon various tooling parameters and materials properties like materials softening, friction parameters, hardness, rate of oxidation and thermal conductivity.

Generally, titanium nitride (TiN) and titanium carbonitride (TiCN) hard coatings are used extensively to improve the hardness, friction and corrosion resistance of surfaces used in tool drilling, cutting, etc. [83]. High resistance to wear makes them excellent tribological coatings. For their excellent tribological behaviour, both TiN- and Ti-based hard coatings (e.g. TiN and TiAlN) are well-known hard coatings in industry applications [84, 85].

As the human lifespan has been significantly extended by considerable advancements in medical technology, the short lifespan of implanted prostheses has become one of the most critical issues in the use of, e.g., hip/knee joint replacements. In this context, due to excellent mechanical properties, high biocompatibility, specific strength and corrosion resistance, titanium alloys (Ti6Al4V) and steel (e.g., biograde SS316L) began to be widely explored for joint replacements [73, 85]. However, the poor wear resistance of Ti6Al4V and SS316L has restricted their usage for the articulating surfaces in total hip and knee joint implants [84].

Due to poor wear resistance, these metals generate wear debris in load-bearing implants. These wear debris lead to localized inflammation reactions causing pain and loosening of implants due to osteolysis. Therefore, it is apparent that the formation of only the ceramic nitride phase along with the multilayers of nitride may not be successful in providing resistance against scratches, sliding wear, corrosion and fatigue [84].

In fact, in developed countries alone, the number of primary total hip replacements was reported to be about 700,000, with a revision surgery number of about 80,000 during the years 2006 to 2008. For the same period, the number of primary total knee replacements was about 400,000 with a revision surgery number of about 20,000 in developed countries [84, 85]. The global number of primary and revision surgeries would be much higher. This fact points towards a huge need for many bioimplants with high reliability. This need, in turn, justifies the large market size. Thus, MLMCHCs, as mentioned earlier, are logically expected to have high hardness, fracture toughness, good tribological behaviour and high corrosion resistance along with a large load-bearing capacity in dynamic conditions. These attributes make it suitable as a potential material for biomedical implants.

7.13.3.3 Recent Developments in MLMCHCs: Surface Engineering Issues

Ti/TiN multilayers offer higher corrosion resistance, compressive strength and fatigue resistance with strong adhesion due to a higher number of interfaces as compared to a monolithic TiN layer [84–86]. Among a handful of synthesis techniques, the most popular is reactive magnetron sputtering (Figure 7.22).

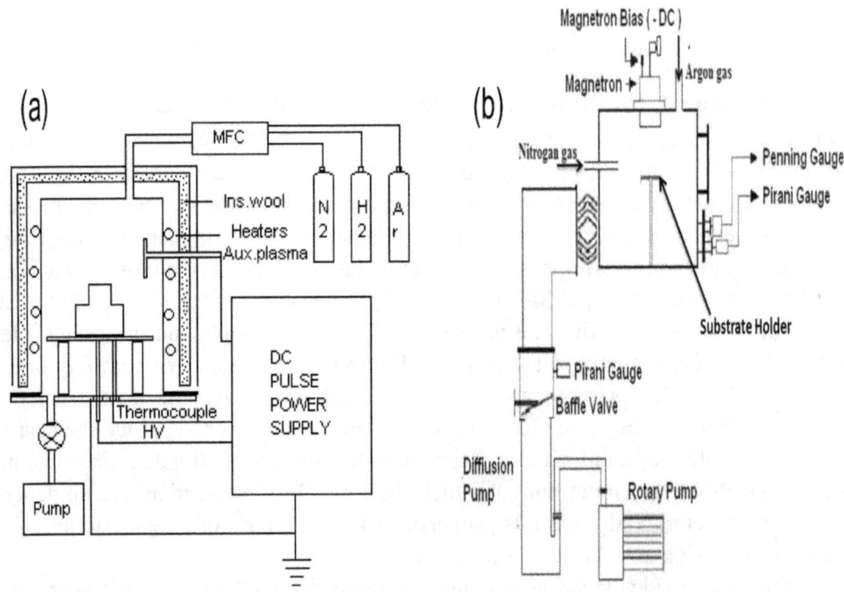

FIGURE 7.22 Reactive magnetron sputtering setup: (a) overview, (b) details of the process.

With reactive magnetron sputtering, TiAlN coating can be deposited on the substrate by TiAl target with approximately 50:50 ratios under an atmosphere of pure argon as a working gas and nitrogen as a reactive gas [72, 87]. Further, the development of complex C/Cr-Ti-N multilayer coatings fabricated through spattering of chromium target and titanium target to form Cr-Ti-N ternary sublayers with a carbon coating on the top is also reported in the literature [79]. The major drawback of other techniques, such as cathodic arc ion plating and plasma-enhanced chemical vapour deposition, is the weak adhesion of the multilayers with the substrate. This happens due to the higher elastic mismatch. Thus, the weak adhesion is the only reason for its failure during circumstances of articulation or any kind of high dynamic mechanical contact.

Thus, in order to overcome such difficulties certain advanced combination surface engineering techniques can be adopted, e.g., plasma nitriding, followed by multilayer coating deposition. The plasma nitriding of Ti provides diffused internal-subsurface layer formation. This layer contains TiN and Ti_2N phases. Such layers are termed the "Embedded Nitrided Layer Zone (ENLZ)" [72, 86–88]. These are ceramic phases. Hence, they are intrinsically hard. As it is just beneath the surface of the substrate, excellent adhesion to the substrate is also guaranteed. Additionally, the deposited coating over the subsurface nitrided layer is also made up of multilayers of Ti/TiN. Thus, such duplex-engineered surfaces with higher elastic modulus compatibility should not only increase the adhesion of the engineered surface but also give resistance towards dynamic fracture. If this promise holds, it should enhance the mechanical and thermal fatigue life.

One additional important point should be kept in mind the surface should have enough cleanness before conducting the experiment in order to control the surface roughness parameters. The cleaning is usually done by inert Ar gas sputtering at ~3.0×10^{-3} to 10^{-5} Pa vacuum. This is achieved by using a diffusion pump supported by a rotary pump. The overall summary of various processes adopted by the researchers [84–86, 89–95] is briefly presented in Table 7.2 [72, 86–88].

TABLE 7.2
Summary of the Methods for Metal–Ceramic Multilayer Composite Structures [84–95]

Composites	Metallic Substrate	Process	Remarks
Ceramic layer over metal	AISI 4340 steel	Plasma nitriding	Mechanical properties improved; fatigue, abrasion and corrosion resistance degraded
Ceramic layer over metal	Stainless steel (SS 304)	Plasma nitriding	Mechanical properties improved; fatigue, abrasion and corrosion resistance degraded
Nitrided ceramic layer TiN, Ti_2N over metal	Ti6Al4V	Plasma nitriding	Mechanical properties improved; fatigue, abrasion and corrosion resistance degraded
Nitrided ceramic layer TiN, Ti_2N over metal	Ti6Al4V	Plasma nitriding	Mechanical properties improved; fatigue, abrasion and corrosion resistance degraded
Nitrided ceramic layer TiN, Ti_2N over metal	Ti6Al4V	Plasma nitriding	Mechanical properties improved; fatigue, abrasion and corrosion resistance degraded
Ti/TiN ceramic bilayers over metal	SS316L	Magnetron sputtering	Static mechanical properties improved; adhesion and fatigue resistance degraded
Ti/TiN multilayers over metal	SS316L	Magnetron sputtering	Static mechanical properties improved; adhesion and fatigue resistance degraded
TiAl/TiAlN multilayer over metal	SS316L	Magnetron sputtering	Static mechanical properties improved; adhesion and fatigue resistance degraded
Nitrided ceramic layer with Ti/TiN multilayers	SS316L, Ti6Al4V	Combined surface engineering	Static and dynamic mechanical properties; adhesion, abrasion and fatigue resistance improved
Nitrided ceramic layer with Ti/TiN multilayers	SS316L, Ti6Al4V	Combined surface engineering	Static and dynamic mechanical properties; adhesion, abrasion and fatigue resistance improved

Such multilayer structures need to be characterized against both static and dynamic contact. Since these are somewhat unique, it is decided to present a brief glimpse of this next.

7.13.3.3.1 Microscratch Resistance

Hardness is a given material resistance against plastic deformation. Therefore, microscratch resistance is measured by the scratch hardness. Scratch hardness is the resistance of a given material against plastic deformation induced by a sharp indenting tip. This technique involves the generation of a controlled scratch with a sharp tip on a selected area of the sample. The tip materials are commonly made up of diamond or hardened metal, which can create a scratch on the surface of the treated/untreated substrate under constant, incremental or progressive load (Figure 7.23a).

The scratch tester can measure the applied normal force, tangential force, penetration depth and acoustic emission. The relevant parameters along with the typical values utilized in actual experiments [72, 85–87] are mentioned in Table 7.3.

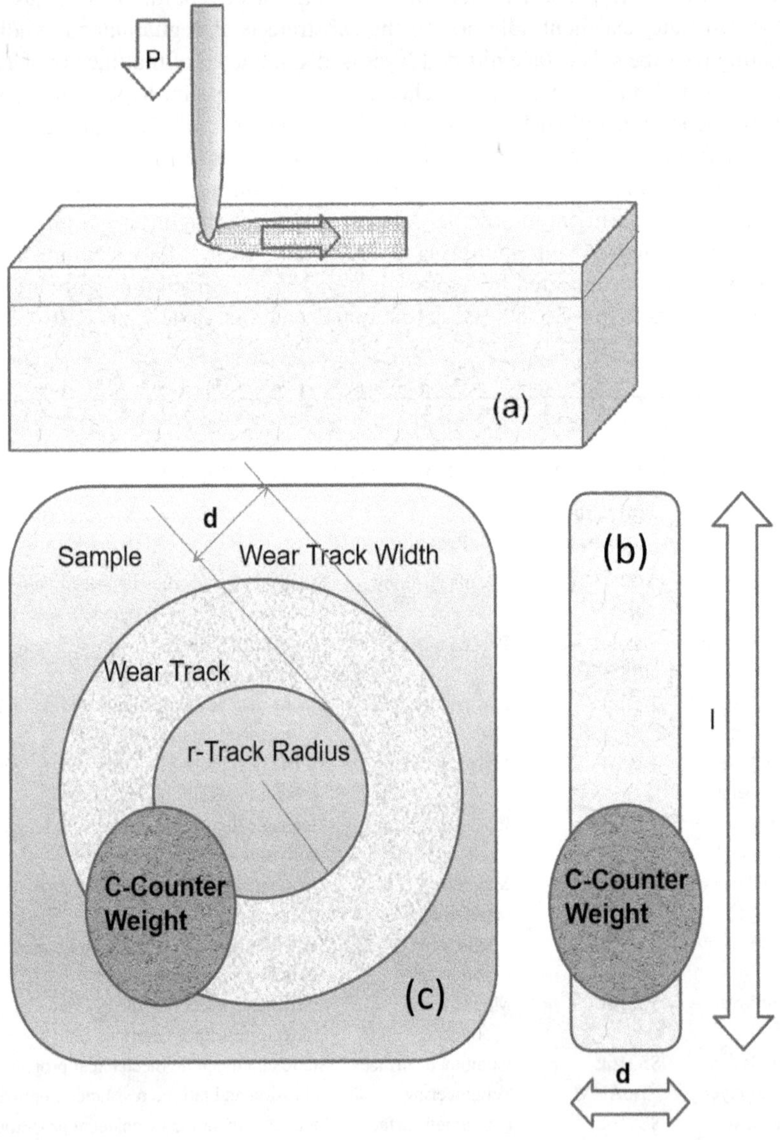

FIGURE 7.23 Tribological experiments: (a) scratch test, (b) ball-on-disk test and (c) sliding wear test.

TABLE 7.3
Typical Parameters for Scratch Teasing [72, 86–88]

Parameter	Value	Measured Parameter
Stroke	Min 1 mm, max 20 mm	Adhesion strength
Normal load	Min 2 N, Max 20	COF and delamination resistance
	Min 20 N, Max 200 N	
Loading rate	Min 0, Max 20 N/mm	Damage resistance
Traction force	Min 0 N, Max 200 N	Measure of surface asperities and damages
Scratching speed	Min 0.1 mm/s, Max 2 mm/s	Measure of adhesion, i.e., scratch resistance

Scratch testing is usually carried out with constant normal load on a specimen or on the reference specimen also using a stylus. The test is used for bulk and coating materials. In such tests, a stylus is moved over a coated specimen surface with a linear increasing load until coating failure occurs at critical loads. Normal force (L) and tangential force (F_t) are recorded. The friction coefficient could be obtained as ($\mu = F_t/L$) at the same time as the tip scratches the sample surface. However, there is also provision for continuously increasing the normal load up to a preset limit when the scratch is conducted [72, 85–87].

Acoustic emission is also measured during the test. In a similar fashion, nanoscratch and nanotribological tests can also be carried out under ambient conditions by using a conventional nanoindenter (e.g., TriboScanner, Hysitron), which is generally equipped with a scanning probe microscope (SPM). The SPM is usually equipped with a dedicated capacitive transducer attached with a conical diamond tip (e.g., 1 μm tip radius with a 90° cone angle). The critical load for the coating failure is a complex function of coating-substrate adhesion, stylus-tip radius, loading rate, thickness of layers, interfacial flaws within the interface and internal stress.

7.13.3.3.1.1 Case Studies Microscratch studies are reported in the literature for the CrN/Cr multilayers deposited onto both silicon and steel substrates. One study investigated the deformation and adhesion behaviour of the composite structure having a CrN/Cr alternate architecture of layers [94]. As expected, the ductile Cr layers fail with a discontinuous chip removal mechanism [94]. On the contrary, the CrN single-layer coating exhibits an abrupt delamination by a flaking mechanism. Scratch tracks in all the multilayers provide evidence for the formation of tensile cracks before the delamination. Further, the adhesion of CrN/Cr multilayers increases if the layer thickness decreases. This is suggested to be due to the relatively lower values of residual stress in comparison to those of the single layers of the CrN and Cr coatings [94].

Similarly, when a series of scratches are made on samples of nickel on silicon and nickel on NiO, no sharp load drop is reported [95] to be observed, whereas in the case of Ni/Al_2O_3, a sharp load drop signifies the start of film delamination. For the Ni/Si sample, the load drop occurs at a load of 34 ± 1 mN, and for the Ni/NiO sample, the load drop occurs at a load of 37 ± 1 mN. The load at which the film delamination begins is referred to as the critical load to failure [95]. The width of the scratch track at the point of film delamination, the area of delamination and the critical normal load to failure are the major factors here [95]. The adhesion strength of the interfacial layers is suggested to depend upon the difference in the modulus of elasticity of the individual layers present at the interface.

The results of progressive scratch tests show [96] that once the CrN/TiN multilayer is deposited on cemented carbide, the interfacial composite is unable to tolerate the deformation. As a consequence, cracks are rapidly initiated and propagated. Generally, when the contact pressure levels are well above the plastic yielding onset of the substrate, then with increasing applied contact load, the tensile radial stresses and strains existing in the vicinity of the residual imprints become large enough to generate the circumferential cracks in the interface layer [96]. The resistance against crack initiation at the interface generally depends upon the value of the (H/E^*) and (H^3/E^{*2}) ratios. Here, the quantity H refers to hardness and the quantity (E^*) refers to the reduced modulus.

Recent experiments [72, 85–87] also report the microscratch properties of Ti/TiN multilayers deposited on SS316L and Ti6Al4V substrates. These situations enhance the elastic mismatch stress. Similar measurements are also made for the same Ti/TiN multilayers deposited by magnetron sputtering on engineered nitrided modified SS316L and Ti6Al4V substrate having elastically matched ceramic layers of FeN and CrN for SS316L, and TiN and Ti_2N for Ti6Al4V [88]. Microscratch studies on duplex surface engineered SS316L (e.g., plasma nitriding and Ti/TiN multilayering) show [88] that the contact stresses (σ_c) and yield stresses of the engineered surfaces are significantly higher than those of the substrate metals. These results [72, 86–88] confirm that such novel surface engineering may help in achieving relatively higher intrinsic strength and higher fracture resistance of the interface layer. The contact stress decreases with an increase in the applied normal load for such intrinsic tough layers [72, 86–88].

7.13.3.3.2 Mechanisms of Sliding Wear

Sliding wear mechanisms in ceramics and composites can be dominated by plastic flow and fracture, depending on the applied normal load and intrinsic mechanical properties of the contacting pairs. Mild wear in ceramics and composites is associated with a low wear rate, smooth plastically deformed surfaces and a relatively smaller magnitude of steady-state friction. Under conditions of mild wear, the predominant mechanism is plastic flow. Hence, the wear debris possesses fine microstructure. On the other hand, the mechanism of severe wear in ceramics and composites is fracture dominated. It leads to a relatively much higher wear rate. The worn surface bears a rough surface. The friction coefficient undulates rapidly prior to reaching a steady-state friction at a later stage when the steady state is reached.

7.13.3.3.2.1 Characterization of Sliding Wear Most sliding wear rigs involve a pair of materials in contact and relative motion. In general, the inputs are the normal load (P), sliding distance (L) and sliding speed (v), and the outputs are the friction coefficient (μ) and wear coefficient (K; in mm^3/N m). There are two types of configurations for sliding wear. One is a linear motion in bidirectional sliding repetitive motion in contact between the sliding couples, one of which is static (e.g., the sample) and the other is dynamic (e.g., the pin). The pin moves as mentioned earlier and repeats the sliding motion as per preset experimental parameters. This is shown schematically in Figure 7.23b. The same setup can be used to make one single pass as well.

The other type of motion is a circular, unidirectional motion, as shown schematically in Figure 7.23c. The circular motion can be made bidirectional by switching the rotation direction (clockwise and counterclockwise) in every cycle. The two most popular test techniques are the pin-on-disc (POD) test to be conducted as per the ASTM G99 standard and the ball-on-disc (BOD) test to be conducted as per the ASTM G133 standard. The configuration consists of a rotating disc of the material to be tested against a stationary pin or a ball as the case may be. The tester consists of a reciprocating horizontally mounted flat specimen that slides against a stationary ball-shaped upper specimen. The test load, speed and sliding distance (in both configurations) are to be determined by the severity of the application or the purpose of the testing. The following case studies are presented to highlight the importance of such characterization techniques from the perspective of sliding wear studies.

7.13.3.3.2.2 Case Studies In the literature, several surface treatment techniques are reported [72–96]. These studies are carried out with the intention of improving the wear resistance of materials by adding alloying elements or designing multilayer composites [72–96]. In the study reported in Ref. [73], TiN and TiAlN coatings are deposited on steel substrates by means of arc ion plating. Then, the tribological measurements are carried out at various loads (1N and 5N). The sliding speeds are varied as 0.1, 0.3 and 0.5 m/s. This study [73] uses an alumina and a steel ball as the counter bodies in a conventional ball-on-disc machine.

Tribological tests performed with the steel ball show adhesive wear for both TiN and TiAlN coatings. When sliding against the steel counter body, wear debris is smeared on the wear track, especially in the case of the TiN coating. However, tribological tests conducted using an alumina ball illustrate an abrasive-like wear behaviour. Further, both coatings exhibit an accumulation of debris at the boundaries of the wear track. Consequently, the TiAlN coatings are reported to be more favourable than TiN coatings for high-speed machining owing to the lower friction coefficient at the highest sliding speed [76]. Further, the Ti/TiN multilayers having a thickness of 4 μm and deposited over steel exhibit better cavitation erosion resistance than stainless steel [91]. It is suggested that the surface roughness of the substrate could play a major role in determining the coating adhesion and hence, its wear resistance [91].

7.13.4 Recent Innovative Developments

Similarly, the thermal expansion coefficient of the substrate SS316L steel is approximately twice as high as that of Ti [72, 85–87]. Thus, the thermal stress generated between the Ti and the TiN layer and the substrate Ti6Al4V provides a direct effect on the adhesion, erosion and wear resistance of the composite interface. In other words, the thermal stress generated is less. Hence, the adhesion of the multilayer coating on the Ti6Al4V is much better [72, 85–87]. This happens as modification of the subsurface is achieved by plasma nitriding.

It is reported [85–87] that the formation of TiN and Ti_2N phases occurs on the subsurface. Therefore, the difference in the thermal expansion coefficient is negligible. As a result, the generation of thermal stress is much less. Consequently, the adhesion and wear resistance of the composite interface become excellent [86–88]. The reduction in microscratch width in sample surfaces engineered by duplex treatment, e.g., by both plasma nitriding and magnetron sputtered Ti/TiN multilayer formation, is evident in the FESEM photomicrographs presented in Figure 7.24 [72, 85–87].

Recent works [85–87] identify also evidence of Ti/TiN multilayer formation on Ti6Al4V substrates by energy dispersive x-ray (EDX) analysis and FESEM techniques (Figure 7.25a, b). Further, the simple plasma nitriding process enhances the nanohardness of SS316L and Ti6Al4V substrates by about 130% (Figure 8.25c) at the interfacial zone (IZ) with respect to that of the bulk metallic layer zone (BMLZ) [72], and by about 120% (Figure 7.25d) with respect to that of the BMLZ in the cross-section of the Ti6Al4V alloy [86–88].

Further, the duplex surface engineered (plasma nitrided along with Ti/TiN multilayered) SS316L-based MCMLHCs have a 70% less wear rate as compared to that of the as-received SS316L [72]. This happens due to the poor adhesion and higher elastic mismatch between the individual components of the composite layers. This situation leads to very poor wear resistance in the Ti/TiN multilayers SS316L [72].

On the contrary, the duplex surface engineered Ti6Al4V has almost a 99% less wear rate as compared to that of normal metallic Ti6Al4V alloy [86–88]. Similarly, the wear resistance of simply plasma nitrided (PN) Ti6Al4V alloy is also much better than that of the as-received Ti6Al4V alloy [85–87]. Thus, both the PN and PN+multilayered Ti6Al4V have greater wear resistance as compared to that of metallic Ti6Al4V alloy. In both of these cases, the TiN, Ti_2N, and α- and β-Ti6Al4V phases are formed. Therefore, there is a good match of elastic moduli and thermal expansion coefficient.

This match reduces the interfacial thermal stress. As a result, only slight plastic deformation of ceramic TiN and Ti_2N layers occurs. The introduction of the ductile Ti layer reduces the possibility of severe brittle fracture. This situation improves not only the microscratch resistance (Figure 7.24) and nanohardness (Figure 7.25) but also the nanoscratch resistance (Figure 7.26) of the duplex surface engineered Ti6Al4V alloy-based MCMLHCs [85–87].

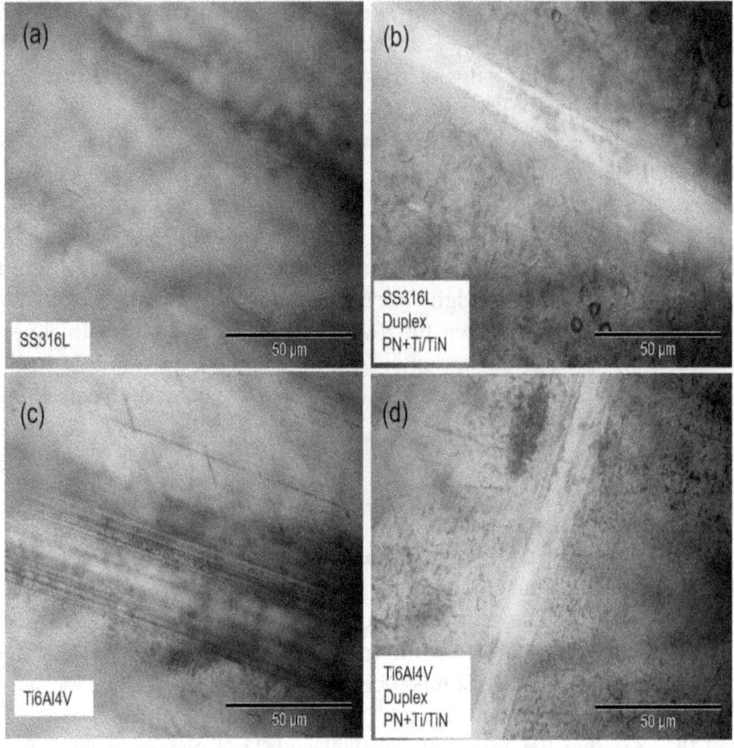

FIGURE 7.24 Microscratch tracks in (a) SS316L, (b) plasma nitrided (PN) SS316L with Ti/TiN MCMLHCs, (c) Ti6Al4V and (d) PN Ti6Al4V with Ti/TiN MCMLHCs architecture design (reprinted with permission from [85,86]).

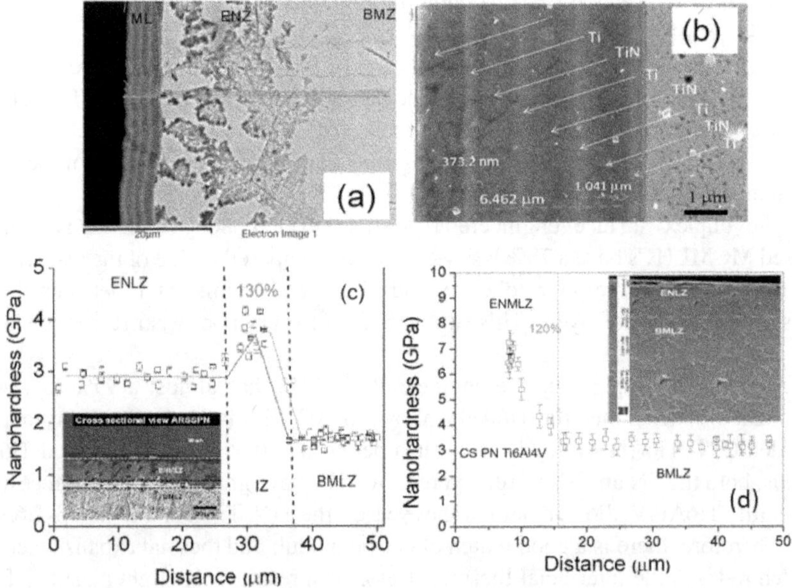

FIGURE 7.25 (a) Microstructure of Ti/TiN layers on PN Ti6Al4V, (b) details of Individual Ti and TiN layers shown in panel a, nanohardness as a function of distance from the surface in the cross-sections of (c) PN SS316L Ti6Al4V and (d) PN Ti6Al4V (reprinted with permission from [85, 86]).

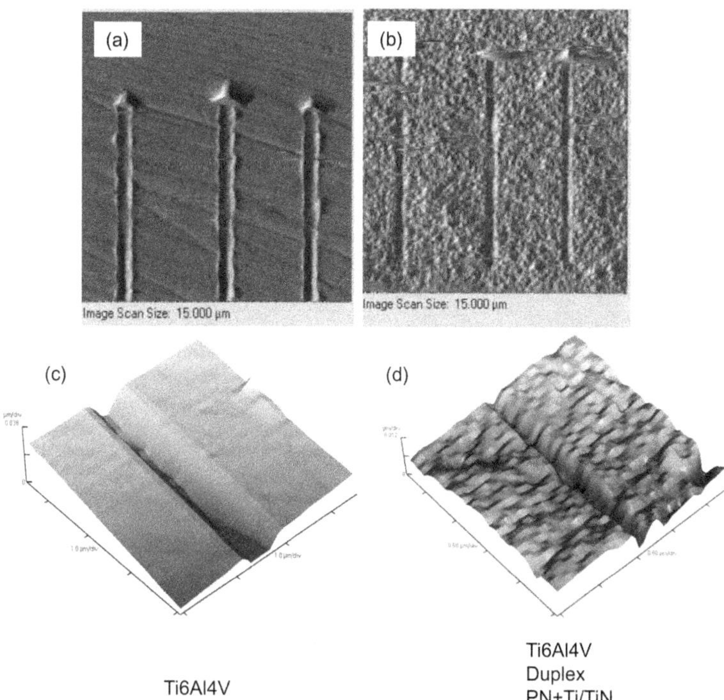

FIGURE 7.26 SPM images of top views of nanoscratch in (a) Ti6Al4V and (b) PN Ti6Al4V with Ti/TiN MCMLHCs design. (c) Topographic view of panel a. (d) Topographic view of panel b.

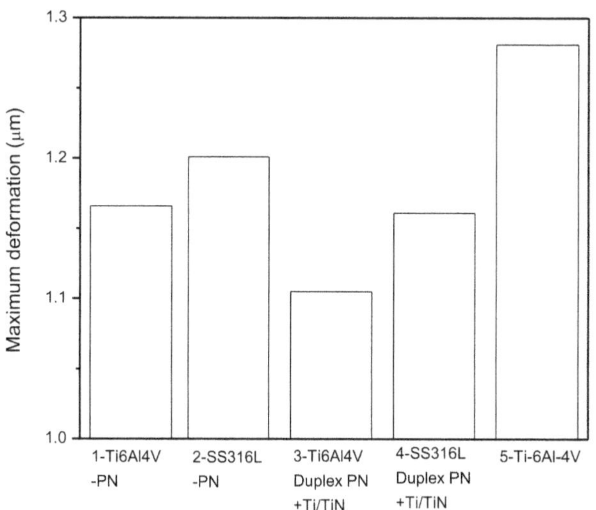

FIGURE 7.27 Maximum deformation recorded during actual cyclic fatigue experiments conducted in simulated body fluid in SS316L, PN SS316L, PN Ti6Al4V with Ti/TiN MCMLHCs design, PN SS316L with Ti/TiN MCMLHCs design, and Ti6Al4V [87].

Further, this unique combination of the beneficial properties led to the highest deformation and fatigue resistance (Figure 7.27) for the femur heads actually manufactured from the duplex surface engineered Ti6Al4V alloy-based MCMLHCs and tested in simulated body fluid under appropriately chosen load in a fatigue tester [87].

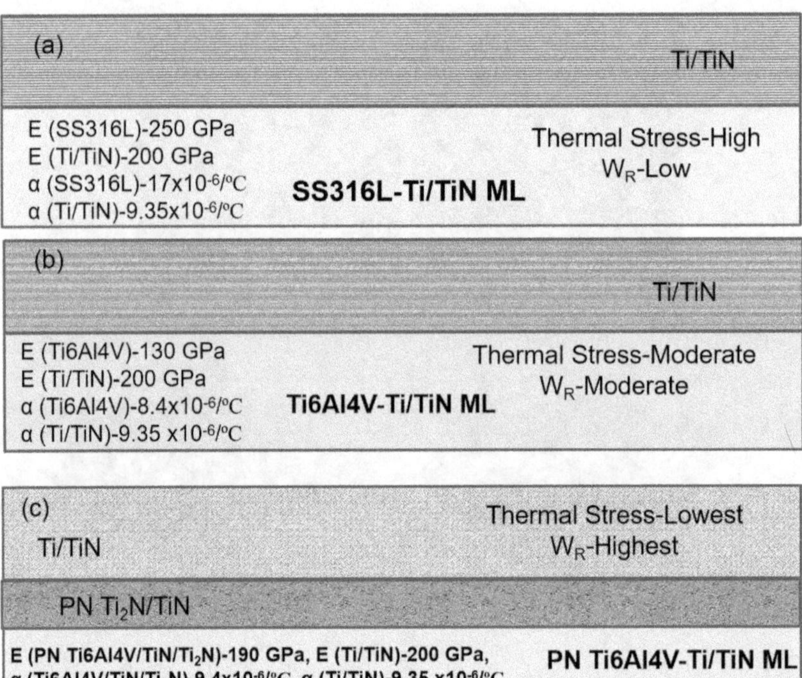

FIGURE 7.28 Schematic highlighting the importance of design philosophy adoptable in MCMLHCs architectures meant to achieve high wear resistance (W_R): (a) SS316L–Ti/TiN MCMLHCs, (b) Ti6A14V–Ti/TiN MCMLHCs and (c) PN Ti6A14V–Ti/TiN MCMLHCs design.

Finally, Figure 7.28 explains the basic concept and physical principles involved in the appropriate design of the various MCMLHCs developed in the recent works [72, 85–87] as detailed earlier.

7.14 SUMMARY AND WAY AHEAD

This chapter provides a brief overview of various conventional composites – MMCs, PMCs, CMCs – and especially the advanced composites. In the case of MMCs, the challenges are identified as the development of biodegradable Mg-based MMCs and interface mechanical properties characterization in FRMMCs, PRMMCs and cold-sprayed/thermally sprayed MMC coatings. In the case of the FRCMCs (GFRPs and CFRPs) as well as PRPMCs, the major challenge is to ensure homogeneous distribution of the reinforcing phases, identification of interface regions through advanced techniques and local mechanical properties characterization of the various interface regions. As far as the CMCs are concerned, the governing processes that determine the adhesion of the reinforcing phase and the matrix phase need to be much better understood. Controlled experiments conducted in situ or ex situ are needed to understand the nanomechanical and/or micromechanical behaviour of the interface region under predetermined loading conditions. In the case of tape cast MLCMCs, the major challenge is to identify the proper choice of the two interfaces, which will help to realize high resistance against failure without seriously compromising the failure strength. However, in the case of MCMLHCs, the major issue of challenge emerges to be the cushioning of residual elastic and thermal mismatch stress that can ensure high resistance against microscratches, nanoscratches, sliding wear and cyclic fatigue. It is shown that when properly designed, such MCMLHCs can be a very promising material for biomedical prosthetic applications.

REFERENCES

1. X. Pang, Y. Song, N. Shi, M. Xu, C. Zhou, J. Chen, Design of zero thermal expansion and high thermal conductivity in machinable xLFCS/Cu metal matrix composites, Compos. B: Eng. 238 2022 109883.
2. A. Adetunla, A. Fide-Akwuobi, H. Benjamin, A. Adeyinka, A. Kolawole, A study of degradable orthopedic implant: An insight in magnesium metal matrix composites, *Heliyon* 8 2022 e10503.
3. G. Luo, J. Guo, J. Hu, P. Li, Y. Sun, Q. Shen, Microstructure and properties of W-Ag matrix composites by designed dual-metal-layer coated powders, Mater. Des. 219 2022 110733.
4. E. Avcu, F. E. Bastan, M. Guney, Y. Y. Avcu, M. A. Ur Rehman, A. R. Boccaccini, Biodegradable polymer matrix composites containing graphene-related materials for antibacterial applications: A critical review, Acta Biomater. 151 2022 1–44.
5. G. R. A. Neto, F. H. M. Matheus, C. A. G. Beatrice, D. R. Leiva, L. A. Pessan, Fundamentals and recent advances in polymer composites with hydride-forming metals for hydrogen storage applications, *Int. J. Hydrog. Energy* 47 2022 34139–34164.
6. J. Zhang, C. Wei, J. Ran, Y. Li, J. Chen, Properties of polymer composite with large dosage of phosphogypsum and its application in pipeline, *Polym. Test.* 116 2022 107742.
7. Z. Fu, A. Pang, H. Luo, K. Zhou, H. Yang, Research progress of ceramic matrix composites for high temperature stealth technology based on multi-scale collaborative design, *J. Mater. Res. Technol.* 18 2022 2770–2783.
8. X. Wang, X. D. Gao, Z. Zhang, L. Cheng, H. Ma, W. Yang, Advances in modifications and high-temperature applications of silicon carbide ceramic matrix composites in aerospace: A focused review, J. Euro. Ceram. Soc. 41 2021 4671–4688.
9. H. Yoon, P. Matteini, B. Hwang, Review on three-dimensional ceramic filler networking composites for thermal conductive applications, J. Non-Cryst. Solids 576 2022 121272.
10. K.Zhong, J. Zhou, C. Zhao, K. Yun, L. Qi , Effect of interfacial transition layer with CNTs on fracture toughness and failure mode of carbon fiber reinforced aluminum matrix composites, *Compos. Part A: Appl. Sci. Manuf.* 163 2022 107201.
11. C. Zweben, Z. Consulting, D. Pennsylvania, *Chapter 10, Mechanical Engineers' Handbook*, Fourth Edition, edited by Myer Kutz, John Wiley & Sons, Inc., 2015 1–37.
12. N. Tsangarakis, J. Nunes, J. M. Slepetz, Fracture toughness testing of metal matrix composites, *Eng. Frac. Mech.* 30 1988 565–577.
13. S. Suresh kumar, S. Thirumalai Kumaran, G. Velmurugan, A. Perumal, S. Sekar, M. Uthayakumar, Physical and mechanical properties of various metal matrix composites: A review, *Mater. Today: Proc.* 50 2022 1022–1031.
14. L. Qian, T. Kobayashi, H. Toda, T. Goda, Z. G Wang, Fracture toughness of a 6061Al matrix composite reinforced with fine SiC particles, *Mater. Trans.*. 43 2002 2838–2842.
15. D. P. Myriounis, S. T. Hasan, N. M. Barkoula, A. Paipetis, T. E. Matikas, Effects of heat treatment on microstructure and the fracture toughness of SiCp/Al alloy metal matrix composites, *J. Adv. Mater. - Covania* 41 2009 18–27.
16. M. A. Maleque, A. A. Adebisi, N. Izzati, Analysis of fracture mechanism for Al-Mg/SiCp composite materials, *IOP Conf. Series: Mater. Sci. Eng.* 184 2017 012031.
17. A. Rabiei, L. Vendra, T. Kishi, Fracture behavior of particle reinforced metal matrix composites, *Compos. Part A* 39 2008 294–300.
18. M. Malaki, A. F. Tehrani, B. Niroumand, Fatigue behavior of metal matrix nanocomposites, *Ceram Int.* 46 2020 23326–23336.
19. S. Chanda, P. Chandrasekhara, R. K. Sarangia, R. K. Nayak, Influence of B4C particles on processing and strengthening mechanisms in aluminum metal matrix composites: A review, *Mater. Today: Proc.* 18 2019 5356–5363.
20. A. Nieto, A. Bisht, D. Lahiri, C. Zhang, A. Agarwal, Graphene reinforced metal and ceramic matrix composites: A review, *Int. Mater. Rev.* 62 2017 241–302.
21. S. C. Tjong, Y. W. Mai, Processing-structure-property aspects of particulate- and whisker-reinforced titanium matrix composites, *Compos. Sci. Technol.* 68 2008 583–601.
22. R. Krishnan, S. Pandiaraj, S. Muthusamy, H. Panchal, M. S. Alsoufifi, A. M. Mahmoud Ibrahim, A. Elsheikh, Biodegradable magnesium metal matrix composites for biomedical implants: Synthesis, mechanical performance, and corrosion behavior: A review, *J. Mater. Res. Technol.* 20 2022 650–670.
23. K. Shirvanimoghaddam, S. U. Hamim, M. K. Akbari, S. M. Fakhrhoseini, H. Khayyam, A. H. Pakseresht, E. Ghassali, M. Zabet, K. S. Munir, S. Jia, J. P. Davim, M. Naebe, Carbon fiber reinforced metal matrix composites: Fabrication processes and properties, Compos. Part A: Appl. Sci. Manuf. 92 2017 70–96.

24. L. He, M. Hassani, A review of the mechanical and tribological behavior of cold spray metal matrix composites, *J. Therm. Spray. Tech.* 29 2020 1565–1608.
25. Q. Guo, Y. Han, D. Zhang, Interface-dominated mechanical behavior in advanced metal matrix composites, *Nano Mater. Sci.* 2 2020 66–71.
26. K. M. Mussert, W. P. Vellinga, A. Bakker, S. V. D. Zwaag, A nano-indentation study on the mechanical behaviour of the matrix material in an AA6061 - Al_2O_3 MMC, *J. Mater. Sci.* 37 2002 789–794.
27. E. Avcu, F. E. Bastan, M. Guney, Y. Y. Avcu, M. A.Ur Rehman, A. R. Boccaccini, Biodegradable polymer matrix composites containing graphene-related materials for antibacterial applications: A critical review, Acta Biomater. 151 2022 1–44.
28. A. Ashothaman, J.Sudha, N.Senthilkumar, A comprehensive review on biodegradable polylactic acid polymer matrix composite material reinforced with synthetic and natural fibers, *Mater. Today Proc.* https://doi.org/10.1016/j.matpr.2021.07.047.
29. P. Kadambi, P. Luniya, P. Dhatrak, Current advancements in polymer/polymer matrix composites for dental implants: A systematic review, *Mater. Today Proc.* 46 2021 740–745.
30. A. K. Sharma, R. Bhandari, C. Sharma, S. K. Dhakad, C. P. Bretotean, Polymer matrix composites: A state of art review, *Mater. Today Proc.* 57 2022 2330–2333.
31. S. Sajan, D. P. Selvaraj, A review on polymer matrix composite materials and their applications, *Mater. Today: Proc.* 47 2021 5493–5498.
32. T. P. Naik, I. Singh, A. K. Sharma, Processing of polymer matrix composites using microwave energy: A review, *Compos. - A: Appl. Sci. Manuf.* 156 2022 106870.
33. A. S. Shinde, I. Siva, Y. Munde, M. Thariq, H. Sultan, L. S. Hua, F. S. Shahar, Numerical modelling of drilling of fiber reinforced polymer matrix composite: A review, *J. Mater. Res. Technol.* 20 2022 3561–3578.
34. L. Tang, J. Zhang, Y. Tang, J. Kong, T. Liu, J. Gu, Polymer matrix wave-transparent composites: A review, *J. Mater. Sci. Technol.* 75 2021 225–251.
35. P. Song, B. Liu, H. Qiu, X. Shi, D. Cao, J. Gu, MXenes for polymer matrix electromagnetic interference shielding composites: A review, *Compos. Commun.* 24 2021 100653.
36. S. Huang, Q. Fu, L. Yana, B. Kasala, Characterization of interfacial properties between fibre and polymer matrix in composite materials: A critical review, *J. Mater. Res. Technol.* 13 2021 1441–1484.
37. P. Davies, B. R. K. Blackman, A. J. Brunner, Standard test methods for delamination resistance of composite materials: Current status, *Appl. Compos. Mater.* 5 1998 345–364.
38. L. B. Ilcewicz, P. E. Keary, J. Trostle, lnterlaminar fracture toughness testing of composite mode I and mode II DCB specimens, *Polym. Eng. Sci.* 28 1988 592–604.
39. V. I. Rizov, Fracture in composites – An overview (part I), *J. Theor. Appl. Mech.* 42 2012 3–42.
40. A. J. Brunner, Fracture mechanics testing of fiber-reinforced polymer composites: The effects of the "human factor" on repeatability and reproducibility of test data, *Eng. Fract. Mech.* 264 2022 108340.
41. A. J. Brunner, Fracture mechanics test standards for fiber-reinforced polymer composites: Suggestions for adapting them to industry 4.0 and the digital age, *Procedia Struct. Integr.* 28 2020 546–554.
42. R. Healey, N. M. Chowdhury, W. K. Chiu, J. Wang, Experimental and numerical determination of mode II fracture toughness of woven composites verified through unidirectional composite test data, *Polym. Polym. Compos.* 27 2019 557–566.
43. R. Naseem, L. Zhao, V. D. Silberschmidt, Y. Liu, Z. Zhang, Characterization of mechanical properties of polymeric stent using nanoindentation, *Procedia Struct. Integr.* 15 2019 51–54.
44. G. Arora, H. Pathak, Nanoindentation characterization of polymer nanocomposites for elastic and viscoelastic properties: Experimental and mathematical approach, Compos. *Part C: Open Access* 4 2021 100103.
45. J. Sun, D. Ye, J. Zou, X. Chen, Y. Wang, J. Yuan, H. Liang, H. Qu, J. Binner, J. Bai, A review on additive manufacturing of ceramic matrix composites, *J. Mater. Sci. Technol.* 138 2023 1–16.
46. W. Wang, L. Zhang, X. Dong, J. Wu, Q. Zhou, S. Li, C. Shen, W. Liu, G. Wang, R. He, Additive manufacturing of fiber reinforced ceramic matrix composites: Advances, challenges, and prospects, *Ceram. Int.* 48 2022 19542–19556.
47. E. Pietrzykowska, M. Małysa, A. Chodara, B. Romelczyk-Baishya, K. Szlazak, W. Swieszkowski, Z. Pakieła, W. Lojkowski, Biodegradable ceramic matrix composites made from nanocrystalline hydroxyapatite and silk fibers via crymilling and uniaxial pressing, *Mater. Lett.* 293 2021 129672.
48. V. N. B. P. Sodisetty, A. K. Singh, H. James, K. Lee, Z. Benedict, Damage evolution in quasi-statically indented alumina based oxide/oxide ceramic matrix composites: An experimental investigation, *Ceram. Int.* 48 2022 32491–32503.

49. J. Yang, J. Chen, F. Ye, L. Cheng, Y. Zhang, High-temperature atomically laminated materials: The toughening components of ceramic matrix composites, *Ceram. Int.* 48 2022 32628–32648.
50. Y. Yan, H. Mei, M. Zhang, Z. Jin, Y. Fan, L. Cheng, L. Zhang, Key role of interphase in continuous fiber 3D printed ceramic matrix composites, *Compos. Part A* 162 2022 107127.
51. P. Makurunje, S. C. Middleburgh, W. E. Lee, Addressing high processing temperatures in reactive melt infiltration for multiphase ceramic composites, *J. Eur. Ceram. Soc.* 2022. https://doi.org/10.1016/j.jeurceramsoc.2022.09.002.
52. Y. Wang, Z. He, J. Yang, X. Ye, C. Yu, S. Qiu, L. Yao, Q. Zeng, D. Jia, Z. Wang, B. Li, X. Pan, Effects of graphene dispersion in hot pressing UO_2-graphene nanosheet ceramic matrix composites, *Ceram. Int.* 48 2022 30779–30787.
53. X. Lv, F. Ye, L. Cheng, L. Zhang, Novel processing strategy and challenges on whisker-reinforced ceramic matrix composites, *Comp. Part A* 158 2022 106974.
54. G. Yu, Y. Ji, C. Xie, J. Du, X. Gao, Y. Song, F. Wang, Transverse tensile mechanical experimental method and behavior of ceramic matrix mini-composites, *Compos. Struct.* 297 2022 115923.
55. Y. Koutsawa, G. Rauchs, D. Fiorelli, A. Makradi, S. Belouettar, A multi-scale model for the effective electro-mechanical properties of short fiber reinforced additively manufactured ceramic matrix composites containing carbon nanotubes, Compos. *Part C: Open Access* 7 2022 100234.
56. J. Du, H. Zhang, Y. Geng, W. Ming, W. He, J. Ma, Y. Cao, X. Li, K. Liu, A review on machining of carbon fiber reinforced ceramic matrix composites, *Ceram. Int.* 45 2019 18155–18166.
57. K. M. Mehta, S. K. Pandey, V. A. Shaikh, Unconventional machining of ceramic matrix composites – A review, *Mater. Today Proc.* 46 2021 7661–7669.
58. X. Zhang, A. Li, J. Chen, M. Ma, P. Ding, X. Huang, T. Yu, J. Zhao, Sustainability-driven optimization of ultrasonic atomization-assisted micromilling process with ceramic matrix composite, *Sustain. Mater. Technol.* 33 2022 e00465.
59. Y. Araia, R. Inoueb, K. Gotoc, Y. Kogo, Carbon fifiber reinforced ultra-high temperature ceramic matrix composites: A review, *Ceram. Int.* 45 2019 14481–14489.
60. X. Wang, X. Gao, Z. Zhang, L. Cheng, H. Ma, W. Yang, Advances in modifications and high-temperature applications of silicon carbide ceramic matrix composites in aerospace: A focused review, *J. Eur. Ceram. Soc.* 41 2021 4671–4688.
61. F. Servadei, L. Zoli, P. Galizia, A. Piancastelli, D. Sciti, Processing and characterization of ultra-high temperature ceramic matrix composites via water based slurry impregnation and polymer infiltration and pyrolysis, *Ceram. Int.* 2022. https://doi.org/10.1016/j.ceramint.2022.09.100.
62. P. Maiti, M. Bhattacharya, P. S. Das, P. Sujatha Devi, A. K. Mukhopadhyay, Indentation size effect and energy balance issues in nanomechanical behavior of ZTA ceramics, *Ceram. Int.* 44 2018 9753–9772.
63. P. Maiti, A. Eqbal, M. Bhattacharya, P. S. Das, J. Ghosh, A. K. Mukhopadhyay, Micro pop-in issues in nanoindentation behaviour of 10 ZTA ceramics, *Ceram. Int.* 45 2019 8204–8215.
64. P. Maiti, M. Bhattacharya, P. S. Das, J. Ghosh, A. K. Mukhopadhyay, A critical note on nanoscale plasticity in 20 ZTA ceramics, *Ceram. Int.* 45 2019 25034–25043.
65. P. Maiti, J. Ghosh, A. K. Mukhopadhyay, New observations and critical assessments of incipient plasticity events and indentation size effect in nanoindentation of ceramic nanocomposites, *Ceram. Int.* 46 2020 3144–3165.
66. S. Sarpure, *Ceramic Nano Composites for Structural Applications, M. Tech.* Dissertation Thesis, MNNIT, Allahabad, India, 2005.
67. A. Sinha, *M. Tech.* Dissertation Thesis, IIEST, Shibpur, Howrah, India, 2005.
68. S. Ghosh, A. Guha, A. K. Mukhopadhyay, H. S. Maiti, Tape cast porous nano alumina multilayer composites, *Trans. Ind. Ceram. Soc.* 64 2005 101.
69. S. Ghosh, A. Guha, A. K. Mukhopadhyay, H. S. Maiti, Sintering and hardness behaviour of nano zirconia tapes, *Trans. Ind. Ceram. Soc.* 64 2005 213.
70. A. K. Mukhopadhyay, *Final Project Report on "Development of laminated ceramic composites for structural applications"*, Department of Science and Technology, Government of India, 2002.
71. S. Ghosh, A. Guha, K. M. Krishna, A. K. Mukhopadhyay, H. S. Maiti, Tape cast multilayer composite of nano zirconia with high toughness, *Mater. Manuf. Process.* 21 2006 662.
72. A. Samanta, R. Rane, B. Kundu, D. K. Chanda, J. Ghosh, S. Bysakh, G. Jhala, A. Joseph, S. Mukherjee, M. Das, A. K. Mukhopadhyay, Bio-tribological response of duplex surface engineered SS316L for hip-implant application, *Appl. Surf. Sci.* 507 2020 145009.
73. M. K. Samani, X. Z. Ding, N. Khosravian, B. Amin-Ahmadi, Yang Yi, G. Chen, E. C. Neyts, A. Bogaerts, B. K. Tay, Thermal conductivity of titanium nitride/titanium aluminum nitride multilayer coatings deposited by lateral rotating cathode arc, *Thin Solid Films* 578 2015 133–138.

74. S. W. Ko, T. Dechakupt, C. A. Randall, S. Trolier-McKinstry, M. Randall, A. Tajuddin, Chemical solution deposition of copper thin films and integration into a multilayer capacitor structure, *J. Electroceram.* 24 2010 161–169.
75. T. Koseki, J. Inoue, S. Nambu, Development of multilayer steels for improved combinations of high strength and high ductility, *Mater. Trans.* 55 2014 227–237.
76. Y. Sahin, The effects of various multilayer ceramic coatings on the wear of carbide cutting tools when machining metal matrix composites, *Surf. Coat. Technol.* 199 2005 112–117.
77. L. Ghalandari, M. M. Moshksar, High-strength and high-conductive Cu/Ag multilayer produced by ARB. *J. Alloys Compd.* 506 2010 172–178.
78. A. A. Voevodin, J. M. Schneider, C. Rebholz, A. Matthews, Multilayer composite ceramic-metal-DLC coatings for sliding wear applications, *Tribol. Int.* 29 1996 559–570.
79. M. P. Schmitt, A. K. Rai, R. Bhattacharya, D. M. Zhu, D. E. Wolfe, Multilayer thermal barrier coating (TBC) architectures utilizing rare earth doped YSZ and rare earth pyrochlores, *Surf. Coat. Technol.* 251 2014 56–63.
80. Y.-L. Shen, S. Suresh, Steady-state creep of metal-ceramic multilayered materials, *Acta Mater.* 44 1996 1337–1348.
81. J. P. Chu, J. E. Greene, J. S. C. Jang, J. C. Huang, Y. -L. Shen, P. K. Liaw, Y. Yokoyama, A. Inoue, T. G. Nieh, Bendable bulk metallic glass: Effects of a thin, adhesive, strong, and ductile coating, *Acta Mater.* 60 2012 3226–3238.
82. A. A. Voevodin, J. M. Schneider, C. Rebholz, A. Matthews, Multilayer composite ceramic-metal-DLC coatings for sliding wear applications, *Tribol. Int.* 29 1996 559–570.
83. K. N. Andersen, E. J. Bienk, K. O. Schweitz, H. Reitz, J. Chevallier, P. Kringhøja, J. Bøttiger, Deposition, microstructure and mechanical and tribological properties of magnetron sputtered TiN/TiAlN multilayers, *Surf. Coat. Technol.* 123 2000 219–226.
84. J. Chen, G. He, Y. Han, Z. Yuan, Z. Li, Z. Zhang, X. Han, S. Yan, Structural toughness and interfacial effects of multilayer TiN erosion-resistant coatings based on high strain rate repeated impact loads, *Ceram. Int.* 47 2021 27660–27667.
85. A. Samanta, M. Bhattacharya, I. Ratha, H. Chakraborty, S. Datta, J. Ghosh, S. Bysakh, M. Sreemany, R. Rane, A. Joseph, S. Mukherjee, B. Kundu, M. Das, A. K. Mukhopadhyay, Nano-and micro-tribological behaviours of plasma nitrided Ti6Al4V alloys, *J Mech Behav Biomed Mater* 77 2018 267–294.
86. A. Samanta, R. Rane, B. Kundu, D. K. Chanda, J. Ghosh, S. Bysakh, G. Jhala, A. Joseph, S. Mukherjee, M. Das, A. K. Mukhopadhyay, Bio-tribological response of duplex surface engineered SS316L for hip-implant application, *Appl. Surf. Sci.* 507 2020 145009.
87. A. Samanta, R. Rane, G. Jhala, B. Kundu, S. Datta, J. Ghosh, A. Joseph, S. Mukherjee, S. Roy, A. K. Mukhopadhyay, Biocompatibility and cyclic fatigue response of surface engineered Ti6Al4V femoral heads for hip-implant application, *Ceram. Int.* 47 2021 6905–6917.
88. A. Amanov, S. Lee, P. Y. Yun, Low friction and high strength of 316L stainless steel tubing for biomedical applications, *Mater. Sci. Eng. C* 71 2017 176–185.
89. K. Shukla, R. Rane, J. Alphonsa, P. Maity, S. Mukherjee, Structural, mechanical and corrosion resistance properties of Ti/TiN bilayers deposited by magnetron sputtering on AISI 316L, *Surf. Coat. Technol.* 324 2017 167–174.
90. W. Yang, G. Ayoub, I. Salehinia, B. Mansoor, H. Zbib, Deformation mechanisms in Ti/TiN multilayer under compressive loading, *Acta Mater.* 122 2017 99–108.
91. A. K. Krella, Cavitation erosion resistance of Ti/TiN multilayer coatings, *Surf. Coat. Technol.* 228 2013 115–123.
92. C. Chen, Q. Li, Y. Leng, J. Y. Chen, P. C. Zhang, B. Bai, N. Huang, Improved hardness and corrosion resistance of iron by Ti/TiN multilayer coating and plasma nitriding duplex treatment, *Surf. Coat. Technol.* 204 2010 3082–3086.
93. E. Martınez, J. Romero, A. Lousa, J. Esteve, Wear behavior of nanometric CrN/Cr multilayers, *Surf. Coat. Technol.* 163–164 2003 571–577.
94. S. K. Venkataraman, D. L. Kohlstedt, W. W. Gerberich, Metal-ceramic interfacial fracture resistance using the continuous micro scratch technique, *Thin Solid Films*, 223 1993 269–275.
95. Y. X. Ou, J. Lin, H. L. Che, W. D. Sproul, J. J. Moore, M. K. Lei, Mechanical and tribological properties of CrN/TiN multilayer coatings deposited by pulsed dc magnetron sputtering, *Surf. Coat. Technol.* 276 2015 152–159.
96. R. D. Jamison, Y. -L. Shen, Delamination analysis of metal–ceramic multilayer coatings subject to nano-indentation, *Surf. Coat. Technol.* 303 2016 3-11.96.

Part 3

Recent Advances in Modeling of Deformation and Fracture in Materials

8 Path-Independent Integrals and Their Applications in Fracture and Defect Mechanics

Anh Tay Nguyen, Shubhada S. Garnaik, Reinhold Kienzler, and Y. Eugene Pak

8.1 INTRODUCTION

The strength of a brittle solid is weakened by defects such as cracks or inclusions. In studying the failure and fracture of materials in the presence of defects, the "material force" or the energy release rate is an important concept to consider. Just as a Newtonian force, which is equal to the negative gradient of the potential energy of a material body with respect to the physical space, the material force (also called configurational force) is defined as the negative gradient of the total energy of the material with respect to the change in position of the defect with respect to the material. Such a force was shown to be a physically meaningful quantity in characterizing various defects in elastic materials. Eshelby [1] was the first to show that the force on an elastic singularity or inhomogeneity can be given as a path-independent integral over any surface enclosing it. This surface integral leads to a conservation law for a homogeneous elastic medium in the absence of defects.

This important conservation law paved the way for the development of other new path-independent integrals and their potential use in fracture mechanics in the decades following. Cherepanov [2] and Rice [3] independently developed the theoretical concept of the path-independent J-integral in order to evaluate the energy release rate for notches and cracks. Günther [4] and Knowles and Sternberg [5] derived conservation laws related to translational, rotational and scaling symmetries of an elastic medium using Noether's theorem [6]. From these findings, Budiansky and Rice [7] introduced the so-called J-, L- and M-integrals and for the first time provided the physical interpretation of the energy release rates per unit cavity translation, rotation and expansion, respectively.

These integrals have been investigated for many different fracture and defect mechanics problems. Freund [8] and Golebiewska-Herrmann and Herrmann [9] gave the relations between the J-, L- and M-integrals and used these integrals to evaluate the stress intensity factor and the energy release rates for a certain class of plane elastic crack problems. Pak et al. [10] evaluated the J-, L- and M-integrals leading to the energy release rates of various defects in an infinite body under various loading conditions. Lazar and Agiasofitou [11] derived the J-, L- and M-integrals of body forces in elasticity and body charges in electrostatics and discussed their physical interpretations. Chen [12] explicitly showed that the M-integral equals twice the change of the total potential energy owed to a single cracking of a central crack in a plane elastic body, although Budiansky and Rice [7] and Golebiewska-Herrmann and Herrmann [9] had implied this conclusion previously in their works. A thorough and unified review of the conservation laws, J-, L- and M-integrals and related concepts with various applications to defect and fracture mechanics is given by Kienzler and Herrmann [13] and Chen [12].

For other applications of the path-independent integrals, various authors [14–18] applied M- and L-integrals in studying interface and microcrack interaction problems. Kienzler and Kordisch [19] illustrated the physical significance of the integrals with a closed-form solution to the hole-dislocation

DOI: 10.1201/9781003359364-11

interaction problem. They further showed that the use of the L- and M-integrals besides J_1 and J_2 is beneficial to the numerical investigation of cracks under mixed-mode conditions. Kienzler [20] studied the changes of energy caused by formation of cavities in stressed elastic plates. The energies of circular holes, elliptical cavities and cracks are calculated with the M-integral. King and Hermann [21] demonstrated an alternative non-destructive experimental measurement technique for evaluating the J- and M-integrals. Suo [22] showed that the energy release rate of a crack created on the interface of a pair of bonded anisotropic solids, also known as Zener's crack, could be expressed by the M-integral. Pak et al. [23] derived a closed-form M-integral expression for a circular inclusion embedded in an infinite matrix and discussed the effect of material inhomogeneity on the energy release rate. Meyer et al. [24] used the M-integral to describe creep closure under various stress states surrounding drainage channels in subglacial hydrologic systems. Recently, Zhang et al. [25] investigated an innovative fatigue model based on the concept of M-integral in notched elastic-plastic material. There are also numerous works [26–29] discussing the energy release rate of nano-defects by considering the surface effects and size dependence in elastic materials.

The purpose of this chapter is to review some recent applications of path-independent integrals to evaluate the energy release rate of various types of defects under different loading conditions. Section 8.2 reviews the definition and the physical significance of the J-, L- and M-integrals with extension for piezoelectric materials. Section 8.3 provides closed-form expressions of J-, L- and M-integrals, the measures of energy release rates, for elliptical and circular cavities as well as for a slit crack embedded in an infinite body under various loading conditions. This can be done by first considering the corresponding expression for an elliptical cavity and then passing to the limit of a vanishingly small semi-axis of the ellipse, i.e., a crack. This approach completely avoids dealing with singular stress fields. Section 8.3 is based on the work by Pak et al. [10]. In Section 8.4, configurational forces acting on two-dimensional (2-D) elastic line singularities are evaluated by path-independent J-, L- and M-integrals in the framework of plane strain linear elasticity. There are similar studies on the in-plane and the out-of-plane line forces [30–32] as well as on 2-D [33] and 3-D nuclei of strain [34]. Here, the elastic line singularities considered are the edge dislocation, the line force, the nuclei of strain and the concentrated couple moment subjected to far-field loads. The interaction forces between two similar parallel elastic singularities are also calculated. It is shown that the self-similar expansion force M evaluated for the line force is exactly the negative of the strain energy pre-logarithmic factor, which is similar to the case of the well-known edge dislocation result by Rice [35]. In addition, the M-integral for the nuclei of strain and the L-integral for the line force yield nonzero expressions under some specific circumstances. Section 8.4 is based on the work by Seo et al. [36]. Finally, the concept of material tractions acting on surfaces and interfaces is introduced for the case of a 2-D circular hollow inclusion embedded in a finite matrix subjected to both internal and external loadings. This matrix-inclusion structure is an elastic system with two external boundaries (on which external loadings can be applied) and an internal stress-generating interface, all of which can be considered as defects. An exact closed-form solution for this particular class of boundary value problem has been derived. It is shown that the self-similar expansion force M corresponds to the energy release rate of the self-similarly expanding defect or the change of the total potential energy of the elastic system with respect to the relative scale change of the defects enclosed within the contour integral.

8.2 PATH-INDEPENDENT INTEGRALS

The J_k-integral vector, M-integral and L-integral are formulated in a two-dimensional field as follows [4, 5, 7]:

$$J_k = \int_{C_J} \left(Wn_k - T_i u_{i,k}\right) ds \tag{8.1}$$

$$M = \int_{C_M} \left(W x_i n_i - T_k u_{k,i} x_i \right) ds \tag{8.2}$$

$$L = \int_{C_I} e_{3ij} \left(W x_i n_i + T_i u_j - T_k u_{k,i} x_j \right) ds \tag{8.3}$$

where W is the strain energy density, x_i is the coordinate direction, u_i is the displacement vector, $T_i = \sigma_{ij} n_j$ is the surface traction vector, n_j is the outward unit normal vector to the contour and C is a contour surrounding a crack tip as shown in Figure 8.1. Here, e_{3ij} is the alternating tensor with the following property:

$$e_{3ij} = \begin{cases} -1 & \text{when } i = 2 \text{ and } j = 1 \\ 0 & \text{when } i = j \\ +1 & \text{when } i = 1 \text{ and } j = 2 \end{cases} \tag{8.4}$$

Rice [3] showed that for a planar crack, the value of the J-integral represents the energy release rate associated with a unit crack tip advancement,

$$J = -\lim_{\Delta a \to 0} \frac{\Delta U - \Delta W}{2\Delta a} \tag{8.5}$$

where ΔU and ΔW are the change in the elastic strain energy of the body and the work done by the loading due to the elongation of the crack, respectively. Alternatively, the J-integral can be viewed as the "material force" or the "crack extension force", which acts on the defect enclosed by the contour C_J. For the case of a defect undergoing self-similar (uniform) expansion characterized by the relative scale change $\Delta a / a$, Budiansky and Rice [7] showed that the M-integral is a measure of the energy release rate of the defect with respect to this change. For a single crack, a simple relationship can be established between M and J [8], i.e.,

$$M = 2aJ \tag{8.6}$$

Equation 8.6 is useful when the J-integral is path-dependent, e.g., if there is loading on the crack faces. In such a case, it is preferable to evaluate the M-integral around a closed contour surrounding the whole crack. It is worth noting that the closed contour C to define the J_k vector is usually

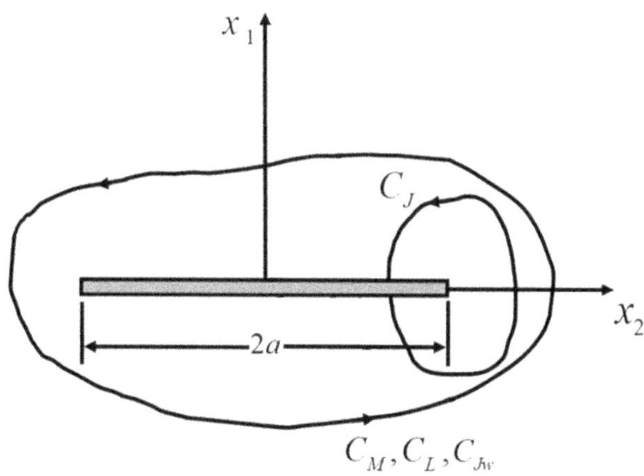

FIGURE 8.1 Paths for the J-, L- and M-integrals for a central crack.

a smooth closed contour surrounding the tip of the crack, while for the M- and L-integrals, the contour C should be chosen to enclose the whole crack completely (Herrmann and Herrmann [9]). However, in Section 8.3, contour C_{Jw}, which completely surrounds the whole crack, is used to evaluate the J-integral to demonstrate that net material force, J_w, acting on the whole cavity is zero under any arbitrary loading.

Herrmann and Herrmann [9] showed that the L-integral is the rotational energy release rate, which is the measure of the energy release rate of a crack per unit rotation. The relationship has the form

$$L = \frac{\partial \Pi}{\partial \varphi} \tag{8.7}$$

where Π is the total potential energy $(U - W)$, and ϕ is a small angle through which the crack rotates with respect to the applied field. In other words, the L-integral can be regarded as the "material moment" acting on the defect. The usual minus sign in Equation 8.7 is omitted because of the way the moment is defined for the L-integral, which is opposite in direction to the traditional definition. This subject has also been discussed in terms of material tractions by Kienzler and Herrmann [13].

The path-independent integrals are useful in fracture mechanics because they can be related to the stress intensity factors. For plane stress conditions and under mode I and mode II loading conditions, one has

$$J = \left(K_I^2 + K_{II}^2\right) / E \tag{8.8}$$

Here, K_I and K_{II} are the stress intensity factors for mode I and mode II fracture, respectively, and E is Young's modulus. In the presence of the far-field loadings σ_{xx}^{∞}, σ_{xy}^{∞} and σ_{yy}^{∞}, K_I and K_{II} are defined as

$$\begin{aligned} K_I &= \sigma_{yy}^{\infty}\sqrt{\pi a} \\ K_{II} &= \sigma_{xy}^{\infty}\sqrt{\pi a} \end{aligned} \tag{8.9}$$

In addition, from the Irwin's relationship, the J-integral can be shown to be equal to the crack extension force G:

$$J = G = \left(K_I^2 + K_{II}^2\right) / E \tag{8.10}$$

By evaluating this integral around the whole crack subjected to far-field loads σ_{xx}^{∞}, σ_{xy}^{∞} and σ_{yy}^{∞}, Herrmann and Herrmann [9] obtained the result

$$L = \frac{2\pi a^2}{E}\sigma_{xy}^{\infty}\left(\sigma_{yy}^{\infty} + \sigma_{xx}^{\infty}\right) \tag{8.11}$$

or

$$L = \frac{2K_{II}}{E}\left(K_I + \sigma_{xx}^{\infty}\sqrt{\pi a}\right). \tag{8.12}$$

The path-independent integrals can be generalized to include the piezoelectric effect. Letting the electric enthalpy density, H, be the Lagrangian density, and differentiating it with respect to the spatial coordinate, x_i, the J-integral for piezoelectric materials, J_p, becomes

$$J_p = \int_{C_J}\left(Hn_x - T_i u_{i,x} + D_i n_i E_x\right) ds \tag{8.13}$$

where D_i and E_i are the electric displacements and the electric fields, respectively. As shown in Pak et al. [37], Equation 8.13 is physically meaningful because it is related to the electroelastic energy release rate for defects in piezoelectric materials. Similarly, the generalized M-integral and L-integral for piezoelectric materials can be written as [10, 42]

$$M_p = \int_{C_M} \left(H x_k n_k - T_k u_{k,i} x_i + D_k n_k E_i x_i \right) ds \tag{8.14}$$

$$L_p = \int_{C_L} e_{3ij} \left(H x_j n_i + T_i u_j - T_k u_{k,i} x_j + D_k n_k E_i x_j \right) ds \tag{8.15}$$

Equation 8.15 has been used to evaluate the rotational energy release rate for cavities embedded in an infinite piezoelectric matrix.

8.3 ENERGY RELEASE RATES FOR VARIOUS DEFECTS UNDER DIFFERENT LOADING CONDITIONS

8.3.1 Validity of the Limiting Procedure

In this section, the method of evaluating path-independent integrals for various defects, e.g., a circular cavity or a crack, is demonstrated (Figure 8.2) by first considering an elliptical cavity and then performing a limiting process. To show the validity of this limiting procedure, the path-independent integrals are evaluated for an elliptical cavity, a circular cavity and a crack in an infinite medium (Figure 8.2) subjected to far-field loads including antiplane shear τ_0 (Figure 8.3d), and combined antiplane shear, τ_0, and in-plane electric field, E_0 (Figure 8.3f) for piezoelectric materials. Similar work done previously [38, 40, 41] is shown in Figure 8.3 and has also been tabulated in Table 8.1 for the cases of (a) remote uniaxial tension, T, at an arbitrary angle β , (b) biaxial tension, T, (c) in-plane shear, S, and (d) uniform heat flow, τ . The M- and L-integrals are first evaluated for an elliptical cavity. The solution for the circular cavity can then be obtained as a special case of the elliptical solution by letting $a \to b$, and for the slit crack, it can be obtained by letting $b \to 0$ (Figure 8.2). The J-integral expression can also be obtained through the M-integral using Equation 8.6. The paths for the integrals are chosen to be along the traction-free boundary, thus eliminating the traction terms, T_i, in the integrand of the integrals. The plane stress elasticity solutions from Timoshenko and Goodier [43] and the piezoelectric solution from Pak [39] are then used to calculate the strain energy density term along the rim. For the circular and the elliptical cavity, the strain energy density only consists of the hoop stress terms.

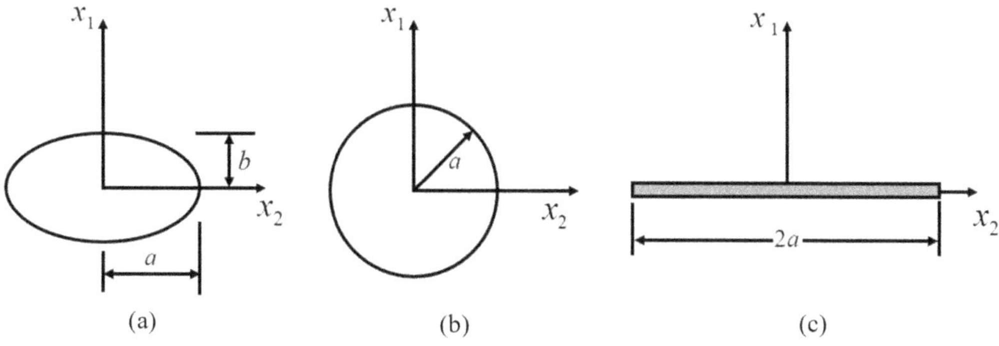

FIGURE 8.2 (a) Elliptical cavity, (b) circular cavity and (c) crack in an infinite medium.

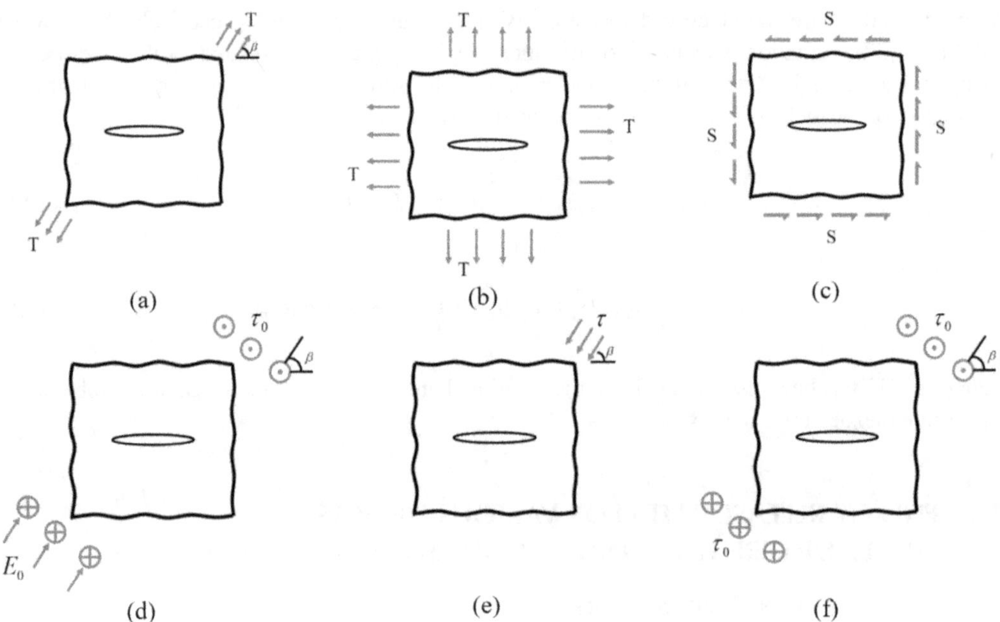

FIGURE 8.3 Elliptical cavity in an infinite medium subjected to (a) far-field uniaxial tension, T, at an arbitrary angle β, (b) biaxial tension, T, (c) in-plane shear, S, (d) antiplane shear τ_0, (e) uniform heat flow, τ, at an arbitrary angle β, and (f) antiplane shear, τ_0, and in-plane electric field, E_0, at an arbitrary angle β.

8.3.2 Discussion

According to Table 8.1, $J_w = 0$ for all defects in all cases of loading. Since the far-field loads are symmetric with respect to the plane $x = 0$ of the defect, the material force acting on the left half and the right half of the defect in the x-direction are equal in magnitude but have opposite signs, resulting in a zero net material force, J_w, acting on the whole cavity. In other words, the change in the total energy of the system due to an infinitesimal translation of the defect in an infinite body under any arbitrary loading is zero. However, the J-integral evaluated around the crack tip is nonzero and validates Equation 8.6, i.e., $M = 2aJ$. For example, the J_p-integral for a crack in an infinite piezoelectric medium obtained by the limiting procedure ($b \rightarrow 0$) on the piezoelectric elliptical cavity solution along with Equation 8.6 matches with the explicit expression obtained by Pak [41],

$$J_p = \frac{\pi a}{E}\left(\tau_0^2 - kE_0^2\right) \tag{8.16}$$

where

$$k = \left(e_{15}^{M^2} + C_{44}^M \varepsilon_{11}^M\right) \tag{8.17}$$

Since the energy of the system changes when a defect expands in a self-similar manner, the self-similar expansion force M is nonzero. Again, the M-integral expressions for a circular cavity ($b \rightarrow a$) and a slit crack ($b \rightarrow 0$) obtained by the limiting method from the embedded elliptical cavity agree with previously calculated results for all loading cases. For example, for a circular cavity, the expression of the M-integral is

$$M = \frac{2\pi a^2}{E}\left(\tau_0^2 - kE_0^2\right) \tag{8.18}$$

TABLE 8.1
Closed-Form Expressions of Path-Independent Integrals under Various Far-Field Loading Conditions

	J_w	M	L
Case 1. Uniaxial Tension at an Arbitrary Angle β (Figure 8.3a)			
Crack	0	$\pi a^2 T^2 (1-\cos 2\beta)/E$	$\pi a^2 T^2 \sin 2\beta / E$
Circular cavity	0	$3\pi a^2 T^2 / E$	0
Elliptical cavity	0	$\pi\left[2a^2 + ab\right]T^2 / E$	$\pi T^2 (a^2 - b^2)\sin 2\beta / E$
Case 2. Biaxial Tension (Figure 8.3b)			
Crack	0	$2\pi a^2 T^2 / E$	0
Circular cavity	0	$4\pi a^2 T^2 / E$	0
Elliptical cavity	0	$2\pi (a^2 + b^2) T^2 / E$	0
Case 3. In-Plane Shear (Figure 8.3c)			
Crack	0	$2\pi a^2 S^2 / E$	0
Circular cavity	0	$8\pi a^2 S^2 / E$	0
Elliptical cavity	0	$2\pi (a+b)^2 S^2 / E$	0
Case 4. Antiplane Shear at an Arbitrary Angle β (Figure 8.3d)			
Crack	0	$\pi a^2 \tau_0^2 (1-\cos 2\beta)/2G$	$\pi a^2 \tau_0^2 \sin 2\beta / 2G$
Circular cavity	0	$2\pi a^2 {\tau_0}^2 / G$	0
Elliptical cavity	0	$\pi (a+b)\tau_0^2\left[(a+b)-(a-b)\cos 2\beta\right]/2G$	0
Case 5. Uniform Heat Flow at an Arbitrary Angle β (Figure 8.3e)			
Crack	0	$\pi E\alpha^2 a^4 \tau^2 \sin^2\beta / 8$	0
Circular cavity	0	$\pi E\alpha^2 a^4 \tau^2 / 2$	0
Elliptical cavity	0	$\pi E\alpha^2 (a+b)^2\left[b^2 + (a-b)^2 \sin^2\beta\right]\tau^2 / 8$	$E\alpha^2 ab(b^2 - a^2)\tau^2 \sin 2\beta / 8$
Case 6. In-plane Electrical and Antiplane Mechanical Loading (Figure 8.3f)			
Crack	0	$\pi a^2\left[\tau_0^2 - kE_0^2\right](1-\cos 2\beta)/2G$	$\pi a^2\left[\tau_0^2 - kE_0^2\right]\sin 2\beta / 2G$
Circular cavity	0	$2\pi a^2\left[\tau_0^2 - kE_0^2\right]/G$	0
Elliptical cavity	0	$\pi (a+b)\left[\tau_0^2 - kE_0^2\right]\left[(a+b)-(a-b)\cos 2\beta\right]/2G$	$\pi (a^2 - b^2)\left[\tau_0^2 - kE_0^2\right]\sin 2\beta / 2G$

where $k = \left(e_{15}^{M^2} + C_{44}^M \varepsilon_{11}^M\right)$

deduced from the elliptical cavity solution for a coupled electromechanical load by letting $b = a$, which is the same solution by Pak [41].

For the cases with loadings at an arbitrary angle, the M-integrals for the elliptical cavity and the crack are maximum when loading is perpendicular to the major axis ($\beta = \pi/2$) of the defects and minimum when it is parallel ($\beta = 0$). However, since a circular cavity expands self-similarly, the energy release rate of a circular cavity is independent of the loading angle.

The material moment, L, acting on the defect is zero when the defect is in equilibrium with respect to the orientation of the external load. Therefore, for the circular cavity ($b \rightarrow a$), $L = 0$ for every loading case, there is no favourable direction for the circular cavity with respect to far-field loading. For a crack ($b \rightarrow 0$) under a uniform thermal load, L is also zero for an arbitrary heat flow angle β . From this result, the crack is predicted to propagate in its own plane when the critical crack extension force is reached at the crack tip [38]. For the elliptical cavity and the crack, the energy release rate per infinitesimal rotation of the defect, i.e., the material moment, is maximum when $\beta = \pm\pi / 4$ in all loading cases. When β is shifted by $\pi / 2$, the elliptical cavity experiences an equal and opposite moment, hence $L = 0$. Here, the rotational material moment tries to orient the defect to maximize the energy (or electric enthalpy for piezoelectric material) release rate, i.e., the material force M. This behaviour is observed in both purely elastic material and piezoelectric material under electromechanical loading. The crack will propagate to a more energetically favourable orientation, at which L would be zero and M would be maximum, under all cases except for the thermal loading case. This behaviour ensures that there would be no crack skewing, i.e., a state of single-mode fracture.

8.4 CONFIGURATIONAL FORCES ON ELASTIC LINE SINGULARITIES

8.4.1 Path-Independent Integrals by Decomposition of Energy Momentum Tensor

For the interacting singularities S and T, the J-integral is obtained by summing up the elastic fields generated by each singularity,

$$
\begin{aligned}
J_k &= \int_C \Big[\frac{1}{2}\left(\sigma_{ij}^{\mathrm{S}} + \sigma_{ij}^{\mathrm{T}}\right)\left(\varepsilon_{ij}^{\mathrm{S}} + \varepsilon_{ij}^{\mathrm{T}}\right)n_k - \left(T_i^{\mathrm{S}} + T_i^{\mathrm{T}}\right)\left(u_{i,k}^{\mathrm{S}} + u_{i,k}^{\mathrm{T}}\right)\Big]dS \\
&= \int_C \left(\frac{1}{2}\sigma_{ij}^{\mathrm{S}}\varepsilon_{ij}^{\mathrm{S}}n_k - T_i^{\mathrm{S}}u_{i,k}^{\mathrm{S}}\right)dS + \int_C \left(\frac{1}{2}\sigma_{ij}^{\mathrm{T}}\varepsilon_{ij}^{\mathrm{T}}n_k - T_i^{\mathrm{T}}u_{i,k}^{\mathrm{T}}\right)dS \\
&\quad + \int_C \Big[\frac{1}{2}\left(\sigma_{ij}^{\mathrm{S}}\varepsilon_{ij}^{\mathrm{T}} + \sigma_{ij}^{\mathrm{T}}\varepsilon_{ij}^{\mathrm{S}}\right)n_k - \left(T_i^{\mathrm{S}}u_{i,k}^{\mathrm{T}} + T_i^{\mathrm{T}}u_{i,k}^{\mathrm{S}}\right)\Big]dS \\
&= J_k^{\mathrm{S}} + J_k^{\mathrm{T}} + J_k^{\mathrm{ST}}
\end{aligned}
\tag{8.19}
$$

where S is the elastic singularity at the origin of the coordinate system and T denotes other interacting singularities or far-field loads. The integrands of the J-integral are then decomposed into the singular self-field and the regular field due to singularity T or the far-field loads. The same approach is also used for the M- and L-integrals, i.e.,

$$
M = M^S + M^T + M^{\mathrm{ST}} \quad \text{and} \quad L = L^S + L^T + L^{\mathrm{ST}} \tag{8.20}
$$

In 2-D elastostatics, the stress singularity near the origin is $1/r$ for the edge dislocation and the line-load [44, 45], and $1/r^2$ for the nuclei of strain and the concentrated couple moment [46].

8.4.2 Configurational Forces Acting on Elastic Line Singularities under Far-Field Loads

The configurational forces for the edge dislocation, the line force, the nuclei of strain and the concentrated couple moment in an infinite solid subjected to far-field loads (Figure 8.4) are calculated using the path-independent integrals. The path-independent integrals are evaluated on a contour C with a vanishingly small radius, r_o, that surrounds the singularity S at the origin. The calculations

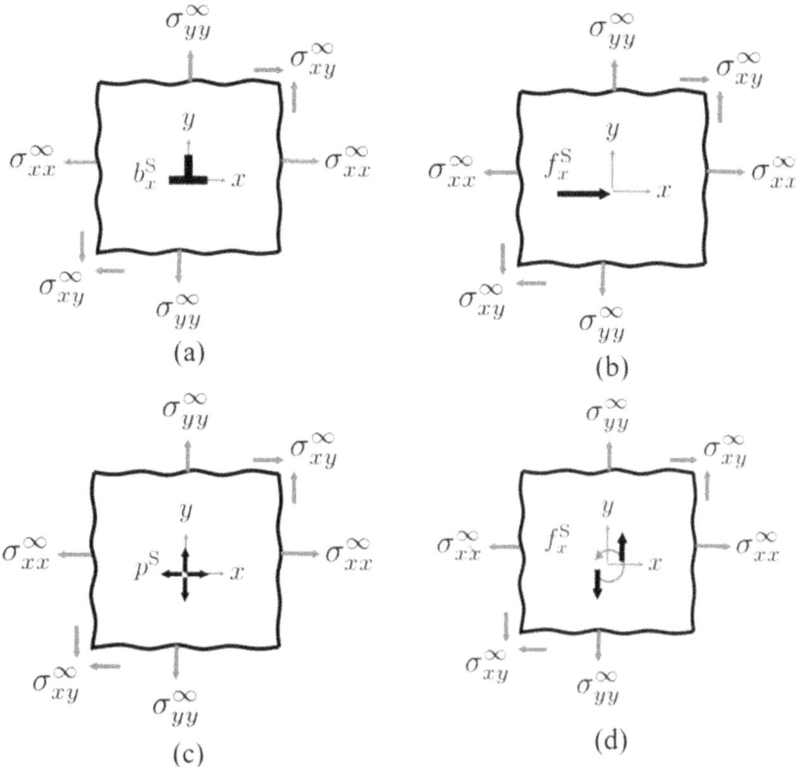

FIGURE 8.4 In-plane elastic singularities: (a) edge dislocation, (b) line force, (c) nuclei of strain and (d) concentrated couple moment, subjected to the far-field loads, σ^{∞}_{ij} , in an infinite medium.

are further simplified by assuming that the elastic field induced by T is constant along the contour C [32]. The results are shown in Table 8.2.

The J-integrals for the edge dislocation are equal to a geometric (Burgers vector) term times a force (far-field load) term. Both the glide and climb forces are correctly obtained by pairing the load with the corresponding Burgers vector (dislocation type). Similarly, the J-integral for the line force, $f^{S}_{k}u^{T}_{k,i}$, is the product of the geometric (strain) term and the force (line force) term which match those given by Cherepanov [2] and Pak [31], if the displacement gradient field due to far-field loads is assumed to exclude the rotation components, i.e., $\omega_{ij} = (u_{i,j} - u_{j,i})/2 = 0$. These translational configurational forces are nonzero because, in the presence of a displacement gradient field, there is finite work done by the line forces. However, the strain energy change of an elastic singularity under a simple translation in a stressed unbound body is zero. Finally, $J = 0$ for the nuclei of strain and the concentrated couple moment, meaning that the translational configurational forces for these line singularities are zero in the presence of a constant stress field. This result agrees with Eshelby's 3-D result [34], which showed that the translational configurational force on a nucleus of strain (centre of dilatation) is nonzero only if there is an applied hydrostatic pressure gradient at the position of the singularity.

The M-integral expressions for the edge dislocation and the line force only include the self-field [22] and is interestingly identical to the negative of the prelogarithmic strain energy factor, $W_{f^{S}_{i}}$,

$$W_{f^{S}_{i}} = \frac{f_i^2(3-4\nu)}{16\pi\mu(1-\nu)}\log\frac{r_2}{r_1} \tag{8.21}$$

TABLE 8.2
Configurational Forces Acting on Elastic Line Singularities Subjected to Far-Field Mechanical Loads

Singularity	J_x	J_y	M	L
Dislocation, b_x^S	$b_x^S\sigma_{xy}^\infty$	$-b_x^S\sigma_{xx}^\infty$	$\dfrac{b_x^{S^2}\mu}{4\pi(1-v)}$	0
Dislocation, b_y^S	$b_y^S\sigma_{yy}^\infty$	$-b_y^S\sigma_{xy}^\infty$	$\dfrac{b_y^{S^2}\mu}{4\pi(1-v)}$	0
Line force, f_x^S	$f_x^S\varepsilon_{xx}^\infty$	$f_x^S\varepsilon_{xy}^\infty$	$-\dfrac{f_x^{S2}(3-4v)}{16\pi\mu(1-v)}$	0
Line force, f_y^S	$f_y^S\varepsilon_{xy}^\infty$	$f_x^S\varepsilon_{yy}^\infty$	$-\dfrac{f_y^{S2}(3-4v)}{16\pi\mu(1-v)}$	0
Nuclei of strain, p^S	0	0	$\dfrac{p^S(1-2v)(\sigma_{xx}^\infty+\sigma_{yy}^\infty)}{2\mu}$	0
Concentrated couple moment, m^S	0	0	0	0

This is similar to the well-known dislocation result given by Rice [35]. However, the M-integral for the nuclei of strain is not related to the coefficient of the strain energy factor

$$W_{p^S} = \frac{p^{S2}(1-2v)^2}{8\pi\mu(1-v)^2}\left(\frac{r_2^2-r_1^2}{r_1^2 r_2^2}\right) \tag{8.22}$$

For the nuclei of strain, the self-similarly expanding force M only exists if the normal stresses (σ_{xx}^∞ and σ_{yy}^∞) are present, because the nuclei of strain induce a state of pure shear in the matrix, which is equivalent to the misfitted cylinder in an infinite matrix, resulting in no volume or scale change. A similar M-integral result was obtained for the case of a three-dimensional misfitted spherical inclusion embedded in an infinite solid [47]. Finally, the M-integral is zero for the concentrated couple moment under far-field loads.

8.4.3 Forces between Two Parallel Elastic Line Singularities

Table 8.3 shows interacting configurational forces between two similar parallel elastic singularities, which are evaluated by the path-independent integrals along the contour C surrounding the elastic singularity S (Figure 8.5). It can be seen that the interaction force for the line force and for the dislocation have opposite signs. Therefore, when there are two parallel line forces in the same direction, they are attracted to each other, which is opposite to the dislocation case [32, 33]. This is because as the line force T approaches the line force S, which induces the displacement in the direction of the line force S, more potential energy drop occurs for the line force T as it gets closer to the line force S.

For the nuclei of strain and the concentrated couple moment, $J=0$ indicating that they do not interact [34]. Similarly, the M value is zero for the case of a concentrated couple

TABLE 8.3
Configurational Forces between Two Parallel Elastic Line Singularities in the Absence of Far-Field Loads

Singularity	J_x	J_y	M	L
b_x^S, b_x^T	$-\frac{b_x^S b_x^T \mu}{2\pi(1-v)a}$	0	$\frac{b_x^{S2}\mu}{4\pi(1-v)}$	0
b_y^S, b_x^T	0	$\frac{b_y^S b_x^T \mu}{2\pi(1-v)a}$	$\frac{b_y^{S2}\mu}{4\pi(1-v)}$	0
b_y^S, b_y^T	$-\frac{b_y^S b_y^T \mu}{2\pi(1-v)a}$	0	$\frac{b_y^{S2}\mu}{4\pi(1-v)}$	0
f_x^S, f_x^T	$\frac{f_x^S f_x^T (3-4v)}{8\pi\mu(1-v)a}$	0	$-\frac{f_x^{S2}(3-4v)}{16\pi\mu(1-v)}$	0
f_y^S, f_x^T	0	$-\frac{f_y^S f_x^T}{8\pi\mu(1-v)a}$	$-\frac{f_y^{S2}(3-4v)}{16\pi\mu(1-v)}$	$\frac{f_y^S f_x^T[1-(3-4v)\log a]}{8\pi\mu(1-v)}$
f_y^S, f_y^T	$\frac{f_y^S f_y^T (3-4v)}{8\pi\mu(1-v)a}$	0	$-\frac{f_y^{S2}(3-4v)}{16\pi\mu(1-v)}$	0
p^S, p^T	0	0	0	0
m^S, m^T	0	0	0	0

moment interacting with another couple moment because the displacement induced by the T field is only in the circumferential direction that may not induce any dilatation. In addition, the M values for the two interacting edge dislocations and line forces are identical to the stand-alone cases under far-field loads. Again, the M-integrals for all line elastic singularities studied here do not recognize any elastic field produced by other interacting defects (or boundaries) that are weaker than $1/r$ singularity at S [22]. Finally, in the absence of normal stresses, the M value for the interacting nuclei of strain is zero as the T field only induces a state of pure shear.

Although not tabulated, interaction forces for the same parallel line singularities in the presence of far-field loads were calculated. The results show that the far-field terms previously obtained for the single-line singularities subjected to far-field loads were picked up by the J-integral. Therefore, the results are simply the sum of the expressions from Tables 8.2 and . The M-integral, however, except for the case of the nuclei of strain, does not recognize far-field loads as expected; thus, the results are the same as those given in Table 8.2.

The L-integral results are zero for all orientations, except for the line forces that are perpendicular to each other. It is speculated that for the arbitrarily positioned and oriented line singularities considered in this study, the L-integrals can be nonzero [48].

The L-integral for the line force f_y^S interacting with f_x^T separated by a distance a showed a nonzero result indicating that there is a configurational moment on f_y^S. This moment is induced because when f_y^S is rotated, its x-component does work through f_x^T-generated displacement, hence, lowering the total energy. Other interacting singularities may also yield nonzero L-integral results if they are arbitrarily positioned and oriented.

8.5 M-INTEGRAL AND ENERGY RELEASE RATE: CASE OF A 2-D CIRCULAR HOLLOW INCLUSION EMBEDDED IN A FINITE MATRIX

8.5.1 Jump in the Material Tractions

In order to introduce the concept of material traction acting on a surface or an interface, let's consider a simple 2-D hollow matrix with inner radius r_i and outer radius r_o, under an internal pressure P_i and external pressure P_o as indicated in Figure 8.6. The textbook solution for displacements and stresses is given as

$$u_r = \frac{r}{4\mu}\left[(\kappa-1)\frac{P_i r_i^2 - P_o r_o^2}{r_o^2 - r_i^2} - 2\frac{(P_o - P_i)r_o^2 r_i^2}{r_o^2 - r_i^2}\frac{1}{r^2}\right]$$

$$\sigma_{rr} = \frac{P_i r_i^2 - P_o r_o^2}{r_o^2 - r_i^2} + \frac{(P_o - P_i)r_o^2 r_i^2}{r_o^2 - r_i^2}\frac{1}{r^2} \tag{8.23}$$

$$\sigma_{\theta\theta} = \frac{P_i r_i^2 - P_o r_o^2}{r_o^2 - r_i^2} - \frac{(P_o - P_i)r_o^2 r_i^2}{r_o^2 - r_i^2}\frac{1}{r^2}$$

where κ is defined as

$$\kappa = \begin{cases} 3-4\nu & \text{for plane strain} \\ \dfrac{3-\nu}{1+\nu} & \text{for plane stress} \end{cases} \tag{8.24}$$

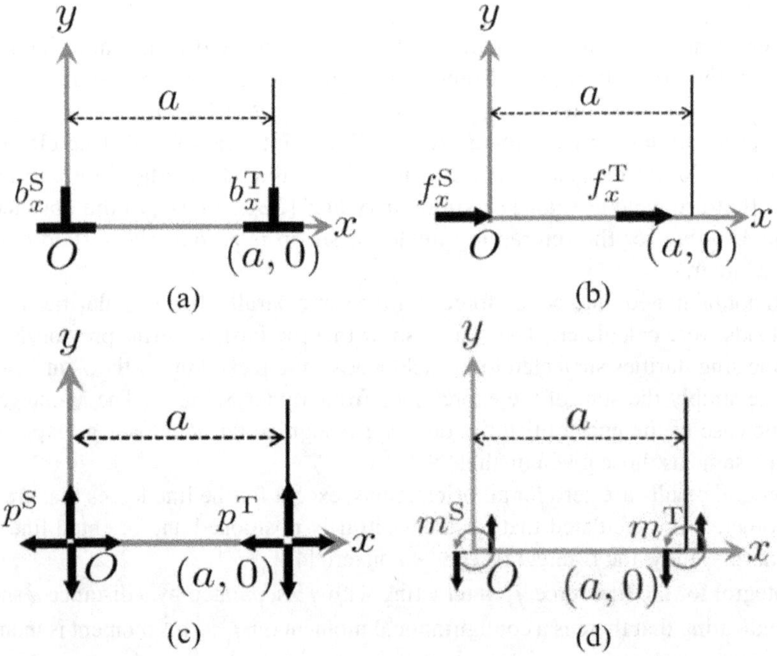

FIGURE 8.5 In-plane elastic singularities (a) edge dislocation, (b) line force, (c) nuclei of strain, and (d) concentrated couple moment, interacting with identical parallel line singularities in an infinite medium.

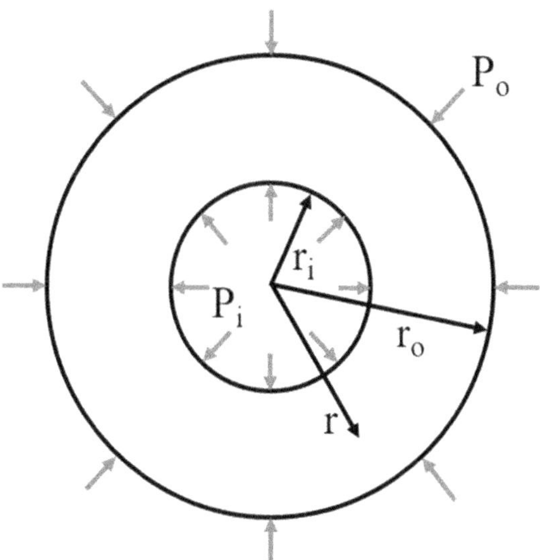

FIGURE 8.6 Hollow 2-D matrix under internal pressure P_i and external pressure P_o.

μ is the shear modulus and ν is the Poisson's ratio. Using the displacements and stresses defined in Equation 8.23, the strain energy and the work done by the external load are given as

$$\begin{aligned} U &= \frac{1}{2}\int_0^{2\pi}\int_{r_i}^{r_o}\left(\sigma_{rr}u_{r,r} + \sigma_{\theta\theta}\frac{u_r}{r}\right) r dr d\theta \\ &= \frac{\pi}{4\mu\left(r_o^2 - r_i^2\right)}\{P_i^2 r_i^2\left[(\kappa-1)r_i^2 + 2r_o^2\right] - 2P_iP_o r_i^2 r_o^2(\kappa+1) + P_o^2 r_o^2\left[(\kappa-1)r_o^2 + 2r_i^2\right]\} \end{aligned} \tag{8.25}$$

$$\begin{aligned} W^{ext} &= W^{r_i} + W^{r_a} = \int_0^{2\pi}\left(-\sigma_{rr}(r_i)u_r(r_i)r_i\right)d\theta + \int_0^{2\pi}\left(\sigma_{rr}(r_o)u_r(r_o)r_o\right)d\theta \\ &= \frac{\pi}{2\mu\left(r_o^2 - r_i^2\right)}\left\{P_i^2 r_i^2\left[(\kappa-1)r_i^2 + 2r_o^2\right] - P_iP_o r_i^2 r_o^2(\kappa+1)\right\} \\ &+ \frac{\pi}{2\mu\left(r_o^2 - r_i^2\right)}\left\{P_o^2 r_o^2\left[(\kappa-1)r_o^2 + 2r_i^2\right] - P_iP_o r_i^2 r_o^2(\kappa+1)\right\} \end{aligned} \tag{8.26}$$

The material traction G_i is defined as

$$G_i = b_{ji}n_i \tag{8.27}$$

where b_{ji} is the Eshelby tensor

$$b_{ji} = W\delta_{ji} - \sigma_{jk}u_{k,i} \tag{8.28}$$

and n_i is the unit normal vector. The jump in the material forces for the M-integral is defined as (see Figure 8.7) [13]

$$j_i = \left[\left[\left(b_{ji}n_j\right)\right]\right] = \left[\left[\left(W\delta_{ji} - \sigma_{jk}u_{k,i}\right)n_j\right]\right] = \left[\left(W\delta_{ji} - \sigma_{jk}u_{k,i}\right)n_j\right]^+ - \left[\left(W\delta_{ji} - \sigma_{jk}u_{k,i}\right)n_j\right]^- \tag{8.29}$$

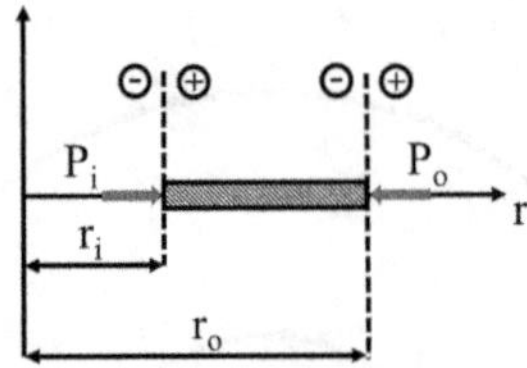

FIGURE 8.7 Physical tractions and jump positions.

Here, in polar coordinates, one has

$$j_r = \frac{1}{2}\left[\left[\left(-\sigma_{rr}u_{r,r} + \sigma_{\theta\theta}\frac{u_r}{r}\right)\right]\right] \tag{8.30}$$

The material jump j_{r_i} at the interior boundary r_i can be expressed as

$$j_{r_i} = \frac{1}{2}\left[\left(-\sigma_{rr}u_{r,r} + \sigma_{\theta\theta}\frac{u_r}{r}\right)\right]_{r_i^+} - \frac{1}{2}\left[\left(-\sigma_{rr}u_{r,r} + \sigma_{\theta\theta}\frac{u_r}{r}\right)\right]_{r_i^-} \tag{8.31}$$

At r_i^-, $\sigma_{rr}^- = -P_i$, $\sigma_{\theta\theta}^- = 0$, and $u_{r,r}^- = u_r^- / r = 0$. Therefore, j_{r_i} becomes

$$j_{r_i} = \frac{1}{2}\left[-\sigma_{rr}(r_i)u_{r,r}(r_i) + \sigma_{\theta\theta}(r_i)\frac{u_r(r_i)}{r_i}\right] = \frac{1}{E^*}\frac{2r_o^2}{(r_o^2 - r_i^2)^2}\left[P_i^2 r_i^2 - P_i P_o(r_i^2 + r_o^2) + P_o^2 r_o^2\right] \tag{8.32}$$

with

$$E^* = \begin{cases} \dfrac{E}{1-\nu^2} & \text{for plane strain} \\ E & \text{for plane stress} \end{cases} \tag{8.33}$$

where E is the Young's modulus of the matrix. Similarly, the material jump j_{r_o} at the outer boundary r_o is given as

$$j_{r_o} = \frac{1}{2}\left[\left(-\sigma_{rr}u_{r,r} + \sigma_{\theta\theta}\frac{u_r}{r}\right)\right]_{r_o^+} - \frac{1}{2}\left[\left(-\sigma_{rr}u_{r,r} + \sigma_{\theta\theta}\frac{u_r}{r}\right)\right]_{r_o^-} \tag{8.34}$$

At r_o^+, $\sigma_{rr}^+ = -P_o$, $\sigma_{\theta\theta}^+ = 0$, and $u_{r,r}^+ = u_r^+ / r = 0$. Therefore, j_{r_o} becomes

$$j_{r_o} = \frac{1}{2}\left[-\sigma_{rr}(r_o)u_{r,r}(r_o) + \sigma_{\theta\theta}(r_o)\frac{u_r(r_o)}{r_o}\right] = \frac{1}{E^*}\frac{2r_i^2}{(r_o^2 - r_i^2)^2}\left[P_i^2 r_i^2 - P_i P_o(r_i^2 + r_o^2) + P_o^2 r_o^2\right] \tag{8.35}$$

The material tractions applied to the hollow matrix are finally obtained as shown in Figure 8.8.

8.5.2 *M*-Integral and the Virial of the Material Traction

The expression for the *M*-integral using the elastic solution obtained in the previous section is

$$M = \frac{1}{2}\int_0^{2\pi}\left(-\sigma_{rr}u_{r,r} + \sigma_{\theta\theta}\frac{u_r}{r}\right)r(rd\theta)$$

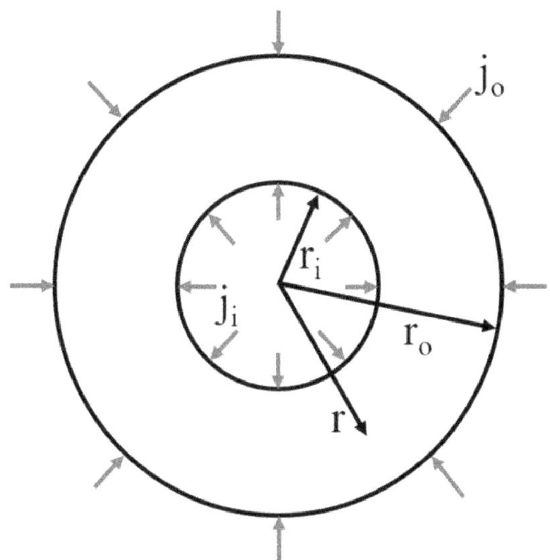

FIGURE 8.8 Material tractions at the inner and outer boundaries.

$$= \frac{4\pi r_i^2 r_o^2}{E^*\left(r_o^2 - r_i^2\right)^2}\left[P_i^2 r_i^2 - P_i P_o\left(r_i^2 + r_o^2\right) + P_o^2 r_o^2\right], \quad r_i^+ \le r \le r_o^- \tag{8.36}$$

The M-integral is also the scalar moment, or the virial, of the material traction. As long as the interaction contour encloses the material traction j_{r_i}, the virial of j_{r_i} is

$$V_{r_i}^{mat} = \int_0^{2\pi} \left(r_i j_{r_i}\right)\left(r_i d\theta\right)$$

$$= \frac{4\pi r_i^2 r_o^2}{E^*\left(r_o^2 - r_i^2\right)^2}\left[P_i^2 r_i^2 - P_i P_o\left(r_i^2 + r_o^2\right) + P_o^2 r_o^2\right] = M, \quad r_i^+ \le r \le r_o^- \tag{8.37}$$

The virial of the material traction j_{r_o} along the outer boundary is

$$V_{r_o}^{mat} = \int_0^{2\pi} \left(r_o j_{r_o}\right)\left(r_o d\theta\right)$$

$$= -\frac{4\pi r_i^2 r_o^2}{E^*\left(r_o^2 - r_i^2\right)^2}\left[P_i^2 r_i^2 - P_i P_o\left(r_i^2 + r_o^2\right) + P_o^2 r_o^2\right] = -M, \quad r = r_o \tag{8.38}$$

The resulting virial calculated along any contour enclosing both boundaries is

$$V^{mat} = V_{r_o}^{mat} + V_{r_i}^{mat} = 0, \quad r_o^+ \le r < \infty \tag{8.39}$$

Therefore, it follows

$$0 \le r < r_i^- : \ M = 0$$

$$r_i^+ \le r < r_o^- : \ M = V_{r_i}^{mat} = M$$

$$r_o^+ \le r < \infty: \quad M = V_{r_i}^{mat} + V_{r_o}^{mat} = 0 \tag{8.40}$$

8.5.3 *M*-Integral and the Energy Release Rate

The *M*-integral can be expressed as the gradient of the total potential energy of the material with respect to the change in the configuration of the defect in the structure. The *M*-integral, however, could be expressed as the variation of the total potential energy with respect to the self-similar expansion of the corresponding defect. In the problem being considered, one has

$$\begin{aligned} &\delta_{r_i}\left(\Pi^{total}\right) = M \\ &\delta_{r_i}\left(\Pi^{total}\right) + \delta_{r_o}\left(\Pi^{total}\right) = 0 \end{aligned} \tag{8.41}$$

where δ_{r_i} and δ_{r_o} are the variational operations with respect to the change in the boundaries at $r = r_i$ and $r = r_o$. The total potential energy is written in terms of the total strain energy and the total work done by the applied loads

$$\Pi^{total} = U - W^{ext} = U - \left(W^{r_i} + W^{r_o}\right) \tag{8.42}$$

During the variation of the work done by the physical traction at the boundary, the applied load, i.e., the physical scalar moment, or physical virial, generated by the traction, must be kept constant. Therefore, W^{r_i} and W^{r_o} can be rewritten as

$$W^{r_i} = \int_0^{2\pi} \left(-\sigma_{rr}(r_i)u_r(r_i)r_i\right)d\theta = 2\pi r_i V_{r_i}^{phys} \tag{8.43}$$

$$W^{r_o} = \int_0^{2\pi} \left(\sigma_{rr}(r_o)u_r(r_o)r_o\right)d\theta = 2\pi r_o V_{r_o}^{phys} \tag{8.44}$$

where $V_{r_i}^{phys}$ and $V_{r_o}^{phys}$ represent the physical virial, or the scalar moment generated by the physical traction at $r = r_i$ and $r = r_o$, respectively. Therefore,

$$\delta_{r_i}\left(W^{ext}\right) = r_i \frac{\partial}{\partial r_i}\left(W^{r_i}\Big|_{V_{r_i}^{phys}=const} + W^{r_o}\Big|_{V_{r_o}^{phys}=const}\right) = r_i\left(2\pi V_{r_i}^{phys}\right) = W^{r_i}$$

$$\delta_{r_o}\left(W^{ext}\right) = r_o \frac{\partial}{\partial r_o}\left(W^{r_i}\Big|_{V_{r_i}^{phys}=const} + W^{r_o}\Big|_{V_{r_o}^{phys}=const}\right) = r_o\left(2\pi V_{r_o}^{phys}\right) = W^{r_o} \tag{8.45}$$

Combining Equations 8.41 and 8.42 and Equation 8.45, the expressions for the *M*-integral in terms of the variation of the total potential energy are obtained

$$\delta_{r_i}\left(\Pi^{total}\right) = \delta_{r_i}\left(U - W^{ext}\right) = r_i \frac{\partial}{\partial r_i}(U) - W^{r_i} = M \tag{8.46}$$

$$\begin{aligned} \delta_{r_i}\left(\Pi^{total}\right) + \delta_{r_o}\left(\Pi^{total}\right) &= \delta_{r_i}\left(U - W^{ext}\right) + \delta_{r_o}\left(U - W^{ext}\right) \\ &= r_i \frac{\partial}{\partial r_i}(U) - W^{r_i} + r_o \frac{\partial}{\partial r_o}(U) - W^{r_o} = 0 \end{aligned} \tag{8.47}$$

The results for the M-integral evaluated by the material tractions, Equation 8.36, and by the energy release rate calculation, Equation 8.46, thus coincide. Please note that Equation 8.46 and Equation 8.47 hold only for the case of applied stress boundary conditions. For the case of applied displacements boundary conditions, the physical virial of the tractions caused by the applied displacements is not constant during the variation of the total potential energy. Therefore, Equation 8.40 and Equation 8.41 simply become

$$\delta_{r_i}\left(\Pi^{total}\right) = r_i \frac{\partial}{\partial r_i}\left(\Pi^{total}\right) = r_i \frac{\partial}{\partial r_i}\left(U - W^{ext}\right) = M \tag{8.48}$$

$$\begin{aligned} \delta_{r_i}\left(\Pi^{total}\right) + \delta_{r_o}\left(\Pi^{total}\right) &= r_i \frac{\partial}{\partial r_i}\left(\Pi^{total}\right) + r_o \frac{\partial}{\partial r_o}\left(\Pi^{total}\right) \\ &= r_i \frac{\partial}{\partial r_i}\left(U - W^{ext}\right) + r_o \frac{\partial}{\partial r_o}\left(U - W^{ext}\right) = 0 \end{aligned} \tag{8.49}$$

8.5.4 Application to a 2-D Circular Hollow Inclusion Embedded in a Finite Matrix

Consider an elastic 2-D hollow inclusion with inner radius a and outer radius b embedded in an elastic 2-D matrix with inner radius b and outer radius c. Suppose that a set of compressive stresses, P_a and P_c, is applied at $r = a$ and $r = c$, respectively (Figure 8.9). The set of continuity and boundary conditions becomes

$$\begin{aligned} u_r^I(r)\Big|_{r=b} &= u_r^M(r)\Big|_{r=b} \\ \sigma_{rr}^M(r)\Big|_{r=b} &= \sigma_{rr}^I(r)\Big|_{r=b} \\ \sigma_{rr}^M(r)\Big|_{r=c} &= -P_c \\ \sigma_{rr}^I(r)\Big|_{r=a} &= -P_a \end{aligned} \tag{8.50}$$

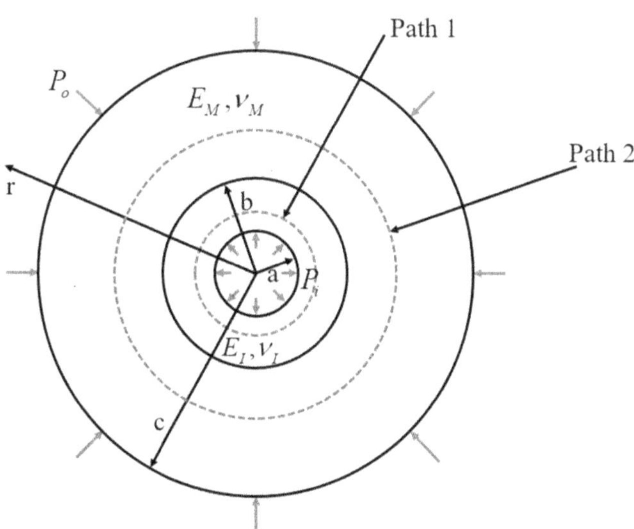

FIGURE 8.9 The hollow inclusion embedded inside the finite matrix subjected to outer and inner external load, i.e., physical tractions.

where $u_r^I(r)$ and $u_r^M(r)$ are the radial displacements of the inclusion and the matrix, respectively. Assuming the plane strain and axisymmetric conditions, the general solution for the displacement fields of the matrix and the inclusion in polar coordinates are given as

$$u_r^M(r) = A^M r + \frac{B^M}{r}$$

$$u_r^I(r) = A^I r + \frac{B^I}{r} \tag{8.51}$$

$$u_\theta^M = u_\theta^I = 0$$

where the superscripts M and I denote the matrix and the inclusion, and A^M, B^M, A^I and B^I are the constants to be determined from the boundary and continuity conditions. Using the set of conditions given in Equation 8.50, the constants A^M, B^M, A^I and B^I are determined as

$$A^M = \frac{\left\{b^2\left[c^2P_c\left(C_2+C_3\right)\right]\frac{C_4}{C_2}+a^2\left[-b^2P_a\frac{C_6}{C_2}C_3+c^2P_c\left(C_3-C_4\right)\right]\right\}}{C_1\left\{b^2\left[b^2\left(C_1-C_2\right)\frac{C_3}{C_1}-c^2\left(C_3+C_2\right)\right]\frac{C_4}{C_2}+a^2\left[b^2\left(C_1+C_4\right)\frac{C_3}{C_1}-c^2\left(C_3-C_4\right)\right]\right\}}$$

$$B^M = \frac{c^2\left\{a^2b^2\left[-P_a\frac{C_6}{C_2}C_1+P_c\left(C_1+C_4\right)\right]-b^4\left[P_cC_4-P_cC_1\frac{C_4}{C_2}\right]\right\}}{C_1\left\{b^2\left[b^2\left(C_1-C_2\right)\frac{C_3}{C_1}-c^2\left(C_3+C_2\right)\right]\frac{C_4}{C_2}+a^2\left[b^2\left(C_1+C_4\right)\frac{C_3}{C_1}-c^2\left(C_3-C_4\right)\right]\right\}}$$

$$A^I = \frac{\left\{b^2c^2P_c\frac{C_5}{C_1}C_4+a^2\left[-b^2P_a\left(C_1+C_4\right)\frac{C_3}{C_1}+c^2P_a\left(C_3-C_4\right)\right]\right\}}{C_2\left\{b^2\left[b^2\left(C_1-C_2\right)\frac{C_3}{C_1}-c^2\left(C_3+C_2\right)\right]\frac{C_4}{C_2}+a^2\left[b^2\left(C_1+C_4\right)\frac{C_3}{C_1}-c^2\left(C_3-C_4\right)\right]\right\}}$$

$$B^I = \frac{a^2\left\{b^4\left[P_aC_1-P_aC_2\right]\frac{C_3}{C_1}-b^2c^2\left[P_a\left(C_3+C_2\right)-P_c\frac{C_5}{C_1}\right]\right\}}{C_2\left\{b^2\left[b^2\left(C_1-C_2\right)\frac{C_3}{C_1}-c^2\left(C_3+C_2\right)\right]\frac{C_4}{C_2}+a^2\left[b^2\left(C_1+C_4\right)\frac{C_3}{C_1}-c^2\left(C_3-C_4\right)\right]\right\}} \tag{8.52}$$

where

$$C_1 = \frac{E^M}{\left(1+\nu^M\right)\left(1-2\nu^M\right)}, C_2 = \frac{E^I}{\left(1+\nu^I\right)\left(1-2\nu^I\right)}$$

$$C_3 = \frac{E^M}{\left(1+\nu^M\right)}, C_4 = \frac{E^I}{\left(1+\nu^I\right)}$$

$$C_5 = \frac{2E^M\left(1-\nu^M\right)}{\left(1+\nu^M\right)\left(1-2\nu^M\right)}, C_6 = \frac{2E^I\left(1-\nu^I\right)}{\left(1+\nu^I\right)\left(1-2\nu^I\right)} \tag{8.53}$$

and E and ν are Young's modulus and Poisson's ratio, respectively. According to Figure 8.9, the M-integral can be evaluated along path 1 in which the void in the inclusion is treated as the defect, and path 2 in which the hollow inclusion containing the void is treated as the defect. Hence, the M-integral on each contour can be expressed as

$$
\begin{aligned}
M_1 &= \frac{1}{2}\int_0^{2\pi}\left(\sigma_{\theta\theta}^{I}\sigma_{\theta\theta}^{I}-\sigma_{rr}^{I}\sigma_{rr}^{I}\right)r^2 d\theta = \pi\left(\sigma_{\theta\theta}^{I}\sigma_{\theta\theta}^{I}-\sigma_{rr}^{I}\sigma_{rr}^{I}\right)r^2, \quad a \le r \le b \\
M_2 &= \frac{1}{2}\int_0^{2\pi}\left(\sigma_{\theta\theta}^{M}\sigma_{\theta\theta}^{M}-\sigma_{rr}^{M}\sigma_{rr}^{M}\right)r^2 d\theta = \pi\left(\sigma_{\theta\theta}^{M}\sigma_{\theta\theta}^{M}-\sigma_{rr}^{M}\sigma_{rr}^{M}\right)r^2, \quad b \le r \le c
\end{aligned}
\tag{8.54}
$$

From Section 8.5.2, the M-integral can also be obtained from the perspective of material tractions. In Figure 8.10, the physical tractions and the jump positions are depicted, whereas in Figure 8.11, the jump of the material tractions is active at the outer boundary $r = c$, inner boundary $r = a$ and the interface $r = b$:

$$
\begin{aligned}
j_a &= \frac{1}{2}\left[\left(-\sigma_{rr}u_{r,r}+\sigma_{\theta\theta}\frac{u_r}{r}\right)\right]_{a^-}^{a^+} = \frac{1}{2}\left[-\sigma_{rr}^{I}(a)u_{r,r}^{I}(a)+\sigma_{\theta\theta}^{I}(a)\frac{u_r^{I}(a)}{a}\right] \\
j_c &= \frac{1}{2}\left[\left(-\sigma_{rr}u_{r,r}+\sigma_{\theta\theta}\frac{u_r}{r}\right)\right]_{c^-}^{c^+} = -\frac{1}{2}\left[-\sigma_{rr}^{M}(c)u_{r,r}^{M}(c)+\sigma_{\theta\theta}^{M}(c)\frac{u_r^{M}(c)}{c}\right]
\end{aligned}
\tag{8.55}
$$

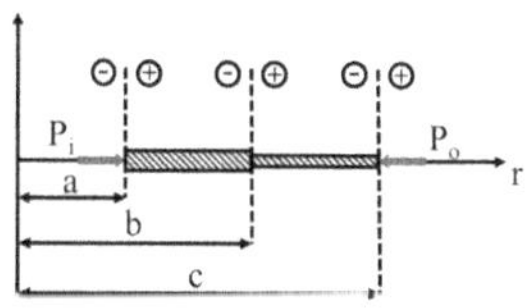

FIGURE 8.10 Physical tractions and jump positions.

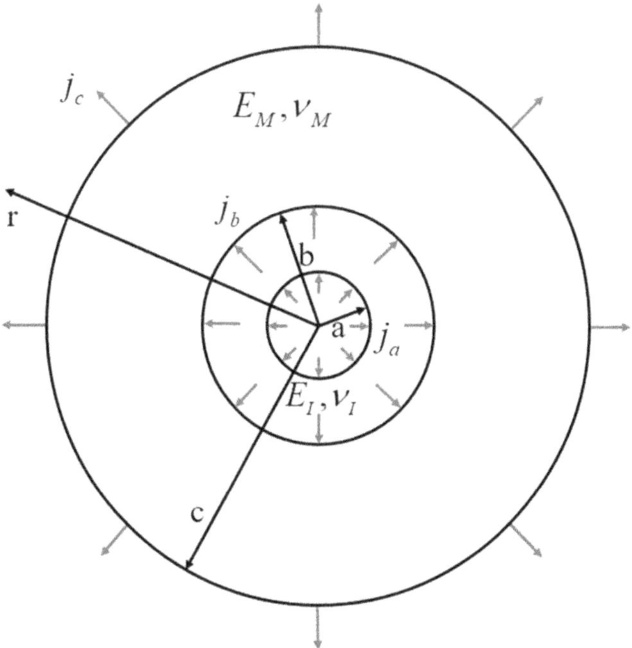

FIGURE 8.11 Material traction acting at the jump position.

The material virials are calculated as

$$V_a^{mat} = \int_0^{2\pi} (aj_a)(ad\theta) = 2\pi a^2 j_a = M_1$$

$$V_b^{mat} = \int_0^{2\pi} (bj_b)(bd\theta) = 2\pi b^2 j_b = M_2 - M_1$$

$$V_c^{mat} = \int_0^{2\pi} (cj_c)(cd\theta) = 2\pi c^2 j_c = -M_2 \quad (8.56)$$

Therefore, the expressions of M-integral in terms of the virial of the material traction for different paths are given as

$$\begin{aligned} &0 \le r < a^-: \quad M = 0 \\ &a^+ \le r < b^-: \quad M = V_a^{mat} = M_1 \\ &b^+ \le r < c^-: \quad M = V_a^{mat} + V_b^{mat} = M_2 \\ &c^+ \le r < \infty: \quad M = V_a^{mat} + V_b^{mat} + V_c^{mat} = 0 \end{aligned} \quad (8.57)$$

In the presence of the applied external field, the total potential energy of the whole structure is

$$\begin{aligned} {}^{total} &= U - W^{ext} = \left(U^I + U^M\right) - \left(W_a + W_c\right) \\ &= \frac{1}{2}\iint_{A_i} \sigma_{ij}^{I}\ {}_{ij}^{I}\, dA + \frac{1}{2}\iint_{A_M} \sigma_{ij}^{M}\ {}_{ij}^{M}\, dA - \left(\int_{S_a} t_i^I(a) u_i^I(a) dS + \int_{S_c} t_i^M(c) u_i^M(c) dS\right) \end{aligned} \quad (8.58)$$

where U^I and U^M are the strain energy of the inclusion and the matrix, while W_a and W_c are the work done by the physical tractions at $r = a$ and $r = c$, respectively. A_I and A_M denote the area of the inclusion and the matrix, and S_a and S_c represent the boundary at $r = a$ and $r = c$, respectively.

In order to obtain the M-integral through the energy release of each defect, as shown in Section 8.5.3, during the variation of the work done by the physical traction at the boundary, the applied load, i.e., the scalar moment or physical virial generated by the physical traction, has to be kept constant. W_a can be rewritten as

$$W_a = \int_{S_a} t_i^I(a) u_i^I(a) dS = \int_0^{2\pi} t_i^I(a) u_i^I(a) a d\theta = 2\pi a V_a^{phys} \quad (8.59)$$

where V_a^{phys} represents the physical virial, or the scalar moment generated by the physical traction at $r = a$. Similarly, $W_c = 2\pi c V_c^{phys}$, where V_c^{phys} is the scalar moment generated by the traction at $r = c$, respectively. Therefore,

$$\delta_a\left(W_{total}^{ext}\right) = a\frac{\partial}{\partial a}\left(W_a\big|_{V_a^{phys}=const} + W_c\big|_{V_c^{phys}=const}\right) = a\left(2\pi V_a^{phys}\right) = W_a$$

$$\delta_c\left(W_{total}^{ext}\right) = c\frac{\partial}{\partial c}\left(W_a\big|_{V_a^{phys}=const} + W_c\big|_{V_c^{phys}=const}\right) = c\left(2\pi V_c^{phys}\right) = W_c \quad (8.60)$$

Combining Equations 8.42, 8.43, 8.58 and 8.60, the expressions for the M-integral in terms of the variation of the total potential energy are obtained:

$$M_1 = \delta_a\left(\Pi^{total}\right) = \delta_a\left(U - W^{ext}\right) = a\frac{\partial}{\partial a}\left(U\right) - W_a \tag{8.61}$$

$$M_2 = \delta_a\left(\Pi^{total}\right) + \delta_b\left(\Pi^{total}\right) = a\frac{\partial}{\partial a}\left(U\right) - W_a + b\frac{\partial}{\partial b}b\left(U\right) \tag{8.62}$$

which match the expressions for the M-integral obtained using Equation 8.56.

8.6 CONCLUSION

In this chapter, the energy release rate of various types of defects and elastic singularities under different loading conditions are evaluated by the path-independent J-, L- and M-integrals. The limiting process is used to obtain the energy release rates for defects such as cracks and circular cavities under various loading conditions using the results for an elliptical cavity. The closed-form expressions of the path-independent integrals for the three types of defects under various loading conditions are reported in a tabulated form in order to compare and understand how different types of loading can affect energy release rates.

Path-independent integrals are also used to evaluate the configurational forces acting on two-dimensional elastic line singularities such as edge dislocation, the line force, the nuclei of strain and the concentrated couple moment subjected to far-field loads. The interaction forces between two similar parallel elastic singularities are also calculated. A self-similar expansion force, M, evaluated for the line force is shown to be the negative of the strain energy pre-logarithmic factor, as in the case of the well-known edge dislocation result previously obtained by Rice [30]. It is also shown that the M-integral for the nuclei of strain is nonzero under far-field normal stresses. This study is useful in understanding how defects in materials interact and influence overall mechanical behaviour.

Finally, the concept of material tractions acting on surfaces and interfaces is introduced for a simple 2-D circular hollow inclusion embedded in a finite matrix subjected to both internal and external loadings. The hollow inclusion allows us to consider both the circular void inside the inclusion and the hollow inclusion itself as defects. A systematic method of calculating the energy release rates via explicit variation of the total potential energy with respect to the self-similar scale change of the defect is demonstrated to be equal to the self-similar expansion force M. Therefore, the M-integral is shown to correspond to the energy release rate, or the change of the total potential energy of the elastic system with respect to the relative scale change of the defects enclosed within the contour integral. It is noted that during the variation of the work done by the traction at the boundary where a load is applied, the applied load must be kept constant.

ACKNOWLEDGEMENTS

This work was supported by a National Research Foundation of Korea grant funded by the Korean Ministry of Science and ICT (No. 2020R1F1A1048208).

REFERENCES

1. J. D. Eshelby (1956). The continuum theory of lattice defects. *Solid State Physics*, 79–144. doi:10.1016/s0081-1947(08)60132-0
2. G. P. Cherepanov (1967). Crack propagation in continuous media. *Journal of Applied Mathematics and Mechanics*, 31(3), 503–512. doi:10.1016/0021-8928(67)90034-2
3. J. R. Rice (1968). A Path-independent integral and the approximate analysis of strain concentration by notches and cracks. *ASME Journal of Applied Mechanics*, 35, 379–386.

4. W. Günther (1962). Über Einige Randintegrale der Elasto-mechanik, Abhandlungen der Braunschweigischen Wissenschaftlichen Gesellschaft, XIV, Verlag Friedr. *Vieweg & Sohn, Braunschweig*, 53–72.
5. J. K. Knowles, and E. Sternberg (1972). On a class of conservation laws in linearized and finite plastostatics. *Archive for Rational Mechanics and Analysis*, 44, 187–211.
6. E. Noether (1971). Invariant variation problems. *Transport Theory and Statistical Physics*, 1(3), 186–207. doi:10.1080/00411457108231446
7. B. Budiansky, and J. R. Rice (1973). Conservation laws and energy release rates, ASME *Journal of Applied Mechanics*, 40, 201.
8. L. B. Freund (1978). Stress intensity factor calculations based on a conservational integral. *International Journal of Solids and Structures*, 14, 241.
9. G. A. Herrmann, and G. Herrmann (1981). On the energy release rate for a plane crack, ASME. *Journal of Applied Mechanics*, 48, 525–528.
10. Y. E. Pak, D. Mishra, and S.-H. Yoo (2012). Energy release rates for various defects under different loading conditions. *Journal of Mechanical Science and Technology*, 26(11), 3549–3554.
11. M. Lazar, and E. Agiasofitou (2018). The J-, M-, and L-integrals of body charges and body forces: Maxwell meets Eshelby. *Journal of Micromechanics and Molecular Physics*. doi:10.1142/s242491301840012x
12. Y. H. Chen (2002). *Advances in Conservation Laws and Energy Release Rates: Theoretical Treatments and Applications*. Kluwer Academic Publishers.
13. R. Kienzler, and G. Herrmann (2000). *Mechanics in Material Space: With Applications to Defect and Fracture Mechanics*. Springer.
14. J. H. Park, and Y. Y. Earmme (1986). Application of conservative integrals to interfacial crack problems. *Mechanics of Materials*, 5(3), 261–276.
15. Y. H. Chen (2001). M-integral analysis for two dimensional solids with strongly interacting microcracks, Part I: in an infinite brittle solids. *International Journal of Solids and Structures*, 38, 3193–3212.
16. Y. Z. Chen (2003). Analysis of L integral and theory of the derivative stress field in plane elasticity. *International Journal of Solids and Structures*, 40, 3589–3602.
17. Y. Z. Chen, and Y. L. Kang (2004). Analysis of M integral in plane elasticity. *ASME Journal of Applied Mechanics*, 71, 572–574.
18. G. A. Herrmann (1983). Description of the plane crack in terms of local fields verses path independent integrals, in: G. C. Sih, E. Sommer, and W. Dahl (Eds.), *Application of Fracture Mechanics to Materials and Structures*, Martinus Nijhoff Publishers, pp. 571–579.
19. R. Kienzler, and H. Kordisch (1990). Calculation of J_1 and J_2 using the L and M integrals. *International Journal of Fracture*, 43, 213–225.
20. R. Kienzler (2008). Energy changes in elastic plates due to holes and cracks. *Proceedings of the Estonian Academy of Sciences*, 57, 26–33.
21. R. King, and G. Herrmann (1981). Nondestructive evaluation of the J and M integrals, ASME *Journal of Applied Mechanics*, 103(1), 83–87.
22. Z. Suo (1999). Zener's crack and the M-Integral. *ASME* Journal of Applied Mechanics, 67(2), 417–418.
23. Y. E. Pak, D. Mishra, and S. H. Yoo (2012). Closed-form solution for a coated circular inclusion under uniaxial tension. *Acta Mechanica*, 223(5), 937–951. doi:10.1007/s00707-012-0617-0
24. C. R. Meyer, J. W. Hutchinson, and J. R. Rice (2016). The path-independent M integral implies the creep closure of englacial and subglacial channels. *Journal of Applied Mechanics*, 84(1), 011006. doi:10.1115/1.4034828
25. Z. Zhang, J. Lv, X. Li, J. Hou, and Q. Li (2021). A fatigue model based on m-integral in notched elastic-plastic material. *International Journal of Solids and Structures*, 111203. doi:10.1016/j.ijsolstr.2021.111203
26. Q. Li, and Y. H. Chen (2008). Surface effect and size dependence on the energy release due to a nanosized hole expansion in plane elastic materials. *Journal of Applied Mechanics*, 75(6), 061008. doi:10.1115/1.2965368
27. T. Hui, and Y.-H. Chen (2010). The M-integral analysis for a nano-inclusion in plane elastic materials under uni-axial or bi-axial loadings. *Journal of Applied Mechanics*, 77(2), 021019. doi:10.1115/1.3176997
28. H. Yi-Feng, and Y. H. Chen (2010). The area contraction and expansion for a nano-void under four different kinds of loading. *Archive of Applied Mechanics*, 81(9), 1323–1331. doi:10.1007/s00419-010-0489-5
29. H. Yifeng, and C. Yi-Heng (2011). Energy release or absorption due to simultaneous expansion of many interacting nano-holes in elastic materials. *Archive of Applied Mechanics*, 82(6), 777–789. doi:10.1007/s00419-011-0591-3
30. G. Cherepanov (1981). Invariant integrals. *Engineering Fracture Mechanics*, 14(1), 39–58.

31. Y. E. Pak (1990). Force on a piezoelectric screw dislocation. *ASME Journal of Applied Mechanics*, 57(4), 863–869.
32. V. A. Lubarda (2016). Determination of interaction forces between parallel dislocations by the evaluation of J integrals of plane elasticity. *Continuum Mechanics and Thermodynamics*, 28(1), 391–405.
33. F. R. N. Nabarro (1985). Material forces and configurational forces in the interaction of elastic singularities. *Proceedings of the Royal Society of London. A. Mathematical and Physical Sciences*, 398(1815), 209–222.
34. J. Eshelby (1951). The force on an elastic singularity. *Philosophical Transactions of the Royal Society of London. Series A, Mathematical and Physical Sciences*, 244(877), 87–112.
35. J. R. Rice (1985). Conserved and integrals and energetic forces. In: B. A. Bilby, K. J. Miller, and J. R. Willis (Eds.), *Fundamentals of Deformation and Fracture*, Cambridge University Press, pp. 33–56.
36. Y. Seo, G.-J. Jung, I.-H. Kim, and Y. Eugene Pak (2018). Configurational forces on elastic line singularities. *Journal of Applied Mechanics*, 85(3), 034501. doi:10.1115/1.4038808
37. Y. E. Pak (1990). Crack extension force in a piezoelectric material. *Journal of Applied Mechanics*, 57, 647–653.
38. Y. E. Pak, G. A. Herrmann, and G. Hermann (1983). Energy release rates for various defects. In: D. O. Thompson, and D. E. Chimenti (Eds.), *Review of Progress in Quantitative Nondestructive Evaluation 2B*. Plenum Publishing Corporation, pp. 1389–1397.
39. Y. E. Pak (1992). Circular inclusion problem in antiplane piezoelectricity. *International Journal of Solids and Structures*, 29, 2403–2419.
40. Y. E. Pak (2010). On the use of path-independent integrals in calculating mixed-mode stress intensity factors for elastic and thermoelastic cases. *Journal of Thermal Stresses*, 33, 661–673.
41. Y. E. Pak (2010). Elliptical inclusion problem in antiplane piezoelectricity: Implication for fracture mechanics. *International Journal of Engineering Science*, 48, 209–222.
42. J. W. Eishen, and G. Hermann (1987). Energy release rates and related balance laws in linear elastic defect mechanics. *Journal of Applied Mechanics*, 109, 388–392.
43. S. P. Timoshenko, and J. N. Goodier (1970). *Theory of Elasticity*, Mc-Graw-Hill.
44. M. Kachanov, B. Shafiro, and I. Tsukrov (2003). *Handbook of Elasticity Solutions*, Kluwer Academic Publishers.
45. A. F. Bower (2009). *Applied Mechanics of Solids*, CRC Press.
46. A. E. Love (1906). *A Treatise of the Mathematical Theory of Elasticity*, Cambridge University Press.
47. S. Y. Seo, D. Mishra, C. Y. Park, and Y. E. Pak (2015). Energy release rate for a misfitted spherical inclusion under far-field mechanical and uniform thermal loads. *European Journal of Mechanics - A/Solids*, 49, 169–182.
48. E. Agiasofitou, and M. Lazar (2017). Micromechanics of dislocations in solids: J-, M-, and L-integrals and their fundamental relations. *International Journal of Engineering Science*, 114, 16–40.

9 Modeling Crack Growth in Materials Using Finite Element Method

Charanjeet Singh Tumrate, Abhishek Rizal, and Dhaneshwar Mishra

9.1 FINITE ELEMENT MODELING APPROACH

Computational modeling approaches are extensively utilized in the field of engineering for analyses of various structures, components, and systems to aid the design process. These approaches are not only helpful in optimizing the structures but also in choosing appropriate materials. Material modeling at different scales under different loadings and loading environments is effective in understanding the material's behavior and underlying mechanics. Defects such as microcracks in materials are very common. These defects do not affect the performance until they reach the critical level. Therefore, it is important to understand the failure behavior of materials with intrinsic defects. Fracture mechanics provides the theoretical basis of defects such as crack initiation and growth in material. There are experimental methods available to evaluate fracture toughness and fracture resistance in materials, but they are expensive. Finite element (FE) analysis-based modeling approaches are extensively utilized to simulate the growth pattern and the material resistance to fracture. There are numerous studies available in the literature on crack growth simulation in materials using the finite element method (FEM) [1–4]. These works range from cracks in an isotropic medium to cracks in a nonhomogeneous medium to interface cracks. There are various techniques available to model and analyze cracks in these different mediums. Figure 9.1 represents the steps considered in general to perform crack growth analysis using finite element-based software.

The required structural geometry is developed at the FE software interface or can be imported from drafting software such as AutoCAD, SolidWorks, or Creo CAD. The material as well as structural properties are applied to the geometry. The mesh elements along with the element size and mesh type (coarse, medium, and fine) are selected. Once the meshing is completed, the boundary conditions (fixed, hinged, roller, etc.) in addition to the external loading are applied to the meshed geometry. Crack properties such as crack tip elements, crack location, and size are allotted along with the declaration of the analysis type (linear, nonlinear, convergence properties). Once all the parameters are complete, the model is run to generate the response. The analysis of the result is performed to determine the critical stresses, remaining resistance of the material, crack propagation, and failure types. In this chapter, various modeling techniques for cracks in an isotropic medium and at the interface have been discussed. Modeling of defect/crack/inhomogeneity by FEM requires a large number of elements (smaller mesh size) to capture the field parameters such as displacements, strain, and stresses. An important development in FEM carried out by Ted Belytschko and his team in 1999 [5] was to correctly analyze defects (discontinuities) such as cracks or inhomogeneities by enriching the nodes near the defects with higher-order polynomials so that the related field variables can be captured correctly. This method is called the extended finite element method (XFEM). In this chapter, the basics of XFEM along with a test case of modeling a functionally graded plate with and without pores using XFEM provides a better understanding of the concept.

 DOI: 10.1201/9781003359364-12

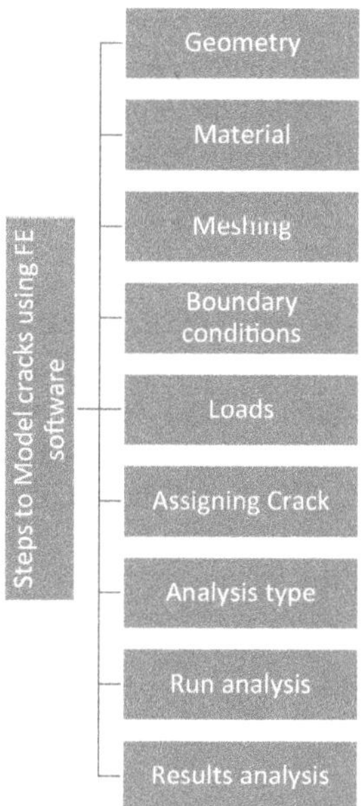

FIGURE 9.1 General steps to analyze cracks in materials using FE software.

9.2 MODELING CRACKS IN AN ISOTROPIC MEDIUM

The properties of the material remain the same in all orientations for an isotropic medium. The characteristics of the material, such as its elastic modulus and fracture toughness, are important in influencing how fractures behave. An isotropic medium can develop cracks, which have linear discontinuities that propagate across the material and deteriorate it while raising the possibility of failure. Additionally, the existence of cracks could result in stress concentration and possibly influence the stress in the surrounding material. The accuracy of the crack propagation analysis in an isotropic medium is affected by the mesh refinement, J-integral domain size [6]. The J-integral denotes the tool to evaluate the strain energy release rate per unit surface area fractures [7]. The application of the finite element method for evaluation of the J-integral by the interaction integral method (M-integral) [8] is based on Equation 9.1:

$$J = \int_{A^*} [\sigma_{ij} \frac{\partial u_i^{aux}}{\partial x_l} - \sigma_{ij}^{aux} \frac{\partial u_i}{\partial x_l} - \sigma_{ij}\varepsilon_{ij}^{aux}\delta_{lj}] \frac{\partial q}{\partial x_j} d" \quad (9.1)$$

where the J-integral (J) is represented by the domain J-integral area (A^*), the stress components σ_{ij}, auxiliary displacement u_i^{aux}, position vector x, strain tensor ε, variation of a function δ, and boundary by Γ. Cracks in an isotropic medium can be minimized with careful material selection, design parameters, construction process, structural health monitoring, and maintenance. The application of non-destructive techniques (NDTs) and artificial intelligence (AI) tools such as machine learning (ML) and computer vision (CV) for monitoring structural health can assist in identifying cracks at an early stage.

9.2.1 Cracks at the Interface

An interface is a zone where any two materials are connected as seen in the layered and composite materials. The most typical interface failure mechanism observed is interface fracture or delamination between the two adjacent layers. The terms in interface analysis such as energy release rate and stress intensity factor (K) illustrate the properties mismatch and its effect on the interface crack propagation [9]. The stress intensity near the crack tip due to residual stresses or remote loads is predicted by the stress intensity factor (K) [10]. To understand the modeling of cracks at the interface, the cohesive zone model and virtual crack closure technique are discussed in Sections 9.2.1.1 and 9.2.1.2.

9.2.1.1 Cohesive Zone Model

The cohesive zone is a region that exists ahead of the crack tip in which multiple microcracks exist generally in quasi-brittle materials. The alliance of these microcracks results in the formation of the main fracture. The traction–separation theory helps in explaining the cohesive zone constitutive behavior to present the link between a constitutive relation and a cracked body [11]. The development of a cohesive zone model (CZM) involves the representation of the fracture process by a local stress–displacement relationship as shown in Figure 9.2. The area under the curve represents Equation 9.2 of fracture energy, G_c [12]:

$$G_c = \int_0^{v_0} \sigma \, dv \tag{9.2}$$

where

v_0= the displacement at final fracture
$\bar{\sigma}$ = maximum stress

Figure 9.2 represents a local stress–displacement relationship where the fracture energy represented by the area under the curve is determined by integration. The fracture calculates the energy required for the microcracks located in the process zone to align into a single crack.

The cohesive zone concept, used for cementitious composites such as concrete, can be effectively applied to a variety of materials. The ability of the cohesive zone model to provide precise predictions for concrete is already proven, along with various notched samples of a glassy polymer (PMMA), certain steels, and concrete [13].

A universal bilinear function has shown to be particularly effective as a softening function for concrete and other cementitious materials. The tensile strength (ft), the abscissa of the centroid of the softening region $(\overline{w})$, the specific fracture energy (GF), and the horizontal intercept (wl) of the

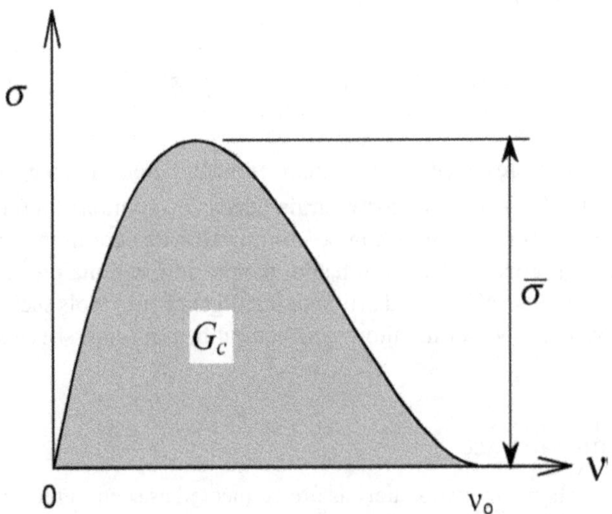

FIGURE 9.2 The cohesive zone law [12].

initial linear approximation are the four parameters that best describe this function, as illustrated in Figure 9.3 [13].

In Figure 9.4, a FEM representation of three different mesh sizes of a notched crack ductile material for a mode I tear experiment is shown. The boundary condition is fixed at nodes B and B′ to avoid translation along the vertical. Nodes A and A′ are subjected to load at constant vertical velocity. The displacement Δ and force P are determined at these nodes. The plasticity zone ahead of the crack tip, i.e., the green area from where the traction–separation curve and fracture energy were calculated, is valid with respect to the experiment [14].

9.2.1.2 Virtual Crack Closure Technique

When implementing the mixed-mode fracture criterion, the mode separation requirement is achieved by the virtual crack closure technique (VCCT), which is widely used for energy release rate analysis based on the outcomes from continuum and solid finite element analyses [15].

According to the VCCT, the quantity of strain energy released for a microscopic crack opening is equal to the amount of work needed to close the crack. Two steps are considered in the analysis of the effort required to close the crack:

i) stress field at the crack tip for crack length
ii) displacement from a to $\Delta\ a$

The expression for two-step VCCT is given by Equation 9.3 [16]:

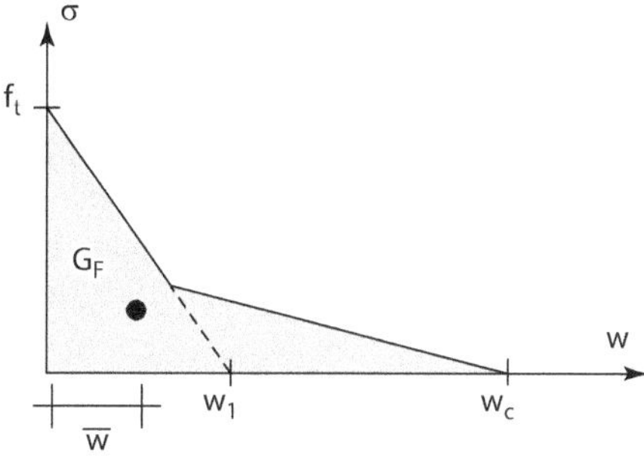

FIGURE 9.3 Illustration of general bilinear softening function [13].

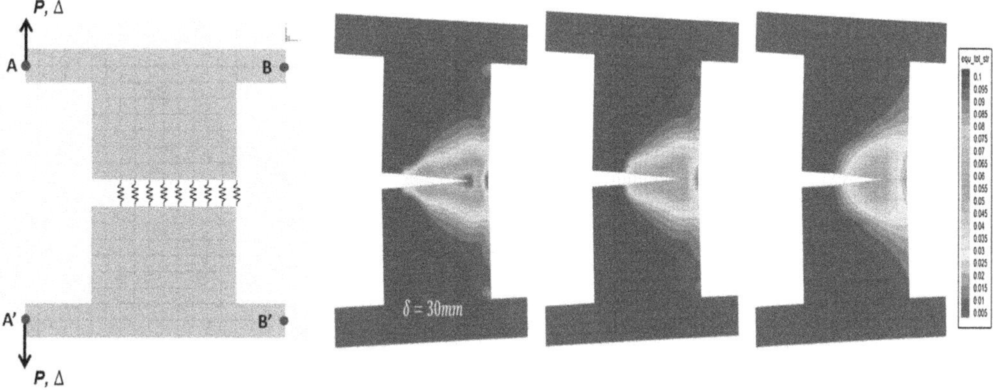

FIGURE 9.4 Finite element model representation and simulation of a performed experiment on a ductile plate for mode I fracture test [14].

$$W = \frac{1}{2}\left(\int_0^{\Delta a} \sigma_{yy}^{(a)}(x)\delta u_y^{(b)}(x)dx + \int_0^{\Delta a} \sigma_{yx}^{(a)}(x)\delta u_x^{(b)}(x)dx + \int_0^{\Delta a} \sigma_{yz}^{(a)}(x)\delta u_z^{(b)}(x)dx\right) \tag{9.3}$$

where

W = work done to close the crack

σ = stress

Δa = distance of crack tip after extension of the crack

Equation 9.3 can be modified, making it simpler to find the strain energy release rate, by assuming that a minuscule crack extension has no effect on the crack font and that both stress and displacement can be obtained by a single analysis represented by Equation 9.4 [16]:

$$W = \frac{1}{2}\left(\int_0^{\Delta a} \sigma_{yy}^{(a)}(x)\delta u_y^{(a)}(x-\Delta a)dx + \int_0^{\Delta a} \sigma_{yx}^{(a)}(x)\delta u_x^{(a)}(x-\Delta a)dx + \int_0^{\Delta a} \sigma_{yz}^{(a)}(x)\delta u_z^{(a)}(x-\Delta a)dx\right) \tag{9.4}$$

where

W = work done to close the crack

σ = stress

Δa = distance of crack tip after extension of the crack

$x - \Delta a$ = displacement

VCCT utilizes fracture mechanics criteria where the strain energy release rate is computed. For the 2D crack problem, an interface element can be implemented, and on one-step analysis 2D VCCT, the approximation of the strain energy release rate is the product of nodal forces and the nodal displacement behind the crack tip, which is given by Equations 9.5 and 9.6 and Figure 9.5 [17].

$$G_I = \frac{F_y \Delta v}{2B\Delta a} \tag{9.5}$$

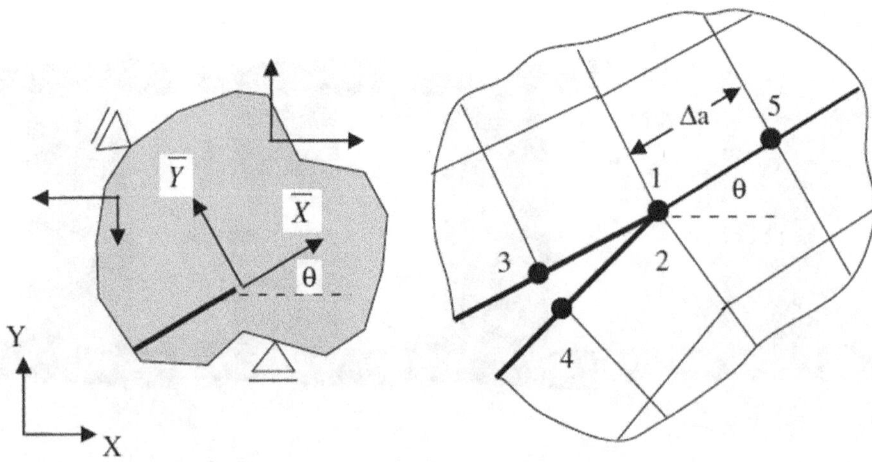

FIGURE 9.5 Interface element (2D) in inclined crack plane [17].

$$G_{II} = \frac{F_x \Delta u}{2B\Delta a} \tag{9.6}$$

The G represents the strain energy release rate, and body thickness and displacements at nodes are given by B and U, respectively. $F_X = k_y(u_2 - u_4)$ and $F_y = k_x(u_1 - u_3)$, where k is the stiffness matrix element. $\Delta v = u_6 - u_8$, $\Delta u = u_5 - u_7$, and $\Delta a = |X_5 - X_1|$. $\overline{XY}$ represents the local coordinate system, where $\sin\theta = \frac{x_5 - x_1}{\Delta_a}$ and $\cos\theta = \frac{y_5 - y_1}{\Delta_a}$.

9.3 EXTENDED FINITE ELEMENT METHOD

Finite element method is extensively used for crack growth modeling. The required structure is divided into regions called elements and are connected through mesh. The complex boundary conditions can easily be handled using FEM, but for complex geometry like 3D models, FEM requires a lot of computation time and becomes expensive [18]. Existing approaches have limitations with crack meshing; however, applying XFEM assists in overcoming these issues. FEM theoretical development can be readily applied to XFEM and are hence the building blocks [19]. XFEM uses enrichment functions such as the partition of unity (PU) technique in the field of discontinuity [19]. The nodes around cracks are enriched as the approximate solution is obtained by the application of higher-order polynomials which results in avoiding the remeshing process. The enrichment technique provides options for tackling issues such as void formation, surface effects in nanomechanics, and subscale models of interface behavior. The XFEM technique has considerably boosted the FEM's power for various problems of interest in mechanics of materials [19, 20]. However, for brittle materials, as shown in Figure 9.5, the results obtained using FEM shows more promising output as compared to XFEM [20].

In Figure 9.6, the red dot denotes the crack initiation location and the crack direction is represented by arrows. A, B, C, and D denote cone crack, critical downward and upward vertical crack, and horizontal crack, respectively. A low-velocity impact was created on a glass silica beam to understand the crack pattern and propagation using FEM, XFEM, DEM (discrete element method), and FEM/DEM. A comparative study of the obtained results was performed. The partition of unity provides a mathematical framework for the development of an enriched solution. The generalization of the standard Galerkin FE method forms the partition of unity FE method [21, 22]. For a scalar function u, the partition of unity approximation is represented by Equation 9.7:

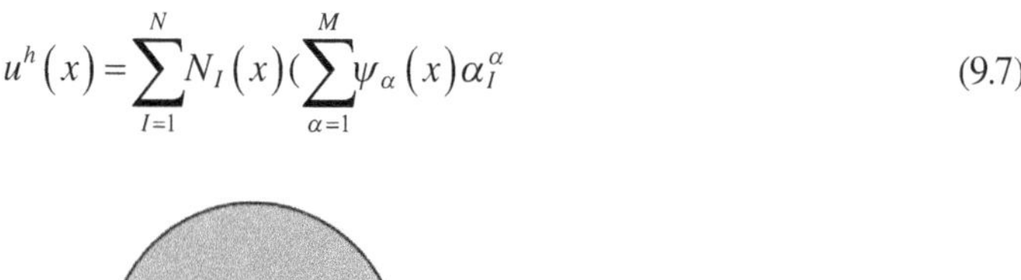

$$u^h(x) = \sum_{I=1}^{N} N_I(x)\left(\sum_{\alpha=1}^{M} \psi_\alpha(x)\alpha_I^\alpha\right) \tag{9.7}$$

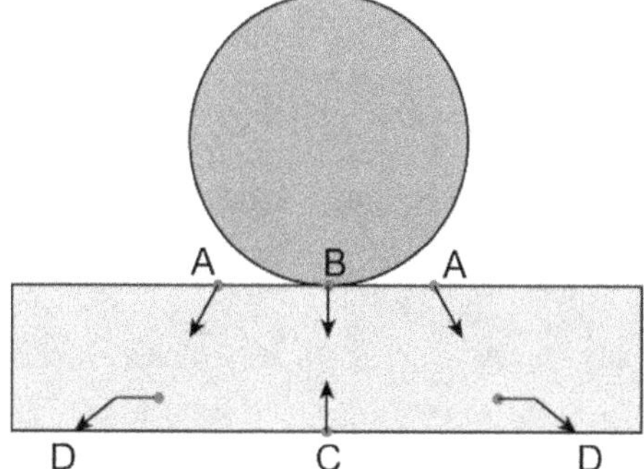

FIGURE 9.6 Illustrations of crack initiation location in a brittle silica glass beam [20].

where u is a scalar-valued function, ψ is the enrichment function, and α_I^{α} is the unknown coefficients related to node (I), a specific geometric entity (cracks), and enrichment function (ψ) [21]. The accuracy of the approximate solutions can be increased in the analytical near crack tip solutions included in the enrichment terms.

9.3.1 Crack Propagation Pattern

From the model, it's clear that cracks originated at point A and the vertical downward cone crack was formed at the time of impact. It was observed that the crack on one face of the cone propagated much faster as compared to the other side in the XFEM model [20]. Unlike XFEM, other models had similar crack patterns and fragmentation to that of glass. Fractured glass pieces detached from the model, discontinuum-based methods resulted in the shattering of glass into small pieces, whereas the continuum method had large pieces of shattered glass. On further simulation, it was found the propagation in XFEM was significantly affected by the meshing strategy. Typical crack patterns like cone crack, critical crack, and horizontal crack were only seen in models other than XFEM and could not predict the cracking pattern of high-velocity impact and instead predicted a single crack. Since the impact energy was released by a rapid fragmentation process and small element size, further cone crack propagation was not effectively contained. When compared to experimental observation, the FEM/DEM model for dynamic fracture of brittle materials, particularly glass, produced the most satisfactory result.

9.3.2 Functionally Graded Materials with Porous Plates

Composite materials such as functionally graded materials (FGMs) have heterogenous properties along a certain direction. In comparison to bulk materials, these materials have better strength and toughness. These materials have a wide range of applications, including bio-implants, automotive, and aerospace, because of the graded qualities at every point in various dimensions. In this study, FGM modeling is done using the commercial FEM software ABAQUS; the crack growth simulation under loading conditions was analyzed using the XFEM technique. Enrichment functions were used for porosity modeling and the pseudo temperature technique was applied for linearly varying material properties, where pseudo temperature is assigned as positions and the corresponding variations in the material properties.

In this study, the y direction represents material gradation. The composite material modeled for this study was a combination of hydroxyapatite and titanium at the upper and lower end, respectively. Crack growths were simulated for shear load and uniaxial tensile load for non-porous and porous plates in isotropic material as well as FGM using XFEM.

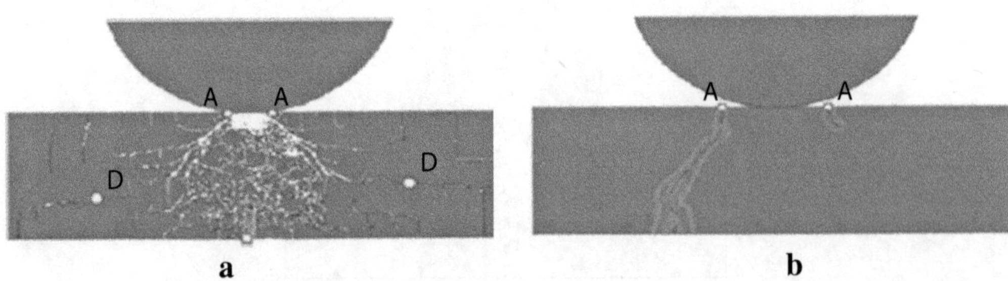

FIGURE 9.7 Crack patterns on a brittle silica glass beam when subjected to impact velocity through steel projectile on a simulated (a) FEM and (b) XFEM model [20].

9.3.3 Crack Growth in Isotopic Materials

For uniaxial tensile and shear loadings, crack development was simulated in isotropic titanium and hydroxyapatite (HAP) plates as shown in Figures 9.8 and 9.9. For the simulation of fracture growth in both porous and non-porous isotropic plates, XFEM was utilized [23].

For non-porous plates, the expansion of the fracture was toward the bottom surface and attributed to loading. In a porous plate, the crack, which was caused by an interior pore and was vulnerable to stress concentration around the pore, expanded toward the right edge with minor deviation toward the bottom. Given the internal pores, the cracks in porous plates developed at a lower load than those in non-porous plates, which was supported by the von Mises stress distribution for porous and non-porous plates. Under tensile loading, the crack pattern for HAP and titanium plate are practically the same, but there is considerable divergence in the von Mises stress distribution [23].

9.3.4 Crack Growth in FGM

The HAP plate and titanium were utilized to simulate the FGM in this work to employ it for bio-implants. Before evaluating the HAP plate, isotropic material crack patterns were examined. Uniaxial

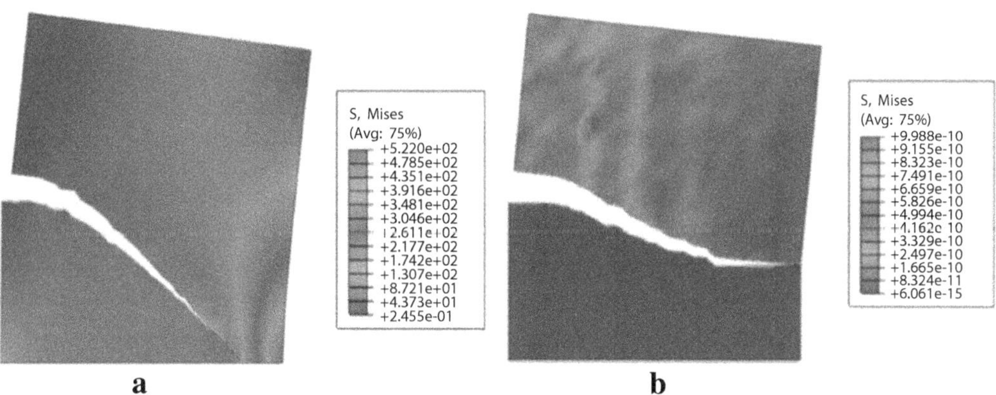

FIGURE 9.8 Crack pattern growth in hydroxyapatite (HAP) plate under shear loading for (a) non-porous and (b) porous plates [23].

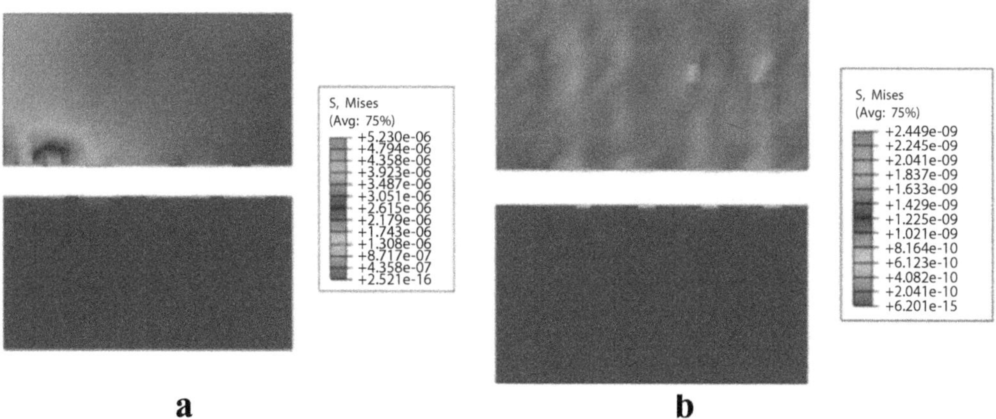

FIGURE 9.9 Crack pattern growth in titanium plate under tensile loading for (a) non-porous and (b) porous plates [23].

tensile and shear loading were the loading conditions employed for the investigation of the fracture growth pattern in the HAP, and the simulation was completed using the XFEM method [23].

The scattered pores nucleate more quickly and change into cracks related to other pores; the crack growth pattern in porous plates has a longer path. In porous plates, the crack growth occurred at a lower rate than that of non-porous plates. When subjected to shear loading, the crack path of the FGM plate was different from that of the isotropic platinum and HAP plate. It was revealed that the FGM plate crack moved more toward the bottom edge than the HAP plate crack did, while the titanium plate crack moved more toward the bottom face than the FGM. This may have been due to titanium's greater resistance to cracking than FGM, whereas HAP's lower resistance to fracture caused a crack to develop along the right edge [23]. Diverse material properties and loading might result in different crack growth and patterns. The investigation enabled an understanding of FGM behavior in fracture patterns and the modeling component of XFEM.

9.3.5 Case Study: Crack Growth Simulation in Reinforced Concrete Using XFEM

Reinforced concrete structures are heterogeneous materials used in the construction industry; they can be either precast or in situ. Concrete structures can undergo cracking due to various factors, and hence understanding and analyzing cracks is vital to building a safer structure. Analysis using the XFEM method and validation with experimental findings and comparison are represented in Figure 9.11 [24].

For simulation of cracks, only half of the beam was simulated because of symmetry. The crack pattern obtained from the XFEM method just before the failure was obtained by three analyses in which the crack initiation location differed for the first two steps, while the third step comprised the first two steps. The crack pattern obtained from the XFEM method agreed with the crack pattern from the test results. Further analysis of structural behavior was done for which the load–displacement curve was generated and compared with the experiment result [24]. All generated graphs were in close alignment with the experimental results.

9.4 CONCLUSION

An in-depth view of the progression of cracks in an isotropic media as well as the interface are provided by crack growth simulation using finite element software. The cohesive zone model and

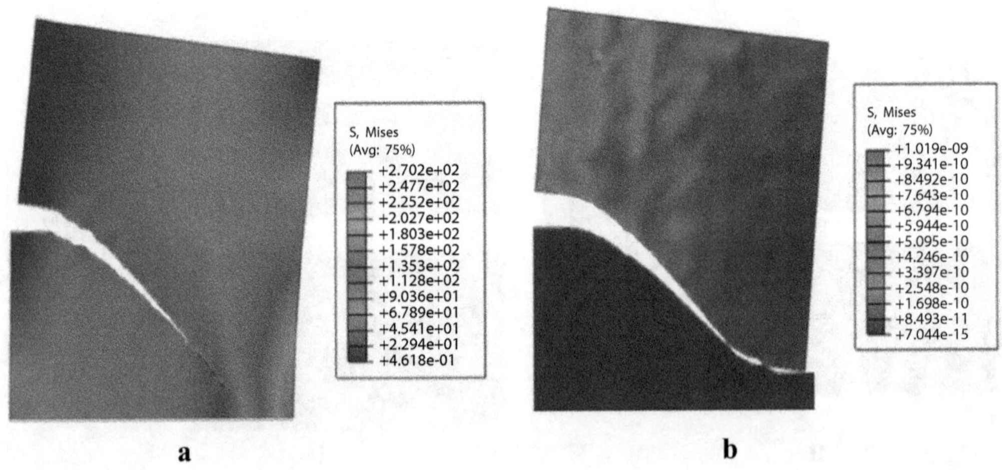

FIGURE 9.10 Crack growth pattern in functionally graded material under shear loading in (a) non-porous and (b) porous plates [23].

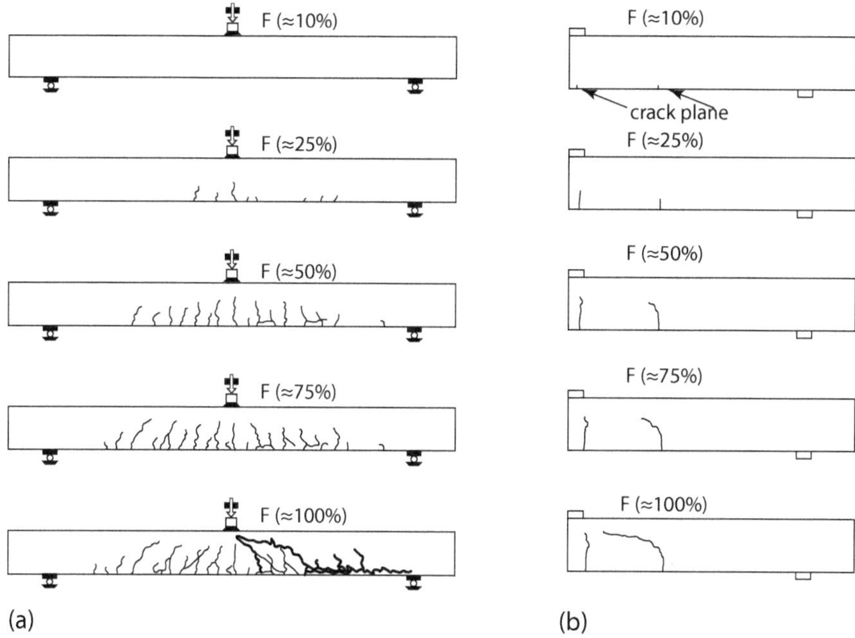

FIGURE 9.11 Crack pattern in reinforced concrete beam under different loading for (a) experimental and (b) simulated results [24].

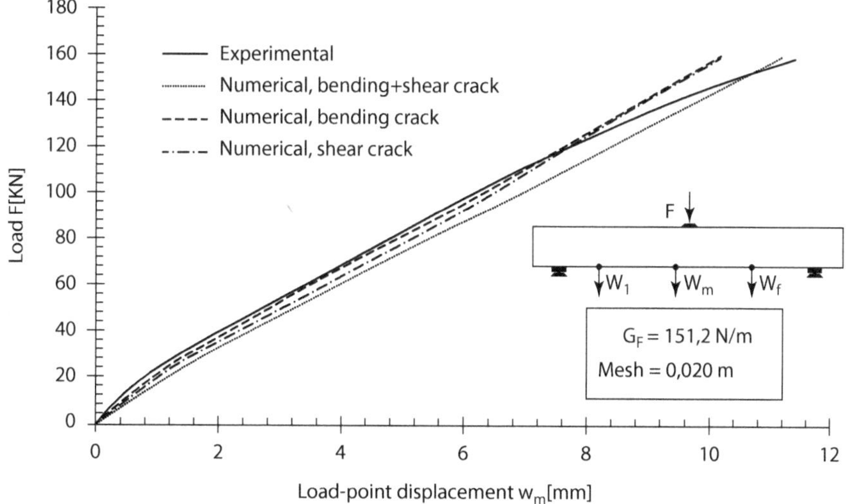

FIGURE 9.12 Load–displacement graph comparison of experimental results and graph generated from XFEM analysis where the numerical graph was generated from the XFEM analysis [24].

virtual crack techniques work extremely well with FEM-assisted simulation software. When using the extended finite element approach, enriched functions such as the PU technique can be used in the field of discontinuities. The pattern of fracture expansion through the materials using functionally graded materials both with and without porous plates can be analyzed using soft computing. A case study on the crack growth simulation in reinforced concrete using XFEM suggests that the results generated from the simulation were in sync with the experimental values. Overall, the simulation results can be used to predict the remaining structural serviceable life for failure prediction analysis.

ACKNOWLEDGEMENTS

The authors would like to acknowledge the high-end computational facility of Manipal University Jaipur, the Multiscale Simulation Research Center (MSRC) for providing computational support.

REFERENCES

1. Swati RF, Wen LH, Elahi H, Khan AA, Shad S. Extended finite element method (XFEM) analysis of fiber reinforced composites for prediction of micro-crack propagation and delaminations in progressive damage: A review. *Microsystem Technologies.* 2019 Mar 4;25:747–63.
2. Xu Y, Yuan H. Computational analysis of mixed-mode fatigue crack growth in quasi-brittle materials using extended finite element methods. *Engineering Fracture Mechanics.* 2009 Jan 1;76(2):165–81.
3. Geelen R, Plews J, Tupek M, Dolbow J. An extended/generalized phase-field finite element method for crack growth with global-local enrichment. *International Journal for Numerical Methods in Engineering.* 2020 Jun 15;121(11):2534–57.
4. Okodi A, Lin M, Yoosef-Ghodsi N, Kainat M, Hassanien S, Adeeb S. Crack propagation and burst pressure of longitudinally cracked pipelines using extended finite element method. *International Journal of Pressure Vessels and Piping.* 2020 Jul 1;184:104115.
5. Moës N, Dolbow J, Belytschko T. A finite element method for crack growth without remeshing. *International Journal for Numerical Methods in Engineering.* 1999 Sep 10;46(1):131–50.
6. Gu J, Yu T, Tanaka S, Yuan H, Bui TQ. Crack growth adaptive XIGA simulation in isotropic and orthotropic materials. *Computer Methods in Applied Mechanics and Engineering.* 2020 Jun 15;365:113016.
7. Bai Q, Bai Y. *Subsea pipeline design, analysis, and installation.* Gulf Professional Publishing; 2014 Feb 18.
8. Mohammadi S. *Extended finite element method: For fracture analysis of structures.* John Wiley & Sons; 2008 Apr 30.
9. Chen L, Liu GR, Nourbakhsh-Nia N, Zeng K. A singular edge-based smoothed finite element method (ES-FEM) for bimaterial interface cracks. *Computational Mechanics.* 2010 Jan;45(2):109–25.
10. Parker, A. P. Stress intensity factors, crack profiles, and fatigue crack growth rates in residual stress fields. *Residual Stress Effects in Fatigue, ASTM STP.* 1982;776: 13–31.
11. Ponnusami SA, Turteltaub S, van der Zwaag S. Cohesive-zone modelling of crack nucleation and propagation in particulate composites. *Engineering Fracture Mechanics.* 2015 Nov 1;149:170–90.
12. Williams JG, Hadavinia H. Analytical solutions for cohesive zone models. *Journal of the Mechanics and Physics of Solids.* 2002 Apr 1;50(4):809–25.
13. Elices MG, Guinea GV, Gomez J, Planas J. The cohesive zone model: Advantages, limitations and challenges. *Engineering Fracture Mechanics.* 2002 Jan 1;69(2):137–63.
14. Woelke PB, Shields MD, Hutchinson JW. Cohesive zone modeling and calibration for mode I tearing of large ductile plates. *Engineering Fracture Mechanics.* 2015 Oct 1;147:293–305.
15. Krueger R. Virtual crack closure technique: History, approach, and applications. *Applied Mechanics Reviews.* 2004 Mar 1;57(2):109–43.
16. Elisa P. Virtual crack closure technique and finite element method for predicting the delamination growth initiation in composite structures. *Advances in Composite Materials-Analysis of Natural and Man-made Materials.* 2011 Sep 9:463–80.
17. Xie D, Biggers Jr SB. Progressive crack growth analysis using interface element based on the virtual crack closure technique. *Finite Elements in Analysis and Design.* 2006 Jul 1;42(11):977–84.
18. Khoei, AR. *Extended finite element method: Theory and applications.* John Wiley & Sons, 2014.
19. Sukumar N, Moës N, Moran B, Belytschko T. Extended finite element method for three-dimensional crack modelling. *International Journal for Numerical Methods in Engineering.* 2000 Aug 20;48(11):1549–70.
20. Wang XE, Yang J, Liu QF, Zhang YM, Zhao C. A comparative study of numerical modelling techniques for the fracture of brittle materials with specific reference to glass. *Engineering Structures.* 2017 Dec 1;152:493–505.
21. Sukumar N, Huang ZY, Prévost JH, Suo Z. Partition of unity enrichment for bimaterial interface cracks. *International Journal for Numerical Methods in Engineering.* 2004 Feb 28;59(8):1075–102.

22. Melenk JM, Babuška I. The partition of unity finite element method: Basic theory and applications. *Computer Methods in Applied Mechanics and Engineering.* 1996 Dec 15;139(1–4):289–314.
23. Singh AP, Tailor A, Tumrate CS, Mishra D. Crack growth simulation in a functionally graded material plate with uniformly distributed pores using extended finite element method. *Materials Today: Proceedings.* 2022 Jan 1;60:602–7.
24. Faron A, Rombach GA. Simulation of crack growth in reinforced concrete beams using extended finite element method. *Engineering Failure Analysis.* 2020 Oct 1;116:104698.

10 Nanoindentation Modeling of Materials Using Finite Element Method

Dhaneshwar Mishra, Harsha Pandey, and Kulwant Singh

10.1 INTRODUCTION

One of the major analysis techniques for investigating the mechanical response of structural materials and other materials at macroscale, microscale, and nanoscale lengths is the finite element method (FEM). It is also one of the most popular techniques for discrete investigation of structural mechanics in various materials. In simple terms, this technique makes the subdivision of a given mathematical model that applies to localized nanoscale deformation and fracture in materials. This process makes very small elements of simple geometry. These elements are non-overlapping. It approximates the actual structure keeping the boundary conditions in mind. These elements are called finite elements. This is done by analytical formation of a huge number of meshes. First, the mechanical response of each finite element is expressed in terms of certain response functions. Each such function has in correspondence the appropriate degree of freedom. These are evaluated at the corresponding nodal points predefined in the model. Ultimately, all the response functions are connected for all the elements. This sum effectively and sufficiently approximates the structural mechanics response of a given material to which the mathematical model is applied. This technique therefore is extendable to many real lives as well as artificial situations. As the finite elements do not overlap, the technique can be utilized for analyzing the macroscale to nanoscale mechanical response of a multitude of systems, e.g., metals, ceramics, thin films, polymers, biomaterials, thick films, coating, composites, and interfaces. In principle, the FEM technique is also applicable to large-scale structures, such as a human bone structure, a building, an aerospace vehicle, or even a bridge. What follows next are some typical examples of the application of the FEM technique and its numerous emerging variants to the nanoindentation of various materials of current relevance from the literature.

Therefore, it comes as no surprise that FEM is applied to analyze phase transformations that occur during nanoindentation of Cu-Ni and NiTiPd shape memory alloys. The results suggest that the deformations at the nanoscale have a strong dependence on the localized interfacial energy [1]. In a very recent work [2], the nanoscale deformation occurring related to both isotropic and kinematic hardening during nanoindentation of a Cu foil used in a composite structure of prepeg/Cu foil/prepeg is analyzed by the FEM technique. Another FEM study [3] reported for thermally oxidized SiC/SiO_2 ceramic structure formed at a temperature of about 1200°C to 1400°C at about 7 KPa pressure in a plasma wind tunnel that the nanoscale deformations are not affected that much by the localized nanoscale delamination but are significantly affected by the residual thermal stress as well as the intrinsic structural stability of the β-SiC material that works as the substrate for the surface SiO_2 thin film that forms during the oxidative exposure process. Furthermore, the recent application of 2-D axisymmetric FEM technique to diamond-like carbon coatings on tool steel [4] confirmed remarkable accuracy in terms of predicting the experimentally measured nanohardness of about 9 GPa, as well as about 94% and 96% accuracies

DOI: 10.1201/9781003359364-13

in predicting the indentation force measured in correspondence during the single nanoindentation and partial load–unload nanoindentation experiments. These facts point to the remarkable accuracy and the ever-increasing need to apply the FEM technique to understand nanoscale deformation and especially fracture initiation behavior of different materials exposed to nanoindentation. The same statement is also in principle valid for other miniaturized test techniques such as micro- and nanopillar compression and microbeam tensile techniques. In addition, the FEM simulations (based on the application of a continuum=based elasto-plastic model) predict [5] all the salient features of localized nanoscale matrix deformation in a unidirectional glass fiber-based thermoplastic polymer matrix composite cross-correlated by experimental data obtained by the novel nano digital image correlation (nano-DIC) spectroscopy technique. Moreover, the application [6] of the FEM-based simulation technique exhibits remarkable efficacy in accurately predicting the experimentally measured nature and magnitude of warpage of a flip chip ball grid array that involves a suitably reinforced polymer matrix molding compound and is an important component of interconnects used in semiconductor devices. FEM coupled with complimentary techniques such as nanoindentation and second harmonic generation microscopy confirm that for collagenous tissue-based biomaterials, e.g., cervix and tendon stiffness, is the highest when the fibrils are oriented in a direction that is parallel to the direction of the nanoindentation axis [7]. In another unique recent effort [8] that pertains to soft hyperelastic biomaterials such as hydrogels and cells, it is shown that FEM-based predictions could successfully verify for two types of nanoindenters (e.g., a parabolic shape and a quartic shape) the solutions of the indentation problem that is accurate up to second order in indentation amplitude profiles. One work [9] combines inverse FEM with nanoindentation and micropillar compression testing for magnetron-sputtered composite Ni/Al nanolaminates (type I, 15–75 nm Ni with fixed 110 nm thick Al; type II, 75 nm fixed Ni and 16–110 nm Al layer thicknesses) and proved that strengthening occurs due to dislocation glide that occurs in a confined manner within the nanosized layers of aluminum. Moreover, it is demonstrated [10] that FEM can successfully analyze cooling processes as well as thermal and nanomechanical responses of TiN/TiSiN multilayer coatings that are used as hard multilayer coatings for protecting the wear of cutting tools.

Nanoindentation is a popular experimental method to characterize the mechanical properties of bulk materials, film on a substrate, coating, etc. [11–20]. The advantage of this method is that the test can be done in a very small region of a sample in place of testing large pieces, and a low number of samples in macroscale testing methods. Finite element analysis-based modeling of nanoindentation experiments is also very popular to characterize materials, as the results obtained through FEM modeling closely follow the experimental results. In addition, FEM-based nanoindentation modeling can also shed light on the mechanics of material deformation and fracture. There are numerous studies available in the literature on nanoindentation numerical experiments that investigate the mechanical properties of various classes of materials [21–52]. The modeling strategies for nanoindentation tests for different classes of materials such as brittle materials, ductile materials, polymeric materials, and composites are different. Brittle materials should ideally have pure elastic deformation followed by fracture when they reach their elastic limit. Practically, there used to be very small plastic deformation leading finally to fracture in brittle materials. However, ductile materials have plastic deformation that must be considered in the nanoindentation modeling scheme. Similarly, for different classes of materials, different responses need special attention while modeling the numerical nanoindentation experiment using the finite element method.

This chapter is focused on discussing the nanoindentation modeling strategy for different classes of materials. The finite element modeling methodology for the nanoindentation test experiment is explained first. This is followed by a test case of modeling thin film on a substrate under misfit strain. This will help readers to understand the usefulness of FEM as a tool in conducting nanoindentation test experiments.

10.2 FINITE ELEMENT MODELING OF NANOINDENTATION EXPERIMENT

A nanoindentation experiment can be modeled using the finite element method. In the literature, there are many works available on nanoindentation modeling to characterize materials and evaluate their mechanical properties such as elastic modulus, hardness, and yield stress [21–50]. The elastic-plastic response of thin film on a substrate system was first developed by Bhattacharya and Nix [52]. A combined experimental nanoindentation and FEM modeling approach yielded the best result as the FEM routines can be utilized to obtain the multitudes of stress and strain experienced by the material under the indenter tip [37–39, 41].

Nanoindentation modeling by finite element method is done in both 2-D and 3-D. In both cases, the indenter is modeled as a rigid body, while the material that has to be characterized is considered as the deformable one. Mostly these 2-D and 3-D models are prepared as axisymmetric models. The results obtained by 2-D models are found to be as good as the results obtained by 3-D models [36]. In all these modeling schemes, the regions in the indentation zone are discretized by a large number of elements in comparison to the other regions. Different cases such as bulk material and film on a substrate use the same concept. In the case of film on substrate, the indentation depth should be smaller than the critical depth ratio (CDR), which is normally 1:10 for indentation depth to film thickness [39] to avoid the effect of the substrate. While modeling nanoindentation numerical experiments, the geometric nonlinearity and materials nonlinearity should be properly taken care of. Therefore, the next section explains the nonlinearity in the analysis covering both geometric and material nonlinearity.

10.3 NONLINEAR ANALYSIS

Loading on any structure/body leads to deformation, which leads to a change in the stiffness. It changes because of the material response, shape/geometry, etc. At a higher deformation level when it reaches the yield limit, the shape of the materials cannot be altered. This causes a change in the mechanical properties. However, we can safely assume that during the deformation process, there is no change in the shape and properties of the material with a minor change in the stiffness. This helps simplify the formulation of the problem and the solution. Finite element solutions are related to the stiffness matrix and unknown displacement with the given nodal load as [39]

$$\{F\} = [K] \times \{d\} \tag{10.1}$$

where $\{F\}$ is the known nodal load vector, $[K]$ is the stiffness, and $\{d\}$ is the unknown nodal displacement vector. When the stiffness of the body/structure changes considerably during the deformation process, linear analysis can't capture the deformation process well. In these cases, the analysis should be made in an iterative fashion that will lead to an increase in the analysis time.

10.3.1 Geometric Nonlinearity

When the stiffness of the continuum body changes due to the change in shape under external loading, it is termed geometric nonlinearity. The nonlinearity in strains and displacements with respect to the stresses and forces is geometric nonlinearity. Examples are large deformation, rotation, and buckling problems such as deformation in soft tissues. Under large deformation, the body/structure takes a different geometry and implies a new stiffness response. The stiffness of the body can change with small deformation as well. An example of this case is a flat membrane under pressure. When the shape of the membrane changes under pressure, a stiffness change can occur. These geometric nonlinearities are important when nanoindentation modeling of soft/complaint body/structure is carried out. It will also be important with a change in the shape of the structure like beams.

10.3.2 Material Nonlinearity

Stiffness changes due to changes in material response are termed material nonlinearity. The material obeying Hook's law gives rise to the linear stress–strain relationship, and the deformation is elastic (temporary). When the load is released, the body regains its original shape. The elasticity-based linear material model is commonly used to determine stress and strain under specific loading conditions. Many materials behave differently when they overtake the yield point beyond which plastic deformation takes place, which is permanent. Therefore, to accommodate the nonlinear material response, one has to consider the plastic deformation part accurately. There are classes of materials available whose response is nonlinear. One such example is metals, which are ductile, leading to plastic deformation taking place beyond the yield point. Similarly, hyperelastic and viscoelastic materials behave differently upon loading, which is nonlinear. Therefore, in the nanoindentation test, the proper material model has to be considered to carry out the numerical experiment for different classes of materials.

10.4 TEST CASE: NANOINDENTATION MODELING OF THIN FILM ON A SUBSTRATE WITH LATTICE AND THERMAL EXPANSION COEFFICIENT MISMATCH

We have implemented the 2-D axisymmetric model for finite element (FE) simulation of elastic-plastic behavior of a specimen in the commercial FE software ABAQUS 6.14. A rigid Berkovich indenter [53] is utilized to apply load, and the indentation depth is measured corresponding to the applied load. The nanoindentation test can be equally simulated by using FEM. The schematic diagram of nanoindentation modeling is illustrated in Figure 10.1a. The load versus displacement diagram is plotted for ascending and descending of both loading cases (Figure 10.2b). The results of load and corresponding displacement are used to calculate various mechanical properties. The literature [54] suggests that there is very little or no difference in results between a 3-D model and a 2-D axisymmetric model. Moreover, a 2-D axisymmetric model requires less computational power and time and is suitable to use.

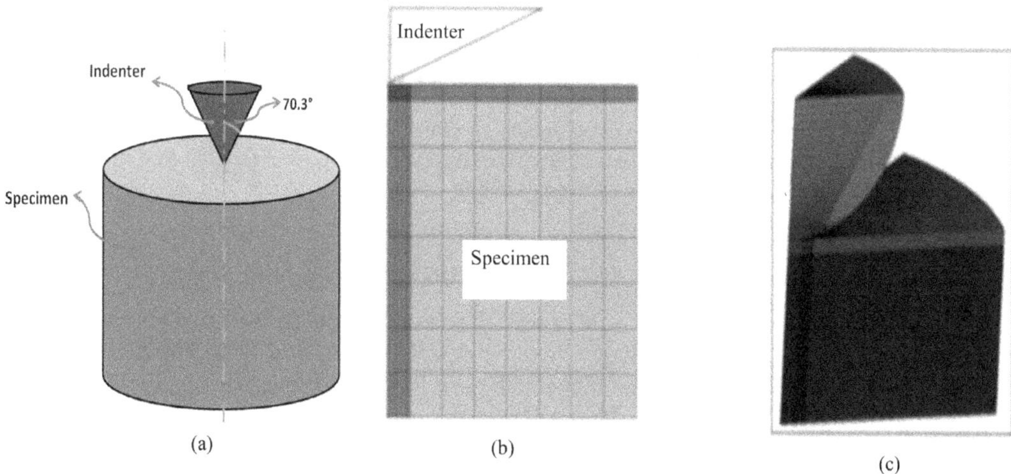

FIGURE 10.1 (a) Schematic representation of nanoindentation test experiment, (b) 2-D axisymmetric nanoindentation model, and (c) 3-D axisymmetric nanoindentation model [36].

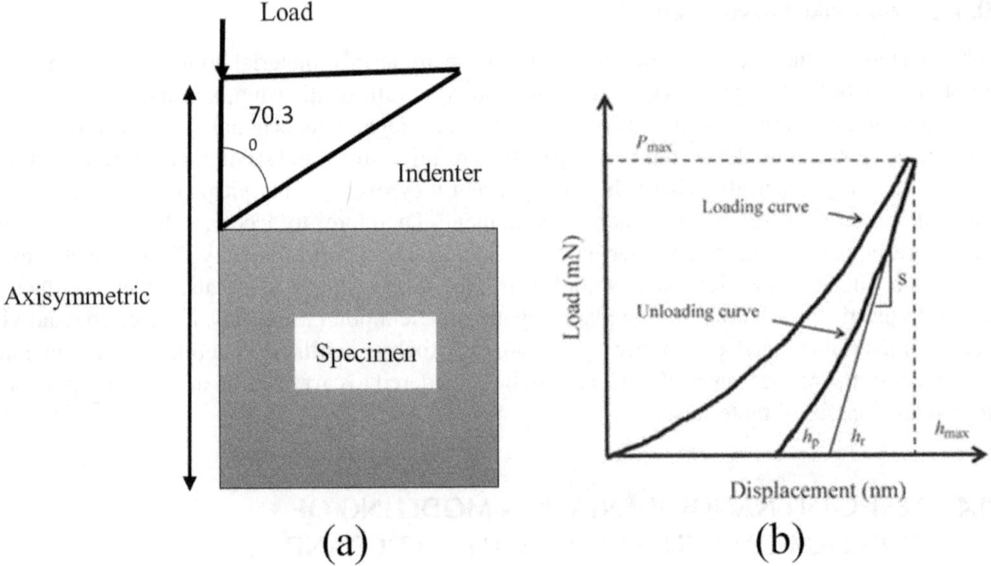

FIGURE 10.2 (a) Schematic diagram of indenter and specimen used for FEM. (b) Load versus displacement curve obtained from nanoindentation analysis.

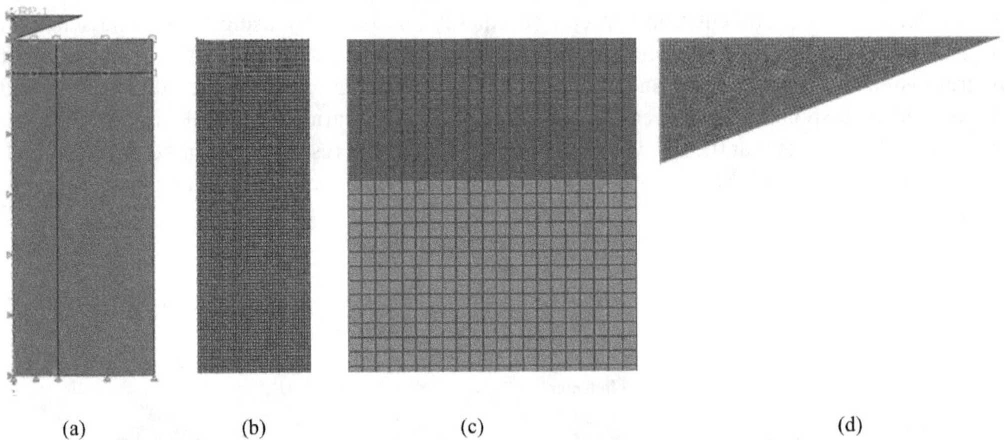

FIGURE 10.3 (a) FEM-based nanoindentation model of a thin film deposited on a substrate. (b) Meshing of the film-substrate system with an axisymmetric element in ABAQUS 6.14 library. (c) Mesh size near the indentation surface. (d) Meshing of Berkovich indenter [35].

10.4.1 Modeling of Gallium Nitride Thin Film on a Sapphire Substrate

In this work, we have investigated the role of the misfit strain on the deformation pattern in a thin film deposited on a thick foreign substrate. We have modeled a thin film 1,000 nm thick deposited on the 10,000 nm sapphire substrate, as shown in Figure 10.3. Both the substrate and the film were considered to be elastic and isotropic. The film was assigned gallium nitride properties, while the substrate was assigned sapphire (Al_2O_3) properties, as mentioned in Table 10.1. The indenter was modeled as a rigid body. The material properties of the indenter are also provided in Table 10.1.

The axisymmetric model of the film–substrate system and the indenter was modeled in ABAQUS 6.14. The indenter was modeled as a rigid body, while the film–substrate system was modeled as an

TABLE 10.1
Material Properties of GaN Film, Sapphire Substrate, and the Indenter

Property	Value	Reference
Ga-N Film		
Elastic modulus	295000 MPa	[60,61]
		[56]
Poisson ratio	0.183	[56]
Expansion coefficient	5.27×10^{-6} K^{-1}	[57, 60]
Sapphire Substrate	345000 MPa	[57]
Elastic Modulus	0.29	[59, 60]
	7.77×10^{-6} K^{-1}	[54]
Poisson ratio	1141000 MPa	[54]
Expansion coefficient	0.07	
Indenter		
Elastic modulus		
Poisson ratio		

elastic material. The indenter was discretized by 5,000 3-node linear axisymmetric triangular elements available in the ABAQUS library, while the film–substrate system was discretized by 18,000 4-node bilinear axisymmetric quadrilateral elements, reduced integration, hourglass control with denser mesh near the contact surface (Figure 10.3c), with relatively coarse mesh away from that. The bottom face of the film–substrate system was applied with displacement constraint in the vertical direction, while the left face was a constraint in the horizontal direction (Figure 10.3a). A reference point was made at the indenter left top corner where we applied the displacement load of 100 nm in a vertically downward direction (Figure 10.3a). The horizontal displacement and the rotation at the reference point were a constraint.

10.4.2 Lattice and Thermal Expansion Coefficient Mismatch Strains

Misfit strains are induced by both lattice mismatch between the substrate and the film, and the thermal expansion coefficient mismatch by inducing thermal strain at the film surface in terms of temperature change. The thermal strain can be calculated as

$$\varepsilon_t = \alpha \Delta T \tag{10.2}$$

where α is the thermal expansion coefficient mismatch and Δ *T* is the temperature change. The misfit strain induced by thermal expansion coefficient mismatch can be calculated by using Equation 10.2. We have applied misfit strain due to thermal expansion coefficient mismatch by taking Δ *T* equal to 100°C, 500°C, and 1000°C. The misfit strain due to lattice mismatch between the substrate and the film was also applied in terms of thermal strain itself by arbitrarily choosing the α and Δ *T* so that the resulting strain comes out to be equal to the magnitude of the lattice misfit strain. We have considered both tensile misfit strain and compressive misfit strain so that the effect of the bigger and smaller film lattice in comparison to the substrate on the nanoindentation deformation behavior can be properly investigated.

10.4.3 Results and Discussions

We would like to understand the deformation behavior of a film–substrate system with prestrained film on a substrate. Therefore, we considered gallium nitride film deposited on a sapphire substrate.

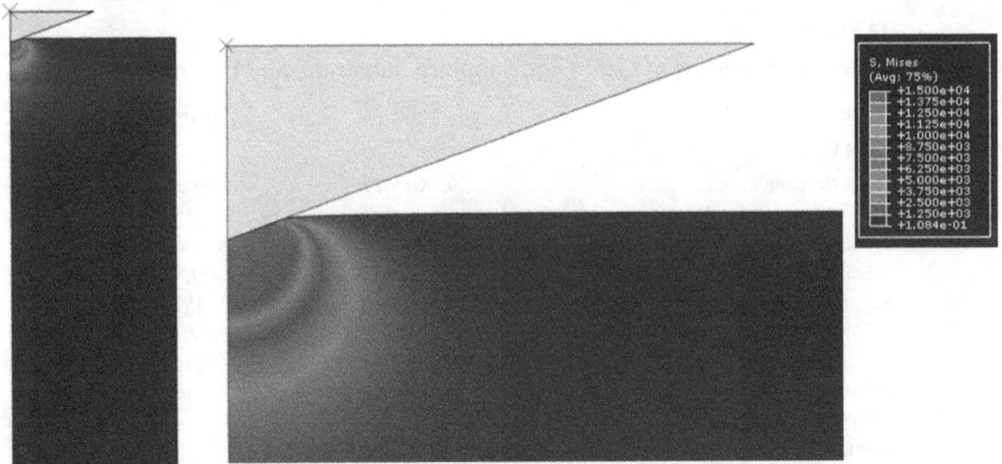

FIGURE 10.4 Nanoindentation deformation in a film substrate system with externally applied misfit strain to the film.

In this case, there is a large amount of lattice mismatch, and thermal expansion coefficient mismatch is present between the substrate and the film. These mismatches can significantly affect the deformation pattern. Therefore, we have carried out nanoindentation-based numerical investigations that could shed light on the deformation pattern of these optoelectronic materials, where a small change in strain can cause considerable changes in optoelectronic performances. The finite element analysis-based nanoindentation studies were carried out to investigate the load-indentation relation at the nanoscale in the presence of misfit strains. Figure 10.4 shows the deformation pattern in the film–substrate system with misfit strain being applied to the film in terms of thermal strain. The following sections discuss the role of thermal mismatch strain and lattice misfit strain on the nanoindentation deformation and the load required to indent.

10.4.3.1 Effect of Thermal Mismatch Strain on Indentation

We have considered three different cases of temperature change to investigate the role of thermal mismatch on the load required to indent 100 nm of the film. As shown in Figure 10.5, the load required to indent 100 nm of the film with temperature change $\Delta T = 100°C$ is 4.28 mN, while at $\Delta T = 1000°C$, it is 4.33 mN. It shows that the film with increased thermal misfit strain causes an increase in the load required for indentation. However, the increase in the load is very small as the effective misfit stain due to the temperature change is minimal (in the range of 0.00527 for 1000°C temperature change). Gallium nitride films are generally deposited on the sapphire substrate at 1000°C. Even at that temperature, the magnitude of the misfit strain due to thermal expansion coefficient mismatch is less, and therefore, there is no significant increase in the load required to deform. One interesting observation, in this case, is the magnitude of elastic deformation is found to increase with an increase in the misfit strain, which requires a higher load to indent the same depth. This signifies that with the increase in temperature, material like gallium nitride behaves more like brittle material and fractures without considerable plastic deformation.

10.4.3.2 Effect of Lattice Mismatch Strain on Indentation

The lattice mismatch between the gallium nitride film and the sapphire substrate is prevalent (in the range of 13% in c-plane growth orientation) and leads to a large amount of dislocation formation [59]. However, there is a limited-slip plane on which dislocations can glide in gallium nitride leading to plastic deformation because of insufficient gliding force. Therefore, it is important to understand the deformation pattern in the presence of lattice misfit between the substrate and the film.

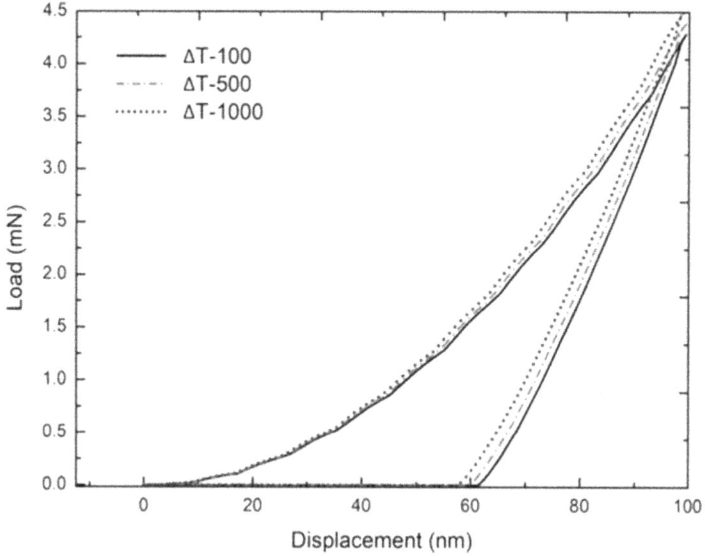

FIGURE 10.5 Load versus indentation plots for various thermal mismatch strains

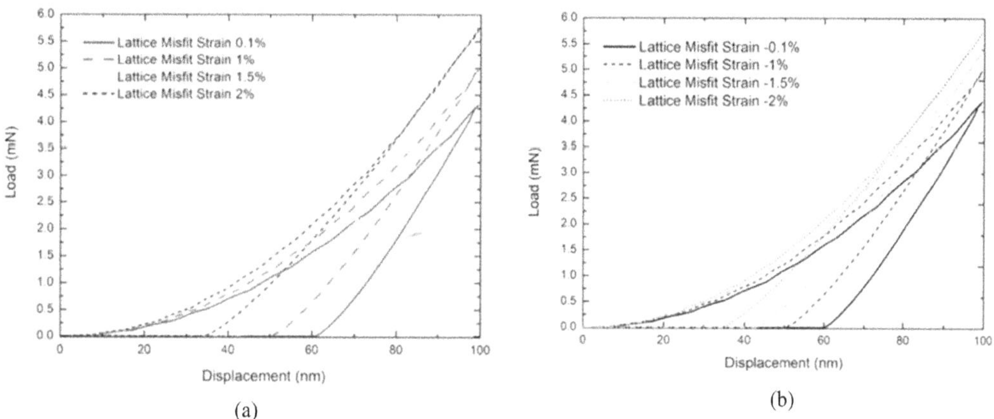

FIGURE 10.6 Load versus indentation depth with varied (a) tensile and (b) compressive lattice misfit strains.

We have considered 0.1%, 1%, 1.5%, and 2% of lattice misfit strain, both tensile and compressive (smaller lattice size of the film than the substrate and vice versa). Figure 10.6 shows the load versus indentation depth plots for tensile misfit strains and compressive misfit strains.

We can observe that the increase in the misfit strain either positive or negative causes an increase in the load required to indent the same depth as in the case of thermal mismatch strain. However, in this case, as the magnitude of strain is higher, the increase in the load required to indent the same depth (100 nm) is considerably high (as at 0.1% misfit strain, the load required is around 4.5 mN, while with 2% misfit strain, it is 5.7 mN). The nature of misfit strain (either tensile or compressive) does not have a considerable effect. The magnitude of elastic deformation is also increasing in this case as in the case of thermal mismatch strain with increasing magnitude of the misfit strain being applied to the film. The increase in the elastic deformation with misfit strain was more evident when we compared the load–deformation plot for the no-misfit case and 2% lattice misfit strain applied to the film (Figure 10.7). This phenomenon can be attributed to the increased brittleness with increasing misfit strain. However, more investigation is required to understand this phenomenon correctly.

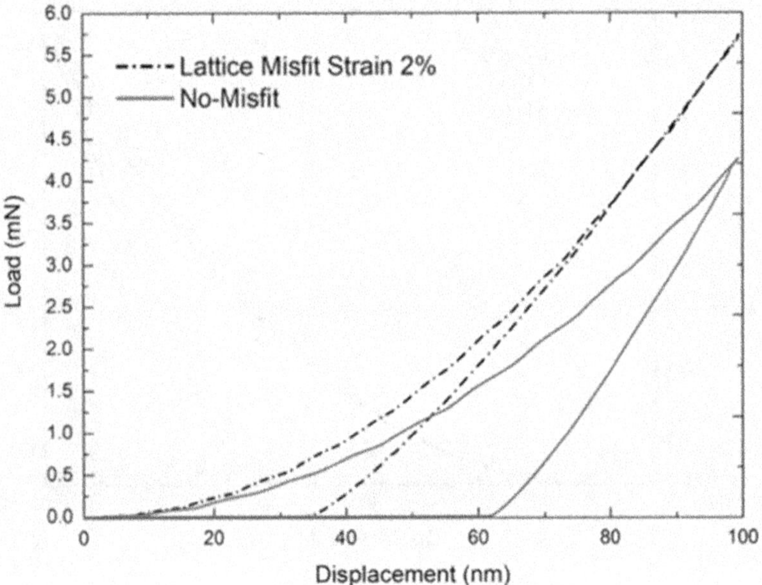

FIGURE 10.7 Comparative load versus indentation depth for no-misfit and 2% lattice misfit strain.

The nanoindentation modeling can also be improved by considering a proper plasticity model with an appropriate hardening coefficient.

10.5 CONCLUSION

Nanoindentation is a useful test method to characterize materials and to evaluate their mechanical properties such as hardness, stiffness, and elastic modulus. The nanoindentation test experiment can also be carried out by modeling through the finite element analysis tool. In fact, FEM modeling-based numerical experiments can shed more light on the underlying mechanics of deformation in these tests, as it can be done iteratively. It also saves lots of time and money, which is required to avail the experimental facility. This chapter provides a critical review of FEM modeling of nanoindentation tests in the literature and issues related to them. It also provides a test case of modeling gallium nitride film deposited on a sapphire substrate for nanoindentation test. The conclusion for the test case is provided in the preceding section.

The role of misfit strain due to lattice and thermal expansion coefficient mismatch between the substrate and the film on indentation deformation for a film deposited on a substrate is investigated. The test case included gallium nitride film deposited on a sapphire substrate. The effect of misfit strain on the indentation deformation was investigated by modeling the film–substrate system with misfit strains being applied in terms of thermal strain to the film using the commercial finite element analysis software ABAQUS 6.14. The 2-D axisymmetric modeling of both the film–substrate system and the indenter was done and the displacement-controlled load was applied. The results concluded that with increased misfit strain (both thermal mismatch and lattice mismatch), the load required to indent the same depth was found to be increasing. The increase in the magnitude of elastic deformation was observed with an increase in misfit strain. This phenomenon can be attributed to the increased brittleness. Future work can include the implementation of varying hardening models and plasticity.

This method is very important in characterizing various classes of materials. Readers can explore the reference literature for further studies on this topic.

ACKNOWLEDGEMENTS

The authors greatly acknowledge the computational support provided by the Multiscale Simulation Research Center (MSRC) at Manipal University Jaipur. The authors also acknowledge the support provided by Prof. A. K. Mukhopadhyay at Sharda University.

REFERENCES

1. M. R. Hajidehi, K. Tůma, S. Stupkiewicz, Indentation-induced martensitic transformation in SMAs: Insights from phase-field simulations, *Int. J. Mech. Sci.*, 245–1, 108100, 2023.
2. C. W. Trost, M. Krobath, S. Žák, R. Hammer, T. Krivec, H.-P. Gänser, T. W. Trost, A. Hohenwarter, M. J. Cordill, Inverse parameter determination for metal foils in multifunctional composites, *Mat. & Des.*, 227, 111711, 2023.
3. L. Yang, Z. Ye, C. Wang, C. Zhao, J. Zhang, Nanoindentation deformation and fracture mechanisms of SiC/SiO_2 thermally oxidized in plasma wind tunnel, *Thin Solid Films*, 768–1, 139715, 2023.
4. V. F. Vieira, Y. Shigaki, P. S. Martins, E. Cheikh, T. Ba, C. A. R. Dias, Nanoindentation test of a DLC coated high-speed steel substrate using a two-dimensional axisymmetric finite element method, *Dia. and Rel. Mat.*, 134, 109792, 2023.
5. N. Klavzer, F. Sarah, M. Gayot, B. Coulombier, B. Nysten, T. Pardoen, Nanoscale digital image correlation at elementary fibre/matrix level in polymer-based composites, *Comp. Part A: App. Sci. and Manufactur.*, 168, 107455, 2023.
6. A. Salahouelhadj, M. Gonzalez, K. Vanstreels, G. Van der Plas, G. Beyer, E. Beyne, Analysis of warpage of a flip-chip BGA package under thermal loading: Finite element modelling and experimental validation, *Microelectronic Eng.*, 271–272, 111947, 2023.
7. A. Ostadi Mghaddam, M. R. Arshee, Z. Lin, M. Sivaguru, H. Phillips, B. L. McFarlin, K. C. Toussaint, A. J. Wagoner Johnson, Orientation-dependent indentation reveals the crosslink-mediated deformation mechanisms of collagen fibrils, *Acta Biomater.*, 158–1, 347–357, 2023.
8. Y. Du, P. Stewart, N. A. Hill, H. Yin, R. Penta, J. Köry, X. Luo, R. Ogden, Nonlinear indentation of second order hyperelastic materials, *J. Mech. Phys. Solids* 171,105139, 2023.
9. S. Shen, H. Li, C. Wang, Y. Liang, N. Feng, N. Zhang, L. W. Yang, Mechanical properties and strengthening mechanism of Ni/Al nanolaminates: Role of dislocation strengthening and constraint in soft layers, *Materials & Design*, 226, 111632, 2023.
10. R. Tu, M. Jiang, M. Yang, B. Ji, T. Gao, S. Zhang, L. Zhang, Effects of gradient structure and modulation period on mechanical performance and thermal stress of TiN/TiSiN multilayer hard coatings, *Mater. Sci. Eng. A*, 866, 144696, 2023.
11. J. A. Rudd, T. R. Jervis, F. Speapen, Nanoindentation of Ag/Ni multilayered thin films. *J. Appl. Phys.* 75–10, 15, 1994.
12. W. C. Oliver, G. M. Pharr, An improved technique for determining hardness and elastic modulus using load and displacement sensing indentation experiments. *J. Mater. Res.* 7, 1564, 1992.
13. M. L. Oyen, Nanoindentation of biological and biomimetic materials, *Expt. Tech.*, 37, 73–87, 2011.
14. M. R. Ayatollahi, E. Alishahi, S. Doagou, S. Shadlou, Tribological and mechanical properties of low content nanodiamond/epoxy nanocomposites, *Compos., Part B*, 43, 3425–3430, 2012.
15. J. L. Cuya, A. B. Mann, K. J. Livi, M. F. Teaford, T. P. Weihs, Nanoindentation mapping of the mechanical properties of human molar tooth enamel, *Arch. Oral Biol.* 47, 281–291, 2002.
16. S. S. R. Koloor, A. Karimzadeh, M. N. Tamin, M. H. A. Shukor, Effects of sample and indenter configurations of nanoindentation experiment on the mechanical behavior and properties of ductile materials, *Metals*, 8, 421, 2018.
17. X. Wang, P. Xu, R. Han, J. Ren, L. Li, N. Han, F. Xing, J. Zhu, A review on the mechanical properties for thin film and block structure characterised by using nanoscratch test, *Nanotechnol Rev*, 8, 628–644, 2019.
18. B. Bor, D. Giuntini, B. Domènech, M. V. Swain, G. A. Schneider, Nanoindentation-based study of the mechanical behavior of bulk super crystalline ceramic-organic nanocomposites, *Journal of the European Ceramic Society*, 39, 3247–3256, 2019.
19. S. A. Epshtein, F. M. Borodich, S. J. Bull, Evaluation of elastic modulus and hardness of highly inhomogeneous materials by nanoindentation, *Appl. Phy.*, 119, 325–335, 2015.
20. Z. Zhang, Z. Wen, Z. Yue, Nanoindentation study and mechanical property analysis of nickel-based single crystal superalloys, *Mater. Res. Express*, 7, 096509, 2020.

21. T. F. Low, C. L. Pun, W. Yan, Theoretical study on nano-indentation hardness measurement of a particle embedded in a matrix, *Philos. Mag.* 95, 1573, 2015.
22. J. A. Knapp, D. M. Follstaedt, S. M. Myers, J. C. Barbour, T. A. Friedmann, Finite-element modeling of nanoindentation, *J. App. Phys.*, 85–3, 1460, 1999.
23. J. Chen, S. J. Bull, On the factors affecting the critical indenter penetration for measurement of coating hardness, *Vacuum*, 83, 911, 2009.
24. C. A. Clifford, M. P. Seah, Modelling of nanomechanical nanoindentation measurements using an AFM or nano indenter for compliant layers on stiffer substrates. *Nanotechnology*, 17, 5283, 2006.
25. P. Duan, S. Bull, J. Chen, Modeling the nanomechanical responses of biopolymer composites during the nanoindentation, *Thin Solid Films*, 596, 277, 2015.
26. K. Inoue, F. Triawan, K. Inaba, K. Kishimoto, M. Nishi, M. Sekiya, K. Sekido, A. Saitoh, Evaluation of interfacial strength of multilayer thin films polymer by nanoindentation technique, *Mech. Eng. J.*, 6, 18-00326, 2019.
27. Y.-P. Cao, K.-L. Chen, Theoretical and computational modelling of instrumented indentation of viscoelastic composites, *Mech. Time-Depend. Mater.*, 16, 1, 2012.
28. J. Chen, G. Lu, Finite element modelling of nanoindentation-based methods for mechanical properties of cells. *J. Biomech.*, 45, 2810–2816, 2012.
29. G. Ananthakrishna, S. Krishnamoorthy, A novel approach to modelling nanoindentation instabilities, *Crystals*, 8–5, 200, 1–26, 2018.
30. A. Karimzadeh, M. R. Ayatollahi, M. Alizadeh, Finite element simulation of nano-indentation experiment on aluminum 110, *Comput. Mater. Sci.*, 81, 595–600, 2014.
31. Y. Li, Y. C. Brinson, Models for nanoindentation of compliant films on stiff substrates. *J. Mater. Res.*, 30, 1747–1760, 2015.
32. M. U. Uysal, Numerical modeling of functional graded TiB coating in nanoindentation on determination of mechanical properties. *Mater. Today Proc.*, 2, 217–223, 2015.
33. L. Bartolomé, E. Oblak, M. Kalin, Mechanical behavior and constitutive models of ZDDP tribofilms on DLC coatings using nano-indentation data and finite element modelling, *Trib. Int.*, 95, 19–26, 2016.
34. A. P. Singh, S. L. Vajire, D. Mishra, Nanoindentation simulation of gallium nitride grown in Different orientations on a sapphire substrate using finite element method, *Mater. Today Proc.*, 79, 282–285, 2022.
35. S. L. Vajire, A. P. Singh, D. K. Saini, A. K. Mukhopadhyay, K. Singh, D. Mishra, Novel machine learning-based prediction approach for nanoindentation load-deformation in a thin film: Applications to electronic industries, *Comput. Ind. Eng.*, 174, 108824, 2022.
36. A. S. Alaboodi, Z. Hussain, Finite element modeling of nano-indentation technique to characterize thin film coatings, *J. King Saud Univ. Eng. Sci.*, 31, 61–69, 2019.
37. X. Chen, I. A. Ashcroft, R. D. Wildman, C. J. Tuck, A combined inverse finite element –elastoplastic modelling method to simulate the size-effect in nanoindentation and characterise materials from the nano to micro-scale, *Int. J. Solids Struct.*, 104–105, 25–34, 2017.
38. A. Karimzadeh, S. S. R. Koloor, M. R. Ayatollahi, A. R. Bushroa, M. Y. Yahya, Assessment of nano-indentation method in mechanical characterization of heterogeneous nanocomposite materials using experimental and computational approaches, *Sci. Rep.*, 9, 15763, 2019.
39. D. Porwal, A. Dey, A. K. Gupta, K. Khan, A. K. Sharma, A. K. Mukhopadhyay, Combined nanoindentation and finite element approach in natural hierarchical structures. In Arjun, D. and Anoop, K. M. (Eds.). *Nanoindentation of Natural Materials* 2018 Sep 3 (pp. 177–202). CRC Press.
40. H. Xiao, X. Wang, C. Long, Theoretical model for determining elastic modulus of ceramic materials by nanoindentation, *Materialia*, 17, 101121, 2021.
41. M. Liu, C. Lu, K. A. Tieu, C. T. Peng, C. Kong, A combined experimental-numerical approach for determining mechanical properties of aluminum subjects to nanoindentation, *Sci. Rep.*, 5(1), 1–16, 2015.
42. S. Guicciardi, C. Melandri, D. Sciti, G. Pezzotti, Analysis of nanoindentation tests on SiC-based ceramics, *Phil. Mag.*, 86(33–35), 5321–5329, 2006.
43. J. Rauker, P. R. Moshtagh, H. Weinans, A. A. Zadpoor, Analytical relationships for nanoindentation-based estimation of mechanical properties of biomaterials, *J. Mech. Med. Biol.*, 14(3), 1430004, 2014.
44. A. Karimzadeh, S. S. Koloor, M. R. Ayatollahi, A. R. Bushroa, M. Y. Yahya, Assessment of nano-indentation method in mechanical characterization of heterogeneous nanocomposite materials using experimental and computational approaches, *Sci. Rep.*, 9(1), 15763, 2019.
45. S. Guicciardi, C. Melandri, D. Sciti, G. Pezzotti, Analysis of nanoindentation tests on SiC-based ceramics. *Phil. Mag.*, 86(33–35), 5321–5329, 2006.

46. J. Chen, H. Li, Finite element modeling of nanoindentation on an elastic-plastic microsphere, 2020. https://doi.org/10.21203/rs.3.rs-25785/v1.
47. J. Duan, Y. Xia, S. Bull, J. Chen, Finite element modeling of nanoindentation response of elastic fiber-matrix composites. *J. Mater. Res.*, 33(17), 2494–2503, 2018.
48. R. Iankov, M. Datcheva, S. Cherneva, D. Stoychev, Finite element simulation of nanoindentation process. In Ivan, D., István, , and Lubin, (Eds.) *Numerical Analysis and Its Applications: 5th International Conference, NAA 2012*, Lozenetz, Bulgaria, June 15–20, 2012, Revised Selected Papers 5 (pp. 319–326). Springer Berlin Heidelberg.
49. J. Chen, S. J. Bull, Modeling of indentation damage in single and multilayer coatings. In *IUTAM Symposium on Modelling Nanomaterials and Nanosystems: Proceedings of the IUTAM Symposium held in Aalborg*, Denmark, 19–22 May 2008 (pp. 161–170). Springer Netherlands.
50. P. Duan, S. Bull, J. Chen, Modeling the nanomechanical responses of biopolymer composites during the nanoindentation, *Thin Solid Films*, 596, 277–281, 2015.
51. M. Simion, M. C. Dudescu, R. Chiorean, Numerical simulation of Vickers micro-indentation test to estimate micro-hardness, *Acta Tech. Napocensis Ser. Appl. Math. Mech. Eng.*, 61(4), 593–6022018.
52. A. K. Bhattacharya, W. D. Nix, Analysis of elastic and plastic deformation associated with indentation testing of thin films on substrates, *Int. J. Solids Struct.*, 24–12, 1287–1298, 1988.
53. A. Karimzadeh, M. R. Ayatollahi, M. Alizadeh, Finite element simulation of nano-indentation experiment on aluminum 1100. *Comput. Mater. Sci.*, 81, 595–600, 2014.
54. M. J. Lichinchi, C. Lenardi, J. Haupt, R. Vitali, Simulation of Berkovich nanoindentation experiments on thin films using finite element method, *Thin Solid Films*, 312(1–2), 240–248, 1998.
55. R. Nowak, M. Pessa, M. Suganuma, M. Leszczynski, I. Grzegory, S. Porowski, F. Yoshida, Elastic and plastic properties of GaN determined by nano-indentation of bulk crystal, *Appl. Phys. Lett.*, 75(14), 2070–2072, 1999.
56. M. Leszczynski, T. Suski, H. Teisseyre, P. Perlin, I. Grzegory, J. Jun, S. Porowski, T. D. Moustakas, Thermal expansion of gallium nitride, *J. Appl. Phys.*, 76(8), 4909–11, 1994.
57. Properties of Sapphire Substrate and Sapphire Wafers. *Properties of Sapphire Wafers, Sapphire Thermal Conductivity*. (n.d.). http://valleydesign.com/sappprop.htm.
58. M. A. Moram, Z. H. Barber, C. J. Humphreys, Accurate experimental determination of the Poisson's ratio of GaN using high-resolution x-ray diffraction, *J. Appl. Phys.*, 102(2), 023505, 2007.
59. D. Mishra, Y. E. Pak, Electroelastic fields for a piezoelectric threading dislocation in various growth orientations of gallium nitride, *Eur. J. Mech. A/Solids*, 61, 279–292, 2017.
60. D. Mishra, S. H. Lee, Y. Seo, Y. E. Pak, Modeling of stresses and electric fields in piezoelectric multilayer: Application to multi quantum wells. *AIP Adv.*, 7(7), 075306, 2017.
61. D. Mishra, Y. H. Cho, M. B. Shim, S. Hwang, S. Kim, C. Y. Park, S. Y. Seo, S. H. Yoo, S. H. Park, Y. E. Pak, Effect of piezoelectricity on critical thickness for misfit dislocation formation at InGaN/GaN interface, *Comput. Mater. Sci.*, 97, 254–262, 2015.

11 FEM-Based Computational Studies on Impression Creep Behavior of Boron-Added P91 Steel

Rajnish Kumar, Raman Bedi, Manoj Kumar, and Akhil Khajuria

11.1 INTRODUCTION

ASTM (A-387) P91 steel is extensively used in process industries and thermal power generation plants due to its good weldability, high corrosion resistance, low thermal expansion, and excellent creep resistance [1–6]. The boron-added version of P91 (P91B) steel has been developed to enhance the creep resistance of P91 because of type IV cracking issues after welding [7–11]. P91B is explored to understand the role of boron in exhibiting improvement in creep resistance [12–17]. Many experimental impression creep studies with a flat-ended circular indenter have been reported in previous literature [1–7, 9–11, 13, 15–17]. The experimental procedure of impression creep is relatively costly considering specimen preparation, specimen cost, and time. However, finite element analysis (FEA) is a less time-consuming and comparatively cheaper method than the experimental. It can predict the behavior of varied types of physical systems such as the deformation of solids and stress distribution. The indentation test using a cylindrical flat indent is now known as the impression creep test [18].

11.1.1 Studies on Impression Creep and Effects of Boron in 9Cr Steels

Hyde et al. (1996) employed P91 steel to observe impression creep behavior by using a rectangular indenter and found primary and secondary creep only. Impression creep data was converted into uniaxial creep data using the reference stress method [19]. Hyde and co-workers (2004) reviewed the impression creep behavior of metals and alloys, weldments, ceramics, and polymers. They observed that a cylindrical indenter with a flat-end was more appropriate for a localized creep test than conical, pyramid, spherical, and trapezoidal indenters [20]. Hyde et al. (2009) studied conversion relationships (called conversion factors) between indentation and impression creep test data of P91 steel. The conversion factor was calculated by the reference stress approach. The influence of specimen geometry on conversion relationships was investigated by comparing the conversion factors derived from 2D and 3D finite element (FE) methods. It was used to convert uniaxial creep data to impression creep data [21]. Baral et al. (2017) studied the behavior of two specimens; one without weld and another with weld over a range of temperatures (600°C–650°C) and stress (50–180 MPa) to find the minimum creep strain rate of modified P91 steel. It was found that the welded specimen had a high creep strain rate at a lower stress region but vice versa in the case of base metal. Also, creep damage (strain) accumulated preferentially around the weld weld-affected zone [22]. Khajuria et al. (2018) investigated the effect of boron in the impression creep test. The specimen materials were P91 steel and P91B steel. In this experiment, steady-state creep was less than P91 steel by approximately 3.5 times. The hydrostatic stress zone was likewise found to be modest in P91 steel, but the deformed

 DOI: 10.1201/9781003359364-14

plastic zone with substantial shear deformation was greater in comparison to P91B steel [23]. Baral et al. (2013) investigated the creep behavior of boron-added P91 steel to encounter type IV cracking in the heat-affected zone [24]. Shrestha et al. (2013) studied the creep deformation behavior of P91 steel at a temperature of 600°C–700°C for an applied stress range of 35–350 MPa. A lower stress regime was observed and dislocation climb and Nabarro–Herring were creep identified [25]. Abe and co-workers (2007) investigated the microstructure of the weld joint and simulated heat-affected zone of 9%–12% Cr steel. Creep and creep rupture tests were carried out for up to 2×10^4 hours at 650°C. Welded joints broke in fine-grained heat-affected zones under modest stresses, indicating type IV fracture, resulting in a shorter creep life than base metal joints. The addition of approximately 100 ppm of boron and 10–20 ppm of nitrogen prevents the formation of fine-grained areas in the heat-affected zone, thus avoiding the formation of type IV fractures [26]. Das et al. (2012) investigated the effects of boron and heat treatment temperature on type IV cracking resistance in modified 9Cr-1Mo steel weldments. For the current study, two unique temperatures of P91 steel were melted: one without boron, known as P91, and one with a controlled addition of boron with extremely little nitrogen, known as P91B. The creep rupture life of P91B weldments made from 1150°C normalized base metal was found to be significantly longer than that of those prepared from 1050°C normalized base metal due to the stabilization of lath martensite by fine $M_{23}C_6$ precipitates [27]. Das et al. (2012) investigated microstructural development of P91B and P91 steels. In the heat-affected zone, P91B austenite grain size was homogeneous and equivalent to that of the base metal. This showed that P91B weldment restored the original grain. It was also found that P91B weldment has a more uniform hardness distribution than P91 weldment, resulting in better type IV fracture resistance [28]. Watanabe et al. (2006) examined the microstructure and creep rupture properties of P91 steel. It was observed that fracture initiation location and crack development route matched the stress triaxiality factor distribution better than the corresponding creep strain distribution [29]. Chu and associates (1977) introduced a new creep test, which was originally the indentation creep test. A flat-ended circular indenter was used so it was called the new indentation creep test hence "impression creep". The steady-state velocity of the indenter was observed for succinonitrile crystals [30]. PM and Ashby (1992) investigated the indentation creep of metals, viz., copper, Armco iron, and non-metals (i.e., germanium, diamond, and silicon). Creep behavior was observed, power law was used, and the stress exponent and activation energy were calculated [31]. Storåkers and Larsson (1994) analyzed the power law behavior of indentation tests from computational and theoretical points of view. A spherical punch was used to know the behavior of creep. In this analysis, strain hardening was modeled [32]. Yu et al. (1985) investigated the impression creep curve of nickel, aluminum copper, and mild steel using a flat-ended circular cylinder. The stress-penetration behavior indicated that an elastic deformation front emerged ahead of the moving indenter, which then gave way to a plastic zone near the elastic limit of the material, which traveled in a steady-state shape as the stress increased [33]. Mathew et al. (2013) carried out an experimental impression creep test of 316LN at different applied stress. The impression creep curve was plotted and it was found that the tertiary stage was absent. Also, impression velocities were found to be in good agreement with coaxial creep data [34]. Eggeler and Wiesner (1993) investigated the creep behavior of circular notched specimens of Nimonic 80A. It was observed that transient creep in brittle materials can disperse stresses more quickly than tertiary-dominated creep in ductile materials and flat-plate specimens had higher maximum axial stresses than cylindrical specimens [35].

11.1.2 Studies on FE Analysis of Impression Creep

Hyde et al. (1993) analyzed the 30 mm diameter of a specimen with a height of 15 mm by using FE analysis. The indenter was used with a diameter of 1 mm and the load was applied to it. From this simulation, they found the impression displacement rate and compared it with uniaxial creep data. Coaxial stress was 0.296 times of impression creep stress [36]. Yang et al. (1995) used ABAQUS FEM software for the impression creep of materials. The Eyring hyperbolic sine constitutive equation was modeled. The conversion factor was reported as 3.5. It was found that for the same striking stress, the imprint velocity was proportional to the punch radius, as demonstrated by a computer simulation using the power law constitutive equation. Also, build-up of material was around the penetration when it was deep [37]. Hyde et al. (2004) performed FE analysis to analyze steady-state creep and damage at the circumferential weld of the P91 steel pipe. FE approach was used to examine differences in creep characteristics among the base, heat-affected zone, and weld material. It was observed that the crack started from the pipe outside diameter in the heat-affected zone. The creep constant was computed based on the results of the impression creep test [20]. Karthik et al. (2012) investigated mechanical properties by using the ball indentation technique. An indentation test was performed by using a tungsten carbide ball with a 0.5 mm radius. The penetrated depth of 100–200μm was modeled using the FE method to evaluate strain hardening and yield stress. It was observed that contact friction between the specimen and indenter affected the load–indenter displacement curve and indentation profile [38]. Lu et al. (2019) performed an indentation test in a self-designed vacuum chamber at high temperatures. Spherical indentation creep tests were evaluated using a systematic FE analysis. Norton power law was used to evaluate creep properties and the creep constant was extracted from the coaxial creep test at different stresses. It was observed that the largest pile-up was observed at a distance of one to twice the radius from the symmetric center [39]. Hyde et al. (1993) created an axisymmetric model to compare uniaxial conventional creep data with impression creep data. The effective gauge length was 0.755 times the indenter diameter, and the coaxial stress was 0.296 times the imprint creep stress. It was seen that the value was near and fitted with uniaxial creep data over impression creep data. In this study, power laws were used to perform creep calculations [36]. Goyal and associates (2013) investigated the creep rupture behavior and steady-state behavior of 2.25Cr–1Mo steel. The 2D axisymmetric model was modeled in the ABAQUS FEM package. The power law was used to analyze the steady-state behavior of the specimen. In FEA, stress distribution was observed. It was found that maximum hydrostatic stress and principal stress shifted from the center of the notch plane [40]. Pétry and Lindet (2009) modeled an axisymmetric model to analyze creep strain, creep behavior, and damage of P92 steel. At different stress levels and different temperatures like 600°C and 650°C, the FE model was developed. Three creep curves were observed and three phases were identified. There was also evidence of steady-state and accelerating creep [41]. Kumar et al. (2015) investigated impression creep behavior and plastic deformation. For this analysis, an axisymmetric model was created with dimensions 10 × 10 mm. An analytical rigid indenter with a radius of 0.5 mm was developed. The evolution of stress beneath the punch with time during plastic, elastic, and creep processes was studied. The produced plastic zone measured nearly 1.5 times the indenter's diameter. Stress inside the plastic zone under the punch is dispersed with creep exposure [42].

11.1.3 Lacuna in Literature and Prime Objective

There are many studies reported on microstructural evolution, FE analysis, and creep behavior of parent 9Cr steels and their welded joints. The heat-affected zone of 9Cr welded joints has been concluded as the region most prone to type IV cracking. To overcome this issue, the introduction of new 9Cr alloys with controlled boron additions proved to be most promising. But the role of boron is still imprecise in 9Cr steel welded joints to clarify enriched creep performance by delaying type IV cracking due to various microstructural mechanisms. In this respect, the following are the key reviewed points from the literature on the creep behavior of P91, P91B, and their welded joints:

(i) The axisymmetric model has been created for FE analysis to calculate the steady-state creep rate.
(ii) A cylindrical flat-ended punch was used for impression creep.
(iii) The conversion factor was used to convert uniaxial creep stress into impression creep stress.

(iv) FE analysis of the boron addition of P91 steel and its creep deformation behavior under the state of high stress and high temperature in P91 steel has not been investigated.

Several studies have been published on standard P91 steel and its welded joints, attempting to explain the underlying microstructural processes of type IV cracking by using creep testing of genuine weldments and simulated specimens. But there are very few studies that have been done on FE modeling of impression creep. On account of the same, the prime objectives of the current work were (i) to create a representative FE model for simulation of the impression creep behavior and (ii) to compare creep strain and steady-state creep rate with experimental data of P91B steel [43].

11.2 METHODOLOGY

11.2.1 Creep Behavior of Isotropic Material and Constitutive Equation

To explain a particular material and its response to the applied charge, a constitutive equation is needed. This equation describes the macroscopic behavior of the internal composition of the material. However, when solid-state material is in consideration, it behaves in complex ways when temperature and deformation of the material are also considered. It is not feasible to write a single equation to describe the material behavior. So there isa need to formulate a separate kind of equation to describe various kinds of ideal material responses. The physical rule should not be affected by the place or arrangement of the entity, i.e., the same physical event is observed from two coordinate systems by two observers. Constitutive equations should be invariant under variation in the frame of reference, i.e., both observers notice equal stress in a loaded material, even if they are in relative motion about each other. The concept of entity objectivity is also understood as the principle of entity frame-indifference [44]. The creep test is performed on a specimen under tension or compression. In most cases, it is carried out in a furnace at a constant temperature T under an invariable load. The length of the specimen is decided as a function of time. It should be noticed that creep is very significant at the temperature of material melting point $0.4T_m$. Herein, only creep under constant uniaxial stress has been considered. Deformation of creep occurs based on three parameters: stress, time, and temperature.

11.2.2 Creep Exponent, Stress Exponent, and Time Exponent Calculation Using Experimental Data

Power law is defined by Equation 11.1 as follows:

$$\dot{\varepsilon}_c = A\sigma^n \exp\left(-\frac{Q}{RT}\right) \tag{11.1}$$

To calculate the stress exponent, a graph is plotted between steady-state rate and stress. The slope of the graph shows the stress exponent. $\dot{\varepsilon}_c$ stands for steady creep rate, A is the creep exponent, σ is stress, and n stands for the stress exponent.

Taking log both sides of Equation 11.1:

$$\log\left(\dot{\varepsilon}_c\right) = \log A + n\log\sigma - \frac{Q}{RT}\exp\left(-1\right) \tag{11.2}$$

$$\log\left(\dot{\varepsilon}_c\right) = \log A + n\log\sigma - \frac{Q}{RT}\frac{1}{2.3026} \tag{11.3}$$

Now plotting $\log\left(\dot{\varepsilon}_c\right)$versus $\log\sigma$, the slope of the graph is called the stress exponent (Figure 11.1).

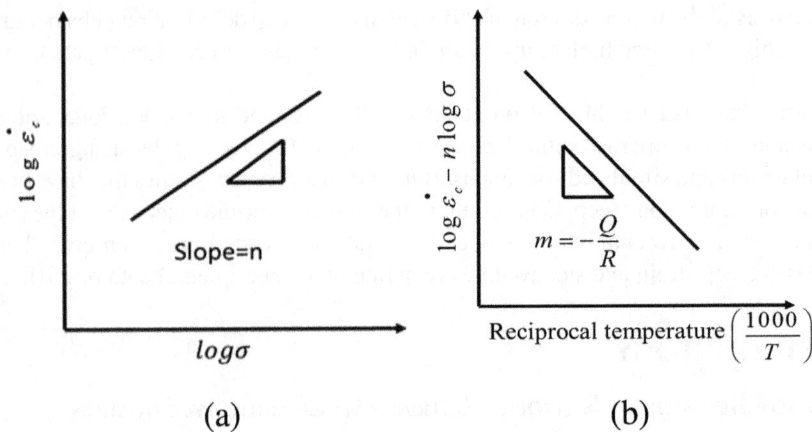

FIGURE 11.1 (a) Plot between steady-state rate versus stress and (b) natural log of creep rate versus reciprocal of temperature.

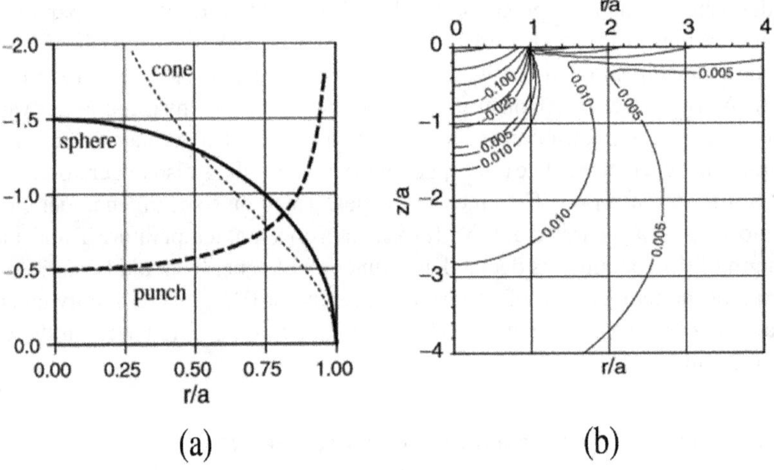

FIGURE 11.2 Normalized contact stress/stress variation σ_z/P_m for cylindrical indenter with flat-ended bottom and Fischer–Cripps contour plot for cylindrical flat punch [46].

To calculate the time exponent, a graph is plotted between the steady-state rate and the reciprocal of absolute temperature. The slope of the graph shows the time exponent, where Q is the activation energy, R is the gas constant, and m is the time exponent.

After arranging Equation 11.3:

$$\log\left(\dot{\varepsilon}_c\right) = n\log\sigma = \log A - \frac{Q}{RT}\frac{1000}{23026} \tag{11.4}$$

The slope of the graph as shown in Figure 11.2a, bis called the time exponent (m), where temperature is in kelvins. For calculating activation energy, Equation 11.5 is used:

$$Q = -m \times R \times 23026 \tag{11.5}$$

$$\text{Creep exponent}\left(\text{A}\right) = \frac{\dot{\varepsilon}_{cr}}{\sigma^n t^m} \tag{11.6}$$

Assumptions:

(i) The applied load is constant throughout the simulation.
(ii) The isothermal process is chosen.
(iii) The material should be isotropic and homogeneous.
(iv) Creep strain is directly proportional to stress.

$$\dot{\varepsilon}_{cr} = \left(Aq^{-n}\left[(m+1)\varepsilon^{-cr} \right]^{m} \right)^{\frac{1}{m+1}} \tag{11.7}$$

where A, n, and m are the creep exponent, stress exponent, and time exponent, respectively, at time t. $\dot{\varepsilon}_{cr}$ is the creep strain rate, q the uniaxial equivalent deviatoric stress, and $\dot{\varepsilon}^{cr}$ the uniaxial equivalent creep strain rate.

11.2.3 Contact Formulation and Stress Distribution underneath the Punch

Stress distribution is determined by the principle of superposition for stress distribution of point contact. For any point of two bodies such as cylindrical indenter with flat punches and spherical punches, the intensity of interpretation of stress distribution of the specimen can be spotted due to point load. The internal stress distribution is determined by the superposition of each point charge to determine the indentation stress field. Hertz was interested in the study of regional deformation and dispensation of the stress of both elastic bodies in contact with each other. He wanted to provide the contact surface with a form that fulfilled specific boundary criteria, such as [45]:

(i) Frictionless contact exists between the bodies.
(ii) The stress must disappear from the contact surface, and the displacement and stress should satisfy the elastic body differential equation in the equilibrium condition.
(iii) The length between the two specimen surfaces is zero inside the circle of contact and larger than zero outside.
(iv) Normal stress on the surface of an object is zero outside the contact circle, and equal and opposite inside.

A flat-ended cylindrical punch, comparable to the Hertzian stress field, is used to create a singularity in the stress region at the periphery of the connection. This causes the specimen or indenter material to distort plastically. However, if the load applied to the punch is not too large, then less plastic deformation at the connection circle's circumference does not affect the elastic stress distribution throughout the specimen material stress field [46].

11.2.4 Material and Its Properties

ASTM (A-387) P91 steel is extensively used in process industries and thermal power generation plants [25, 47]. After many years of use, creep damage and rupture buildup due to the deterioration of P91 metallic structural additives at elevated temperatures. There is a big difference in the characteristics between the weld region and the parent metal. It was revealed to be the root of most carrier issues, i.e., the heat-affected zone. For this FEA, the creep constant of P91B was taken as shown in Table 11.2 from the literature [22]. Elastic properties as shown in Table 11.1 were also taken from the literature [48].

11.2.5 Evaluation of Impression Creep Parameters

The conversion factor was taken as 3.3 for this FEA, which was found by Yu and Li in 1977 [49]. The conversion factor was used to balance the stress level between uniaxial stress (σ_{uni}) and impression

TABLE 11.1
Elastic Properties of P91B Steel [48]

Young's Modulus of Elasticity	Poisson's Ratio
162 GPa	0.3

TABLE 11.2
Experimental Creep Parameters of Coaxial Creep of Parent Metal, Weld Metal, and Composite Weld Metal at 650°C [22]

Creep Parameter	Base Metal	Weld Metal	Composite Weld Metal
A	1.83E–9	1.48E–11	8.46E–15
n	2.90	4.35	5.948
m	–0.679	0.65	0.59

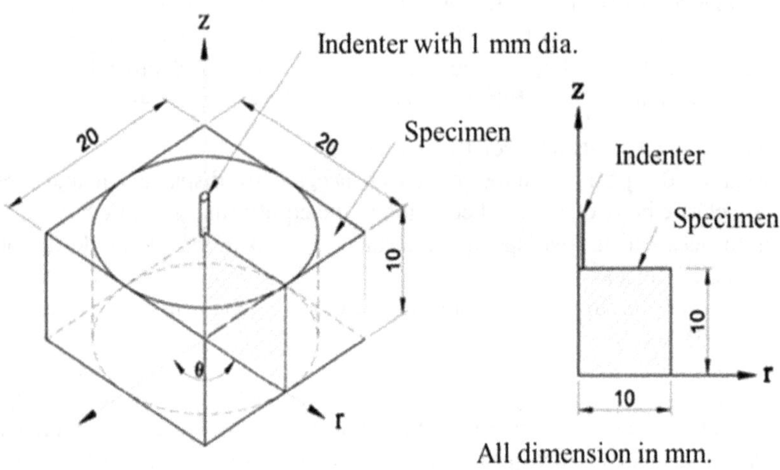

FIGURE 11.3 Schematic diagram of the axisymmetric analysis [42].

stress (σ_{imp}) techniques that are suitable for most materials. Therefore, four stress levels of 165, 231, 396, and 594 MPa were used for the impression creep test at 650°C. The stress level values used in the impression creep test in this research work correspond to equivalent uniaxial tensile stresses of 50, 70, 120, and 180 MPa.

11.2.6 Finite Element Modeling

For FE simulation of impression creep test using ABAQUS FEA software, an axisymmetric FE model of a cylindrical indenter with a flat-end diameter of 1 mm [46] in contact with an indenter with a flat-end specimen of 10 mm × 10 mm was created (Figure 11.3). The ABAQUS assembly interface allows developing of FE mesh using a structural system that meets actual assembly. The assembled part is called the instance part. Anything in a component, case, or assembly is exclusive to that part, instance, or assembly. This means that node/element identifiers and names do not have to be unique throughout the whole model; they only have to be unique within the component, instance, or assembly where they are specified in Figure 11.4.

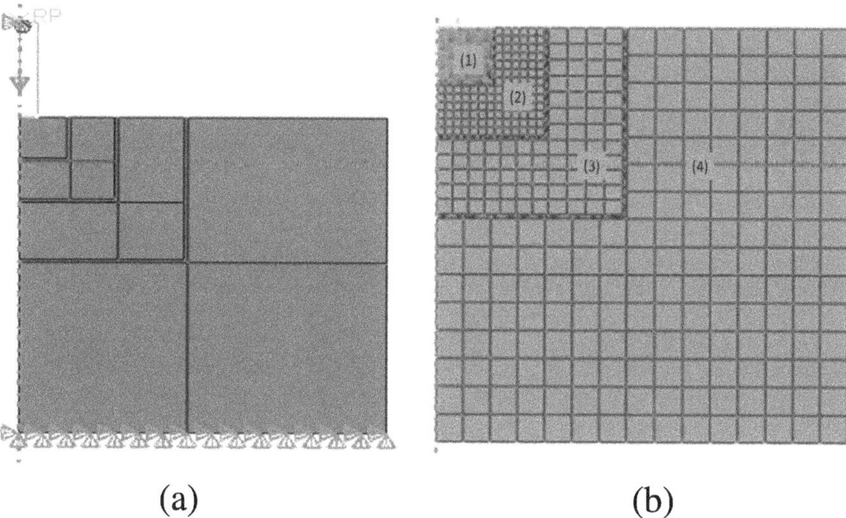

FIGURE 11.4 Applied load and boundary conditions and meshing of model.

TABLE 11.3
Different Sizes of Meshing and Number of Elements Used for Mesh Convergence

Seed Size	Region 1	Region2	Region 3	Region 4	Number of Elements
1	30 μm	200 μm	400 μm	700 μm	2603
2	15 μm	100 μm	200 μm	350 μm	9425
3	7.5μm	30 μm	100 μm	175 μm	36365

11.2.7 Interaction and Boundary Condition

In this model, the specimen is selected as a slave surface and the punch is selected as an ace surface. The surface-to-surface method was used to make contact between the punch and specimen. The discretization method was used as the node to surface. The finite sliding tracking approach was used because it allows arbitrary separation, rotation of the surfaces, and sliding. The contact interaction property was taken as a frictionless surface, which was used in specimen and indenter contact. In the assembly, some boundary conditions were applied. The movement of the indenter was restricted in the longitudinal direction, and the rotational degree of freedom was also restricted. The moment was permitted in the y-direction of the indenter. The longitudinal and lateral motion of the specimen was also restricted as shown in Figure 11.4. Different impression stress was applied to the reference point of the punch.

11.2.8 Type of Analysis and Meshing

Each continuous object has an infinite number of degrees of freedom (DOF); human or machine calculations are impossible to solve. As a result, a mesh is generated that divides the domain into a predetermined number of elements, each of which is utilized to compute the solution. The data is then interpolated throughout the whole domain. The mesh refinement process involves three types. Herein, only the H and P types have been discussed. In H type, there is an increase in the number of elements and therefore it decreases the length while maintaining the polynomial order of the form characteristic constant [50]. In P type mesh generation, only change in order of shape function is considered as shown in Equation 11.8 [50]. There are four sizes of mesh used in this model as shown in Table 11.3. In this model, a 30 μm local mesh size was employed in contact, followed by 200 μm, then 400 μm. The global mesh size was kept as 700 μm. From the element library, an implicit solver was used. For this model, two types of the element are used: CAX4R and CAX8R. The full explanation of CAX4R is continuum, axisymmetric non-linear deformation, four-node, reduced integration with one integration point or Gauss point and four Gauss points, respectively [51, 52].

$$ax + b \tag{11.8}$$

11.3 RESULTS AND DISCUSSION

For FE analysis, creep parameters and material properties were taken from the literature as described earlier. Base metal, weld metal, and composite weld metal simulations were run at four stresses (i.e., 165, 231, 396, and 594 MPa) to calculate the minimum creep strain rate. Base metal, weld metal, and composite weld metal simulation results were compared with experimental uniaxial creep results as reported in the literature at stresses 50 MPa and 70 MPa for 10000 hours and 200 hours, respectively. Weld metal and composite weld metal simulation results were analogized with experimental coaxial creep results as reported in the literature at stress 70 MPa for 200 hours only.

11.3.1 Mesh Convergence for Base, Weld, and Composite Weld Metal Simulation

The accuracy, convergence, and computation speed of a FE model are affected by the mesh size. A mesh convergence test was performed in the present study before FE simulation. As demonstrated in Figure 11.5, Figure 11.6, Figure 11.7, and Figure 11.8, variations in von Mises stress, depth of indentation, and creep strain are insignificant when the number of elements increases or mesh size decreases. An element's size changes to the optimal size of 30 μm at the contact area of the penetrating specimen.

Figure 11.5 depicts von Mises stress variation of base metal at 231 MPa and 165 MPa for 200 hours and 10000 hours, respectively. As shown in Figure 11.5, von Mises stress variation is studied along with the coaxial direction during creep. The highest stress is found around the circumference of the indenter rather than the core. Because of hydrostatic stress, there is less stress present near the

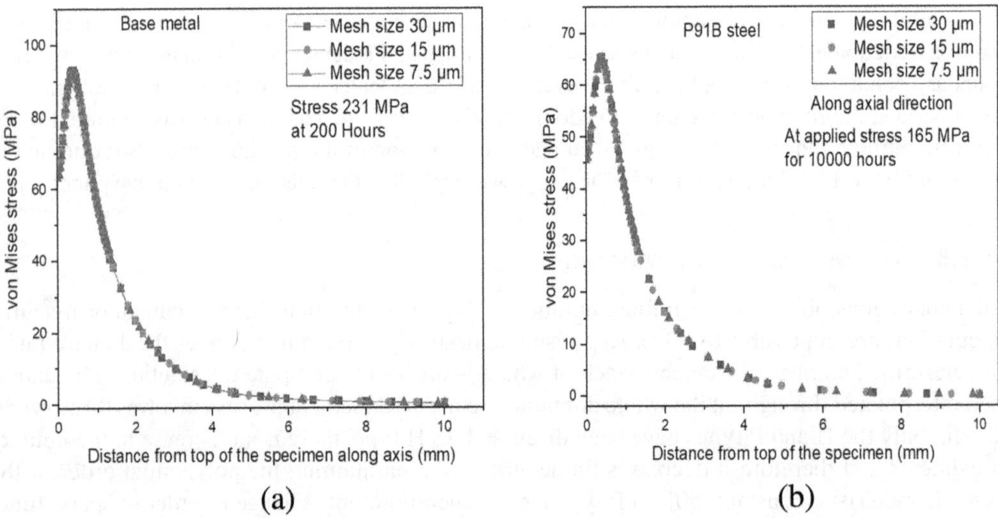

FIGURE 11.5 Variation of von Mises stress profile along the axis at 231 and 165 MPa.

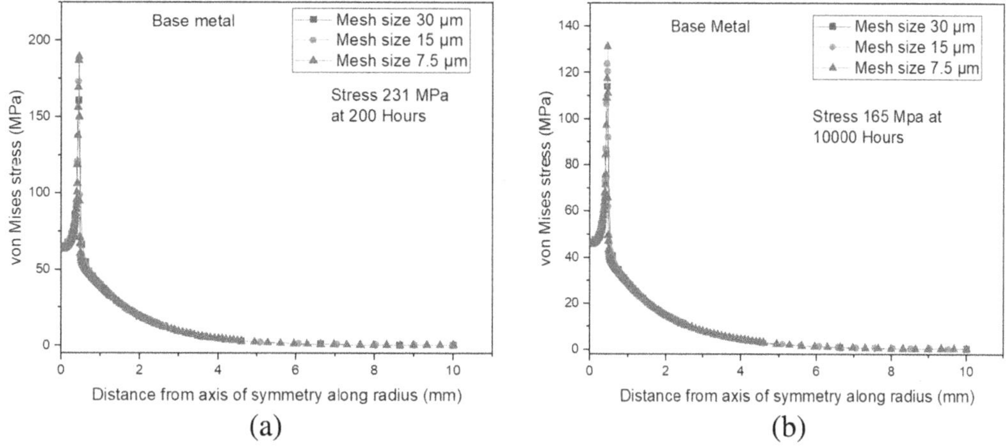

FIGURE 11.6 Variation of von Mises stress profiles in radial direction at 231 MPa and 165 MPa.

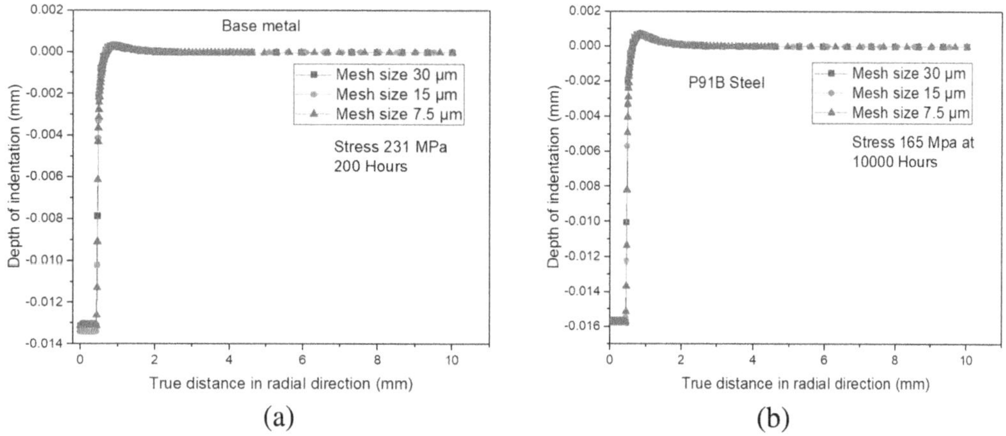

FIGURE 11.7 Comparison of depth profiles for different element sizes at 165 and 231 MPa.

center of the indenter [53]. Maximum von Mises was seen at the circumference of the punch. This phenomenon was also seen by Yu and Li [49].

Figure 11.6 depicts the onset of creep, the evolution of the plastic region, and the duration of the plastic region. A graph was plotted between von Mises stress and radial distance from the axis of the indenter of base metal at applied stress 165 MPa and 231 MPa for 10000 hours and 200 hours, respectively. It showed that maximum stress developed at the girth of the indenter as shown in Figure 11.6.

The typical depth of indentation profile at the various element sizes is depicted in Figure 11.7, and their profile differences are negligible. The mesh size near the indentation specimen's contact area is tuned at 30 μm to decrease calculation time, while the mesh size is gradually kept thicker in areas far from the indentation specimen's contact area.

In this model, 2603 elements were used for further simulation. Depth of indentation versus true distance in radial distance is plotted at stress 165 MPa and 231 MPa for 10000 hours and 200 hours, respectively at different sizes of the element as shown in Figure 11.7. The creep strain versus time plot for element sizes 30, 15, and 7.5 μm at 165 and 231 MPa, respectively, is depicted in Figure 11.8a, b.

Similar results were obtained for mesh convergence of weld and composite weld metal simulation.

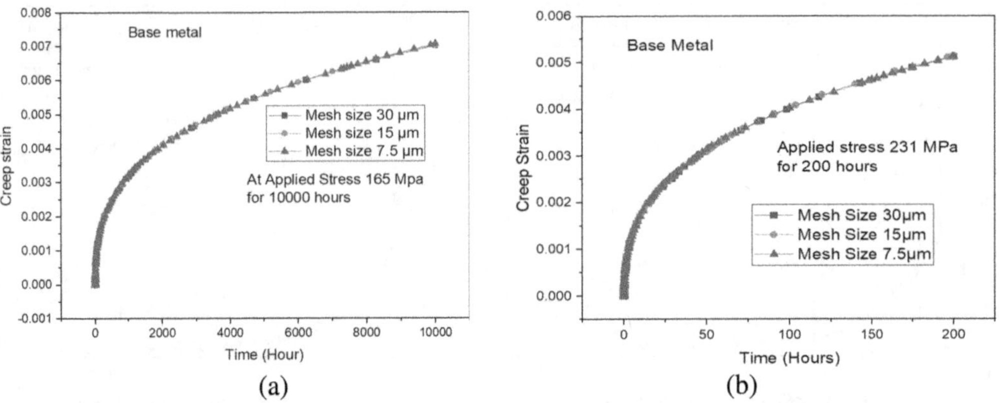

FIGURE 11.8 Creep strain versus time plot for element sizes 30, 15, and 7.5 μm at 165 and 231 MPa, respectively.

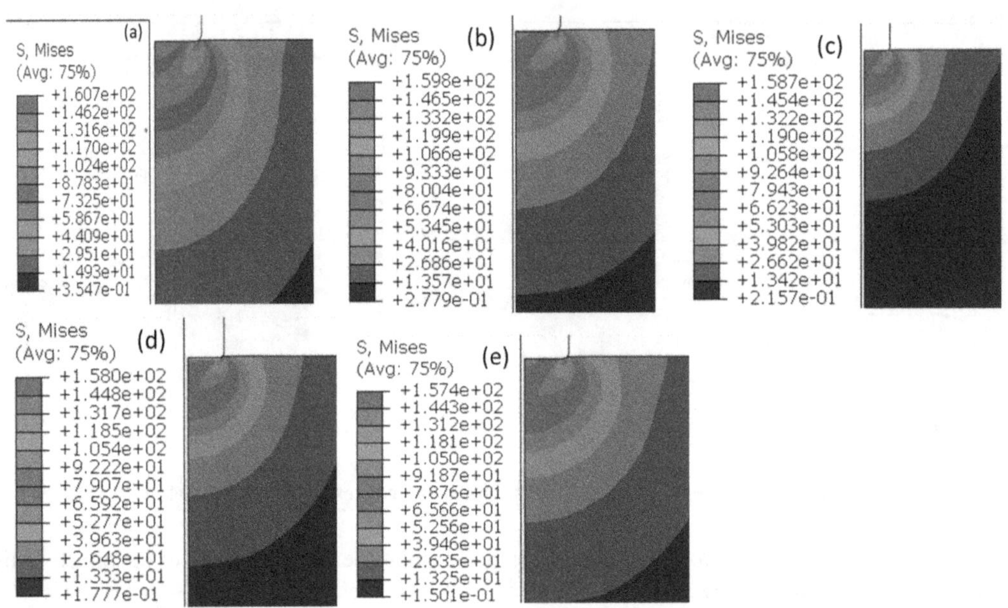

FIGURE 11.9 Growth of von Mises stresses for (a) 200 hours, (b) 500 hours, (c) 1000 hours, (d) 1500 hours, and (e) 2000 hours at 231 MPa.

11.3.2 FE Simulation of P91B Base Metal

11.3.2.1 Base Metal Impression Creep Behavior at Different Applied Stresses

The development of von Mises stress during impression creep simulation is depicted in Figure 11.9. Von Mises stress, present on the punch's circumference, decreases with creep time 200, 500, 1500, and 2000 hours at applied stress 231 MPa. The von Mises stress at the indenter corner reduced to 160.7, 159.8, 158.7, and 157.4 MPa. As a result, compression inside the plastic zone under the punch is redistributed during creep. During creep exposure, a similar redistribution of stress around the notch has been observed [35, 40].

Contour plots of base metal are shown in Figure 11.10 at stresses 165, 231, 396, and 594 MPa 200 hours. When applied stress increases, plastic deformation spreads out toward the bottom of the

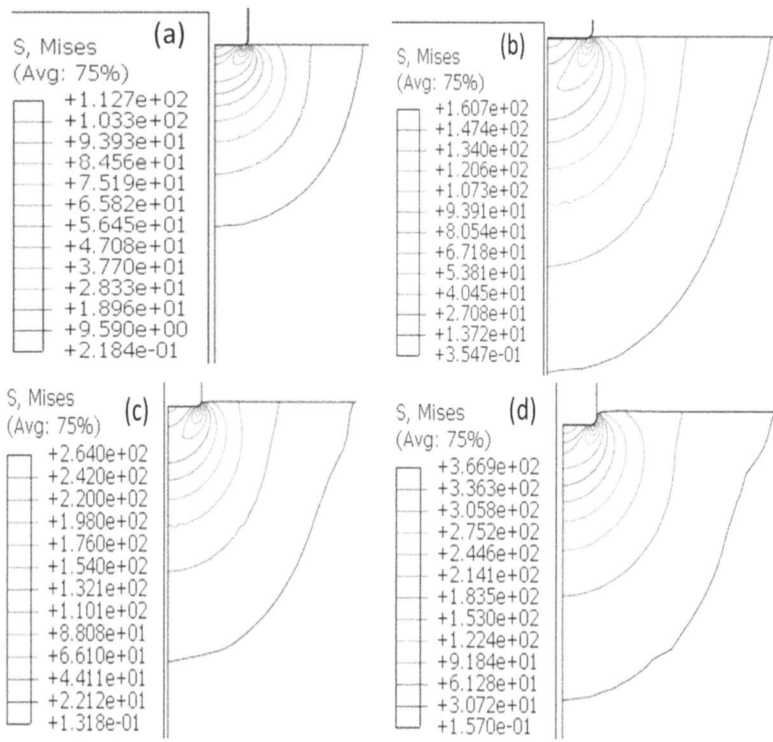

FIGURE 11.10 Contour plots of base metal at (a) 165 MPa, (b) 231 MPa, (c) 396 MPa, and (d) 594 MPa.

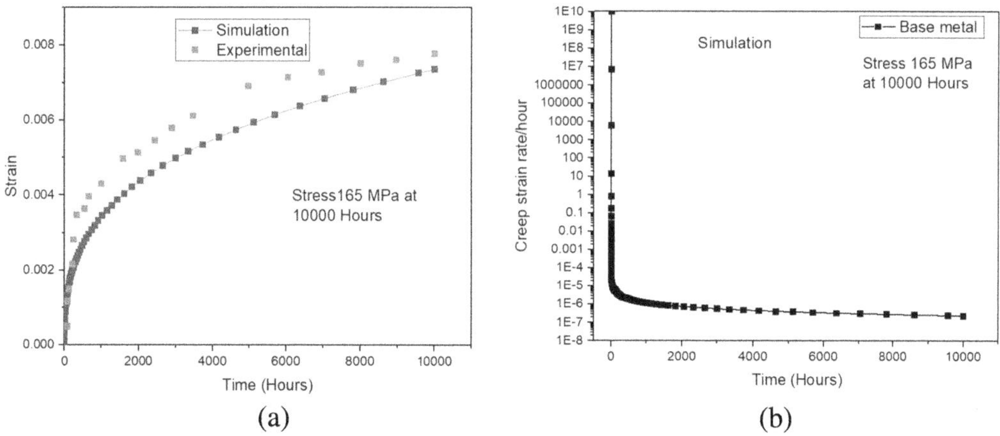

FIGURE 11.11 Creep strain and minimum strain rate at 165 MPa for 10000 hours.

indenter and moves away from the indenter. It created a hemispherical shape around the indenter during the impression. The same incident was also reported by Naveena [42].

Figure 11.11 depicts creep strain versus time plots for base metals at 165 MPa for 10000 hours. Maximum creep strain and creep strain rate were found to be 0.00736 and 2.337×10^{-7} per hour at 650°C. The corresponding experimental uniaxial strain was 0.00777 for 10000 hours at applied stress 50 MPa. However, because of the compression character of the load in an impression creep test, this curve only provided primary and secondary creeping stages. Figure 11.12 exhibits creep strain versus time and minimum strain rate versus time of base metal. Creep strain and minimum

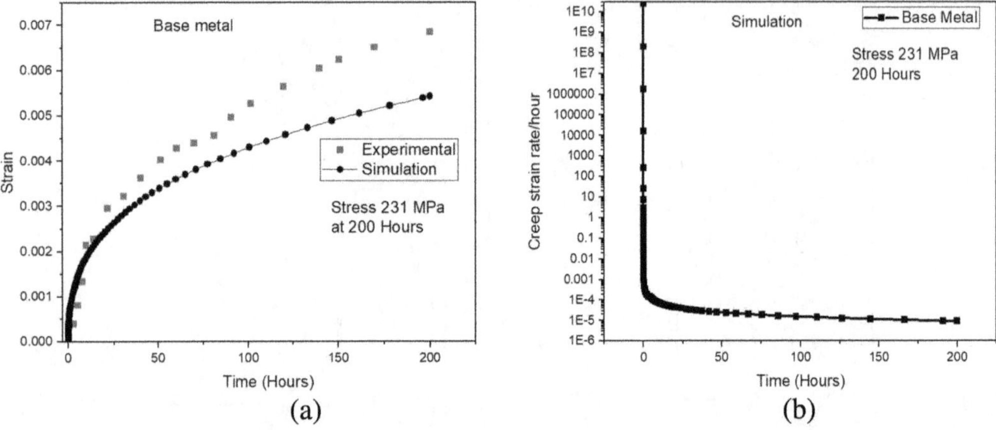

FIGURE 11.12 Creep strain and minimum strain rate at 231 MPa for 200 hours.

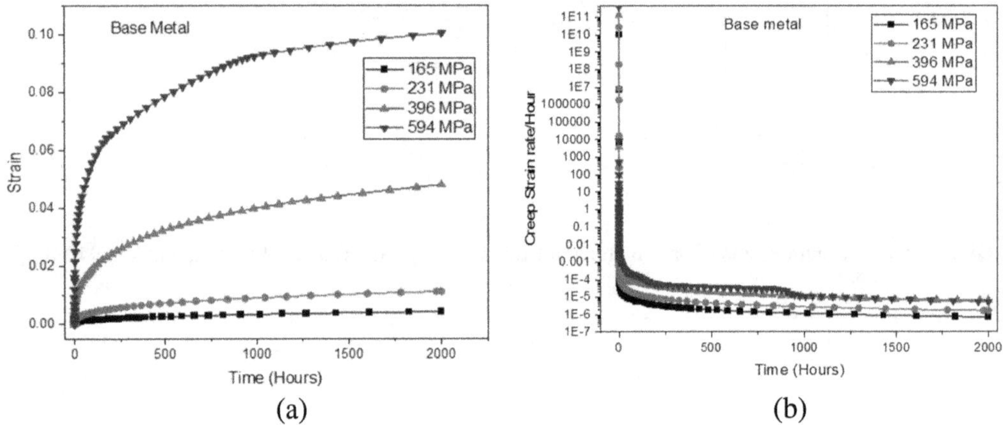

FIGURE 11.13 Creep strain versus time plot and steady-state rate versus time plot at 165, 231, 396, and 594 MPa for 2000 hours.

strain rate were found to be 0.00543 and 9.065×10^{-6} per hour at stresses 231 MPa for 200 hours. The corresponding experimental uniaxial strain was 0.00685 for 200 hours at an applied stress of 70 MPa for base metal.

Figure 11.13 depicts the creep strain versus time plot of base metal at stresses 165, 231, 394, and 594 MPa. Maximum strains were observed as 0.00434, 0.01127, 0.04811, and 0.10041, respectively. The creep curve showed only two stages of the standard creep curve because compressive stress was applied (no necking zone); hence, the accelerating or tertiary creep stage was not present. At the initial stage, the creep strain rate was very high. As time increased, the creep strain rate decreased and became approximately constant (Figure 11.13). Different steady creep rates were found as 7.360×10^{-7}, 1.695×10^{-6}, 6.667×10^{-6}, and 5.306×10^{-6} at stresses of 165, 231, 394, and 594 MPa, respectively. Initially, the strain rate was very high due to material resistance, and the strain rate decreased due to strain hardening.

When it entered into a steady state, it became constant. The transient creep of base metal P91B exhibits strain hardening owing to the intrinsic reluctance of material distorted at the beginning of an impression creep test; the latter base metal enters into the steady-state secondary creep stage in a very short time [54]. Due to work hardening induced by dislocation multiplication and related interactions, the strain rate slowed during transient creep. In the second creep stage, the strain rate became stable when work hardening backwashed and with recovery mechanisms, equilibrium approached, i.e., rearrangement annihilation and rearrangement [25, 54]. Figure 11.14 depicts the variation of von Mises stresses with increasing simulation time. The von Mises stresses were found as 160.7, 159.8, 158.7, 158, and 157.4 MPa for simulation time 200, 500, 1000, 1500, and 2000 hours for the same applied stress of 231 MPa. It may be observed that von Mises stress decreased with increasing time. During creep exposure, comparable observations of stress redistribution around the notch were obtained [35, 40]. The depth of indentation versus distance along the radius of the specimen is shown in Figure 11.15 at stress 231 MPa for simulation times 200, 500, 1000, 1500, and 2000 hours. It was found that the height and width of the pile in base metal increased with an increase in penetration depth, and the same was reported by Naveena [42].

Figure 11.15b also depicts a comparison of experimental uniaxial minimum strain rate and FE simulation of base metal at stress 231 MPa for 200 hours. The creep strain rate of experimental

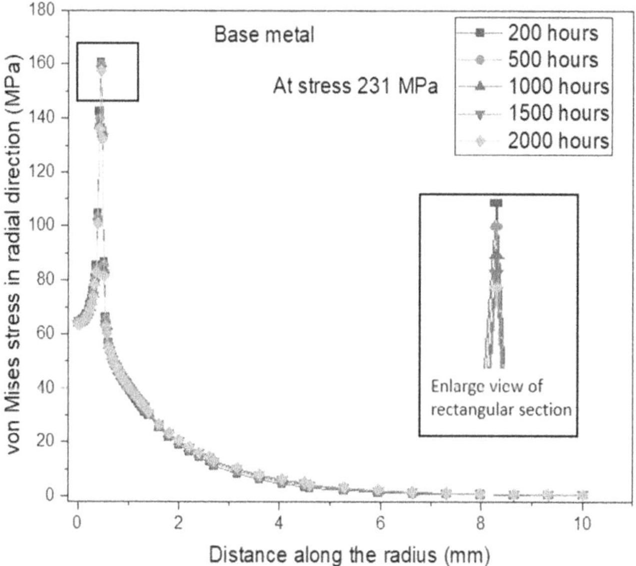

FIGURE 11.14 Von Mises stress distribution with varying time at 231 MPa.

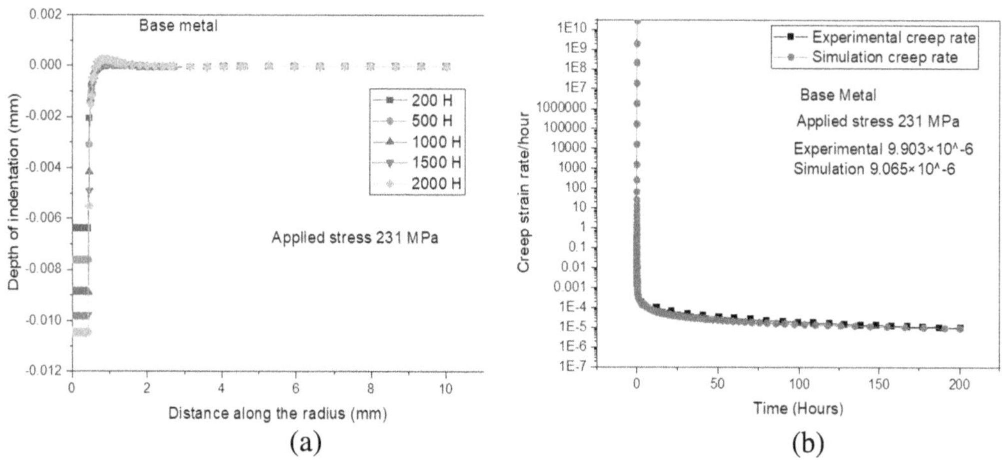

FIGURE 11.15 Depth of indentation for 200, 500, 1000, 1500, and 2000 hours at 231 MPa and steady-state strain rate of experimental and simulation at 231 MPa for 200 hours.

uniaxial creep strain and simulated impression creep were 9.903×10^{-6} per hour and 9.065×10^{-6} per hour. Initially, the creep strain rate was very high due to material resistance and after strain hardening; it became constant with time.

11.3.3 FE Simulation of P91B Weld Metal

11.3.3.1 Impression Creep Behavior of Weld Metal at Different Applied Stresses

The elaboration of von Mises stress during the impression creep process is given in Figure 11.16. It decreased at the punch circumference regarding creep time, as observed earlier for base metal. After 200, 500, 1500, and 2000 hours, von Mises stress decreased to 95.28, 93.78, 91.75, and 91.22 MPa, respectively, at the punch corner at 231 MPa. As a result, tensions inside the plastic zone under the punch possibly redistribute during creep. Earlier too, a similar advertence of stress dispensation around the notch during creep exposure was observed [35, 40]. Contour plots for weld metal are shown in Figure 11.17 at stresses 165, 231, 396, and 594 MPa for 200 hours. It may be noticed that as stress increases, plastic deformation spreads toward the bottom of the indenter and away from the indenter, thus forming a hemispherical shape as reported by Naveena [42].

Creep curves at 231 MPa with 650°C for 200 hours are shown in Figure 11.18. The peak creep strain and creep strain rate were observed as 0.03412 and 5.4955×10^{-5} per hour at 650°C. The experimental uniaxial strain was 0.0302 for 200 hours at 70 MPa. In Figure 11.19, the creep strain versus time graph at stresses 165, 231, 394, and 594 MPa is shown. Peak strain was found as 0.01793, 0.06263, 0.18318, and 0.20778, respectively, for these stress values. Further, it can be viewed thatinitially the strain rate was very high. However, as time increased,the creep strain rate decreased and became consistent. Steady creep rates were found as 3.648×10^{-6}, 1.279×10^{-5}, 7.2×10^{-6}, and 6.452×10^{-6} at different applied stresses of 165, 231, 394, and 594 MPa, respectively. For weld metal, the transient creep stage, which exhibits strain hardening due to the inherent resistance of metal to deform during initiation of an impression creep test, is large, whereas weld metal takes a comparatively short time to enter the steady-state secondary creep stage [23]. The depth of indentation versus distance along the radius of the specimen is shown in Figure 11.20 at stress 231 MPa for

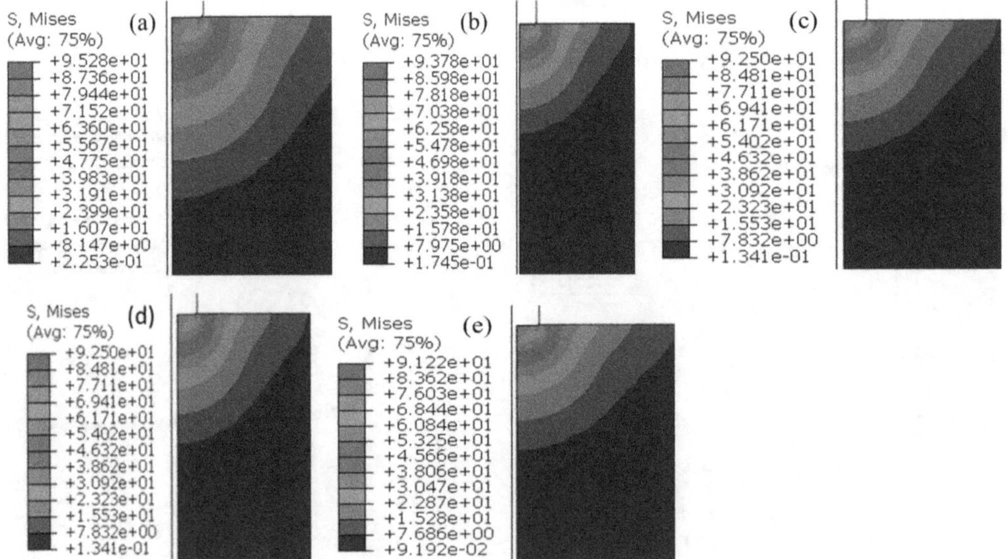

FIGURE 11.16 Von Mises stresses in the area of punch during impression creep: (a) 200 hours, (b) 500 hours, (c) 1000 hours, (d) 1500 hours, and (e) 2000 hours at 231 MPa.

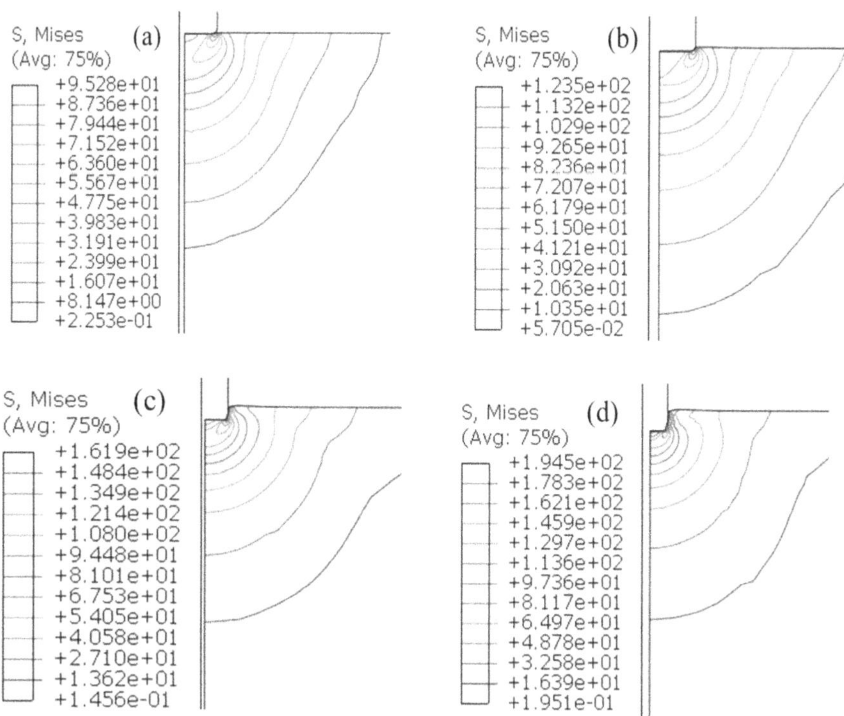

FIGURE 11.17 Contour plots of weld metal for 200 hours at (a) 165 MPa, (b) 231 MPa, (c) 396 MPa, and (d) 594 MPa.

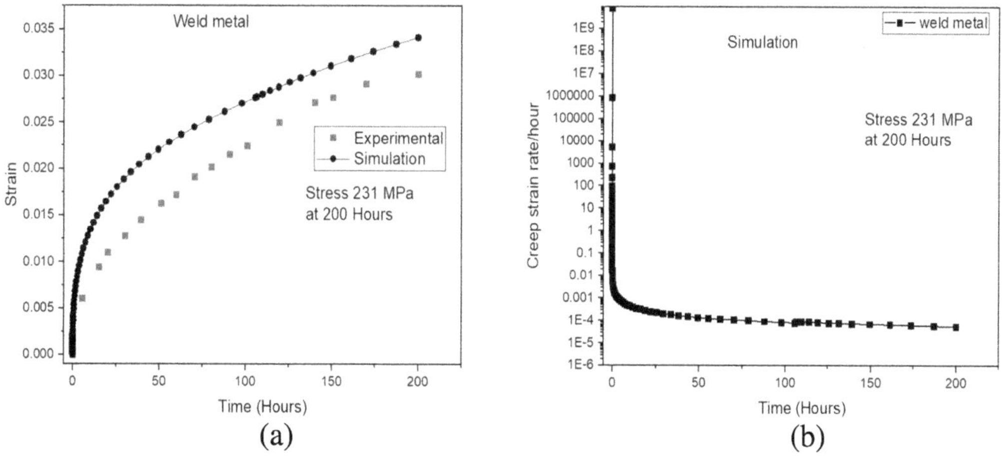

FIGURE 11.18 Creep strain and minimum creep strain rate at 231 MPa for 200 hours.

simulation times 200, 500, 1000, 1500, and 2000 hours. The height and width of the pile of weld metal increased with an increase in penetration depth as reported by Naveena [42]. Figure 11.20 depicts a variation of von Mises stresses with increasing simulation time. The von Mises stresses were found as 123.5, 119.8, 116.218, 116.1, and 115.2 MPa for simulation times of 200, 500, 1000, 1500, and 2000 hours for the same applied stress of 231 MPa. The von Mises stress decreased with increasing time. During creep exposure, comparable observations of stress redistribution around the notch were obtained earlier too [35, 40].

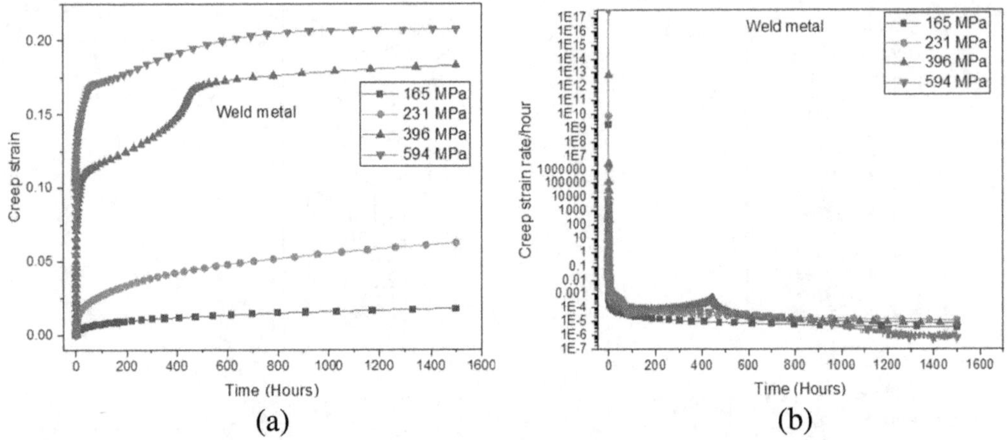

FIGURE 11.19 Creep strain and steady-state strain rates of weld metal at stresses 165, 231, 396, and 594 MPa for 1500 hours.

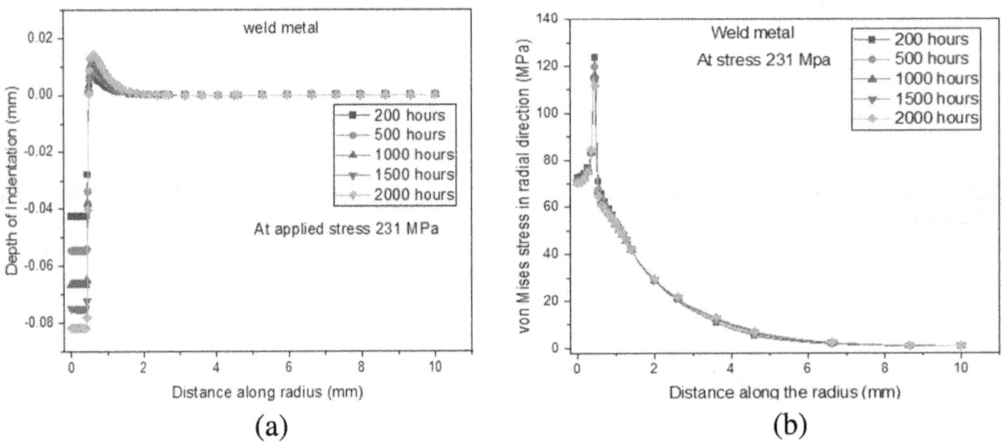

FIGURE 11.20 Depth of indentation and von Mises stresses distribution of weld metal for 200, 500, 1000, 1500, and 2000 hours at 231 MPa.

Figure 11.21 depicts an equivalent experimental uniaxial creep rate at 70 MPa and FE simulation of weld metal at 231 MPa for 200 hours. The creep strain rate of experimental uniaxial data and simulated impression creep were 7.65581×10^{-6} per hour and 5.4955×10^{-6} per hour. In the early stage, the creep strain rate was very high due to material resistance; however, after strain hardening, it became constant.

11.3.4 FE Simulation of P91B Composite Weld Metal

11.3.4.1 Impression Creep Behavior of Weld Metal at Different Applied Stresses

The behavior of von Mises stress during creep exposure is depicted in Figure 11.22 .It decreased at punch girth as regards to time. After 200, 500, 1500, and 2000 hours, stress at the punch corner decreased to 87.75, 86.54, 85.41, 84.72, and 84.16 MPa, respectively, with applied stress of 231 MPa. It resulted in stress redistribution inside the plastic zone as also observed in previous studies [35, 40, 54]. The contour plot of composite metal is shown in Figure 11.23 at different applied stresses of 165, 231, 396, and 594 MPa for 200 hours. When stress increases, plastic deformation spreads out

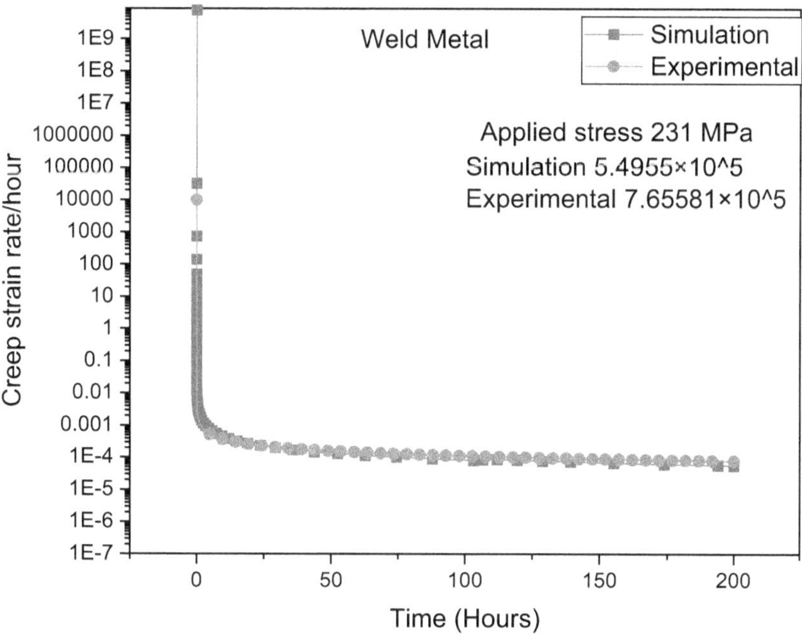

FIGURE 11.21 Comparison of experimental and simulation creep rate at 231 MPa for 200 hours.

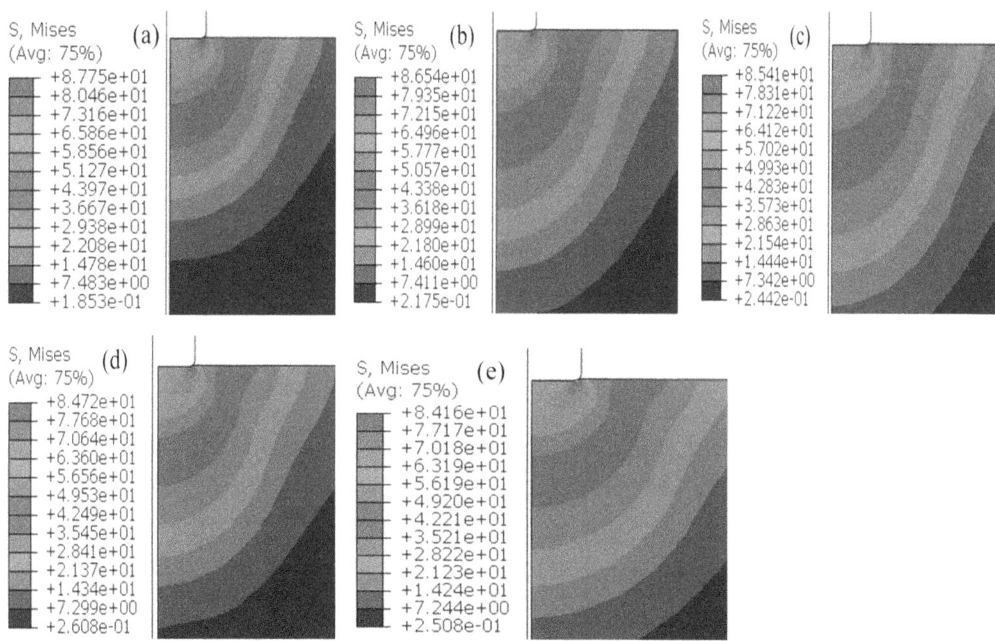

FIGURE 11.22 Von Mises stresses in the area of punch during creep at 165 MPa for (a) 200 hours, (b) 500 hours, (c) 1000 hours, (d) 1500 hours, and (e) 2000 hours.

toward the bottom of the indenter and away from the indenter eventually generating a hemispherical shape around the indenter during the impression as noted by Naveena [42].

The impression creep curve for composite weld metal at 231 MPa in the form of creep strain versus time and steady-state creep rate versus time at 650°C for 200 hours is shown in Figure 11.24.

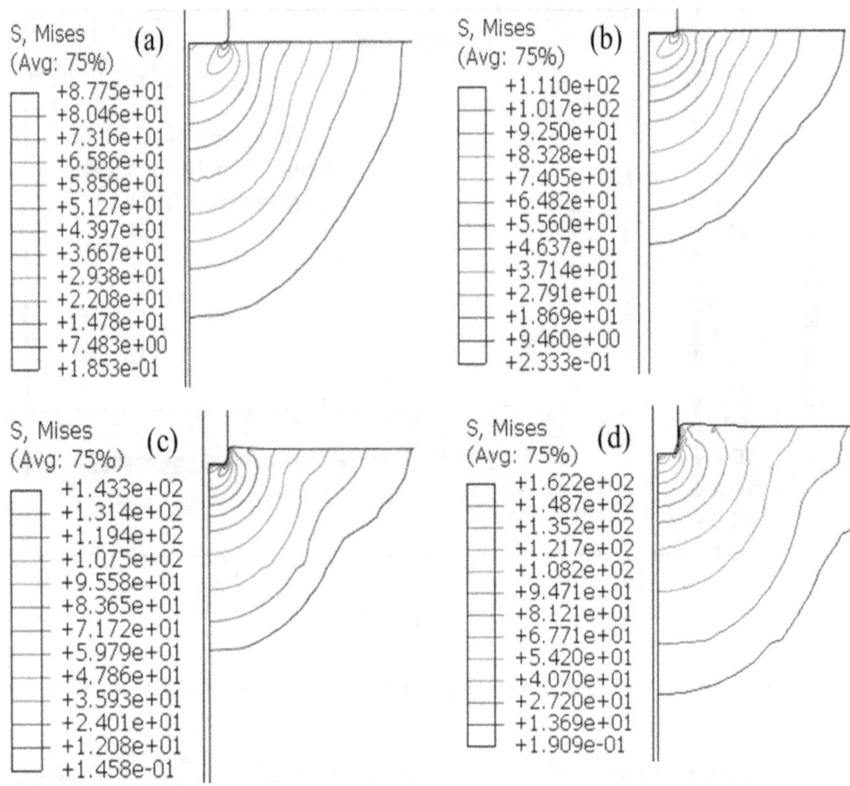

FIGURE 11.23 Contour plots of composite weld metal for 200 hours at (a) 165 MPa, (b) 231 MPa, (c) 396 MPa, and (d) 594 MPa,

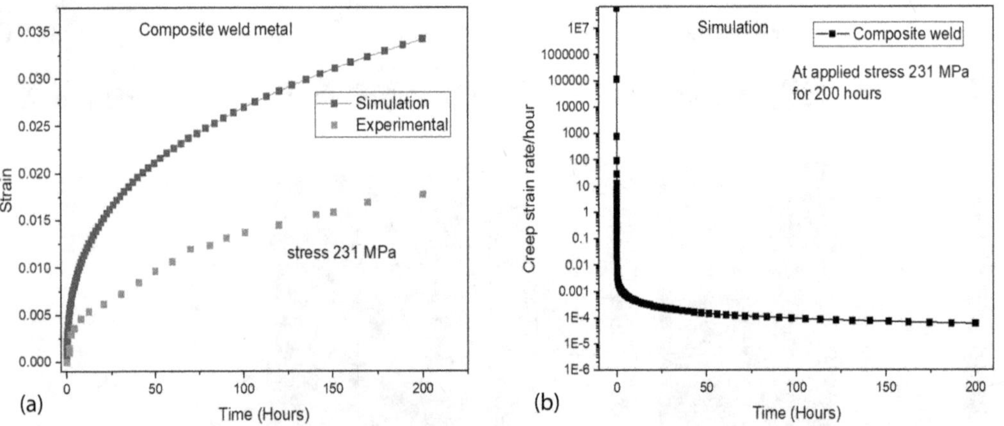

FIGURE 11.24 Creep strain and steady-state strain rate at 231 MPa for 200 hours.

The peak creep strain and creep strain rate were indicated as 0.03421 and 5.8435×10^{-5} per hour at 650°C at 231 MPa. The corresponding experimental uniaxial strain was 0.017 for 200 hours at 70 MPa. As depicted in Figure 11.25, the creep strain versus time graph of composite weld metal is plotted at different applied stresses of 165, 231, 394, and 594 MPa, and maximum strain was found as 0.01432, 0.07417, 0.22298, and 0.26929, respectively.

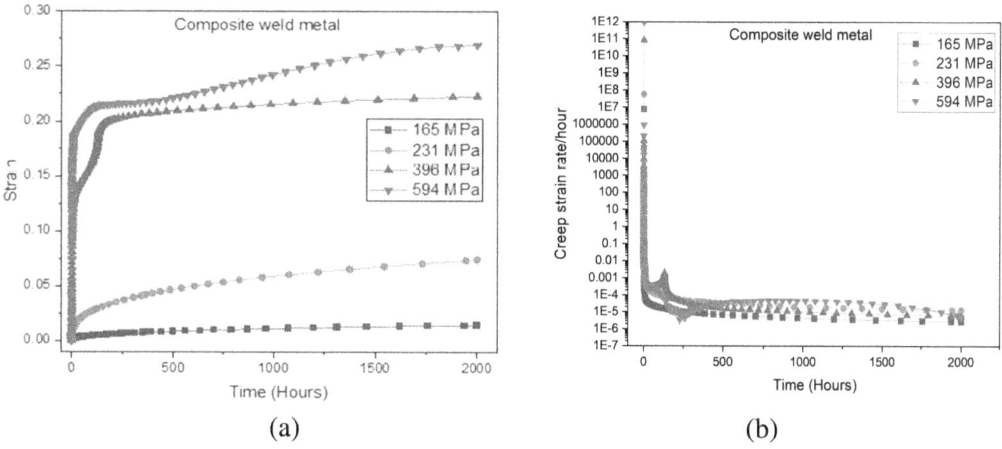

FIGURE 11.25 Creep strain and steady-state strain of composite weld metal at 165, 231, 396, and 594 MPa for 2000 hours.

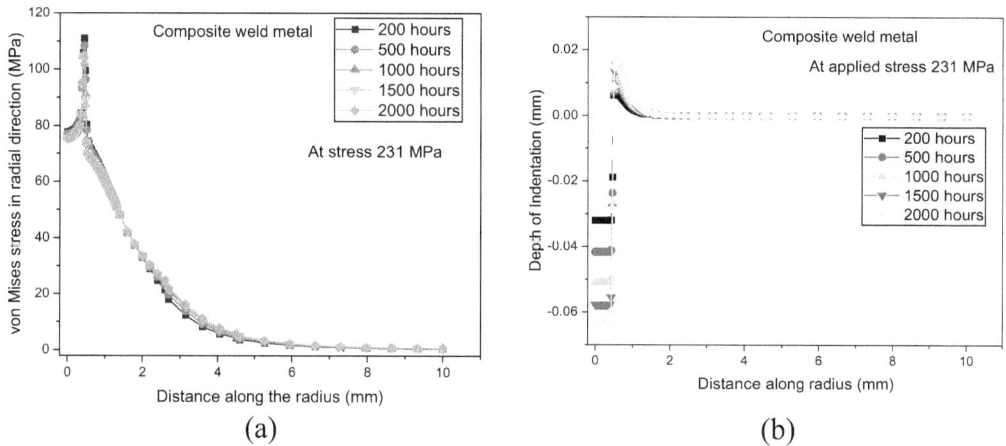

FIGURE 11.26 Von Mises stresses distribution and depth of indentation of composite weld metal for varying time at 231 MPa.

A different steady creep rate was found as 2.671×10^{-6}, 1.255×10^{-5}, 5.556×10^{-6}, and 5.125×10^{-6} at different applied stresses 165, 231, 394, and 594 MPa, respectively. The deformation mechanisms could be explained as earlier outlined in the case of base and weld metal FE simulations. Figure 11.26a depicts the variation of von Mises stresses with increasing simulation time. These stresses were found as 110.95, 108.368, 105.316, 103.467, and 101.9 MPa for simulation times of 200, 500, 1000, 1500, and 2000 hours for stress 231 MPa. It was decreasing with increasing time. During creep exposure, comparable observations of stress redistribution around the notch were obtained [35, 40, 54]. The depth of indentation versus distance along the radius of the specimen is shown in Figure 11.26b at stress 231 MPa for simulation times 200, 500, 1000, 1500, and 2000 hours. It was found that the height and width of the pile of composite weld metal increased with an increase in penetration depth as reported by Naveena [42].

Figure 11.27 shows a comparison of the experimental uniaxial creep rate at applied stress 70 MPa and FE simulation of composite weld metal at 231 MPa for 200 hours. The creep strain rate of experimental uniaxial creep strain and simulated impression creep are 3.922×10^{-5} per hour and 5.843×10^{-3} per hour.

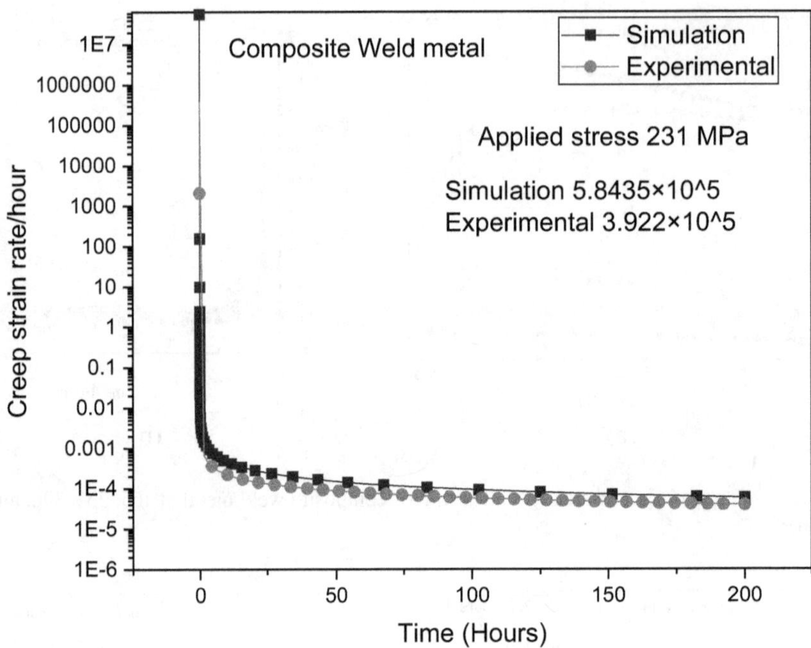

FIGURE 11.27 Comparison of experimental and simulation creeping rates of composite weld metal at 231 MPa for 200 hours.

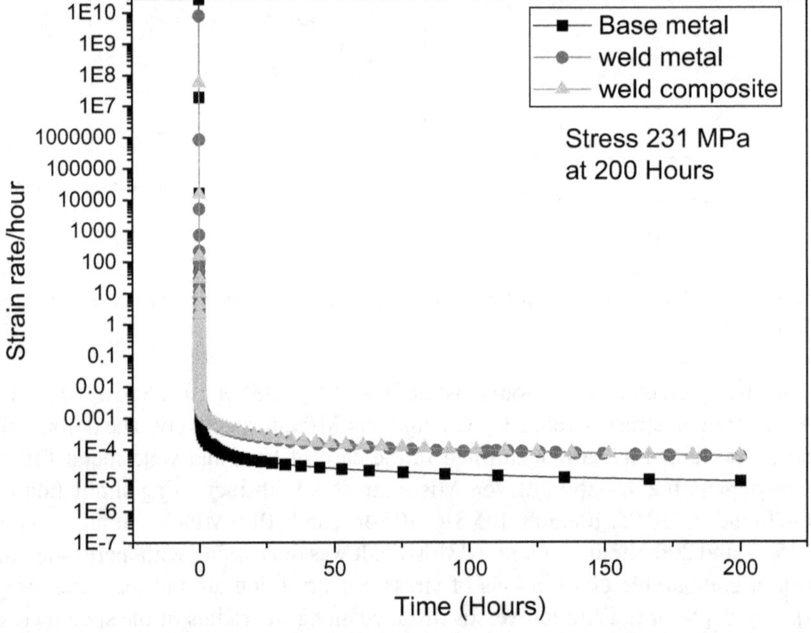

FIGURE 11.28 Comparison at 231 MPa of steady-state creep rate for P91B base, weld, and composite weld.

11.3.5 Comparison of FE Simulation in Terms of Steady-State Creep Rate

The steady-state creep rate of P91B base, weld, and, composite weld comparison was done using simulation results. The steady-state strain rate of base metal, weld metal, and composite weld metal were found as 9.903×10^{-6}, 5.4955×10^{-5}, and 5.8435×10^{-5}, respectively, at applied stress of 231 MPa for 200 hours as shown in Figure 11.28.

Base metal minimum creep strain rate performance was 5.55 times better than weld metal. Base metal minimum creep strain rate performance was 5.9 times better than composite weld metal and weld metal minimum creep strain rate performance was 1.06 times better than composite weld metal.

11.4 SUMMARY

An extensive computational FE-based simulation on ABAQUS was performed to understand the deformation and impression creeping behavior of P91B steel. The material constant was obtained from experimental uniaxial creep and was successfully utilized to simulate the finite element model for impression creep. One of the most significant observations from FE analysis was the similar behavior of impression creep and experimental uniaxial creep. FE simulation results of P91B steel base, weld, and, composite weld in terms of creep strain as well as creep steady-state strain rate matched well with experimental uniaxial creep data. It was found that von Mises stress of P91B steel base, weld, and composite weld decreased for simulation times of 200, 500, 1000, 1500, 2000, and 10000 hours. Further, less stress appeared underneath the punch than the circumference of the indenter for P91B steel base, weld, and composite weld because of the development of hydrostatic stress below the punch. It was also noticed that radial stress was tending to zero concerning radial distance. Moreover, the height and width of the pile of P91B base metal, weld metal, and composite weld metal increased with an increase in penetration depth. Also, von Mises stress for P91B weld metal was less than P91B base metal and composite weld metal for the same applied stress for simulation time. After comparison of the steady-state creep rate of P91B base metal and P91B weld metal, it was also found that the steady-state creep rate of P91B base metal was 5.55 times less than the steady-state creep rate of P91B weld metal.

ACKNOWLEDGEMENTS

The authors extend their gratitude to the Department of Mechanical Engineering, and Director Dr B R Ambedkar, National Institute of Technology, Jalandhar, Punjab, India for their timely support that helped to complete this research work.

REFERENCES

1. A. Khajuria, M. Akhtar, and R. Bedi, "Boron addition to AISI A213/P91 steel: Preliminary investigation on microstructural evolution and microhardness at simulated heat-affected zone," *Mater. Werkst.*, vol. 53, no. 10, pp. 1167–1183, Oct. 2022. https://doi.org/10.1002/mawe.202100152
2. M. Akhtar and A. Khajuria, "The synergistic effects among crystal orientations, creep parameters, local strain, macro–micro deformation, and polycrystals' hardness of Boron Alloyed P91 steels," *Steel Res. Int.*, vol. 93, no. 9, p. 2100819, Sep. 2022. https://doi.org/10.1002/srin.202100819
3. A. Khajuria, M. Akhtar, and R. Bedi, "A novel approach to envisage effects of boron in P91 steels through Gleeble weld-HAZ simulation and impression-creep," *J. Strain Anal. Eng. Des.*, vol. 57, no. 8, pp. 647–663, Nov. 2022. https://doi.org/10.1177/03093247211061943
4. A. Khajuria, R. Kumar, R. Bedi, J. Swaminanthan, and D. Kumar, "Impression creep studies on simulated reheated HAZ of P91 and P91B steels," p. 7. https://modtech.ro/international-journal/vol10no12018/07_Khajuria_Akhil.pdf
5. A. Khajuria, R. Bedi, and R. Kumar, "Investigation of Impression Creep Deformation Behavior of Boron-Modified P91 Steel By High-End Characterization Techniques," in *Manufacturing Engineering*, V. S. Sharma, U. S. Dixit, and N. Alba-Baena, Eds. Singapore: Springer Singapore, 2019, pp. 137–150. https://doi.org/10.1007/978-981-13-6287-3_10
6. M. Akhtar, A. Khajuria, R. Kumar, and R. Bedi, "*Metallurgical Investigations on Dual Heat Cycled Boron Alloyed P91 Ferritic/Martensitic*," p. 10, 2017. http://dx.doi.org/10.13140/RG.2.2.18467.30241/2
7. M. Akhtar and A. Khajuria, "Probing true creep-hardening interaction in weld simulated heat affected zone of P91 steels," *J. Manuf. Process.*, vol. 46, pp. 345–356, Oct. 2019. https://doi.org/10.1016/j.jmapro.2019.08.029

8. M. Akhtar, A. Khajuria, and R. Bedi, "Effect of Re-normalizing and Re-tempering on Inter-critical Heat Affected Zone(S) of P91B Steel," in *Manufacturing Engineering*, V. S. Sharma, U. S. Dixit, K. Sørby, A. Bhardwaj, and R. Trehan, Eds. Singapore: Springer Singapore, 2020, pp. 255–270. https://doi.org/10.1007/978-981-15-4619-8_20
9. M. Akhtar and A. Khajuria, "Probing true microstructure-hardening relationship in simulated heat affected zone of P91B steels," *Metallogr. Microstruct. Anal.*, vol. 8, no. 5, pp. 656–677, Oct. 2019. https://doi.org/10.1007/s13632-019-00573-w
10. A. Khajuria, M. Akhtar, R. Bedi, R. Kumar, M. Ghosh, C.R. Das, S.K. Albert, "Influence of boron on microstructure and mechanical properties of Gleeble simulated heat-affected zone in P91 steel," *Int. J. Press. Vessels Pip.*, vol. 188, p. 104246, Dec. 2020. https://doi.org/10.1016/j.ijpvp.2020.104246
11. A. Khajuria, R. Kumar, and R. Bedi, "Effect of boron addition on creep strain during impression creep of P91 steel," *J. Mater. Eng. Perform.*, vol. 28, no. 7, pp. 4128–4142, Jul. 2019. https://doi.org/10.1007/s11665-019-04167-z
12. A.Khajuria, M. Akhtar, R. Bedi, R. Kumar, M. Ghosh, C.R. Das, S.K. Albert , "Microstructural investigations on simulated intercritical heat-affected zone of boron modified P91-steel," *Mater. Sci. Technol.*, vol. 36, no. 13, pp. 1407–1418, Sep. 2020. https://doi.org/10.1080/02670836.2020.1784543
13. M. Akhtar, A. Khajuria, M. K. Pandey, I. Ahmed, and R. Bedi, "Effects of boron modifications on phase nucleation and dissolution temperatures and mechanical properties in 9%Cr steels: Alloy design," *Mater. Res. Express*, vol. 6, no. 12, p. 1265k3, Feb. 2020. https://iopscience.iop.org/article/10.1088/2053-1591/ab6db7
14. M. Akhtar and A. Khajuria, "Effects of prior austenite grain size on impression creep and microstructure in simulated heat affected zones of boron modified P91 steels," *Mater. Chem. Phys.*, vol. 249, p. 122847, Jul. 2020. https://doi.org/10.1016/j.matchemphys.2020.122847
15. M. Akhtar, A. Khajuria, J.K. Sahu, J. Swaminathan, R. Kumar, R. Bedi, S.K. Albert , "Phase transformations and numerical modelling in simulated HAZ of nanostructured P91B steel for high temperature applications," *Appl. Nanosci.*, vol. 8, no. 7, pp. 1669–1685, Oct. 2018. https://doi.org/10.1007/s13204-018-0854-1
16. M. Akhtar, A. Khajuria, V. S. Kumar, R. K. Gupta, and S. K. Albert, "Evolution of microstructure during welding simulation of boron modified P91 steel," *Phys. Met. Metallogr.*, vol. 120, no. 7, pp. 672–685, Jul. 2019. https://doi.org/10.1134/S0031918X19070056
17. A. Khajuria, R. Kumar, and R. Bedi, "*Characterizing Creep Behaviour of Modified 9Cr1Mo Steel by Using Small Punch Impression Technique for Thermal Powerplants*," Oct. 2018. https://doi.org/10.5281/ZENODO.1453768
18. O. Khokhar, A. Khajuria, and R. Bedi, "An initial study on the impression creep behaviour of stir-cast GNP reinforced AA6061 composite (AMMC)," *IOP Conf. Ser. Mater. Sci. Eng.*, vol. 1248, no. 1, p. 012086, Jul. 2022. https://iopscience.iop.org/article/10.1088/1757-899X/1248/1/012086/meta
19. T. H. Hyde, W. Sun, and A. A. Becker, "Analysis of the impression creep test method using a rectangular indenter for determining the creep properties in welds," *Int. J. Mech. Sci.*, vol. 38, no. 10, pp. 1089–1102, Oct. 1996. https://doi.org/10.1016/0020-7403(95)00112-3
20. T. H. Hyde, W. Sun, A. A. Becker, and J. A. Williams, "Creep properties and failure assessment of new and fully repaired P91 pipe welds at 923 K," vol. 218, p. 12, 2004. https://journals.sagepub.com/doi/pdf/10.1177/146442070421800305
21. T. H. Hyde and W. Sun, "Evaluation of conversion relationships for impression creep test at elevated temperatures," *Int. J. Press. Vessels Pip.*, vol. 86, no. 11, pp. 757–763, Nov. 2009. https://doi.org/10.1016/j.ijpvp.2009.07.001
22. J. Baral, J. Swaminathan, D. Chakrabarti, and R. N. Ghosh, "Effect of welding on creep damage evolution in P91B steel," *J. Nucl. Mater.*, vol. 490, pp. 333–343, Jul. 2017. https://doi.org/10.1016/j.jnucmat.2017.04.056
23. A. Khajuria, M. Akhtar, R. Kumar, J. Swaminanthan, R. Bedi, and D. K. Shukla, "*Effect of Boron Modified Microstructure on Impression Creep Behaviour of Simulated Multi-Pass Heat Affected Zone of P91 Steel*," 2018. http://dx.doi.org/10.13140/RG.2.2.34633.24164/2
24. J. Baral, J. Swaminathan, and R. N. Ghosh, "Creep behaviour of 9CrMoNbV (P91) steel having a small amount of boron," *Procedia Eng.*, vol. 55, pp. 88–92, 2013. https://doi.org/10.1016/j.proeng.2013.03.224
25. T. Shrestha, M. Basirat, I. Charit, G. P. Potirniche, and K. K. Rink, "Creep rupture behavior of grade 91 steel," *Mater. Sci. Eng. A*, vol. 565, pp. 382–391, Mar. 2013. https://doi.org/10.1016/j.msea.2012.12.031
26. F. Abe, M. Tabuchi, M. Kondo, and S. Tsukamoto, "Suppression of type IV fracture and improvement of creep strength of 9Cr steel welded joints by boron addition," *Int. J. Press. Vessels Pip.*, vol. 84, no. 1–2, pp. 44–52, Jan. 2007. https://doi.org/10.1016/j.ijpvp.2006.09.013

27. C. R. Das, S. K. Albert, J. Swaminathan, S. Raju, A. K. Bhaduri, and B. S. Murty, "Transition of crack from type IV to type II resulting from improved utilization of boron in the modified 9Cr-1Mo steel weldment," *Metall. Mater. Trans. A*, vol. 43, no. 10, pp. 3724–3741, Oct. 2012. https://doi.org/10.1007/s11661-012-1179-4
28. C. R. Das, S. K. Albert, A. K. Bhaduri, B. Raj, J. Swaminathan, and B. S. Murty, "Improvement in creep resistance of modified 9Cr-1Mo steel weldment by boron addition," *Weld. World*, vol. 56, no. 7–8, pp. 10–17, Jul. 2012. https://doi.org/10.1007/BF03321360
29. T. Watanabe, M. Tabuchi, M. Yamazaki, H. Hongo, and T. Tanabe, "Creep damage evaluation of 9Cr–1Mo–V–Nb steel welded joints showing Type IV fracture," *Int. J. Press. Vessels Pip.*, vol. 83, no. 1, pp. 63–71, Jan. 2006. https://doi.org/10.1016/j.ijpvp.2005.09.004
30. S. N. G. Chu and J. C. M. Li, "Impression creep; a new creep test," *J. Mater. Sci.*, p. 9. https://doi.org/10.1007/BF00552241
31. P. M. Sargent and M. F. Ashby, "Indentation creep," vol. 8, p. 8, 1992. https://doi.org/10.1179/mst.1992.8.7.594
32. B. Storåkers and P.-L. Larsson, "On Brinell and Boussinesq indentation of creeping solids," *J. Mech. Phys. Solids*, vol. 42, no. 2, pp. 307–332, Feb. 1994. https://doi.org/10.1016/0022-5096(94)90012-4
33. H. Y. Yu, M. A. Imam, and B. B. Rath, "Study of the deformation behaviour of homogeneous materials by impression tests," *J. Mater. Sci.*, vol. 20, no. 2, pp. 636–642, Feb. 1985. https://doi.org/10.1007/BF01026536
34. M. D. Mathew, Naveena, and D. Vijayanand, "Impression creep behavior of 316LN stainless steel," *J. Mater. Eng. Perform.*, vol. 22, no. 2, pp. 492–497, Feb. 2013. https://doi.org/10.1007/s11665-012-0290-4
35. G. Eggeler and C. Wiesner, "A numerical study of parameters controlling stress redistribution in circular notched specimens during creep," *J. Strain Anal. Eng. Des.*, vol. 28, no. 1, pp. 13–22, Jan. 1993. https://doi.org/10.1243/03093247V281013
36. T. H. Hyde, K. A. Yehia, and A. A. Becker, "Interpretation of impression creep data using a reference stress approach," *Int. J. Mech. Sci.*, vol. 35, no. 6, pp. 451–462, Jun. 1993. https://doi.org/10.1016/0020-7403(93)90035-S
37. F. Yang, J. C. M. Li, and C. W. Shih, "Computer simulation of impression creep using the hyperbolic sine stress law," *Mater. Sci. Eng. A*, vol. 201, no. 1–2, pp. 50–57, Oct. 1995. https://doi.org/10.1016/0921-5093(95)09763-5
38. V. Karthik, P. Visweswaran, A. Bhushan, D.N. Pawaskar, K.V. Kasiviswanathan, T. Jayakumar, B. Raj, "Finite element analysis of spherical indentation to study pile-up/sink-in phenomena in steels and experimental validation," *Int. J. Mech. Sci.*, vol. 54, no. 1, pp. 74–83, Jan. 2012. https://doi.org/10.1016/j.ijmecsci.2011.09.009
39. W. Lu, X. Ling, and S. Yang, "A modified reference area method to estimate creep behaviour of service-exposed Cr5Mo based on spherical indentation creep test," *Vacuum*, vol. 169, p. 108923, Nov. 2019. https://doi.org/10.1016/j.vacuum.2019.108923
40. S. Goyal, K. Laha, C. R. Das, S. Panneerselvi, and M. D. Mathew, "Finite element analysis of effect of triaxial state of stress on creep cavitation and rupture behaviour of 2.25Cr–1Mo steel," *Int. J. Mech. Sci.*, vol. 75, pp. 233–243, Oct. 2013. https://doi.org/10.1016/j.ijmecsci.2013.07.005
41. C. Pétry and G. Lindet, "Modelling creep behaviour and failure of 9Cr–0.5Mo–1.8W–VNb steel," *Int. J. Press. Vessels Pip.*, vol. 86, no. 8, pp. 486–494, Aug. 2009. https://doi.org/10.1016/j.ijpvp.2009.03.006
42. J. Naveena, G. Kumar, and M. D. Mathew, "Finite element analysis of plastic deformation during impression creep," *J. Mater. Eng. Perform.*, vol. 24, no. 4, pp. 1741–1753, Apr. 2015. https://doi.org/10.1007/s11665-014-1225-z
43. J. Betten, *Creep Mechanics*, 3rd ed. Berlin Heidelberg: Springer, 2008.
44. S. Murakami and N. Ohno, Eds., *IUTAM Symposium on Creep in Structures: Proceedings of the IUTAM Symposium Held in Nagoya*, Japan, 3–7 April 2000, vol. 86. Dordrecht: Springer Netherlands, 2001. https://doi.org/10.1007/978-94-015-9628-2
45. A. C. Fischer-Cripps, *Introduction to Contact Mechanics*, 2nd ed. New York: Springer, 2007.
46. P. Wriggers and U. Nackenhorst, Eds., *Analysis and Simulation of Contact Problems*. Berlin: Springer-Verlag, 2006.
47. B. Arivazhagan and M. Kamaraj, "Metal-cored arc welding process for joining of modified 9Cr-1Mo (P91) steel," *J. Manuf. Process.*, vol. 15, no. 4, pp. 542–548, Oct. 2013. https://doi.org/10.1016/j.jmapro.2013.07.001
48. T. H. Hyde, I. A. Jones, S. Peravali, W. Sun, J. G. Wang, and S. B. Leen, "Anisotropic creep behaviour of Bridgman Notch Specimens," *Proc. Inst. Mech. Eng. Part J. Mater. Des. Appl.*, vol. 219, no. 3, pp. 163–175, Jul. 2005. https://doi.org/10.1243/146442005X10364

49. H.-Y. Yu and J. C. M. Li, "Computer simulation of impression creep by the finite element method," *J. Mater. Sci.*, vol. 12, no. 11, pp. 2214–2222, Nov. 1977. https://doi.org/10.1007/BF00552243
50. M. G. K. Ronald and L. Huston, "*Mesh Refinement in Finite Element Analysis by Minimization of the Stiffness Matrix Trace*," Ohio: University of Cincinnati Cincinnati, USAAVSCOM Technical Report 89-C-019 AD-A219 303, Nov. 1989.
51. "Abaqus/CAE User's Manual," , p. 1174.
52. "Abaqus Theory Manual," ABAQUS, Inc., Rising Sun Mills, 166 Valley Street, Providence, RI 02909, USA, p. 1172.
53. D. R. Hayhurst and J. T. Henderson, "Creep stress redistribution in notched bars," *Int. J. Mech. Sci.*, vol. 19, no. 3, pp. 133–146, Jan. 1977. https://doi.org/10.1016/0020-7403(77)90073-X
54. Y. L. Zhang and Y. Z. Wang, "Systematic study on the competition between α-decay and spontaneous fission of superheavy nuclei," *Nucl. Phys. A*, vol. 966, pp. 102–112, Oct. 2017. https://doi.org/10.1016/j.nuclphysa.2017.06.005.

12 Channelling Deformation-Induced Electric Field Property of Polymer Hybrid Nanocomposite for Energy Harvesting

Arpana Pal Sharma, Uvais Valiyaneerilakkal, Kulwant Singh, and Saleem Khan

12.1 INTRODUCTION

In this chapter, we channel the piezoelectric property of polymer hybrid nanocomposite to use them as energy harvesting material. The development of sustainability is the key demand of the world. Energy consumption is increasing with the population and so is the exploitation of fossil fuels. Harvesting energy from nonrenewable resources has certain limitations, hence researchers are looking towards materials which require human activities for synthesizing energy harvesting devices. Harnessing energy at the nanoscale is the need of the hour. The device which can generate electricity at nanoscale and power electronic devices is called a nanogenerator (NG). There are three types of NGs: piezoelectric [1–9], triboelectric [10–15], and pyroelectric, based on the principle of their working. The piezoelectric NG, also known as PENG, was first introduced by Wang in 2006 and progress has been ongoing in the field [16].

12.2 WORKING PRINCIPLE OF PENG

Electric charge generation on the surface of the piezoelectric material is due to mechanical deformation. When the stress is applied on the surface of the material, electric charges are induced resulting in polarization of the material. The charges induced depend on the stress applied, piezoelectric coefficient of the material, dimension, or area of contact of the piezoelectric layer with electrodes. The polarized charges give rise to an electric field and develop a potential on the two sides of the electrodes. For the displacement current $D = \in E$, where ε is the electrical permittivity and E is the electric field generated due to surface charge density.

12.3 MATERIALS

The material used for the fabrication of the nanodevice plays a vital role in its performance. The generation of power and the stability of the device are major concerns. Several organic materials, such as graphene and polymer, and inorganic materials, such as ZnO and CdS [17], have been used. ZnS has been used to successfully create the Schottky contact between the electrodes and the structures in PENG. Structures like nanorods, nanowires [17, 18], and nanoflowers can significantly vary the output parameters of NGs. Polymer nanocomposite formation is an integral part of the

DOI: 10.1201/9781003359364-15

polymer nanotechnology industry. When the nanoparticles are dispersed in the polymer matrix, the efficiency of the nanogenerators is enhanced. Low dosage of the material significantly modifies the intrinsic properties of polymers. The high surface-to-volume ratio affects the mechanical, thermal, electrical, physical, and chemical properties.

Polycrystalline material is cheap, flexible, and accessible. When they are subjected to stress, the dipoles are formed, which give rise to an induced electric field. Due to the formation of dipoles in the material, an electric field is induced. These polymers crystallize when subjected to cooling and form crystallites or grains. The crystallites are microscopic and have different orientations which show various phase formations in the matter. The boundary where two crystallites meet is known as the grain boundary having various phase formation on either side.

It has been found that the polyvinylidene fluoride polymer (PVDF) and its copolymers are piezoelectric and ferroelectric in nature. PVDF is a semicrystalline polymer of (–CH_2–CF_2–,) chains and crystallizes in five phases: α, β, γ, δ, and ε [19, 20]. The non-polar phase α has a $TGTG^-$ conformation which is easily attained when the polymer is dissolved in an organic solvent, while the β phase is the polar phase in which an all-trans (TTTT) conformation is present as shown in Figure 12.1 and 12.2. Research shows that the γ, δ, and ε phases are also polar [21] due to the T_3GT_3G conformation [22]. Among these polymorphs of PVDF, the β phase exhibits spontaneous polarization and the highest piezoelectric, ferroelectric, and pyroelectric properties [23–28]. Polarization in piezoelectric material is directly proportional to the stress applied. It is the measure of dipole moment per unit volume. To enhance the piezoelectricity of PVDF, one of the hydrogens in PVDF is replaced with fluorine creating the copolymer poly(vinylidene-trifluoride) P(VDF–TrFE), which has high piezoelectric and ferroelectric properties [29, 30]. Also, the degree of β crystallization is more than PVDF. Due to the highly compact size, it has a relatively large dipole moment due to the presence of one extra electronegative fluorine, which makes it chemically more stable.

12.3.1 Neodymium Manganite ($NdMnO_3$)

Neodymium manganite ($NdMnO_3$) is a rare earth-based oxide whose structure is like gadolinium orthoferrite ($GdFeO_3$). When the perovskite structure crystals are incorporated into the polymer,

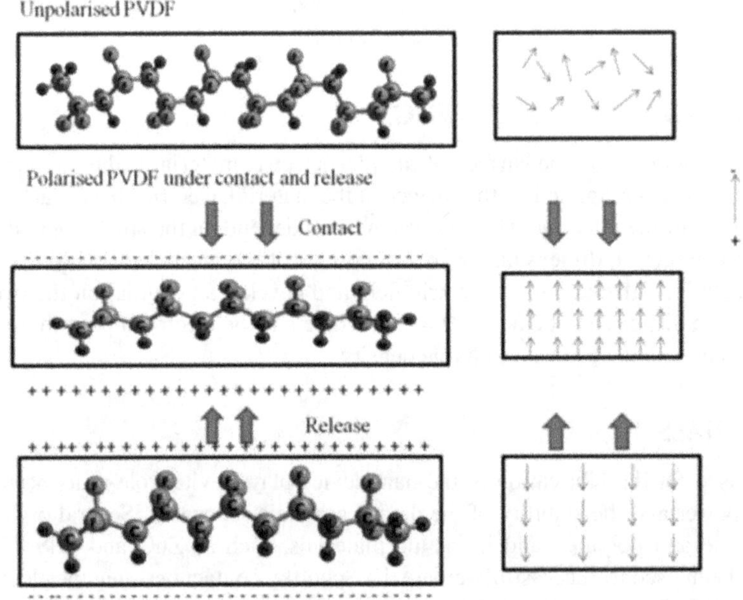

FIGURE 12.1 Dipole formation in PVDF during press and release [25].

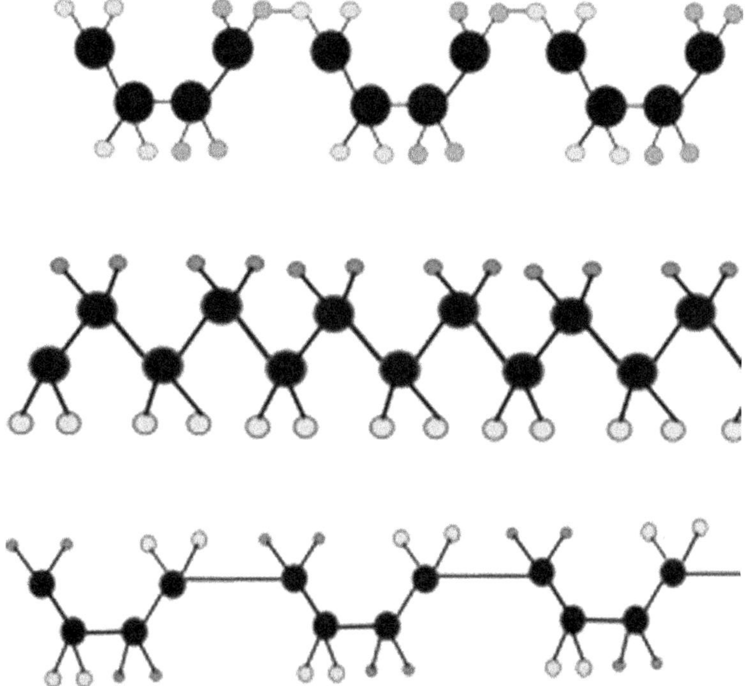

FIGURE 12.2 Alpha, beta, and gamma phases of PVDF.

electrical conductivity increases [31, 32]. Perovskite ceramics like $BaTiO_3$, $SrTiO_3$, and $PbTiO_3$ when dispersed in polymers have drawbacks like environmental issues, flexibility, and strength. The simple unit cell of $NdMnO_3$ has Nd atoms at the corners, oxygen atoms at the midpoints of the edges, and a manganese atom at the centre. It is a multiferroic composite which shows electrical and magnetic properties. The cause of multiferroicity is the structural disorder, which is induced by the rotation of MnO_6 octahedra and the decrease in Mn–O–Mn bond angles. The doping of cations, i.e., Nd and Mn, improve the electrical properties [31].

12.3.2 Reduced Graphene Oxide (rGO)

Graphene is made up of carbon atoms arranged in a hexagonal lattice one atomic layer thick. It is an excellent conductor of heat and electricity. Graphene oxide (GO) is made by treating graphene with strong oxidizers. This process adds oxygen atoms on the surface due to which GO becomes hydrophilic. The ratio of carbon and oxygen atoms in GO decides the conductivity of the material. When the oxygen atoms are removed from the surface of GO using a strong reducing agent, the final product is reduced graphene oxide (rGO). There are various techniques like laser scribing, and chemical, thermal, and electrochemical reduction processes by which GO can be converted to rGO [33].

12.4 SAMPLE PREPARATION

$NdMnO_3$ is synthesized by the sol–gel method as it is a low cost and feasible process to produce a wide range of nanostructures. This method includes hydrolysis and polycondensation of the metal alkoxide solution to form sol, which is further heated to form gel [2]. The process is shown in Figure 12.3.

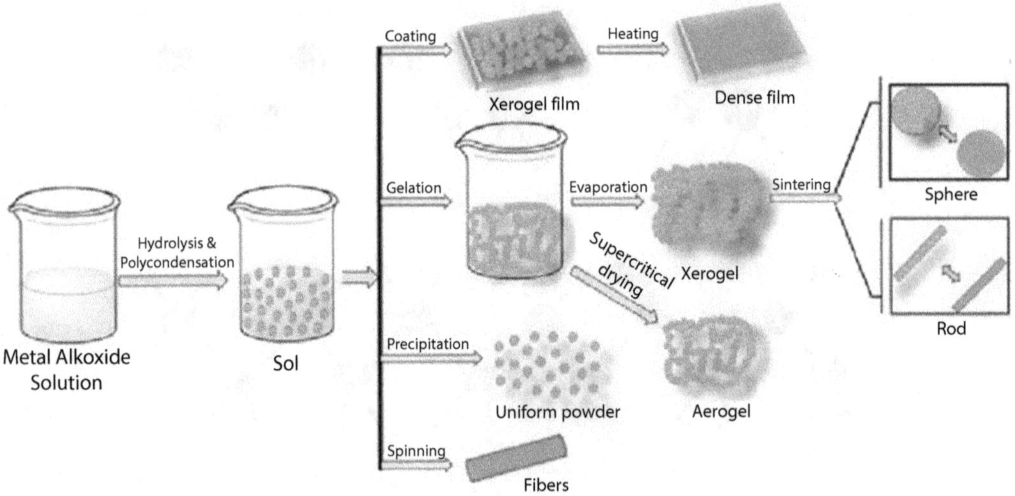

FIGURE 12.3 Pictorial representation of sol–gel process [2].

The oxide precursors Nd_2O_3 and MnO_2 were taken in stoichiometric ratio, then transformed to a colloidal solution by a chelating agent and dried to obtain a homogeneous gel. rGO is synthesized by the modified Hummers method. This method is divided into two stages. In the first stage, graphite is processed to form graphene oxide by using a strong oxidizing agent. In the second stage, graphene is prepared from graphene oxide, which is further reduced to form reduced graphene oxide.

There are various methods by which a hybrid piezoelectric nanogenerator can be fabricated. Films are used for characterization.

12.5 CHARACTERIZATION TECHNIQUES

12.5.1 Fourier Transform Infrared (FTIR) Characterization

Fourier transform infrared spectroscopy (FTIR) is a method of getting mathematical formulation from an interferogram. The principle is based on the Michelson interferometer experiment, which uses infrared rays of the electromagnetic spectrum. When infrared rays incident upon the sample, it absorbs or transmits rays and a spectrum is obtained. The spectrum is used to identify the organic and inorganic functional groups. The spectrum has two regions based on the wavenumber, e.g., from 400 to 1500 cm^{-1} is the fingerprint region, which is difficult to interpret. The presence of peaks at 840, 1280, and 1400 cm^{-1} signifies the β phase in the sample. The second region is the functional group region, from 1500 to 4000 cm^{-1}. The presence of nanofillers can be justified by the FTIR spectrum. The ATR-transmittance graph for powder samples and thin films shows the presence of the Mn-O bond. Also shifting the peak from 600 cm^{-1} to 550 cm^{-1} will show how the deformation takes place in the sample [31, 32].

12.5.2 X-Ray Diffraction (XRD) Characterization

X-ray diffraction (XRD) is a nondestructive technique by which the structure of the crystals can be determined. When a beam of X-rays is incident on the surface of the crystals, it scatters the beam in different directions. The scattered rays from the periodic array produce diffraction peaks which are caused by interference of the scattered waves. The powder diffraction method can be used to identify the orthorhombic structure of $NdMnO_3$. Also, the particle size can be deduced using the

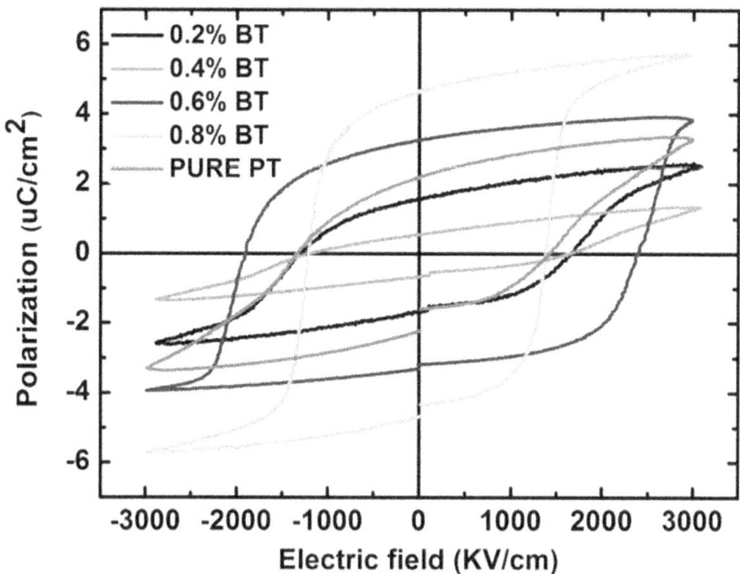

FIGURE 12.4 Polarization versus electric field analysis for PVDF and PVDF–BaTiO3 nanocomposite [30].

Debye–Scherrer formula, which is $t = \frac{0.9\lambda}{\sqrt{B_m^2 - B_s^2}\cos\theta}$, where t is the crystallite size, λ is the wavelength of X-rays, and Θ is the Bragg angle. The phase identification, degree of crystallinity, and orientation can also be determined in doped and undoped thin films of P(VDF–TrFE) [34]. Since the sample is polycrystalline with thousands of crystallites, the diffraction will take place for every set of parallel planes.

12.5.3 Field-Emission Scanning Electron Microscopy Analysis

To study the surface morphology of the thin films formed, the field-emission scanning electron microscopy (FESEM) technique is used. The formation of grain boundaries on thin films shows the crystallization of the polymer, which should be consistent with the XRD and FTIR results.

12.5.4 Polarization versus Electric Field Analysis

The impact of the electric field on the polarization of dipoles in PVDF–TrFE has been studied by various researchers. Ferroelectric material and other dielectric material properties are such that once the electric field is removed, ferroelectrics retain their polarization, whereas dielectrics do not. Like ferromagnetic materials, ferroelectric materials also show a hysteresis loop. Unlike ferromagnetic material, the hysteresis loop is due to deformation happening due to an electric field. Hysteresis properties like electric coercivity (E_C), remnant polarization (P_r), and saturation polarization (P_s) explain the impact of polarization/deformation due to an electric field. A study by [30] explains the change in the hysteresis loop of P(VDF-TrFE) polymer by the addition of $BaTiO_3$ nanoparticles into the matrix (Figure 12.4). The corresponding remnant polarization (P_r), coercive voltage (V_c), and saturation polarization (P_s) for each concentration of $BaTiO_3$ in PVDF–TrFE are given in Table 12.1.

12.6 CONCLUSION

In this chapter, we presented case studies related to the piezoelectric nanogenerator. Deformation in material induces an electric field which generates electrical signals. In this work, thin films of neat

TABLE 12.1
Ferroelectric Properties of PVDF–TrFE/$BaTiO_3$ Nanocomposites [30]

Sample	P_r ($\mu C/cm^2$)*	V_c (V)†	P_s ($\mu C/cm^2$)‡
Pure PT	2.18	142	3.27
0.2% BT	1.59	166	2.59
0.4% BT	0.54	164	1.33
0.6% BT	3.36	239	3.83
0.8% BT	4.67	138	5.72

*Remnant polarization.
†Coercive voltage.
‡Saturation polarization.

and doped P(VDF–TrFE) were proposed, and the nanofillers $NdMnO_3$ and Rgo, when added to the polymer matrix, show a significant increase in the polar active β phase. The result is confirmed by FTIR and FE-SEM results. The peaks prominent for the polar phase at 840, 1280, and 1400 cm^{-1} are seen in the ATR–FTIR graphs, and the degree of crystallinity is depicted on the surface morphology of the samples with an increase in the content of rGO.

ACKNOWLEDGEMENTS

The authors are grateful to SAIF and CAF labs, Manipal University Jaipur, for characterization of the samples.

REFERENCES

1. Elham Maghsoudi Nia, N. A. (2017). A review of walking energy harvesting using piezoelectric materials. *International Conference on Architecture and Civil Engineering (ICACE 2017)*, *291*, 012026. https://doi.org/10.1088/1757-899X/291/1/012026
2. Ani Melfa Roji, M. J. G. (2017). A retrospect on the role of piezoelectric nanogenerators in the development of green world. *RSC Advances*, *7*, 33642. https://doi.org/10.1039/c7ra05256a
3. Dipti Dhakras, S. O. (2016). High performance organic-inorganic hybrid piezo-nanogenerator via interface enhanced polarization effects for self-powered electronic systems. *Advanced Material Interfaces*, *3*, 1600492. https://doi.org/10.1002/admi.201600492
4. Hyeoung Woo Kim, A. B. (2004). Energy harvesting using a piezoelectric "cymbal" transducer in dynamic environment. *Japanese Journal of Applied Physics*, *43*(9A), 6178–6183. https://doi.org/10.1143/JJAP.43.6178
5. Long Gu, J. L. (2020). Enhancing the current density of a piezoelectric nanogenerator using a three-dimensional intercalation electrode. *Nature Communications*, *11*(1), 1030. https://doi.org/10.1038/s41467-020-14846-4
6. Hemalatha Parangusan, D. P. (2019). Toward high power generating piezoelectric nanofibers: Influence of particle size and surface electrostatic interaction of Ce–Fe2O3 and Ce–Co3O4 on PVDF. *ACS Omega*, *4*, 6312–6323. https://doi.org/10.1021/acsomega.9b00243
7. Ramesh, M. (2020). *Electrically conductive self-healing materials: Preparation, properties, and applications* (p. 817354). Elsevier. https://doi.org/10.1016/B978-0-12-817354-1.00001-6
8. Nyugen, V., Kelly, S., & Yang, R. (2017). Piezoelectric peptide-based nanogenerator enhanced by single-electrode triboelectric nanogenerator. *APL Materials*, *5*, 074108. https://doi.org/10.1063/1.4983701
9. Peng-Kun Yang, X.-L.-.Q.-C.-P. (2020). Skin-inspired electret nanogenerator with self-healing abilities. *Cell Reports Physical Science*, *1*, 100185. https://doi.org/10.1016/j.xcrp.2020.100185
10. Dzhardimalieva, G. I., Yadav, B. C., Lifintseva, T. V., & Uflyand, I. E. (2020). Polymer chemistry underpinning materials for triboelectric nanogenerators (TENG's): Recent trends. *European Polymer Journal*, *142*, 110163. https://doi.org/10.1016/j.eurpolymj.2020.110163

11. Giyoung Song, Y. K.-O.-H. (2015). Molecularly engineered surface triboelectric nanogenerator by self-assembled monolayers (METS). *Chemistry of Materials*, *27*(13), 4749–4755. https://doi.org/10.1021/acs.chemmater.5b01507
12. Kyung-Eun Byun, M.-H. L.-G.-J. (2017). Potential role of motion for enhancing maximum output energy of triboelectric nanogenerator. *APL Materials*, *5*, 074107. https://doi.org/10.1063/1.4979955
13. Simiao Niu, Y. L. (2013). Theory of sliding-mode triboelectric nanogenerators. *Advanced Materials*, *25*(43), 6184–6193. https://doi.org/10.1002/adma.201302808
14. Wanchul Seung, H.-J. Y.-H.-W. (2017). Boosting power-generating performance of triboelectric nanogenerators via artificial control of ferroelectric polarization and dielectric properties. *Advanced Energy Materials*, *7*, 1600988. https://doi.org/10.1002/aenm.201600988
15. Xin Chen, X. M. (2020). A triboelectric nanogenerator exploiting the Bernoulli effect for scavenging wind energy. *Cell Reports Physical Science*, *1*(9), 100207. https://doi.org/10.1016/j.xcrp.2020.100207
16. Zhong Lin Wang, G. Z. (2012, December). Progress in nanogenerators for portable electronics. *Materials Today*, *15*(12), 532–543. https://doi.org/10.1016/S1369-7021(13)70011-7
17. Yi-Feng Lin, J. S.-Y. (2008). Piezoelectric Nanogenerators Using CdS Nanowires. *Applied Physics Letters*, *92*, 022105.
18. Long Lin, C.-H. L. (2011). High output nanogenerator based on assembly of GaN nanowires. *Nanotechnology*, *22*, 475401. http://dx.doi.org/10.1088/0957-4484/22/47/475401
19. Lovinger, A. J. (1982). Poly(vinylidene fluoride). In D. Bassett (Ed.), *Developments in crystalline polymers* (Vol. 33, p. 195). London: Springer.
20. Subash Cherumannil Karumuthil, S. P. (2019). Electrospun Poly(vinylidene fluoride-trifluoroethylene)-based polymer nanocomposite fibers for piezoelectric nanogenerators. *ACS Applied Materials and Interfaces*, *11*, 40180–40188. http://dx.doi.org/10.1021/acsami.9b17788
21. Tashiro, K. (1995). Crystal structure and phase transitions of PVDF and related copolymers. In H. S. Nalwa & M. Dekker (Eds.), *Ferroelectric polymers* (p. 63). Scientific Research Publishing, New York.
22. Gaur, A. K. (2017). Induced Piezoelectricity in poly(vinylidene fluoride) hybrid as efficient harvester. *Chemistry Select 2*, 8278–8287. http://dx.doi.org/10.1002/slct.201701780,
23. Ueberschlag, P. (2001). PVDF piezoelectric polymer. *Sensor Review*, *21*, 118–125.
24. Sreenidhi Prabha Rajeev, S. C. (2017). Nanoscale static voltage generation and its surface potential decay using scanning probe microscopy. *Micro & Nano Letters*, 1–6. https://doi.org/10.1049/mnl.2017.0236
25. Sreenidhi Prabha Rajeev, S. C. (2018). α- & β-crystalline phases in polyvinylidene fluoride as tribo-piezo active layer for nanoenergy harvester. *High Performance Polymers*, 1–15. https://doi.org/10.1177/0954008318796141
26. Siuk Cheon, H. K.-J.-W. (2017). High-performance triboelectric nanogenerators based on electrospun polyvinylidene fluoride–silver nanowire composite nanofibers. *Advanced Functional Materials*, 1703778. https://doi.org/10.1002/adfm.201703778
27. Zagni, N., Pavan, P., & Alam, M. A. (2020, October 15). A memory window expression to evaluate the endurance of ferroelectric FETs. *Applied Physics Letters*, *117*(15), 152901. https://doi.org/10.1063/5.0021081
28. Zerun Yin, B. T. (2019). Characterization and application of PVDF and its copolymer films prepared by spin-coating and Langmuir–Blodgett method. *Polymers*, *11*(12), 2033. https://doi.org/10.3390/polym11122033
29. Uvais Valiyaneerilakkal, S. C. (2021). High-performance P(VDF–TrFE)/BaTiO3 nanocomposite based ferroelectric field effect transistor (FeFET) for memory and switching applications. *Nano Select*, 1–7. https://doi.org/10.1002/nano.202100081
30. Valiyaneerilakkal, U., & Varghese, S. (2013). Poly (vinylidene fluoride-trifluoroethylene)/barium titanate nanocomposite for ferroelectric nonvolatile memory devices. *AIP Advances*, *3*, 042131. https://doi.org/10.1063/1.4802980
31. Somvanshi, A., & Husain, S. (2017). Study of structural, dielectric and optical properties of NdMnO3. In *2nd International Conference on Condensed Matter and Applied Physics (ICC 2017)* (pp. 030242-1–030242-4). America: AIP Publishing. https://doi.org/10.1063/1.5032577
32. Assirey, E. A. (2019). Perovskite synthesis, properties and their related biochemical and industrial application. *Saudi Pharmaceutical Journal*, *27*(6), 817–829. https://doi.org/10.1016/j.jsps.2019.05.003
33. Karan, S. K., Mandal, D., & Khatua, B. B. (2015). Self-powered flexible Fe-doped RGO/PVDF nanocomposite: An excellent material for a piezoelectric energy harvester. *Nanoscale*, *7*(24), 10655–10666. https://doi.org/10.1039/C5NR02067K
34. Dipankar Mandal, K. J. (2012). Simple synthesis of palladium nanoparticles, β-phase formation, and the control of chain and dipole orientations in palladium-doped poly(vinylidene fluoride) thin films. *Langmuir*, *28*, 10310–10317. https://doi.org/10.1021/la300983x

13 Atomistic Modelling and Molecular Dynamics Simulation for Elastic Deformation in Nanocomposites

Neetu Chaudhary, Jash Rana, and Mithilesh K. Dikshit

13.1 INTRODUCTION

In order to build incredibly high-performance aircraft components during the Cold War, high-performance composites were developed. The use of composites has increased significantly due to their high-performance characteristics, low-cost manufacturing processes, and increased level of user confidence. These days, composites are employed in a variety of industries, including sports, medicine, and the military. Strength, stiffness, fatigue life, corrosion and wear resistance, temperature-dependent behaviour, thermal insulation, thermal conductivity, and acoustic insulation are some of the features of materials that can be enhanced by creating composite materials. Composites have weight and economic advantages in addition to the aforementioned qualities.

In numerous defence applications, such as anticorrosive coatings, polymer membranes, laminate structures, and electronic devices, chemically cross-linked thermosetting polymer networks are used. These networks can be found in composite protective materials, nanocomposites, anticorrosive coatings, and more. Epoxy is one of the resins that is utilized most frequently in the polymer matrix of fibre-reinforced composites. In this context, epoxy is typically cross-linked with amine curing agents in order to produce a network that is conducive to the adhesion of fibres. In addition to that, epoxy is frequently utilized in the production of adhesives, varnishes, and paints. Epoxy-based networks can range from glassy structural resins to loosely cross-linked elastomers and gels as a result of the high degree of mechanical tuning that is possible with these networks. Binders made of flexible polymer networks are essential components of fibres, energy materials, biological tissue surrogates, robotics, and wearable electronics. The rising interest in biopolymer networks can be attributed to the improved energy-dissipating qualities that these networks possess. These properties are essential for the construction of armour. Carbon nanotubes (CNTs) have demonstrated that they have a significant amount of potential for enhancing the material properties of polymers [1]. In addition to the enhancement of electrical and thermal conductivity, the enhancement of mechanical properties is also of particular interest. This stems from the recent discovery that combining CNTs with brittle materials may impart some of the desirable mechanical properties of the CNTs to the resulting composites, making CNTs an excellent candidate for use as reinforcement in polymeric materials [2, 3].

13.2 EPOXY RESINS

Epoxy resin is a molecule that has been defined as having more than one epoxide group within its structure. Thermosetting polymers have various applications such as adhesives, high-performance

 DOI: 10.1201/9781003359364-16

coatings, and encapsulating materials. These resins have excellent electrical properties; low shrinkage; good adhesion to a wide variety of metals; resistance to moisture, thermal shock, and mechanical shock; and low shrinkage. The viscosity, epoxide equivalent weight, and molecular weight of epoxy resins are the three most important properties.

13.2.1 Types of Epoxy Resins

Epoxy resins can be divided into two primary groups: glycidyl epoxy resins and non-glycidyl epoxy resins. Glycidyl epoxies can be further broken down into the categories of glycidyl ether, glycidyl ester, and glycidyl amine. Epoxy resins that do not contain glycidyl are classified as either aliphatic or cycloaliphatic depending on their chemical structure. Glycidyl epoxies can be made by condensing an appropriate dihydroxy compound, a dibasic acid, or a diamine with epichlorohydrin in a reaction. Through the process of peroxidation of olefinic double bonds, non-glycidyl epoxies can be produced. Glycidyl-ether epoxies, such as diglycidyl ether of bisphenol A (DGEBA) and novolac epoxy resins, are the types of epoxies that are utilized most frequently.

13.2.1.1 DGEBA

In the presence of a basic catalyst, bisphenol A and epichlorohydrin can be reacted to produce the epoxy resin known as DGEBA. This epoxy resin is typically used in commercial applications. An atomistic model of DGEBA is shown in Figure 13.1.

The number of repeating units, also referred to as the degree of polymerization, has a direct bearing on the characteristics of the DGEBA resins. The stoichiometry of the synthesis reaction determines how many repeating units there will be in total. The degree of polymerization in many commercial items normally ranges from 0 to 25.

13.2.1.2 Epoxy Novolac Resins

Epoxy novolac resins are glycidyl ethers that are derived from phenolic novolac resins. For the production of phenolic novolac resin, an excessive amount of phenols are reacted with formaldehyde in the presence of an acidic catalyst. To create epoxy novolac resins, phenolic novolac resin is reacted with epichlorohydrin in the presence of sodium hydroxide, which acts as a catalyst. In most cases, multiple epoxide groups can be found within epoxy novolac resins. The number of epoxide groups that are present in a molecule is determined by the number of phenolic hydroxyl groups that are present in the starting phenolic novolac resin, the degree to which these groups react, and the degree to which low-molecular-weight species are polymerized during the synthesis process. Because of the multiple epoxide groups, these resins are able to achieve a high cross-link density, which results in superior resistance to temperature extremes, chemicals, and solvents. Because of their superior performance at elevated temperatures, excellent moldability, mechanical properties, superior electrical properties, and resistance to heat and humidity, epoxy novolac resins are frequently used in

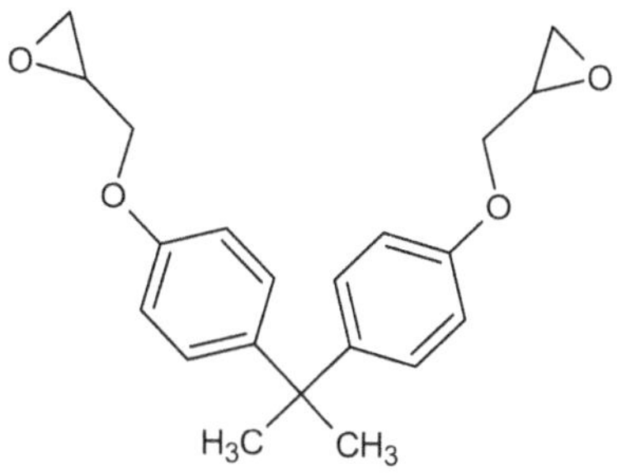

O O O O H_3C CH_3

FIGURE 13.1 Line representation of DGEBA.

the formulation of moulding compounds for microelectronics packaging. This is due to the fact that these resins are easy to mould and have excellent mechanical properties.

13.2.1.3 Curing of Epoxy Resins

During the curing process, an epoxy resin undergoes a chemical reaction in which the epoxide groups in the resin combine with a curing agent (also known as a hardener) to produce a highly interconnected and three-dimensional network. It is necessary to cure the epoxy resin with a hardener in order to transform it into a material that is hard, infusible, and rigid. This can be accomplished by using an epoxy mould. Epoxy resins undergo a rapid and simple curing process at virtually any temperature between 5 and 150 degrees Celsius, depending on the curing agent that is used. Epoxy resins can be cured with a large number of different types of agents, each of which has its own unique set of properties and applications. Epoxies are cured with a variety of different agents, the most common of which are amines, polyamides, phenolic resins, anhydrides, isocyanates, and polymercaptans. The molecular structure of the hardener determines not only the cure kinetics but also the glass transition temperature (T_g) of a system after it has been cured. When selecting a resin and hardener to use, it is important to consider the application, the process that will be used, and the properties that are desired. The stoichiometry of the epoxy-hardener system is another factor that influences the properties of the material after it has cured. The structure can be changed by employing a wide variety of hardeners, both in terms of the types used and the amounts used, which have the tendency to control the cross-link density. For the curing of epoxy resins, amine and phenolic resin-based curing agents are frequently used as curing agents.

13.2.1.3.1 Amine-Based Curing Agents

In order to cure epoxy, amines are the curing agents that are utilized most frequently. Epoxy is known to provoke strong chemical reactions in primary and secondary amines. Tertiary amines are typically put to use in the curing process in the capacity of a catalyst, also known as an accelerator. The use of an excessive amount of catalysts can result in faster curing; however, this typically comes at the expense of the catalyst's working life as well as its thermal stability. The mechanism of action of the catalysts has an impact on the final polymer's physicochemical properties.

13.2.1.3.2 Phenolic Novolac Resins

When cured with phenolic hardener, epoxy resins provide exceptional adhesion, strength, chemical resistance, and flame resistance. Because they have low water absorption, excellent heat resistance, and good electrical resistance, phenolic novolac-cured epoxy systems are the ones that are most commonly used for encapsulation. In order for complete healing to take place, an accelerator is required.

13.3 CARBON NANOTUBES

Iijima, a Japanese scientist, made a discovery in 1991 regarding a new type of carbon known as graphitic tubules. He also made a hypothesis regarding the possible structure of these tubes, which

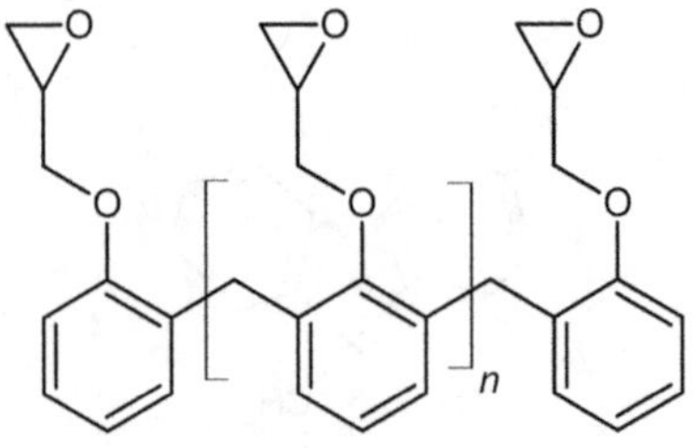

FIGURE 13.2 Line structure of epoxy novolac resin.

turned out to be correct. Iijima was the first person to discover and describe carbon nanotubes (CNTs), and because the walls of these CNTs were constructed from two to fifty layers of graphene, the term "multiwalled carbon nanotube" (MWCNT) was coined to describe them. Single-walled carbon nanotubes (SWCNTs) were later found to have been discovered [4].

CNTs can be conceptualized as sheets of graphite that are rolled into seamless cylinders. These carbon nanotubes are referred to as single-walled or multiwalled carbon nanotubes (Figure 13.3), depending on the number of sheets they contain. The distance between two graphite sheets in MWCNTs is typically 0.340 nm. This is a slightly greater distance than the distance between two consecutive graphite sheets in a single crystal graphite, which is 0.335 nm. This difference may be explained by the specific cylindrical geometry of MWCNTs [5].

Metallic and semiconducting properties have both been seen in carbon nanotubes [6]. The fact that carbon nanotubes have a hollow interior, in addition to their unique mechanical qualities, metallic and semiconducting properties, and other desirable characteristics, makes them particularly appealing to scientific researchers. On the other hand, due to the chemical structure of carbon nanotubes, it is not possible for these nanotubes to react with or form complexes with a large number of different chemical components. As a consequence of this, the functionalization of carbon nanotubes has developed into an extremely significant area of chemistry. CNTs that are functional and have been supplemented with other physicochemical features bring up new opportunities for the technological applications of these nanomaterials [7]. Different strategies for the synthesis of nanotubes and their functionalization have been developed over a couple of decades. The synthesis, purification, and characterization of nanotubes' chemical and physical properties used to be the primary focus of research. However, as a result of recent advancements in nanotube synthesis, the emphasis of research has shifted to nanotube functionalization and applications.

13.4 MOLECULAR DYNAMICS

In molecular dynamics (MD) simulation techniques, the system's governing equations of motion are numerically solved to determine the time evolution of a number of atoms. The forces between the atoms are often calculated using either quantum mechanical ab initio calculations or classical interatomic potentials. The terms classical and quantum molecular dynamics, respectively, are used to describe these two main strategies. If a simulation includes atoms that move at energies more than a few hundred electron volts (eV), then one must also include a description of the energy loss of the energetic atoms to electrons in the medium, commonly known as electronic stopping power [6]. In a typical MD simulation, the only physical inputs that are provided are those pertaining to the

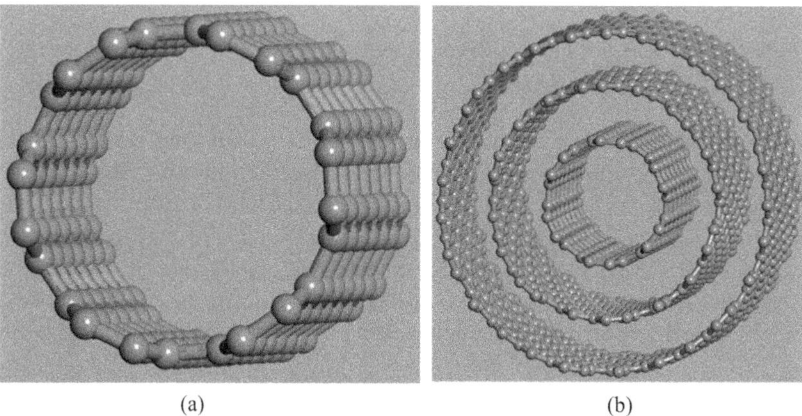

FIGURE 13.3 (a) Atomistic model of single-walled carbon nanotubes and (b) atomistic model of multiwalled carbon nanotubes.

interatomic forces and the electrical stopping power. Therefore, provided that both are thoroughly understood, MD approaches have the potential to be utilized to examine ion implantation processes in a manner that is realistic by simulating the time development of the atomic collision processes that are induced during the process of implantation. The investigation of structures with the lowest possible potential for energy loss can yield a great deal of insightful knowledge. On the other hand, these structures are just models; in reality, molecules are flexible structures that are prone to thermal motion. Molecular dynamics is a technique that can be utilized to simulate the thermal motion of a structure as a function of time. In this technique, the motion of the structure is driven by the forces that are acting on the atoms. Studies of the time evolution of a variety of systems at nonzero temperatures, such as biological molecules, polymers, or catalytic materials, in a variety of states, such as crystals, aqueous solutions, or in the gas phase, can benefit greatly from the use of dynamics simulations. These simulations can be extremely helpful. The most common uses of molecular dynamics include conducting conformational searches, producing statistical ensembles, and researching the motions of molecules.

13.4.1 General Principles of MD Simulations

The time evolution of a system of atoms is computed by numerically solving the equations of motion in molecular dynamics simulations. These simulations model molecular dynamics. As a result of the fact that the motion of every single atom that is involved in a collision cascade can be tracked in MD simulations, these simulations provide the most realistic method for analyzing the formation of defects during ion implantation. One of the very first applications of molecular dynamics methods [8] was the simulation of collision sequences in metals. Ever since then, MD simulations have been used to study a wide range of phenomena that occur in collision cascades. The force acting on each atom in the system is first calculated as part of the molecular dynamics modelling procedure. A suitable algorithm is used to solve the system's equations of motion [9, 10]. The result reveals how the atom locations, velocities, and accelerations change during a limited amount of time [21]. The solution of the equations of motion will be more precise the lower the time step. The simulation proceeds by recalculating the forces in the new positions after the atoms have been relocated. Typically, an orthogonal simulation cell contains the atoms that are part of the calculation. Both classical and quantum mechanical calculations can be used to determine the forces governing the simulation. Despite recent promising developments in tight-binding molecular dynamics (TBMD) techniques, quantum molecular dynamics techniques still take too long to simulate complete collision cascades [11]. The force F_i acting on an ith atom in the system is estimated using traditional MD techniques and the Newtonian formalism as [21]:

$$\bar{F}_i(\bar{r}_i) = \sum_{j \neq i} \bar{F}_{ij}(\bar{r}_{ij}) = -\sum_{j \neq i} \nabla V_{ij}(r_{ij}) \tag{13.1}$$

where r_{ij} is the distance between the atoms, F_{ij} is the force operating between the atoms, and $V_{ij}(r_{ij})$ is a potential energy function. All atoms whose interaction with the ith atom is stronger than a threshold value are included in the sum over j [7]. The cut-off radius for the interactions is often defined in order to execute the threshold in practise. Repulsive contact can be thought of as essentially Coulombic at very close proximity between the nuclei. The electron clouds separate the nuclei from one another when they are farther apart. In order to characterize the repulsive potential, the Coulombic repulsion between nuclei can be multiplied by a screening function $\phi(r)$ [21]:

$$V(r) = \frac{1}{4\pi \epsilon_0} \frac{Z_1 Z_2}{r} \phi(r) \tag{13.2}$$

The interacting nuclei's charges are Z_1 and Z_2, and their separation from one another is measured by r. Over the years, many various repulsive potentials and screening functions have been put forth, some derived from theoretical calculations and others from semi-empirical research. The repulsive potential provided by Ziegler, Biersack, and Littmark, sometimes known as the ZBL repulsive potential, is one that is frequently employed. It was created by applying a universal screening function to theoretically predict potentials for a wide range of atom pair combinations [6]. Self-consistent total energy calculations utilizing density-functional theory and the local density approximation (LDA) for electronic exchange and correlation can yield a more precise repulsive potential. This method determines the total energy as a function of the distance between the atoms in a dimer bond, which includes both the electronic component and the inter-nuclear Coulomb repulsion. The interactions regulating the system must be described in a way that takes into consideration the bonds' angular dependence when one wants to describe a solid with covalent bonds. Utilizing potentials that depend on the locations of three or more atoms can do this [12, 13].

13.4.2 Case Study: DGEBA–DETDA

To further understand the atomistic models, it is useful to first study the synthesis of DGEBA and its DETDA-cured polymeric forms. DGEBA monomer production from bisphenol A and an epoxide is shown in Figure 13.4.

Only the para-para structural form is possible for the reactant bisphenol A and the final product, DGEBA monomer: the other two forms are ortho-ortho and para-ortho. In order to create a polyether product during the synthesis, a freshly generated DGEBA monomer can also engage in the polyaddition reaction with bisphenol A, as shown in Figure 13.1. This reaction can be accelerated by temperature and a suitable catalyst [14]. Figure 13.5 depicts the interaction of an amine group of diethyl toluene diamine (DETDA) with an epoxide group of the DGEBA monomer.

Due to the fact that there are still three hydrogen atoms between the two amine groups, the final product may react with up to three additional epoxide groups and begin further cross-linking, which will eventually result in the production of a networked polymer. In conclusion, decisions must be made about the following while creating the molecular models of cross-linked DGEBA–DETDA:

- DGEBA monomers' ortho/para structural conformation
- Percentage of polyether species in existence
- Cross-linking intensity and molecular topology

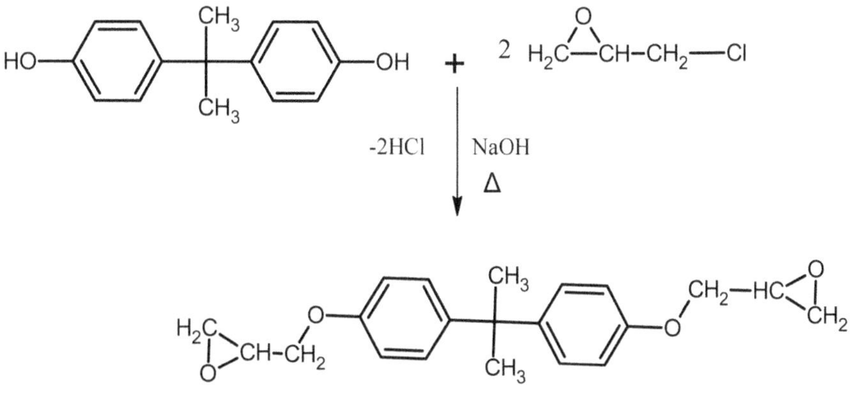

FIGURE 13.4 Synthesis of DGEBA monomer from bisphenol A and an epoxide.

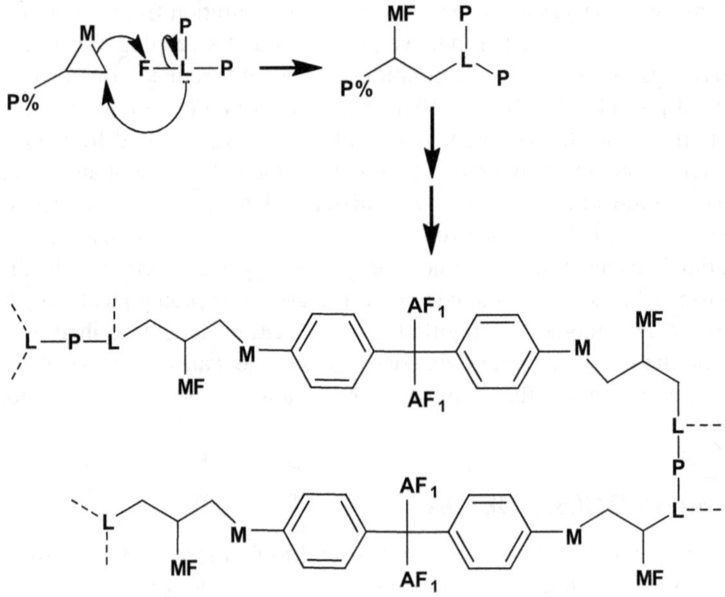

FIGURE 13.5 Reaction mechanism of DGEBA monomer with amine group.

13.5 ATOMISTIC MODEL BUILDING AND MD SIMULATIONS

13.5.1 Atomistic Model Building

For running the MD simulations based on the aforementioned literature, a representative composite model is required. Epoxy DGEBA and hardener DETDA are taken into account for model creation and MD simulations in the current chapter. According to product information from a producer of DGEBA–DETDA resins, the usual weight ratio of the final resin's monomer and curing agent is 100/26.4 [15]. According to this, there are 2.15 regular DGEBA monomers for every DETDA molecule. Figure 13.6 depicts the atomistic triangular model that has been developed within that restriction. Seven DGEBA monomers and three DETDA molecules were used to create the model [16].

The creation of an atomistic model of a variety of amorphous systems begins with the selection of the required number and kind of epoxy molecules, which is followed by the stochastic packing of those molecules into a periodic simulation volume shown in Figure 13.7a. The amorphous cell with periodic boundary conditions is represented by Figure 13.7a, which is also referred to as the representative composites. To improve the cell properties and make them more like those of the real composite, the amorphous cell must be simulated using MD simulations.

13.5.2 MD Simulations

A smart algorithm that combines the conjugate gradient algorithm and the steepest descent method for energy minimization has been employed after creating the initial coordinates for the amorphous cell. It was used for 10,000 steps, then four stages of MD (every 10,000 steps), with velocity-rescaling applied at each time step using progressively bigger time steps (up to 1 fs time step size) to carefully remove any residual initial poor contacts. The initially produced amorphous cell has flaws including vacuum gaps and randomly arranged molecules. A geometric relaxation was applied to the structure to remove this kind of stability. Forcite dynamic simulation has been used to attain the necessary density of the amorphous cell after the structure has been relaxed. At a pressure of 0.5 GPa, isobaric–isothermal (NPT) ensembles have been conducted. The duration of the simulation

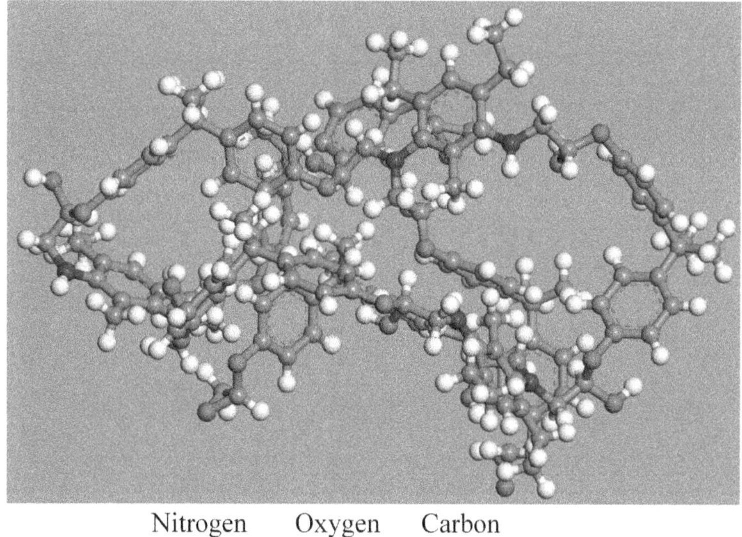

FIGURE 13.6 Triangular atomistic model of DGEBA cured with DETDA.

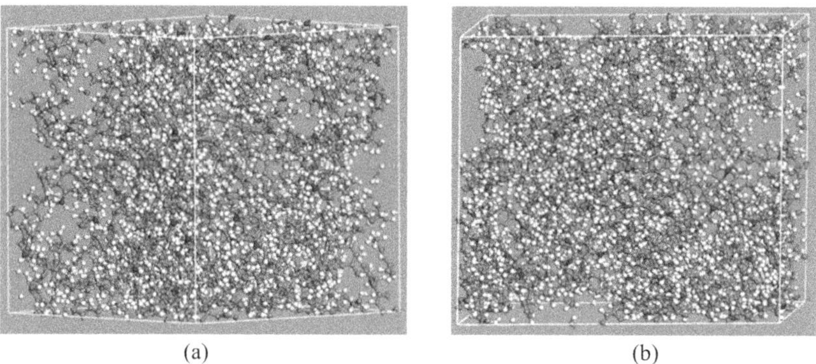

FIGURE 13.7 (a) Amorphous cell before simulation and (b) amorphous cell before simulation.

was 5 ps, with a time step of 1 fs [16]. For further processing, it is crucial to ensure the model's stability. The model is shown to be stable enough to move forward according to the dynamic energy simulation, one of the criteria. The model's kinetic energy should be lower, and the total, non-bond and potential energies shouldn't fluctuate during the course of the simulation. Additionally, at ambient temperature, the density of the simulated amorphous cell must be comparable to that of the actual composite. According to Figure 13.8, the density in this instance must be 1.2 g/cm^3 [17].

13.5.3 Reinforcement of SWCNTs in Epoxy Matrix

In general, CNTs are thought to have great potential as fillers to enhance the material properties of polymers. The stiffness and strength of CNTs are incredibly great [18, 19]. CNTs have a high heat conductivity and an electrical conductivity that can either be semiconducting or conducting due to their graphitic structure or having the appearance of metal. In the epoxy (DGEBA) matrix, the armchair SWCNT (10, 10) is strengthened. By establishing the isosurface in a periodic boundary cell, the DGEBA monomers cured with DETDA are closely packed outside the SWCNT. The epoxy's molar ratio is assumed to be 10 [16].

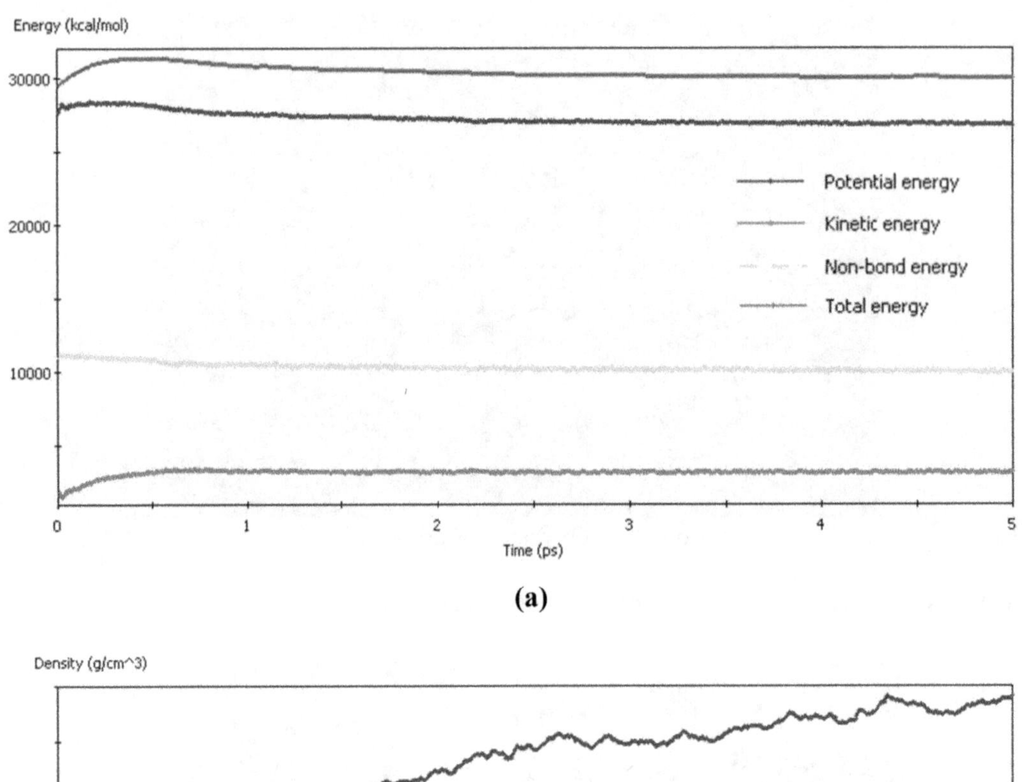

(a)

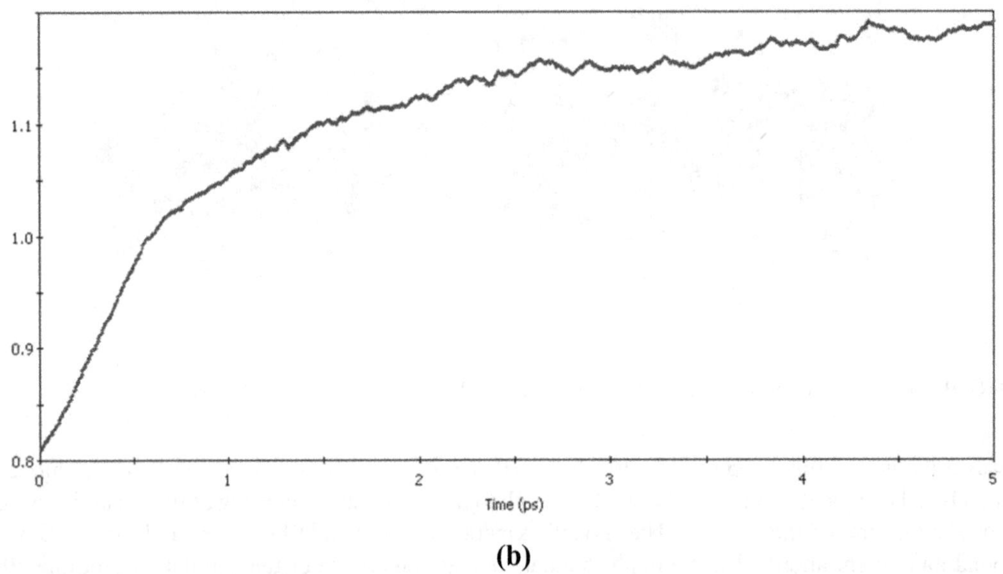

(b)

FIGURE 13.8 (a) Simulated energies and (b) density at room temperature.

The volume fraction (f_{NT}) of the reinforced CNT may be obtained using the following relation:

$$f_{NT} = \frac{\grave{A}\left(R_{NT} + \left(h_{eq}/2\right)\right)^2}{A_{cell}} \tag{13.3}$$

A_{cell} is the cross-sectional area of the unit cell, and h_{eq} is the van der Waals equilibrium separation distance between the CNTs and the epoxy matrix. The interaction between epoxy and CNTs at the interface determines the van der Waals separation distance. In the illustration, the van der Waals separation is depicted.

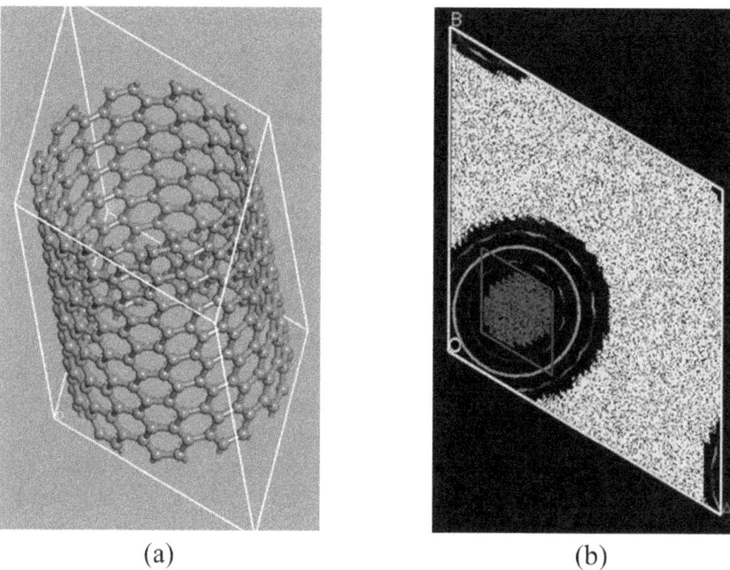

FIGURE 13.9 (a) Armchair (10, 10) CNT and Connolly and (b) isosurface and segregated view of CNT.

13.5.4 Calculation of Elastic Properties of Isotropic Materials

COMPASS (Condensed-phase Optimized Molecular Potential for Atomistic Simulation Studies), the first ab initio-based force field to have been parameterized using extensive data for molecules in the condensed phase, is the force field widely used for MD simulations. It is based on earlier class II CFF9 and PCFF force fields. The integration time step plays an important factor in MD simulations, which was set to 1 fs. Van der Waals and Coulomb forces were combined using an atom-based technique (the cut-off, spline width, and buffer width were 9.5, 1.0, and 0.5 Å, respectively). Boltzmann distributions were used to generate random initial velocities. The temperature was set to 298 K, and Berendsen control was used with collision frequency steps every 690 fs. The Parrinello

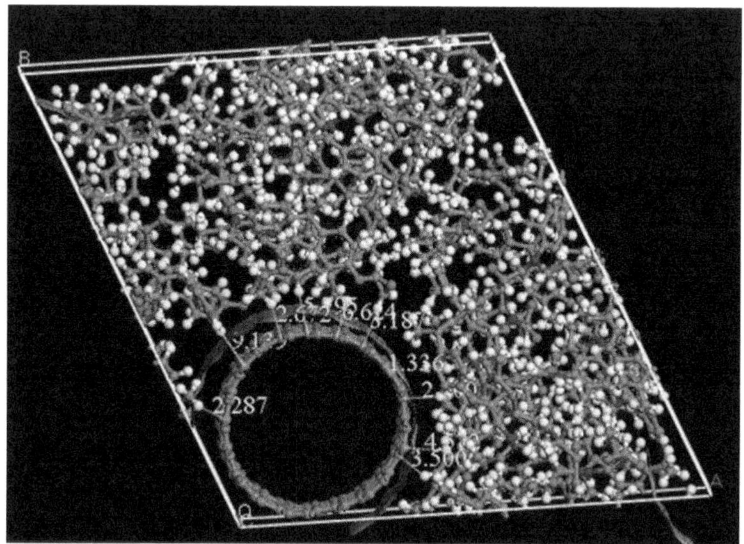

FIGURE 13.10 Van der Waals separation of CNT and epoxy.

method was used to manage pressure. The MD trajectory was created by storing the coordinates and velocities of all atoms at each 100 fs. The cell mass was set to 20.00 a.m.u (i.e., output a frame every 100 fs). The analysis module of the system estimated the static elastic properties, and the data from the relevant trajectory was then analyzed to determine the values of strain under various stressors. Five increments of 2.0% each were applied in the longitudinal direction during the MD simulation [16]. For every unit of applied deformation, a uniform strain was imparted to the entire specimen. In other words, each increment was achieved by uniformly enlarging the cell along its length and rescaling the atoms' new coordinates to meet the new dimensions. With 20 ps (20,000 steps) and a 10 ps NVT ensemble, the system was minimized after each 1.5% increment of strain was added (10,000 steps). The atoms were permitted to equilibrate within the revised MD cell dimensions during minimization and NVT. For temperatures of 298 K and 350 K, respectively, Young's modulus of the pure DGEBA/DETDA was predicted to be 2.429 GPa and 2.184 GPa, while the obtained Poisson's ratios were 0.42 and 0.40, respectively [16]. Young's modulus at room temperature was calculated to be extremely close to the observed value of 2.71 GPa [20]. As can be seen from Figure 13.11a and b, Young's modulus shows a clear temperature influence. As the simulation temperature rises, Young's modulus falls (Figure 13.11a). After being reinforced with CNT, the DGEBA's Young's modulus and stress–strain relationship improved, as illustrated in Figure 13.11c and d. According to Figure 13.11c, at a very low deformation of 0.003 strain amplitude, the Young's modulus increased to 13.1 GPa.

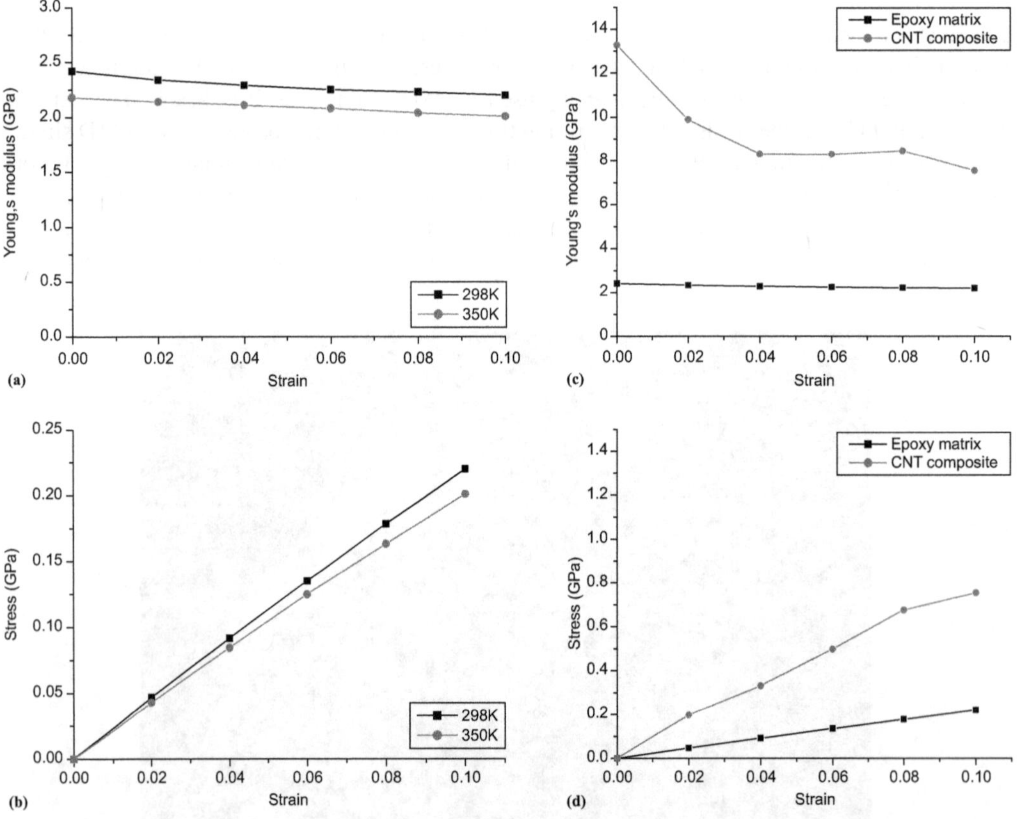

FIGURE 13.11 (a) Young's modulus versus strain and (b) stress versus strain for pure DGEBA/DETDA at 298 and 350 K. (c) Young's modulus versus strain and (d) stress versus strain of CNT-reinforced DGEBA/DETDA at 298 K.

13.6 CONCLUSIONS

The MD simulation is an effective technique for predicting the many properties of the various types of materials, including mechanical, thermal, electrical, etc. The mechanical characteristics and deformation of the CNT-reinforced DGEBA–DETDA nanocomposite were examined in the current chapter. The investigation's findings support the following judgement: the two composite concepts under consideration are CNT-reinforced composite and epoxy matrix. Young's modulus consistently decreases as temperature increases, according to the results of the MD simulations for the epoxy DGEBA–DETDA. Young's moduli (slopes of the stress–strain curves) of epoxy DGEBA treated with DETDA decrease as strain levels increase. The MD simulation findings demonstrate that Young's modulus (slope of the stress–strain curve) of epoxy or reinforced composite matrix materials diminishes as strain levels rise (from 0.003 to 0.1). The CNT-reinforced epoxy composite is five times stiffer than the pure epoxy matrix at any given strain level, which is the most important discovery from the MD data.

REFERENCES

1. Fitria, T. N. (2021). QuillBot as an online tool: Students' alternative in paraphrasing and rewriting of English writing. *Englisia: Journal of Language, Education, and Humanities*, *9*(1), 183–196.
2. Zhu, R., Pan, E., & Roy, A. K. (2007). Molecular dynamics study of the stress–strain behavior of carbon-nanotube reinforced Epon 862 composites. *Materials Science and Engineering: A*, *447*(1–2), 51–57.
3. Mokashi, V. V., Qian, D., & Liu, Y. (2007). A study on the tensile response and fracture in carbon nanotube-based composites using molecular mechanics. *Composites science and technology*, *67*(3–4), 530–540.
4. Iijima, S., & Ichihashi, T. (1993). Single-shell carbon nanotubes of 1-nm diameter. *Nature*, 363(6430), 603–605.
5. Ajayan, P. M. (1999). Nanotubes from carbon. *Chemical Reviews*, *99*(7), 1787–1800.
6. Dresselhaus, G., Dresselhaus, M. S., & Saito, R. (1998). *Physical properties of carbon nanotubes*. World Scientific.
7. Balasubramanian, K., & Burghard, M. (2005). Chemically functionalized carbon nanotubes. *Small*, *1*(2), 180–192.
8. Pagona, G., & Tagmatarchis, N. (2006). Carbon nanotubes: Materials for medicinal chemistry and biotechnological applications. *Current Medicinal Chemistry*, *13*(15), 1789–1798.
9. Lide, D. R. (2008). *CRC Handbook of chemistry and physics*, 88th ed. The Chemical Rubber Company.
10. Ebbesen, T. W., & Ajayan, P. M. (1992). Large-scale synthesis of carbon nanotubes. *Nature*, *358*(6383), 220–222.
11. Wong, E. W., Sheehan, P. E., & Lieber, C. M. (1997). Nanobeam mechanics: Elasticity, strength, and toughness of nanorods and nanotubes. *Science*, *277*(5334), 1971–1975.
12. Nieto, A., Agarwal, A., Lahiri, D., Bisht, A., & Bakshi, S. R. (2021). *Carbon nanotubes: Reinforced metal matrix composites*. CRC Press.
13. Yin, K., Zou, D., Zhong, J., & Xu, D. (2007). A new method for calculation of elastic properties of anisotropic material by constant pressure molecular dynamics. *Computational Materials Science*, *38*(3), 538–542.
14. Kobayashi, M., Sanda, F., & Endo, T. (1999). Application of phosphonium ylides to latent catalysts for polyaddition of bisphenol A diglycidyl ether with bisphenol A: Model system of epoxy-novolac resin. *Macromolecules*, *32*(15), 4751–4756.
15. Tack, J. L., & Ford, D. M. (2008). Thermodynamic and mechanical properties of epoxy resin DGEBF crosslinked with DETDA by molecular dynamics. *Journal of Molecular Graphics and Modelling*, *26*(8), 1269–1275.
16. Dikshit, M. K., & Engle, P. E. (2018). Investigation of mechanical properties of CNT reinforced epoxy nanocomposite: A molecular dynamic simulations. *Materials Physics & Mechanics*, *37*(1), 7–15.
17. Dikshit, M. K., Nair, G. S., Pathak, V. K., & Srivastava, A. K. (2019). Investigation of Mechanical Properties of Graphene Reinforced Epoxy Nanocomposite Using Molecular Dynamics. *Materials Physics & Mechanics*, *42*(2), 224–233.
18. Thostenson, E. T., & Chou, T. W. (2003). On the elastic properties of carbon nanotube-based composites: Modelling and characterization. *Journal of Physics D: Applied Physics*, *36*(5), 573.

19. Yu, M. F., Files, B. S., Arepalli, S., & Ruoff, R. S. (2000). Tensile loading of ropes of single wall carbon nanotubes and their mechanical properties. *Physical Review Letters*, *84*(24), 5552.
20. Qi, B., Zhang, Q. X., Bannister, M., & Mai, Y. W. (2006). Investigation of the mechanical properties of DGEBA-based epoxy resin with nanoclay additives. *Composite Structures*, *75*(1–4), 514–519.
21. Nordlund, K. H. (1997). Molecular dynamics simulations of atomic collisions for ion irradiation experiments, Acta Polytechnica Scandinavica, Applied Physics Series, vol. 202, p. 1, 1995, PhD thesis, University of Helsinki.

14 Stress–Strain Response of Graphene-Reinforced Aluminium Composite

A Molecular Dynamics Study

Ashish Kumar Srivastava, Ramanpreet Singh, and Vimal Kumar Pathak

14.1 INTRODUCTION

Graphene sheet (GS), a plate-like crystal lattice structure with sp^2-hybridized carbon atoms, possesses unusually high mechanical/electrical properties [1]. Theoretically, GSs have been explored for approximately 80 years [2, 3]. Though GSs were known as an inherent part of 3-dimensional (3D) engineering materials, the presumption was that the GSs do not last in the free state and are assumed to be unstable because of the curved structures of carbon atoms, e.g., fullerenes and nanotubes [4]. In 2004, Novoselov et al. [5] experimentally developed naturally happening GSs and unfolded a new branch of nanoscience. The strength and stiffness of GSs are found nearly the same as that of sp^2-hybridized, 1D structure carbon atoms, i.e., carbon nanotubes (CNTs) [6]. Whereas the larger contact surface area of GSs with matrix material and lesser production cost than CNTs makes GSs a favoured reinforcing nanofiller [7].

Characterization of nanoscale GS-embedded composites is a challenging task analytically and experimentally. Molecular dynamics (MD) simulations play an important role in the prediction of mechanical properties of nanocomposites [8–10]. The 2D-plate-like structure of GSs has enhanced mechanical connectivity with matrix owing to the enlarged interfacial area of GS and provides the improved stiffness and strength properties of the composite materials in comparison to CNT composites [11–13]. Sharma et al. [14] made a comparative analysis between the elastic properties of CNT and GS-reinforced copper composite. It was observed that for a reinforcement of 8% by volume fraction, the elastic modulus of CNT and GS composite is enhanced by 13% and 18%, respectively. The shear stiffness of GS and GS-embedded composites are also analyzed in the literature employing MD computations. In 2011, Min and Aluru [15] examined the temperature and chirality-dependent shear stiffness properties of GS by MD computation and established that the shear stiffness of GS reduces after 800 K. It was also found that the zigzag GS offers marginally enhanced shear modulus in comparison to armchair GS. Sakhaee-Pour [16] employed molecular mechanics to explore the shear stiffness of GS and found likely results, but the diversion between the shear stiffness of GS in the armchair and zigzag direction was not significant. Fereidoon et al. [9] used Materials Studio software to estimate the elastic and shear stiffness of GS-embedded rutile composite.

Research articles on the mechanical properties of the GS-embedded aluminium (Al) composite are very rare. Therefore, we are reporting here our molecular dynamics simulation on the GS-embedded Al composite. We characterize the composite material in terms of stiffness and

DOI: 10.1201/9781003359364-17

strength properties under tensile and shear loading using MD simulations. This study aims to demonstrate the effect of GS reinforcement on the stiffness and strength properties of GS-embedded Al composite.

14.2 MODELLING OF COMPOSITE REPRESENTATIVE UNIT CELL (RUC)

14.2.1 Simulation Model

The MD-based model of the GS-Al composite representative unit cell (RUC) is represented in Figure 14.1. To represent the nanoscale geometry within the GS-embedded Al composite, periodic boundary conditions (PBCs) are utilized in x, y and z orientations. The simulation model has the dimensions 50.5 Å × 50.5 Å × 28.35 Å (i.e., $v_f = 12\%$).

14.2.2 Potentials

Inter-atomic potentials utilized in the present article are:

1. Embedded-atom method (EAM)
2. Adaptive intermolecular reactive empirical bond order (AIREBO) potential
3. Lennard-Jones (LJ) 12-6 potential for Al-Al, C-C and Al-C interactions, respectively.

The EAM potential, incorporating many-body potential, proposed by Winey et al. [17] is given as:

$$E = \sum_{i \neq j} \phi\left(r_{ij}\right) + F\left(\bar{\rho}\right),$$

$$\bar{\rho} = \sum_{i \neq j} f\left(r_{ij}\right), \tag{14.1}$$

where $\phi\left(r_{ij}\right)$ and $F\left(\bar{\rho}\right)$ represent the pairwise potential and embedding energy for an atom, offering constraints that are used in determining $\phi\left(r_{ij}\right)$ and $f\left(r_{ij}\right)$ by experimental data. Expressions for $\phi\left(r_{ij}\right)$ and $f\left(r_{ij}\right)$ are given as:

$$\phi(r) = A\left(r - r_m\right)^4 \left(1 + d_1 r + d_2 r^2 + d_3 r^3 + d_4 r^4 + d_5 r^5 + d_6 r^6 + d_7 r^7\right), \tag{14.2}$$

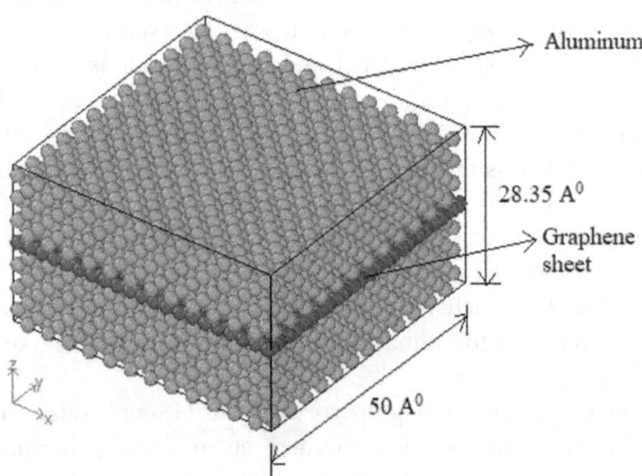

FIGURE 14.1 Atomic model for MD simulations, showing the GS-embedded Al composite, $v_f = 12\ \%$.

and

$$f(r)=\exp\left[-\frac{(r-b_1)^2}{b_2}\right](r-r_m-1)$$

$$=\frac{1}{2}\{1+\cos[\pi(r-r_m+1)]\}\exp\left[-\frac{(r-b_1)^2}{b_2}\right](r_m-1<r<r_m) \tag{14.3}$$

$$=0\,(r>r_m),$$

where the cut-off radius r_m is in angstroms. The splines are employed to offer qualitatively right $f(r)$ and $\varphi(r)$ functions for small r values that are scarcely tried by Monte Carlo computations. A, b_1, b_2, d_1, d_2, d_3, d_4, d_5, d_6 and d_7 are the parameters for EAM function.

To mimic the carbon–carbon atomic bond in GS, AIREBO [18] potential consisting of long-range van der Waals (vdW) and torsion interactions [19] is selected due to its preciseness to describe the mechanical responses of CNT/GS. The AIREBO potential possesses reactive empirical bond order (REBO) and the torsional interaction potential LJ and is represented as:

$$E_C=\frac{1}{2}\sum_i\sum_{j\neq 1}\left[E_{ij}^{REBO}+E_{ij}^{LJ}+\sum_{k\neq i}\sum_{l\neq i,j,k}E_{kijl}^{TORSION}\right] \tag{14.4}$$

where

E_{ij}^{REBO}= covalent bonding forces

E_{ij}^{LJ}= intermolecular interaction

$E_{kijl}^{TORSION}$= dihedral angle effect

To confirm the cut-off radius of AIREBO potential and also to avoid truncation errors, the cut-off radius is considered 3 $\sigma_{Al\text{-}C}$. For the interaction between the carbon atoms of GS and Al atoms, vdW bonding is taken and represented by LJ 12-6 potential [18] represented as:

$$E_{Al-C}=4\varepsilon_{Al-C}\left[\left(\frac{\sigma}{r}\right)^{12}-\left(\frac{\sigma}{r}\right)^{6}\right] \tag{14.5}$$

ε and σ show the deepness of the potential curve and the radius (r) at zero inter-atomic potential. LJ parameters for carbon and aluminium atom interaction, the widely used Lorentz–Berthelot (LB) principle is employed [20, 21] and represented as:

$$\sigma=\frac{\sigma_C+\sigma_{Al}}{2} \tag{14.6}$$

$$\varepsilon=\sqrt{\varepsilon_C\varepsilon_{Al}} \tag{14.7}$$

Representing values of these LB potential parameters are given in Table 14.1.

Thus, the combined energy of the RUC is given as [23]:

$$E_{total}=E_{Al}^{EAM}+E_C^{AIREBO}+E_{LJ}^{Al-C} \tag{14.8}$$

Simulations for MD models are accomplished by utilizing the open-source code LAMMPS (31 May 2019 version) originated by Sandia National Laboratories [24], employing its distributed memory message-passing parallelism characteristic.

TABLE 14.1
Lennard-Jones (LJ) Potential Parameters for Atomic Bonding [22]

Atom	Al-Al	C-C	Al-C (Using LB Rule)
ε (eV)	0.4157	2.8440	0.03438
σ (Å)	2.6200	3.4000	3.0100

ε: depth of the LJ 12-6 well; σ: vdW radius; cut-off distance: $3 \times \sigma_{Al\text{-}C} = 9.03$ Å.

14.2.3 Simulation Conditions

Ahead of performing the numerical experiments on the RUC, to predict the stress–strain response of the GS-reinforced Al composite, it is important to equilibrate RUC models at 300 K by utilizing the micro-canonical ensemble. Time-step considered for the equilibrium process is 0.5 femtosecond (fs) [25–27]. The equilibrium curve representing the convergence of the potential energy of RUC at 300 K is shown in Figure 14.2. After the equilibration process of total potential energy, the RUC is subjugated to the isothermal–isobaric ensemble, or the NPT ensemble, to hold the pressure of RUC at zero bar to remove the internal stresses. In the LAMMPS code, the axes of composite RUC were set in the directions of the rectangular restricting box by employing the region "block" command. Thenceforth, to estimate the elastic and shear properties of composite RUC, the RUC was applied with a constant strain rate of 0.05/picosecond (ps) in the zigzag/armchair direction of GS.

The stress in the RUC is computed by the negative pressure in the armchair/zigzag directions by utilizing the following equation [28]:

$$P_{ij} = \frac{\sum_{k}^{N} m_k v_{ki} v_{kj}}{V} + \frac{\sum_{k}^{N} r_{ki} f_{kj}}{V} \tag{14.9}$$

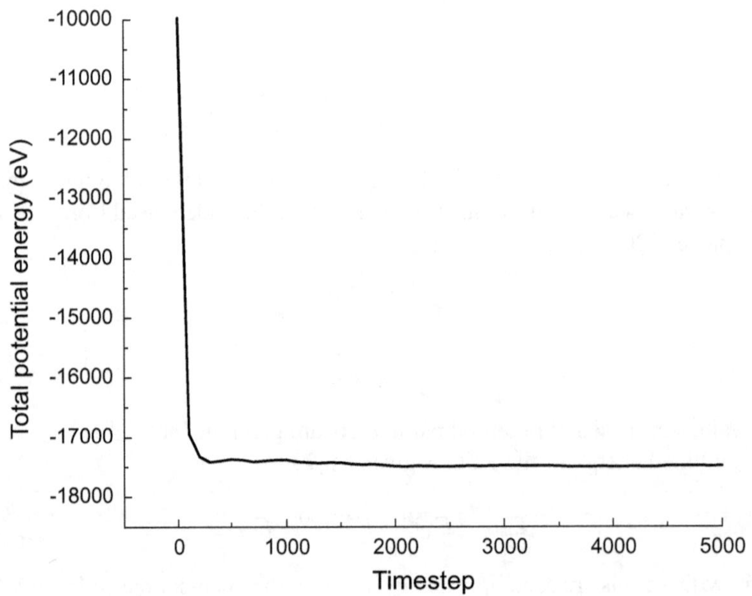

FIGURE 14.2 Total potential energy equilibrium curve for GS-Al RUC.

where v_{ki} and v_{kj} are the instantaneous velocities in the x and y directions, respectively, of the k_{th} atom; m_k represents the mass of the k_{th} atom; r_{ki} represents the instantaneous position vector of the k_{th} atom; f_{kj} shows the total force working on the k_{th} atom; and V is the system volume. Also, the longitudinal strain is computed as:

$$\varepsilon_{x/y} = \left(L_{x/y} - L_{x0/y0}\right) / L_{x0/y0}, \tag{14.10}$$

where the strain in the loading direction is $\varepsilon_{x/y}$, and the terminal and initial length of RUC are $L_{x/y}$ and $L_{x0/y0}$, respectively, in the armchair/zigzag direction.

The unitless shear strain was defined as offset/length, where length is the RUC length perpendicular to the shear direction (e.g., y box length for x-y deformation) and offset is the displacement distance in the shear direction (e.g., x direction for x-y deformation) from the unstrained RUC.

14.3 VALIDATION

To check the correctness of the molecular dynamics simulation approach employed in the present study, the obtained stress–strain response of pure Al RUC is equated with the results reported in the literature [22]. It is found from Figure 14.3 that the stress–strain response obtained in the current study is in good agreement with the published strengthening curve in the literature.

14.4 RESULTS AND DISCUSSION

After equating the MD simulation approach utilized in the present study, a similar methodology is employed to predict the stress–strain behaviour of GS-reinforced Al RUC. The elastic and shear moduli of GS-reinforced RUCs are estimated from the linear region of the stress–strain response using regression analysis. In the range of $0 < \varepsilon \leq 0.02$, the behaviour of the stress–strain curve is determined to be nearly linear [22]. The tensile stress–strain response of GS-Al composite RUC in the armchair and zigzag direction and shear stress–strain response in the -xy, -xz and -yz planes are compared in Figures 14.4 and 14.5, respectively. Corresponding values of tensile and shear mechanical properties for pure Al and composite are also compared in Table 14.2. It can be seen

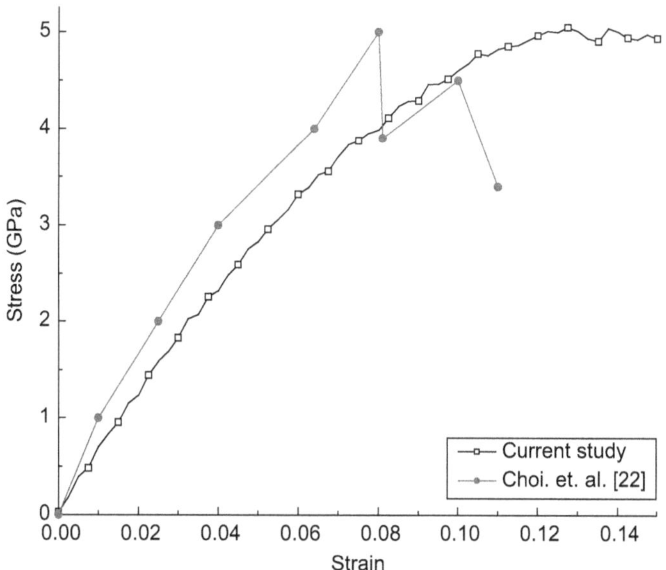

FIGURE 14.3 Stress–strain response of pristine Al strain rate 0.05/picosecond.

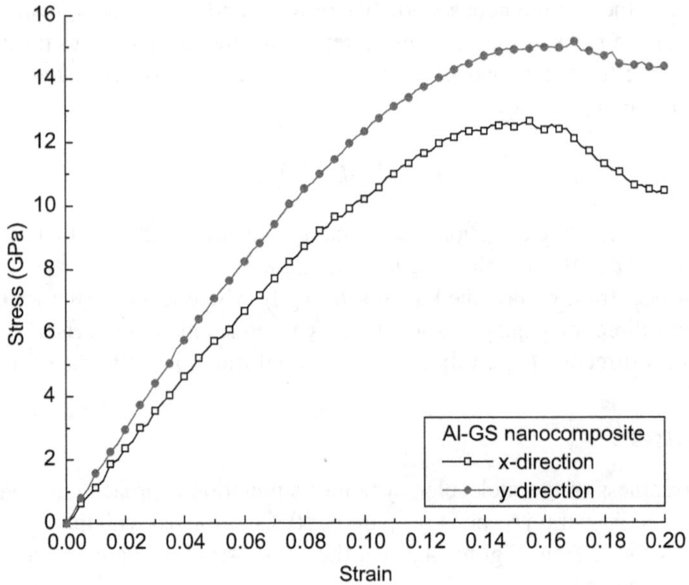

FIGURE 14.4 Stress–strain response of GS-reinforced Al composite.

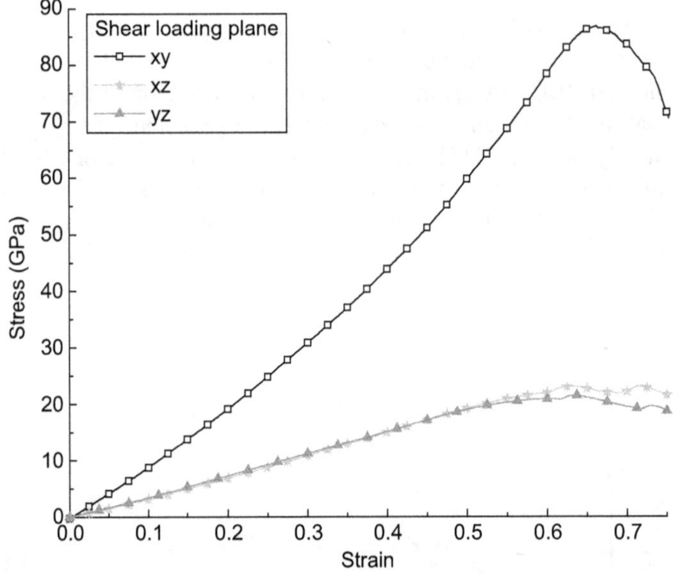

FIGURE 14.5 Shear stress–strain response of GS-reinforced Al composite.

from Figure 14.4 that the elastic modulus and strength of the composite are higher in the zigzag direction than armchair direction. The elastic modulus of RUC, for 12% of GS reinforcement, is enhanced by 76.04% and 121.87% in the armchair and zigzag directions, respectively. The tensile strengths of GS-embedded composite in armchair and zigzag directions are reported to be 12.1 and 14.2 GPa, respectively. This finding is in good agreement with the results published by Mortazi et al. [29]. It can be observed from Figure 14.4 that in both directions, i.e., armchair and zigzag, GSs carry higher loads than the Al matrix, resulting in better tensile modulus and strength properties.

The impression of the shear strain on the stress–strain response of GS-embedded RUC is shown in Figure 14.5. It can be observed from Figure 14.5 that the GS-embedded Al RUC takes the highest

TABLE 14.2
Elastic and Shear Moduli and Strength (in GPa) of GS-Embedded Al Composite

Matrix Material	Tensile Property in Armchair Direction	Tensile Property in Zigzag Direction	Shear Property in *x-y* Plane	Shear Property in *x-z* Plane	Shear Property in *sy-z* Plane
Pure Al: Modulus $\left(E_{Al}\right)$	64.9200	64.9200	24.9692	24.9692	24.9692
Pure Al: Strength $\left(S_{Al}\right)$ $\dot{\varepsilon}^{cr}$	4.9000	4.9000	10.7768	10.7768	10.7768
RUC: Modulus $\left(G_{RUC}\right)$	114.2857	142.8571	98.3471	40.8759	40.2679
RUC: Strength $\left(Sh_{RUC}\right)$	12.1000	14.2000	88.5672	21.9681	19.5573

load in in-plane shear loading (i.e., *x-y* direction) condition. Application of fixed in-plane shear strain, γ_{xy}, in the armchair and zigzag direction on the GS-embedded RUC results in shear modulus, G_{xy}, of 98.3471 GPa. On the other hand, if the RUC is applied with out-of-plane shear strains, γ_{xz} and γ_{yz}, the predicted shear moduli, G_{xz} and G_{yz}, are 40.8759 and 40.2679 GPa, respectively. Thus, it can be concluded from Table 14.2 that the chirality of GS affects both the tensile and shear moduli of composite RUC. The deviation in the shear moduli for out-of-plane shear strain loading conditions (i.e., γ_{xz} and γ_{yz}) on the RUC is not found to be significant. This finding is concurrent to the published findings for GS-embedded composite that 2D periodically distributed GSs embedded composite extend weak out-of-plane mechanical properties [30, 31].

14.5 CONCLUSION

In the current study, MD computer simulations are performed to mechanically characterize the GS-embedded Al RUC under tensile and shear loading conditions. It is found that the overall tensile/shear moduli and strength of nanocomposite RUC increased considerably with the addition of GS as a nanofiller. Based on the present study, the following important conclusions can be drawn:

- GSs can considerably enhance the tensile modulus and strength of Al composites. This shows that there is an effectual load transfer among high-strength GSs and Al matrix.
- The chirality of GS impresses the tensile/shear moduli and strength of the RUC. The maximum elastic/shear moduli are reported when the RUC is loaded in the zigzag direction only.
- The GS-Al composite RUC carries maximum load if it is applied with in-plane shear strain loading, and therefore in-plane shear moduli (i.e., G_{xy})are considerably more than out-of-plane shear moduli (i.e., G_{yz}).

ACKNOWLEDGEMENTS

The present research study was performed by employing the computation research facilities available at the Multiscale Simulation Research Center (MSRC), Manipal University Jaipur (MUJ), India.

REFERENCES

1. A.K. Geim, K.S. Novoselov, The rise of graphene, Nature. 6 (2007) 183–191.
2. P.R. Wallace, The band theory of graphite, Phys. Rev. 71 (1947) 622–634.

3. M.P. Sharma, L.G. Johnson, J.W. McClure, Diamagnetism of graphite, Phys. Lett. A. 44 (1973) 445–446.
4. E. Fradkin, Critical behavior of disordered degenerate semiconductors. I. Models, symmetries, and formalism, Phys. Rev. B. 33 (1986) 3257–3262.
5. K.S. Novoselov, A.K. Geim, S.V. Morozov, D. Jiang, Y. Zhang, S.V. Dubonos, I.V. Grigorieva, A.A. Firsov, Electric field effect in atomically thin carbon films, Science 306 (2004) 666–669.
6. I. Ovid'ko, Mechanical Properties of Graphene, Rev. Adv. Mater. Sci. 34 (2013) 1–11.
7. T. Kuilla, S. Bhadra, D. Yao, N.H. Kim, S. Bose, J.H. Lee, Recent advances in graphene based polymer composites, Prog. Polym. Sci. 35 (2010) 1350–1375.
8. S. Bashirvand, A. Montazeri, New aspects on the metal reinforcement by carbon nanofillers: A molecular dynamics study, Mater. Des. 91 (2016) 306–313.
9. A. Fereidoon, S. Aleaghaee, I. Taraghi, Mechanical properties of hybrid graphene/TiO2 (rutile) nanocomposite: A molecular dynamics simulation, Comput. Mater. Sci. 102 (2015) 220–227.
10. A. Singh, D. Kumar, Effect of temperature on elastic properties of CNT-polyethylene nanocomposite and its interface using MD simulations, J. Mol. Model. 24 (2018) 177–178.
11. S. Stankovich, D.A. Dikin, G.H.B. Dommett, K.M. Kohlhaas, E.J. Zimney, E.A. Stach, R.D. Piner, S.T. Nguyen, R.S. Ruoff, Graphene-based composite materials., Nature. 442 (2006) 282–286. h
12. H. Kim, C.W. Macosko, Processing-property relationships of polycarbonate/graphene composites, Polymer (Guildf). 50 (2009) 3797–3809.
13. A. Srivastava, D. Kumar, A continuum model to study interphase effects on elastic properties of CNT/GS-nanocomposite, Mater. Res. Express. 4 (2017) 025036.
14. S. Sharma, P. Kumar, R. Chandra, Mechanical and thermal properties of graphene–carbon nanotube-reinforced metal matrix composites: A molecular dynamics study, J. Compos. Mater. 51 (2017) 3299–3313.
15. K. Min, N.R. Aluru, Mechanical properties of graphene under shear deformation, Appl. Phys. Lett. 98 (2011) 1–3.
16. A. Sakhaee-Pour, Elastic properties of single-layered graphene sheet, Solid State Commun. 149 (2009) 91–95.
17. J.M. Winey, A. Kubota, Y.M. Gupta, Thermodynamic approach to determine accurate potentials for molecular dynamics simulations: Thermoelastic response of aluminum, Model. Simul. Mater. Sci. Eng. 18 (2010).
18. B.K. Choi, G.H. Yoon, S. Lee, Molecular dynamics studies of CNT-reinforced aluminum composites under uniaxial tensile loading, Compos. Part B. 91 (2016) 119–125.
19. S.K. Deb Nath, S.-G. Kim, Study of the nanomechanics of CNTs under tension by molecular dynamics simulation using different potentials, ISRN Condens. Matter Phys. 2014 (2014) 1–18.
20. A. White, *Intermolecular Potentials of Mixed Systems: Testing the Lorentz-Berthelot Mixing Rules with Ab Initio Calculations*, Australia: Defense Science and Technology Organization 2000.
21. A. Srivastava, D. Kumar, Post-buckling behaviour of carbon-nanotube-reinforced nanocomposite plate, Sadhana - Acad. Proc. Eng. Sci. 42 (2017) 129–141.
22. B.K. Choi, G.H. Yoon, S. Lee, Molecular dynamics studies of CNT-reinforced aluminum composites under uniaxial tensile loading, Compos. Part B Eng. 91 (2016) 119–125.
23. Y. Rong, H.P. He, L. Zhang, N. Li, Y.C. Zhu, Molecular dynamics studies on the strengthening mechanism of Al matrix composites reinforced by graphene nanoplatelets, Comput. Mater. Sci. 153 (2018) 48–56.
24. https://lammps.sandia.gov.
25. K. Gordiz, S.M. Vaez Allaei, F. Kowsary, Thermal rectification in multi-walled carbon nanotubes: A molecular dynamics study, Appl. Phys. Lett. 99 (2011) 251901.
26. X. Chen, L. Zhang, M. Zheng, C. Park, X. Wang, C. Ke, Quantitative nanomechanical characterization of the van der Waals interfaces between carbon nanotubes and epoxy, Carbon 82 (2015) 214–228.
27. M. Kirca, Design and analysis of sandwiched fullerene-graphene composites using molecular dynamics simulations, Compos. Part B Eng. 79 (2015) 513–520.
28. S. Subedi, L.S. Morrissey, S.M. Handrigan, S. Nakhla, The effect of many-body potential type and parameterisation on the accuracy of predicting mechanical properties of aluminium using molecular dynamics, Mol. Simul. 46 (2020) 271–278.
29. B. Mortazavi, Y. Rémond, S. Ahzi, V. Toniazzo, Thickness and chirality effects on tensile behavior of few-layer graphene by molecular dynamics simulations, Comput. Mater. Sci. 53 (2012) 298–302.
30. A. Srivastava, D. Kumar, A continuum model to study interphase effects on elastic properties of CNT/GS-nanocomposite, Mater. Res. Express. 4 (2017) 025036.
31. A. Kumar Srivastava, V. Kumar Pathak, Elastic properties of graphene-reinforced aluminum nanocomposite: Effects of temperature, stacked, and perforated graphene, Proc. Inst. Mech. Eng. Part L J. Mater. Des. Appl. 234 (2020) 1218–1227.

Part 4

Progress in Experimental Approaches

Case Studies

15 Physics of Deformation Behaviour in Nickel-Based Super Alloys

Monika Srivastava, Sangeeta Rawal, and Anoop Kumar Mukhopadhyay

15.1 INTRODUCTION

The superalloys are alloys of nickel (Ni) or nickel–iron (Ni-Fe) or cobalt (Co). They have some characteristic features. They maintain structural stability up to elevated temperatures. They preserve their mechanical integrity up to high temperatures at high stress. They can also preserve their surface condition when exposed to high stress and temperature under severe environments. We shall confine our discussion to the Ni-based superalloys.

As such, the Ni-based superalloys are very complex. These superalloys find maximum usage as the hottest part components. About 50% by weight of advanced aircraft engines comprises these Ni-based superalloys. The major component of all such alloys is the face-centred cubic (FCC) Ni matrix. This matrix can preserve its structural stability up to temperatures as high as about 70% to 80% of its melting temperature. The big advantage of Ni-based superalloys is that they can be strengthened by direct means as well as by indirect means. The choice to strengthen the surface stability is opted by using chromium (Cr) or aluminium (Al) as the alloying element in the Ni-based superalloys. The application possibilities of Ni-based superalloys are very rich. These alloys are mostly used for making gas turbine engine components in military aircraft. Even the gas turbine engine components in civilian aircraft use these superalloys.

These superalloys also find many applications in power generation industries. Another major area of application for Ni-based superalloys is marine propulsion, which involves a corrosive atmosphere. Almost all types of oil and gas industries deal with components made from Ni-based superalloys. This information suggests the suitability of these superalloys for harsh environments. Further, Ni-based superalloys are the most preferred material for spacecraft engine components.

Furthermore, these alloys are in heavy demand for applications in nuclear reactors and submarines. They also constitute vessels used in the chemical processing industries involving highly oxidizing and/or nitriding atmospheres. They are also the most preferred material for application in tubes of heat exchangers. Today there are many generations of Ni-based superalloys available. However, such choices are dependent on the target applications.

All such applications in high temperature, high stress and harsh conditions need the mechanical integrity of Ni-based superalloys to be preserved. This is what makes the study of deformation and fracture physics of Ni-based superalloys a very attractive domain for research and further development. In what follows, the basic aspects of Ni-based superalloys are touched upon. Next, the various types of Ni-based superalloys and their application prospects are briefly discussed. In addition, focus is put on the important mechanical properties of these materials and the common techniques utilized to measure them. Moreover, a detailed survey of the research on deformation and fracture reported in the literature is presented. This also helps in understanding the research gaps. Further, based on these research gaps, a map is charted for future research that needs to be conducted for

DOI: 10.1201/9781003359364-19

further development of these materials. Finally, the key aspects of the knowledge gathered are summarized.

15.2 BASIC ASPECTS OF NI-BASED SUPERALLOYS

Superalloys are heat-resisting alloys that are based on group VIII A elements such as nickel, cobalt, and iron, and they exhibit excellent resistance to mechanical and chemical degradation. They exhibit stability in the structure, surface and property even at elevated temperatures, high stress and severe environments. This description accurately defines existing superalloys and also includes the properties required to include new materials such as titanium and aluminium [1, 2]. The structure of superalloys typically has a matrix with an austenitic FCC crystal structure. A superalloy is a metallic alloy which is resistant to high temperatures, creep, oxidation and corrosion, which is why it is well suited for many applications such as gas turbines, aeroengine applications, coal conversion plants and chemical process industries. The superalloys are categorized as:

1) Nickel-based superalloys
2) Iron-based superalloys
3) Cobalt-based superalloys

Figure 15.1 represents the classification of the superalloys [3]. The iron-based superalloys evolved from austenitic stainless steels, combining a closed-packed FCC matrix with both, precipitate forming elements and solid solution hardening elements. Austenitic alloys are more difficult to forge than the martensitic grades [4]. However, cobalt-based superalloys are not as strong as nickel-based superalloys, but they possess their strength up to higher temperatures. They derive their strength largely from a distribution of refractory metal carbides.

Nickel-based superalloys are the most complex but most widely used for all the applications [2, 3, 5]. They currently constitute over 50% of the weight of advanced aircraft engines. Nickel as an alloy has a high FCC phase stability and the capability to strengthen by indirect or direct means. Further, chromium and aluminium ions improve the surface stability of nickel. Nickel-based superalloys possess high-temperature mechanical properties much better than other superalloys. It is very difficult to cut nickel-based superalloys as they have high values of hardness, strength and work hardening [6].

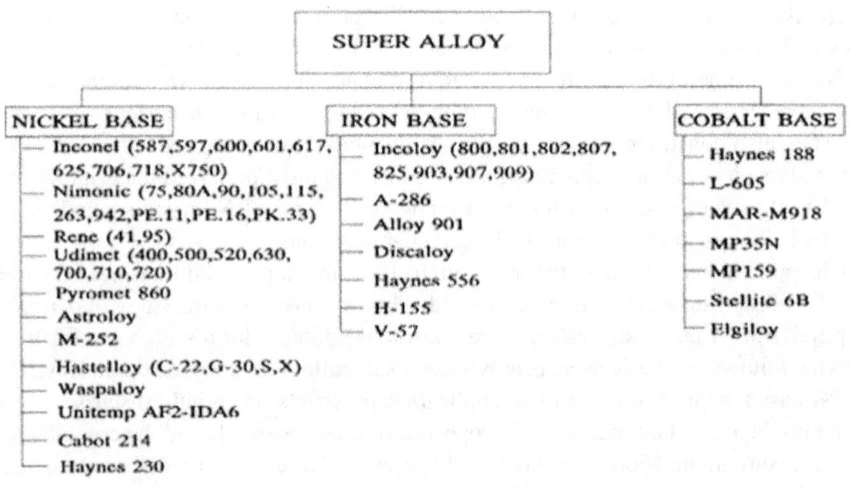

FIGURE 15.1 Classification of superalloys [3].

The major phases in most nickel superalloys are the gamma (γ) phase, gamma prime (γ ′) phase and carbides phase. In addition, there are the topologically closed-packed phases. The gamma (continuous matrix) phase is an FCC nickel-based austenitic phase which is mainly composed of Ni and other solid solution elements like cobalt (Co), chromium (Cr) and molybdenum (Mo).

The gamma prime (γ ′) phase is the primary strengthening phase of nickel-based superalloys and it is represented by $Ni_3(Al,Ti)$ [5, 7]. The chemical compatibility and the compatibility in the matrix (γ) and precipitate (γ ′) lattice parameter (~0%–1%) allows the γ ′ to precipitate throughout the matrix. This phase is responsible for the long-term stability. As the temperature increases up to 650°C, the flow stress of γ ′ is elevated. In addition, γ ′ is quite ductile. Thus, it imparts strength to the matrix without lowering the fracture toughness of the alloy.

The carbide phases are formed when 0.05–0.2 wt.% carbon is added to the reactive elements, such as titanium, tantalum and hafnium. This leads to formation of TiC, TaC and HfC as carbides. After heat treatment, these carbides are reduced to lower carbides which have fcc lattice and they form grain boundaries. Now these carbides at the grain boundaries are responsible for the high values of the rupture strength at high temperatures. Therefore, the carbides are beneficial for the nickel-based superalloys [5]. The equations showing the decomposition of carbides into lower carbides are given as:

$$MC + \gamma = M_6C + \gamma' \quad (15.1)$$

$$MC + \gamma = M_{23}C_6 + \gamma' \quad (15.2)$$

$$M_6C + M' = M_{23}C_6 + M'' \quad (15.3)$$

where M is Ti, Ta or Hf; and M′ and M″ is Cr, Co, Ni or Mo.

The TCP phases, or topologically close-packed phases, are the undesirable brittle phases that can form during heating. TCP phases have a plate-like structure which adversely affects the mechanical properties such as ductility and creep rupture. TCPs are potentially damaging for two reasons: (1) they initiate the formation of cracks as they are brittle in nature, and (2) they do not combine with the γ and γ ′ strengthening elements in any useful forms and thus they decrease the creep strength.

The nickel-based superalloys have solutes such as aluminium and/or titanium, with a total concentration that is typically less than 10 atomic percent. Figure 15.2 represents the phase diagram

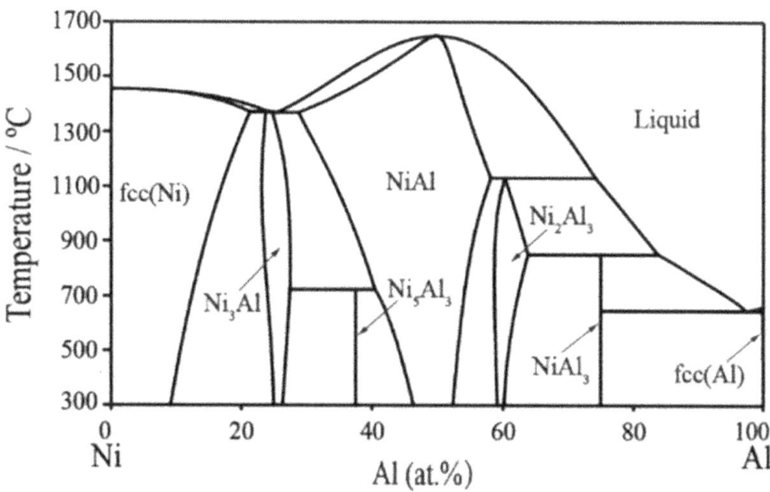

FIGURE 15.2 Phase diagram of Ni-Al [8].

of binary Ni-Al [8]. This gives rise to a two-phase equilibrium structure, which consists of both of γ and γ' phases. As mentioned earlier, the γ' phase is largely responsible for the strength of the material at high temperatures and its creep deformation resistance. The amount of γ' depends on two important factors: the chemical composition and temperature.

15.3 PROCESSING OF SUPERALLOYS

The processing of Ni-based superalloys initializes with the preparation of large metal ingots. Ingots are fabricated by vacuum induction melting (VIM) in a refractory crucible to consolidate elemental and/or revert materials to form a base alloy. The formation of ingots is achieved by three different processing routes: (1) remelting and subsequent casting, (2) remelting and wrought processing and (3) remelting followed by formation of superalloy powder, which is further subjected to wrought processing operations. These different processes are applied to remove segregation or dispersion of elements; to eliminate tramp elements such as oxygen, sulphur and nitrogen; and to obtain an ingot with uniform microstructure [9]. The ingots that are formed are now annealed to further reduce the segregation of impurities at grain boundaries and triple junctions. However, different processing approaches affect the properties of the superalloys which are formed by various processing routes, as expected. Some of the processing approaches are explained here.

15.4 CAST PROCESSING OF SUPERALLOYS

In the casting process, fabrication of superalloy components of complex shapes, including blades and vanes, is performed. In this process, ceramic moulds of alumina, silica and/or zirconia are used for the fabrication of ingots. The ceramic moulds are embedded in the wax to have internal cooling structures. The wax is removed by using thermal cycles. Meanwhile, the superalloy is preheated in the vacuum chamber and then the remelted superalloy is filled in the mould and left to cool at room temperature. Three types of casting can be done: equiaxed, columnar grained and single crystal. The solidification of the superalloy is uniform through equiaxed casting. The casted superalloy then undergoes cycles of heat treatments to reduce segregation and establish γ' precipitation. Figure 15.3 represents the casting process of superalloys.

15.5 WROUGHT PROCESS

Wrought alloys are fabricated by remelting VIM ingots. This process forms a secondary ingot for subsequent deformation processing. The secondary ingots are again melted, as a secondary melting process is required for the wrought alloys. This is so because the structural properties of Ni-based superalloys at elevated temperatures are highly sensitive towards microstructural variations and chemical inhomogeneities. As ingot sizes increase, VIM at high temperatures often results in macrosegregation or the formation of large shrinkage cavities during solidification.

During the solidification of the VIM ingots, the heat transfer is limited due to the low thermal conductivity of the solidifying mass. The deformation of the superalloys by mechanical processes such as cast and wrought processing becomes difficult as more alloying elements are added to the nickel-based superalloys for higher strength of the resulting superalloys. Although certain elements such as W, Mo, Ti, Ta and Nb increase the strength of the superalloy, the resultant superalloy exhibits macrosegregation at solidification. Also, the low ductility of high-strength alloys is highly prone to the formation of cracks due to thermally induced stress during cooling. Additionally, the limited ductility of high-strength alloys renders the ingot susceptible to cracking as thermally induced stresses evolve during cooling. This has led to the use of an alternative route, e.g., powder processing [3, 9].

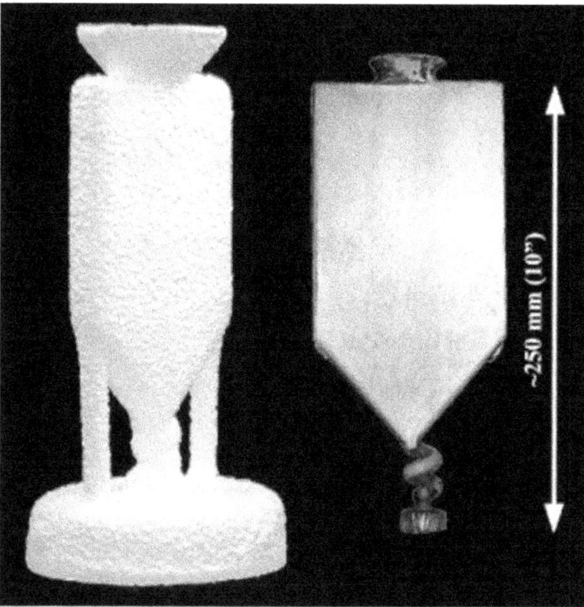

FIGURE 15.3 Ceramic investment casting mould with single crystal starter at the bottom of the plate and single crystal plate following directional solidification and removal of the ceramic mould [2].

15.6 POWDER METALLURGY

Powder processing has been developed to overcome the difficulties of melt-related processes. It is able to produce advanced high-strength polycrystalline superalloys. In this process, vacuum atomization of VIM ingot is performed. This step is followed by rapid solidification of the fine powder. Due to this rapid solidification, the macrosegregation within the alloy is suppressed. As high-strength superalloys have low ductility, they are usually very sensitive to initial flaw sizes. Thus, the atomized powders are separated based on particle size. Standard 150 or 270 size meshes are used to separate the powders into sizes >100 μm and >50 μm, respectively. Although finer powder sizes are desired to minimize initial defect sizes, costs increase substantially as yields are substantially reduced [9, 10].

15.6.1 Vacuum Induction Melting (VIM)

VIM is considered the primary melting process in cast and wrought processing routes. In this process, liquid metal is produced under vacuum in an induction heated crucible. Before heating, the sample is purified and refined so that the composition of the resultant melt can be controlled. Vacuum or inert gas atmosphere is preferred to prevent the reactions of the sample with the atmospheric oxygen and nitrogen.

15.6.2 Vacuum Arc Remelting (VAR)

Vacuum arc remelting (VAR) is a secondary melting technique. It converts the VIM-processed electrodes into ingots with improved chemical and physical homogeneity. In this process, a stub welded to one end of an electrode is suspended over a water-cooled copper crucible. An arc is struck between the end of the electrode and the crucible bottom. The arc generates the heat required to melt the electrode, which drips into the crucible and the melt is then poured into moulds.

15.6.3 Electroslag Remelting (ESR)

Electroslag remelting (ESR) is also a secondary melting technique. In this process, remelting is not processed by striking an arc under vacuum. Here, a consumable electrode is melted by immersing it in a slag which is superheated by means of resistance heating. The molten sample is then poured into a water-cooled mould. The whole process is conducted in ambient air rather than using vacuum conditions. During melting, metal droplets fall through the molten slag, and chemical reactions reduce sulphur and non-metallic inclusions.

15.7 VARIOUS TYPES OF NI-BASED SUPERALLOYS AND THEIR APPLICATION PROSPECTS

There are three different classes of nickel-based superalloys. The first class consists of elements having an FCC austenitic matrix, which are the elements from periodic table groups V, VI and VII, i.e., nickel, cobalt, iron, chromium, molybdenum, tungsten and vanadium. The second class of elements, which are from groups III and IV and have an odd-sized diameter, form the γ' precipitate in Ni_3Al. The third group of elements, namely, B, C and Zr, segregate to grain boundaries [11]. Nickel-based alloys can be either of two types. One type is strengthened by solid solution. The other type is strengthened by precipitate formation.

For those applications in which modest strength is required, the solid solution process is used. Most nickel-based alloys contain 10%–20% Cr; 8% Al and Ti; 5%–10% Co; and small amounts of B, Zr and C. Other common additives such as Mo, W, Ta, Hf and Nb are also used. In broad terms, the combinations of elemental additions in the formation of Ni-based superalloys can be categorized as follows:

(i) γ formers, which participate in the matrix. These are the elements that tend to partition into the matrix, e.g., Co, Cr, Mo, W and Fe. The difference in the atomic diameter of γ formers is only slightly different from the primary matrix element Ni.
(ii) γ' formers are the elements that partition to the γ' precipitate, e.g., Al, Ti, Nb, Ta and Hf.
(iii) Carbide formers.
(iv) Elements that segregate to the grain boundaries [4, 11]. Some of the nickel-based superalloys with their composition and application are described as follows:

a) Inconel alloy 600 (76Ni-15Cr-8Fe): It is an alloy that is a standard material for the construction of nuclear reactors. It is also used in the chemical industry in heaters, stills, evaporator tubes and condensers.
b) Nimonic alloy 75: It contains about 80 wt.% Ni and about 20 wt.% Cr with additions of Ti and C. It is used in gas turbine engineering, furnace components and heat-treatment equipment.
c) Alloy 601: It contains a relatively lower Ni content with aluminium and silicon additions for improved oxidation resistance. It is used for chemical processing, pollution control, aerospace and power generation.
d) Alloy X750: In this alloy composition, the addition of Al and Ti is for age hardening. It is primarily used in gas turbines, rocket engines, nuclear reactors, pressure vessels, tooling and aircraft structures.
e) Alloy 718: It has a composition of about 55 wt.% Ni, about 21 wt.% Cr, 5 wt.% Nb and 3 wt.% Mo. Here, Nb is added to prevent the formation of cracks during the welding process. It is used in aircraft and land-based gas turbine engines and cryogenic tanks.
f) Alloy X: It has a wt.% composition of about 48% Ni, 22% Cr, 18% Fe and 9% (Mo + W). This alloy is used in high-temperature flat-rolled products for aerospace applications.
g) Waspaloy: It has a wt.% composition of about 60% Ni, 19% Cr, 4% Mo, 3% Ti and 1.3% Al. It is used for jet engine applications.

h) ATI 718 Plus: It exceeds the operating temperature capability of standard 718 alloy by about 55 C°. It also allows engine manufacturers to improve fuel efficiency and it is cost effective.
i) Nimonic 90: It has a wt.% composition of about 54% Ni, 18%–21% Cr, 15%–21% Co, 2%–3% Ti and 1%–2% Al. It is used for turbine blades, discs, forgings, ring sections and hot-working tools.
j) René N6: It has a wt.% composition of about 4% Cr, 12% Co, 1% Mo, 6%W, 7% Ta, 5.8% Al, 0.2% Hf, 5% Re and balance Ni. It is used in jet engines [2].

15.8 IMPORTANT MECHANICAL PROPERTIES OF NI-BASED SUPERALLOYS

The most important mechanical properties of Ni-based superalloys are tensile strength, ultimate tensile strength, compressive strength, creep resistance, high-temperature strength and stress rupture. The other properties of importance include hardness, Young's modulus and shear modulus [1–11].

15.9 STRENGTH BEHAVIOUR OF NI-BASED SUPERALLOYS

The mechanical properties of nickel-based superalloys have a strong dependence on their microstructure, which in turn is governed by the chemical composition and the processing conditions of the superalloy. The nickel-based superalloy has a remarkable characteristic of yield stress. It is important to note that it is the stress required to initiate the plastic deformation by dislocation flow. As compared to other alloy systems, in the case of Ni-based superalloys, the yield stress does not greatly decrease with increasing temperature. The variation of the yield stress of Ni-based superalloys with temperature is shown in Figure 15.4 [12].

To analyze the curve, further study was performed on the yield properties of Ni-Al binary alloy between –200°C and 800°C [13]. The result of this study is shown in Figure 15.5. Two types of alloys were taken in the study: Ni–8%Al, and Ni–14%Al. The Ni–8%Al alloy is insufficient for the precipitation of the γ phase. Thus, it benefits only from a solid solution strengthening. Depending upon the temperature, the strengthening is reported [13] to be between 50 and 100 MPa.

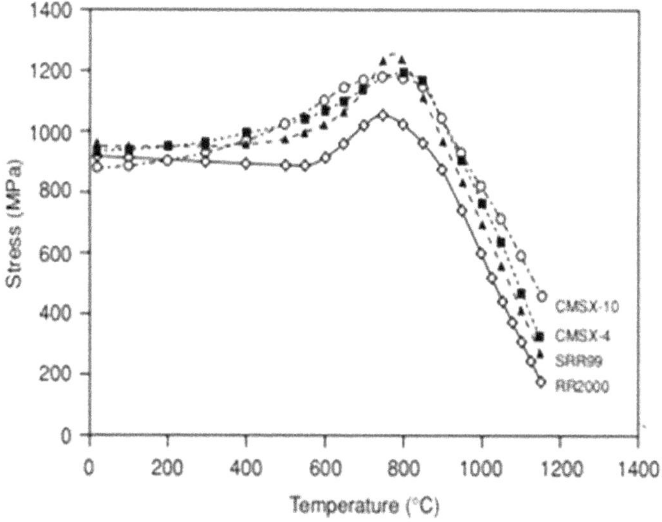

FIGURE 15.4 The variation of yield stress with temperature [12].

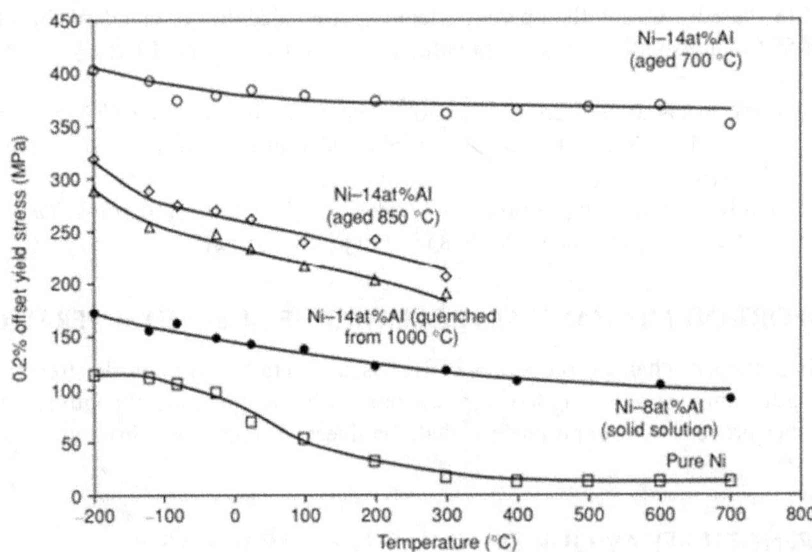

FIGURE 15.5 Data for the yield stress of alloys in the binary Ni–Al system as a function of composition and temperature. Adapted from [13].

On the other hand, the Ni–14%Al alloys displayed precipitation strengthening. In this case, however, the extent of strengthening is reported [13] to be dependent upon the fraction, size and distribution of the γ' particles, and hence on the heat treatment applied. It is analyzed [13] that the constant yield stress displayed by the Ni–14%Al alloy heat treated at 700°C is due to microscopic deformation of the γ' phase. It is further noted [13] that at and beyond the peak stress, the behaviour of the alloy is determined by the strength of the γ' phase. As the temperature increases, the γ' imparts a fraction of the strength to the alloy. The analysis [13] also indicates that for the weak-coupling mechanism, hardening of the Ni-based superalloy is approximately proportional to the square root of the particle radius (r), and for the strong coupling mechanism, it is inversely proportional to r. Thus, hardening is reported [13] to be maximized for a particle size at an intermediate situation between the two extremes.

15.10 CREEP BEHAVIOUR OF NI-BASED SUPERALLOYS

Creep resistance means the resistance to the tendency of a material to slowly deform over a long period of exposure to high levels of stress. The creep behaviour of the nickel-based superalloys depends strongly on the grain size, phase composition, particle size and the concentrations of alloying additions [14]. For polycrystalline nickel, a fifth power law dependence of the applied stress on the creep behaviour is also observed. Creep resistance is reported [14] to be the most potent when the atomic size misfit of solute and solvent is large.

Further, the γ phase fraction is found [14] to have a profound influence on creep life. The creep cavitation occurs at γ grain boundaries. This creep damage reduces with the increase in grain size. Thus, the maximum influence happens when the grain size is large. The creep strengthening in polycrystalline nickel-based superalloys can occur due to both, solid solution strengthening due to the presence of solute atoms and precipitation hardening due to phases such as the $\acute{\gamma}$ phase. Furthermore, the amount of creep strain is reported [14] to increase in the order rhodium (4%), iridium (5%), palladium (6%) and gold (6%). These values are slightly higher than that (e.g., about 2%) due to Ni. The creep strain is reported [14] to be not so strongly dependent on the stress level applied.

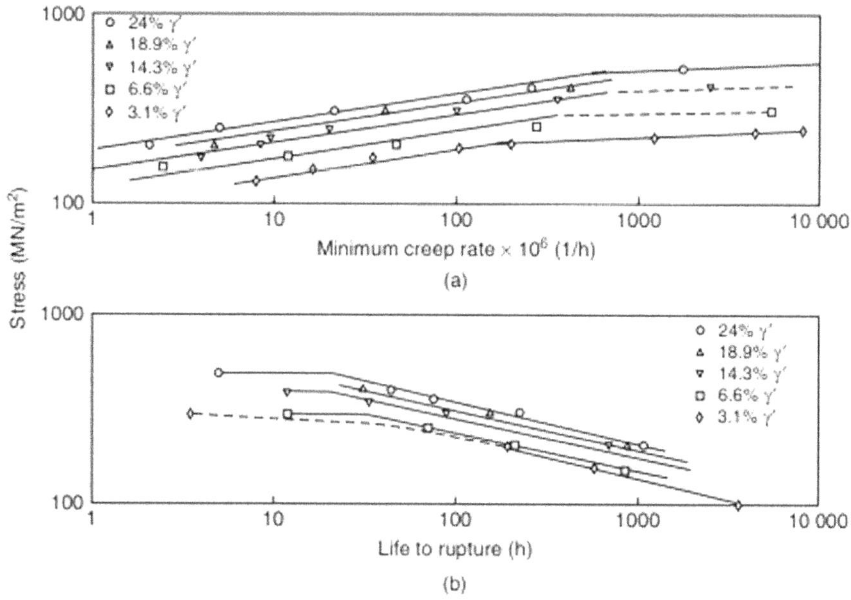

FIGURE 15.6 Variation of the (a) minimum creep rate and (b) life to rupture for a number of polycrystalline superalloys with varying fractions of the γ phase [14].

However, the presence of the γ ′ phase, which is promoted by alloying with Al, Ti and Ta, has a profound influence on the creep resistance of nickel-based superalloys (Figure 15.6). If the stress level is not too high, the creep behaviour improves substantially. Also, nickel-based superalloys possess outstanding stress rupture strength as compared with other materials, e.g., Mg, Ti and Al alloys, used in the structures of aircraft [15]. Therefore, nickel-based superalloys can resist creep deformation even at 850°C, while the other elements exhibit rapid creep deformation even at relatively lower temperatures of 100°C, 150°C and 350°C [16]. The compositions of the Ni-based superalloys Haynes-556, Haynes HR-120, Hastelloy X and Haynes-230 are shown in Figure 15.7. The variations of ultimate tensile strength, 0.2% yield strength, elongation and stress rupture at 1000 h as a function of temperature change are shown in Figure 15.8. These data confirm that nickel-based superalloys with a maximum composition of Fe, Cr and Co give better mechanical properties when compared to other alloys even at relatively higher temperatures.

Both strain rate enhancement and grain size reduction, especially at the nano-metric size, can enhance the yield strength, ultimate tensile strength and fracture strain in Ni-Al intermetallic materials [17–19]. Thus, the importance of alloying Ni with other elements for mechanical properties enhancement is shown in Table 15.1.

15.11 ELASTIC BEHAVIOUR OF NI-BASED SUPERALLOYS

A turbine blade is often made from Ni single crystal (NiSC) or columnar-grained Ni. However, NiSC is elastically anisotropic. The elastic moduli tensor is a fourth-order tensor, as both stress and strain tensors are second-order tensors [14]. Hence, the stiffness depends on the crystallographic orientation relative to the loading configuration. Thus, the specimens machined at random from a large single crystal display different elastic properties in different directions. For a given cubic single crystal with orthogonal axes 1, 2 and 3 lying along the [100], [010] and [001] directions, the anisotropic elastic properties are reported [14]. However, for polycrystalline Ni-based superalloys, such as wrought superalloys with equiaxed castings, the elastic properties are expected to be isotropic. Hence, they are invariant with respect to the various crystallographic directions. The elastic

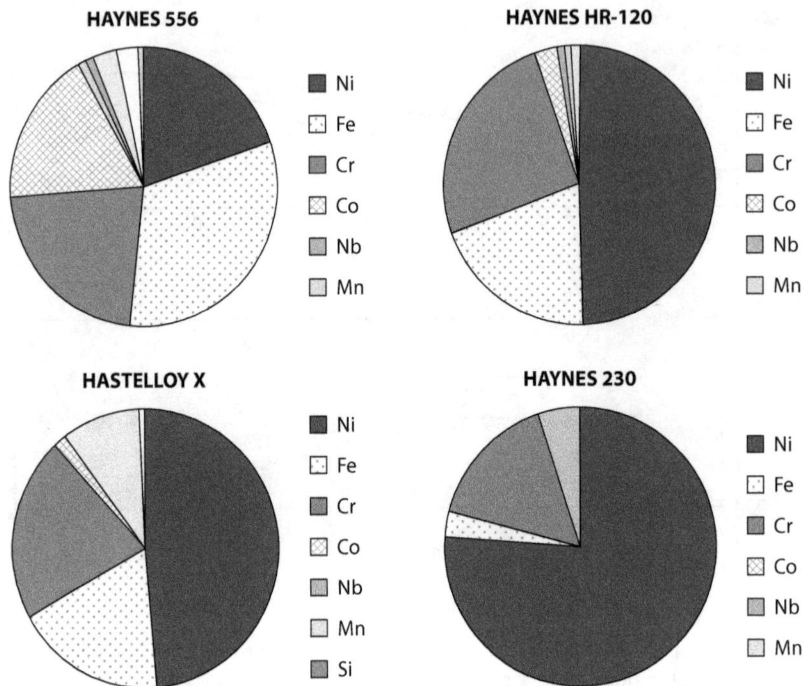

FIGURE. 15.7 Composition of various Ni-based superalloys [16].

modulus of pure Ni along the [100] direction is about 125 GPa which is much less than that (E about 207 GPa, G about 80 GPa) of the polycrystalline Ni-based superalloys [5]. When designing single crystal blades from nickel-based superalloys, it is important to recognize this fact.

15.12 TECHNIQUES TO EVALUATE THE MICROSTRUCTURE AND THE MECHANICAL PROPERTIES OF NI-BASED SUPERALLOYS

There are a number of techniques [20–23] utilized to measure the mechanical properties of Ni-based superalloys. In one of the studies, the tensile strength and fracture of cast nickel-base superalloy K445 in a temperature range of 25°C–1000°C was measured [20]. This alloy is used in the high-pressure turbine blades used in industrial gas turbines. Although it has good high-temperature properties, it still has low ductility.

The brittleness is mainly due to the presence of W, Mo, Ta and Nb. The high-temperature behaviour of the alloy depends strongly on the γ' precipitates. To obtain the desired γ' precipitation structure, the alloy is first melted in a vacuum induction furnace and then cast in a round diameter of 16 mm and length of 130 mm. Further, the preheating and moulding temperatures are maintained at 950°C and 1430°C, respectively. The investigated samples are then heat treated to obtain the desired γ' precipitation microstructure. The microstructure and fracture surfaces of the nickel-based superalloy K445 are observed by using well-known techniques such as optical microscopy (OM), scanning electron microscopy (SEM) and transmission electron microscopy (TEM). The results reveal that an anomalous yield strength phenomenon exists in the alloy at medium-high temperatures.

For another Ni-based superalloy, Inconel 718, tensile tests are reported [21] to be performed with an initial strain rate of 5×10^{-5} to 1.04×10^{-4} s^{-1} on a universal AG-250KNE test machine in air at a temperature range of 25°C–1000°C. The sample for tensile testing is induction heated. For

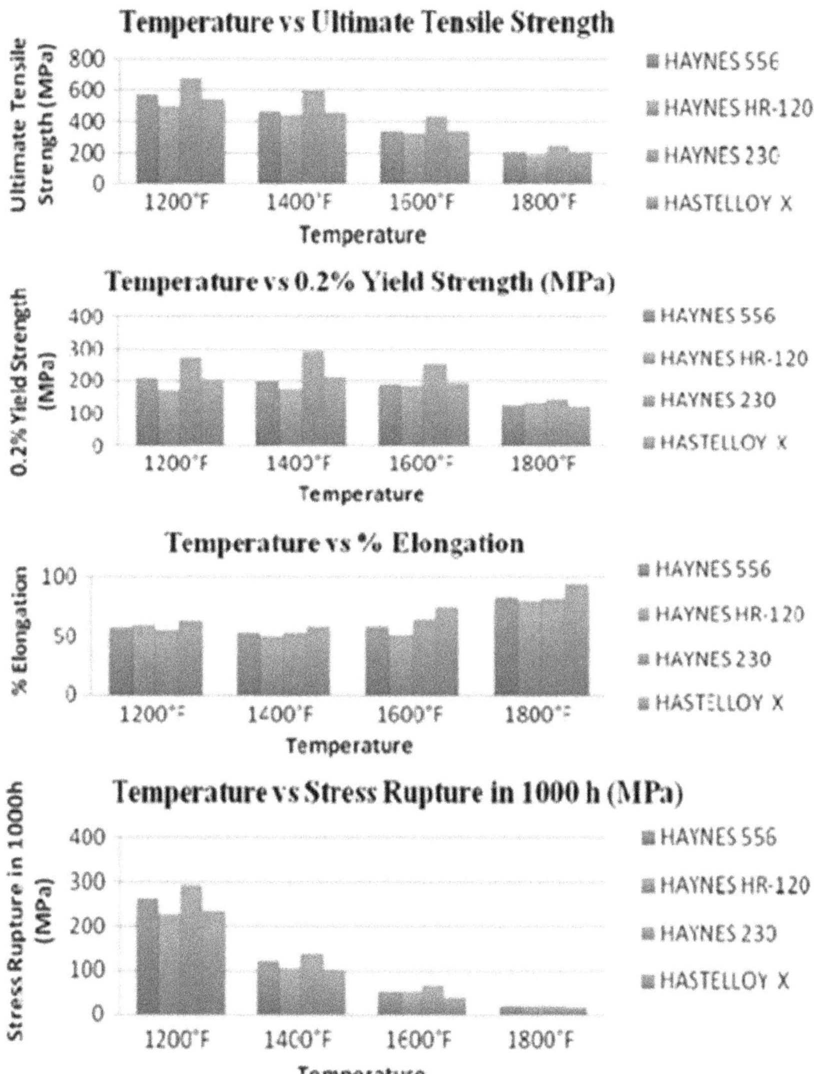

FIGURE 15.8 Variations of ultimate tensile strength, 0.2% yield strength, elongation and stress rupture at 1000 h as a function of the temperature change [16].

TABLE 15.1
Importance of Alloying Ni with Other Elements for Mechanical Properties Enhancement

Nickel When Alloyed With	Provides
Chromium, molybdenum, iron, tantalum, tungsten	Higher strength
Boron, zirconium, carbon	Higher creep resistance
Chromium, aluminium, tantalum	Higher oxidation resistance
Aluminium, titanium	Larger high temperature strength
Hafnium	Higher ductility at intermediate temperatures

microstructural observation, the specimens are reported [21] to be cut from the as-cast and heat-treated samples. The microstructure and fracture surface are reported [21] to be examined using the conventional techniques of OM, SEM, TEM and EDAX (energy-dispersive X-ray spectroscopy).

Further, the fracture analysis is reported [21] to be done by in situ digital image correlation (DIC) and synchrotron radiation X-ray tomography (SRXT) techniques. For DIC, the sample is machined to 5 mm in gauge width and 12 mm in gauge length using electrical discharge machining (EDM). Similarly, for SRXT experiments, the samples are reported [21] to be cut into dimensions of 300 μm in gauge width and 3 mm in gauge length. This is done by using the aforementioned EDM machine. All specimens are polished with 2000 grit SiC papers to reduce the chances of edge fracture and to reduce the impact of the EDM process. The samples are ultrasonically cleaned prior to the commencement of all testing processes [21].

To study the tensile and fracture properties of the Ni-based superalloy DZ951G [22], it is also initially melted in a VIF furnace and then cast into a polycrystalline ingot of about 5 kg. To obtain a directionally solidified superalloy, the polycrystalline ingot is remelted and poured into a ceramic mould at a temperature of 1550°C. The incipient melting temperature is reported [22] to be precisely measured by the differential thermal analysis (DTA) technique. The samples are finally tested at a temperature up to 1100°C by using an AG-250KNE mechanical test machine.

The as-cast, heat-treated and tensile-ruptured samples are reported [22] to be etched by chemical etching and electrolytic etching. The microstructure and phases of the sample are analyzed using the field-emission scanning electron microscopy (FESEM) technique. After the tensile test, for TEM analysis, the thin foils with a thickness of 500 μm are horizontally cut about 5 mm away from the fracture surfaces. These foils are subsequently ground down to about 50 μm and finally punched by an ion milling process prior to insertion in the TEM chamber. [22].

For the creep test of the Ni-based superalloy M951G [23], it is heat treated at three temperatures: 1210°C, 1100°C and 870°C. The samples are kept for 4 h at each temperature. Then they are air cooled to room temperature. Next, the creep tests are conducted under applied stresses from 240 MPa to 400 MPa at 900°C in air. The phase identification is done by the X-ray diffraction analysis (XRD) technique, while microstructural and fracture surface observations are conducted by the OM, SEM and TEM techniques. Further, the volume fractions of carbides and γ' phase are calculated by using an image processing software such as Image-Pro Plus software (IPP) applied to the relevant images.

Generally, the elastic properties of several monocrystalline Ni-based superalloys are measured [24] as a function of orientation. In addition, the elastic properties of Ni-based superalloys are also reported [23] to be measured as a function of texture. Thus, to determine the elastic behaviour of anisotropic solids, knowledge of both the orientation and texture is required.

Recent work [24] on the γ' precipitate-hardened NiSC samples IN 738 LC, SRR 99, CMSX4 and CMSX-6 uses the cylindrical specimens (ϕ ~4–5 mm, L ~40–50 mm). The orientations of these single crystal specimens are distributed statistically in the standard stereographic triangle. Through the orientation distribution function (ODF), the texture of polycrystalline materials is described. A modified Forster resonance technique is used for the determination of the elastic properties from resonant frequency measurements.

Further, piezoelectric transducers are reported [24] to be used to excite vibrations in the aforementioned specimens and the resonant frequencies are measured. The measurements are reported [24] to be carried out in vacuum in the 4 to 150 kHz frequency range between 20°C and 1250°C. Three vibrational modes, i.e., the flexural, longitudinal and torsional vibrations, are excited. Flexural and longitudinal vibrations supply information about Young's modulus E (T) as a function of temperature (T). Similarly, the torsional vibrations provide information about the shear modulus G (T) as a function of temperature (T). The temperature is reported [24] to be controlled by a thermocouple located 1 mm away from the middle of the specimen. Thus, the experimentally measured data plotted in Figure 15.9 show the variations of E(T) and G(T) as a function of the orientation parameter.

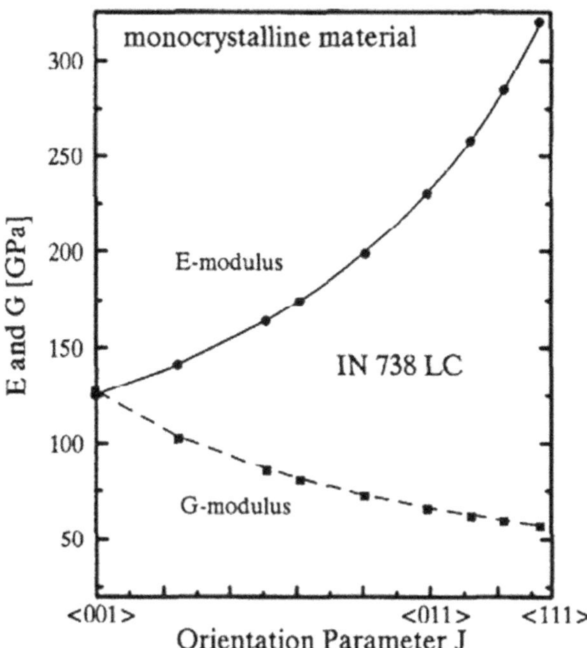

FIGURE 15.9 Young's modulus (E) and shear modulus (G) of monocrystalline IN 738 LC as a function of the orientation parameter, describing the orientation [24].

15.13 LITERATURE REVIEW ON MECHANICAL PROPERTIES OF NI-BASED SUPERALLOYS

As already mentioned, the research domains on superalloys in general and Ni-based superalloys in particular are very attractive for researchers worldwide [25–33] simply because of their huge application prospects in a wide variety of technologies including aeroengines, gas turbines, spacecraft, nuclear power industry, nuclear reactor industry and conventional power sector industries [25–32].

Before we go into the details of the literature review, it needs to be noted once again that the main solute in nickel-based superalloys is titanium or aluminium which generates a two-phase equilibrium microstructure (gamma and gamma prime). Gamma (γ) is a solid solution with a cubic F lattice consisting of a random distribution of atoms. Gamma prime (γ ′) has a cubic P or primitive cubic lattice with Ni atoms at the face centres and Ti or Al atoms at the cube corners. It has the chemical formulae Ni_3Ti, Ni_3Al or $Ni_3(Al,Ti)$. The amount of γ ′ depends on the temperature and the chemical composition. It is the γ ′ phase which is responsible for the high temperature strength and high resistance to creep deformation properties of Ni-based superalloys, as already mentioned. For a given chemical composition, the fraction of γ ′ decreases with an increase in temperature. The γ phase forms the matrix in which the γ ′ precipitates. Further, the γ ′ phase precipitates in a cube–cube orientation with respect to γ phase, which makes its (i.e., γ ′) cell edges parallel to the edges of the γ phase.

There is only a small misfit between γ and causes low $\frac{\gamma}{\gamma'}$ interfacial energy, and precipitate coarsening is exclusively driven by the minimization of total interfacial energy. However, an interface whether coherent or non-coherent makes the microstructure stable. The misfit value can be altered with the chemical composition, especially by altering the ratio of aluminium to titanium. The sign of the misfit is also important. Thus, the misfit is said to be positive if γ ′ has larger lattice parameter values than those of the γ phase. On the other hand, the misfit is said to be negative if γ ′ has a smaller lattice parameter value than those of the γ phase.

An earlier modelling effort [33] made about a decade back proposes a physical model for the analysis of creep deformation of Ni-based single crystal superalloy. It is proposed that no empirical parameter is required for this model and the relative effect of certain parameters like γ' volume fraction on macroscopic properties (such as strain rate) is performed. It is assumed dislocations that are responsible for strengthening can be either (a) trapped at the $\frac{\gamma}{\gamma'}$ interfaces or (b) free to slip in the γ matrix. Further, work [34] confirms that the climbing rate of the trapped dislocations at the interface is controlled by the vacancy absorption and emission. The kinetics of γ channel thickening/thinning is related to the relocation of the vacancy emission/absorption fluxes at the $\frac{\gamma}{\gamma'}$ interfaces [34].

Furthermore, it is noted [34–36] that the microstructure and mechanical properties of Ni-based superalloys are greatly affected by heat treatment. The study of these changes helps in predicting and forecasting the service life of such superalloys. As already discussed, the Ni-based superalloys are made up of γ and γ' phases but with the addition of other alloying elements (such as Cr, Co, Nb and Ti), β phase, carbides and α-Cr precipitates are also found. The heat treatment of these superalloys leads to the dissolution of carbides and γ' precipitates and further cooling or subsequent ageing causes reprecipitation in the matrix [34–36]. It needs to be recognized that the first report on the synthesis of Ni-based superalloys was done in 1980 [37, 38]. Since then, efforts have been ongoing to understand the structure–microstructure–mechanical properties correlation of Ni-based superalloys and their further improvements.

In addition, it is found that the reprecipitation of γ' is practically impossible, and the microstructure of the superalloys depends on the cooling rate. A slow cooling rate favours the formation of a higher volume fraction of longer, bigger γ' precipitates. A high ageing temperature causes the formation of secondary phases and dissolution of unstable precipitates, but this effect of ageing temperature is not constant for all superalloys and is sensitive to the initial microstructure, ageing parameters and alloying composition [34]. As such, innumerable efforts are there in the literature. To totally enumerate all of them is beyond the scope of the present chapter. Therefore, an attempt is made to provide a brief overview of the literature.

Recent work reported [39] about two decades back illustrates that the addition of Al or Cr increases the density and strength of the pure Ni foam. Creep resistance also increases with an increase in density and hardening of the γ matrix [39]. In the case of the GTD-111 superalloy, abnormal fluctuation in ductility and tensile strength are reported when measured as a function of temperature, 25°C to 900°C [40]. In addition, the thermal exposure of alloy 263 to 700°C results [41] in the precipitation of $M_{23}C_6$ carbides. The carbides are reported [41] to be small, bulky and acicular type. These carbides help to increase tensile strength. If the temperature is raised to 750°C, it results [41] in the precipitation of the η phase along with the γ' particle depletion zone and enhances the creep rupture life and tensile strength at high temperatures. Due to the very good combination of mechanical properties at high temperatures, the UDIMET 500 Ni-based superalloy is reported [42] to exhibit high strength, stability and good hot corrosion resistance. These properties make it more suitable for application in second-stage blades and other hot components of gas turbines [42]. A faster cooling rate enhances the amount of inhomogeneous precipitation in the microstructure. As a result, the alloy consists of coherent and ordered γ' precipitates, and carbides dispersed in the matrix and austenitic γ matrix [42]. Earlier work also confirms that the cooling rate also affects the shape, volume fraction and size of the γ' particles and an increase in cooling rate causes a decrease in the volume percent and size of the γ' particles [41, 42].

For important practical applications, often superalloy-based joints are needed [41, 43]. The materials used for making such joints are reported [42–45] to affect their microstructure, morphology, fracture and tensile strength properties. Moreover, the wielding speed, impulse frequency and wielding current are the main parameters [41, 43] that govern the structural health of superalloy-based joints.

Particularly, in the case of Ni-based superalloys, it is found [41, 43] that if the welding current is lower, the root of the welding joint is unwelded so that the elongation and strength of the welded joint becomes inferior. If the welding current is too high, then a more brittle and harder phase starts precipitating in the welding zone. A high welding current also causes the formation of welding defects in the welding zone. The strength of the welded joint increases with a decrease in the impulse frequency [41, 43].

Earlier research [44] elaborates on how the microstructure affects the crack propagation rate in GH169 Ni-based superalloys. Further work [45] examines the mechanical and microstructural properties of TMW-4, which is also a Ni-based superalloy. TMW-4 alloy is reported [45] to be heat treated at temperatures ranging from 700° to 1220°C for various time periods ranging from 1 to 1000 h. It is identified [45] that in the subsolvus heat-treatment samples, grain size, fraction of secondary gamma prime, tertiary gamma prime and secondary gamma prime precipitate size increase with increasing solution temperature. Grain growth is reported to be very sluggish due to primary gamma prime grain boundary pinning. Further, the supersolvus heat treatments provide a large, grained microstructure with all of the gamma prime phases precipitated inside the gamma phase grains. The eta phase is reported to be present only in samples homogenized at a temperature of 1220°C and aged at a temperature range of 1000°C to 1175°C [45]. The highest yield strength and elongation are noted [45] to occur at 750°C with a larger grain size of 10.7 μm and very small secondary γ' precipitate size (e.g., 8 Å). Furthermore, the creep mechanisms at applied stress of 630 MPa at an intermediate temperature of 725°C are reported [46] for TMW-4M3 superalloys.

In the case of powder metallurgy route processed superalloys, e.g., U720LI, it is noted [47] that the size and volume fraction of gamma prime precipitates are strongly sensitive functions of the cooling rate. For instance, for cooling rates in the range of 1 to 167°C.min^{-1}, the gamma prime precipitates are evaluated [47] to be about 0.15 to 0.35 μm, respectively. Further, both the cooling rate and ageing treatment affect the tensile strength. The tensile strength is reported [47] to increase from about 90 to 130 MPa with ageing treatment. This behaviour is suggested to be due to the formation of very fine gamma prime precipitates.

In the case of a Co–Ni–Al–Ti superalloy, however, a related work [48] reports the effects of the relative content of Al and Ti on the microstructure and mechanical properties. It also concerns how the addition of 2 at.% Cr or Mo to partially replace Co can affect the microstructure and mechanical properties of this Co–Ni–Al–Ti superalloy. The results show that γ'-$(Co,Ni)_3(Al,Ti)$ precipitates are set aside on the γ-Co matrix. The precipitates exhibit a dendrite-like morphology. It is reported further that ageing causes the precipitation of uniformly nanosized γ'-$(Co,Ni)_3(Al,Ti)$ on the γ-Co matrix. On the other hand, the Ni, Al, Mo and Ti atoms are segregated preferentially onto the γ' phase. However, the Cr and Co atoms are found [48] to segregate preferentially to the γ matrix phase. The addition of 2 at.% Mo or Cr along with an increase in Ti content increases the average size and volume fraction of γ' phase. At the constant addition level of 2 at.% Mo, the lattice misfit strain increases with the increase in concentration of Ti. On the contrary, at the constant addition level of 2 at.% Cr, the lattice misfit strain decreases with the increase in concentration of Ti. Ageing times from 20 to 100 h leads to an increase in the lattice misfit strain and size of γ' phase. But the γ' phase volume fraction increases initially with ageing time up to the maximum at 100 h prior to decreasing with a further increase in ageing time. The lattice misfit strain plays an important role in enhancing the Vickers hardness for an ageing time of 20 to 100 h. However, for still higher ageing periods, the γ' phase volume fraction plays a critical role in affecting the Vickers hardness [48].

It is interesting to note that the relationship between the initial δ phase versus fracture characteristics and tensile deformation behaviours in uniaxial tensile tests at 920°C–1010°C at a varying strain rate (0.01–0.001 s^{-1}) are also reported [49] for a Ni-based superalloy with approximate wt% composition of Ni-53, Cr-19, Nb-5, Mo-3, Ti-1, Al-0.6, C-0.03, Co-0.03 and balance Fe. This work finds [49] that microvoid coalescence and localized necking cause the ductile fracture of specimens for various initial contents of δ phase. Further, in the solution-treated specimens, the microvoids present mostly nucleate through the interfacial debonding of carbides and γ phase in the matrix. However, in the aged specimens, the microvoid formation is initiated by the precipitation of δ phase [49]. It is suggested that to enhance the plastic deformation capability, it is important to reduce the formation of microvoids in both types of specimens.

A solo but important attempt is also made [50] to understand how the prolonged 10000 h exposure at 871°C to 982°C affects the eta phase formation in GTD 111Ni-based superalloys. The η phase is reported [50] to be formed below 927°C and precipitation of the σ phase occurs at ~871°C. Further, it is noted that the increase in temperature leads to both the dissolution of η within the γ-γ' eutectic and the precipitation of $\gamma'+M_{23}C_6$ films along the grain boundary. The results obtained from the creep tests (440 MPa/810°C and 210 MPa/927°C) and tensile tests confirm that the rupture life and creep ductility are not affected by the change of primary γ' size or η phase [50]. It is also important to note that the presence of the suitable microstructure, grain size, morphology and distribution of the γ' precipitates as well as the heat treatment and post-heating ageing schedule need to be optimized to attain high fatigue resistance in the Ni-based superalloys [51].

In addition, it is reported [52] that the ductility, tensile life and stress-rupture life of the conventional casting route-based Ni superalloys dramatically decrease with the increase of nitrogen content from about 22 to 30 ppm in the melt. However, in the longitudinal direction of directionally solidified Ni superalloys, ductility, tensile life and stress-rupture life are independent of nitrogen within the contents studied. Many collections [53] report the development pathway of various advanced structural materials, e.g., Ni-based superalloys. It is also shown [54] that the well-known high temperature nanoindentation technique can be used to study the nanohardness and Young's moduli of Ni-based superalloys, e.g., Rene95, IN718, IN100, CMSX-3 and ME3, as a function of temperature, 30°C to 400°C.

A remarkable recent review [55] looks into the design aspect of NiSCs for advanced applications. Generally speaking, the microstructure, and mechanical and thermal stability of Ni-based single crystalline superalloys depend on the type and concentration of the elements present. The alloying elements can be categorized in four ways depending upon their effect on the properties [56]. These are base elements, oxidation resistance elements, mechanical strengthening elements and long-term stability elements. The mechanical strengthening elements can be categorized further as solid solution, precipitation and grain boundary elements. The solid solution elements in Ni-based superalloys are generally the elements of higher atomic diameter, e.g., Co, W, Re, Cr and Ru. Hence, lattice distortion affects the properties of the alloy.

The presence of these elements decreases the stacking fault energy (SFE) of the superalloy. The decrease in SFE helps to obstruct the dislocation to cross slip. Even the thermal stability of the superalloy is influenced by the elements. For instance, the presence of the elements Ru, Co, W and Re increases the processing temperature of NiSCs. Similarly, the presence of the elements Ti, Ta, Cr, Mo and Al decreases the processing temperature of the NiSC. It is reported further that [55] the elements Al, Ta, Ti and W help to form the requisite precipitate phase, i.e., the γ' phase. For 60%–75% volume fraction of γ' phase, the Ni-based superalloy gives maximum service life. The grain boundary elements are B, C and Hf. These elements decrease the casting pores of the alloys [55].

The tensile and yield strength of the Ni-based superalloy DZ951 are reported [56] to reach a minimum value of around 700°C prior to an increase in temperature, with a maximum of about 760°C. The maxima are followed by sharp decreases with further temperature rises. Confirmatory studies by SEM and STM (scanning tunnelling microscopy) techniques reveal that the defects in the γ matrix are full dislocations and match those present in traditional solid solution FCC superalloys. At relatively lower temperatures (to 750°C), aspect ratios of both the phases (γ and γ' phases) remain almost unchanged, which indicates that plastic strain in the superalloy is mainly caused by localized shearing in both phases. Deformation of γ' phase in the superalloy is suggested [56] to be dominated by anti-phase boundary (APB) shearing, and APB and SF (stacking fault) shearing. The dominance of APB shearing increases with temperature.

It is reported further that [56] when the temperature is increased above 850°C, plastic strain occurs by uniform deformation of both phases. But the deformations which were earlier accompanied by the vertical channels are now dominated by the horizontal channels. Once the vertical channels are narrowed enough, the deformation of γ ′ phase starts, and this deformation is mainly due to the motion of superdislocation and SF shearing. Furthermore, it is reported that [56] when the temperature is raised above 1000°C, most dislocations get deposited at the interface. At such high temperatures, the dislocation densities decrease for both phases. Deformation of the γ ′ phase generally involves motion of the superdislocations.

In this connection, it is important to note that recent work [57] also proposes a non-Schmid model to analyze the temperature-dependent yield anisotropy and orientation of the Ni-based superalloys. The effect of various microstructural features such as matrix channel width and precipitate volume fraction are simulated by the model [57]. Further, a single-stage ageing procedure is successfully achieved for precipitation-strengthened Ni-based superalloys. The ageing treatment is reported to enhance the high-temperature mechanical properties of the superalloy [58]. The same researchers also report [59] the in situ nano-deformation performed on nanopillars of additively manufactured Ni-based superalloys. On the other hand, the critical influence of cobalt, tantalum and tungsten on the elevated temperature mechanical properties of NiSCs is also reported [60]. Thus, truly wide aspects of the mechanical properties of Ni-based superalloys are studied by various researchers [25–60] in search of developing a better understanding of the structure–microstructure–mechanical properties interrelationship.

15.14 FUTURE RESEARCH NEEDS

A multitude of advanced research areas [61–80] are opening in various aspects of research and development in Ni-based superalloys, especially in the domain of their mechanical properties including both deformation and fracture. However, the most important horizon of emerging research is in additive manufacturing technologies for the fabrication of complex, near net shaped parts of Ni-based superalloys for applications in high-temperature gas turbines, spacecraft, nuclear power plants, reactors and other demanding harsh environments. The emphasis in some recent reviews [61–62] is to understand how the processing parameters and resulting microstructures affect the mechanical properties of Ni-based superalloys.

For instance, attempts [63] are made to comprehend how the addition of yttrium can change the microstructure and affect the high-temperature mechanical properties of Ni-based superalloys developed by selective laser melting, which is one of the major new routes under exploration through AM technology. In a similar way, recent attempts are reported [64] on how the evolution of the morphologies of the γ/γ ′ phases affect the mechanical properties of NiSCs developed by AM technology which utilizes laser metal deposition to fabricate the single crystal. Another recent effort [65] compares the mechanical properties of additively manufactured Hastelloy, Inconel 718 and Inconel 625 superalloys. This work also utilizes well-known AM technologies, such as laser powder bed fusion (L-PBF) and laser powder direct energy deposition (LP-DED). The comparison of fatigue lives of Ni-based alloys reveals very interesting features. The results confirm that at both strain amplitudes of 0.01 and 0.005 $mm.mm^{-1}$, the L-PBF/LP-DED IN 625 and LP-DED/L-PBF Hastelloy X superalloys sustain a higher number of cycles to failure in comparison to those sustained by the LP-DED/L-PBF IN 718 superalloys during the low cycle fatigue failure tests. It is suggested that the comparatively better performances of IN625 and Hastelloy could be linked to their better ability to undergo plastic deformation. This study is unique in the sense that no similar work has yet been reported. Further, this area demands huge attention in the future.

In a very recent unique effort [66], two lasers – one with low-power beams and another with high-power beams – are utilized in the AM technology of the twin laser-based LP-PBF process to develop the layer-by-layer 3D-printed heterogeneous microstructure of IN718 alloy that is tailored to deliver desired mechanical properties. These results open a new horizon of microstructurally tuned additively manufactured NI superalloy development to attain desired mechanical properties. This effort will certainly frame a major area of future research [66].

The well-known spherical nanoindentation technique provides a Young's modulus estimate of about 200 GPa in a selective laser-melted IN738LC Ni-based superalloy [67]. It is reported that the indentation stress–strain (ISS) curve compares well with those obtained from the micropillar compression tests. Further, the indentation stress is noted to be highly sensitive to precipitate distribution. Furthermore, the grain boundaries mediate the strain-hardening behaviour when the nanoindentation marks are made in a region close to the grain boundary region [67]. This approach provides the idea of microstructural length scale mechanical properties of the Ni-based superalloys and should constitute a major domain of future research.

Another recent work [68] explores the microstructure–mechanical properties correlation in crack-free Ni-based GH3536 superalloys. This superalloy is also formed by another prospective AM technology, namely, laser solid forming. Thus, a major chunk of recent emerging areas focuses on the mechanical properties of unconventional, AM technology-based Ni-based superalloys [61–68].

A unique recent attempt [69] utilizes ultrasonic wave agitation during the solidification process of the 4716MA0 Ni-based superalloy. The results confirm enhancement in tensile strength behaviour. This is suggested [69] to be linked to a reduction in grain size, inhibition of segregation, changing of the carbide distribution, and gamma as well as gamma prime eutectic distribution. This work opens up a new way to improve the mechanical properties of Ni-based superalloys and hence it is likely to enjoy huge prospects in future research. In a similar way, the reduction in carbon content from the standard composition in Nimonic alloy 105 leads to enhancement in creep resistance due to a large, grained microstructure [70]. This important attempt deserves a new look to explore further optimization of conventional Ni-based superalloy compositions in terms of carbon content [70].

Another important major work [71] explores the effect of the ultrasonic surface rolling process (USRP) and post-treatment on the microstructure and elevated temperature tensile properties of ATI 718Plus Ni-based superalloys. This deliberate creation of a gradient in the heterogeneous microstructure leads to about a 12% enhancement in yield strength (YS) and ultimate strength (UTS) at 704°C. The improvement is noted to occur in comparison to the untreated sample tested at the same temperature. This work shows that there are many opportunities to explore even with conventionally processed Ni-based superalloys.

In tune with such efforts, another recent attempt [72] explores the development of a new alloy composition, Ni–25Co-7(Al+Ti)-4(W+Mo)–10Cr-0.7Nb (wt%) through VIM, ESR and VAR, followed by homogenization and ageing. The alloy is reported [72] to exhibit YS of about 1.2 GPa at room temperature, which very slightly reduces to about 1 GPa at a temperature as high as 750°C. Similarly, it exhibits UTS of about 1.6 GPa at room temperature, which very slightly reduces to about 1.2 GPa at a temperature as high as 750°C. These excellent high-temperature mechanical properties are suggested [72] to be linked to the favourable interactions among dislocations, SFs and nanotwins. This work marks an important milestone to illustrate that even with conventional processing, decent high-temperature mechanical properties are attainable through composition optimization and heat treatment optimization. Such efforts need to be pursued in future research.

In a very encouraging recent effort [73], the role of the γ/γ' interfacial misfit dislocation network on the mechanical behaviour and the creep characteristics of Ni-based single crystal superalloys is reported. This work reveals the effect of the temperature, stress and Re atoms on the evolution of the interfacial dislocation network by utilizing the MD simulations on the (100), (110) and (111) phase interface models. Intelligent heat treatment and optimized ageing leads [74] to better tensile strength in a new conventionally wrought cast γ'-hardened Ni-based superalloy after thermal exposure at 650°C for 500 h, 1000 h and 2300 h.

Similarly, Ni-based superalloys processed [75] by conventional techniques, such as VIM and ESR, followed by a combination of cold rolling, intelligent heat treatments and optimized recrystallization by appropriate ageing treatment can show both high strength and ductility (e.g., 1386 MPa and 20.5%). It is reported [75] that the dislocations and deformation nanotwins induced by cold rolling and γ' particles lead to high strength. Further, the fine-grained microstructure, annealing twins and γ' particles coupled with a high density of intersected SFs are suggested [75] to lead to the desired combination of high strength and high ductility. It is expected that many more future works will pursue this domain to gain further improvement in the properties of Ni-based superalloys. Other researchers also reported interesting observations [76]

A major area of emerging importance is the application of machine learning (ML) and deep learning (DL) in the development of Ni-based superalloys [77, 78]. For instance, a very important recent work [76] uses ML to quickly obtain the optimized composition of Ni-based superalloys. This work starts with a numerical inverse method. This method is used to massively calculate the multi-element diffusion coefficients. This is work done based on an accurate atomic mobility database. Next, it uses these coefficients to refine the physical models with two objectives: (a) to tune the creep rates at high temperatures and (b) to confirm the structural stability of the superalloy at high temperatures. Next, the technique of unsupervised machine learning is used to find the range of compositions able to deliver optimal performance. Finally, two sets of compositions are zeroed in on and their performance prediction is also experimentally validated. Finally, it needs to be appreciated that a new cluster formula approach [79] is also reported. Such an approach is used for the optimization of the creep resistance of Ni-based single crystal superalloys.

Thus, additive manufacturing, conventional processing, unconventional complementary processing, applications of ML and DL, and the cluster formula approach appear to be some of the major areas of emerging importance. The pursuit of these areas will help to mitigate current and new challenges in the deformation and fracture of Ni-based superalloys.

15.15 SUMMARY AND CONCLUSION

This chapter focuses on the physics of deformation and fracture in Ni-based superalloys. First, the importance of the Ni-based superalloys is outlined briefly to highlight their importance as an advanced structural material suitable for numerous high-temperature and room-temperature applications. This is followed by a brief overview of the fundamental aspects of Ni-based superalloys to emphasize the relationship between different processing techniques and the resultant properties that are achieved. Further, the important mechanical properties of Ni-based superalloys are identified. Furthermore, the most important mechanical properties – strength and creep resistance – are briefly discussed. This is followed by a detailed overview of the literature on mechanical properties of Ni-based superalloys. Finally, emerging areas of importance are highlighted.

ACKNOWLEDGEMENTS

The authors acknowledge the kind support and encouragement received from various directors of the Council of Scientific and Industrial Research (CSIR)–Central Glass and Ceramic Research Institute (CGCRI), Kolkata, India, during the tenures of whom parts of the research work were carried out. They also gratefully acknowledge the various infrastructural support received from various divisions of CSIR-CGCRI, Kolkata, India. The financial support received from Indian sponsoring agencies, including CSIR, UGC, DST(SERB), DAE-BRNS, and IPR, of the Government of India (GOI) in terms of various project sponsorships and fellowships is gratefully acknowledged by the authors. Finally, the author AKM acknowledges the kind and continued support received from the authorities of Sharda University, Greater Noida, Uttar Pradesh, India.

REFERENCES

1. A Kracke, A Allvac, Superalloys, the most successful alloy system of modern times-past, present and future, *Proceedings of the 7th International Symposium on Superalloy*, Volume 718, 6 October 2010, pages 13–50.
2. E Akca, A Gürsel, A review on superalloys and IN718 nickel-based INCONEL superalloy, *Periodicals of Engineering and Natural Sciences (PEN)*, Volume 3(Part 1), 26 June 2015, https://www.researchgate.net/publication/296013959.

3. GR Thellaputta, PS Chandra, CS Rao, Machinability of nickel-based superalloys: A review, *Materials Today: Proceedings*, Volume 4(Part 2), 1 January 2017, pages 3712–3721, https://doi.org/10.1016/j.matpr.2017.02.266.
4. CT Sims, NS Stoloff, WC Hagel, *Superalloys II*, New York: Wiley, August 1987.
5. W Betterodge, J Heslop, *Nimonic Alloys and Other Nickel-Based High-Temperature Alloys*, 2 ed. Edward Arnold; London, United Kingdom, 1974.
6. SH Chen, SC Su, PC Chang, SY Chou, KK Shieh, The machinability of MAR-M247 superalloy, *Advanced Engineering Forum*, Volume 1, 2011, pages 155–159. Trans Tech Publications Ltd, https://doi.org/10.4028/www.scientific.net/AEF.1.155.
7. K Ankamma, AK Singh, KS Prasad, GC Reddy, M Komaraiah, NE Prasad, Effects of aging and sheet thickness on the room temperature deformation behavior and in-plane anisotropy of cold rolled and solution treated Nimonic C-263 alloy sheet, *International Journal of Materials Research*, Volume 102(Part 10), 1 October 2011, pages 1274–1285, https://doi.org/10.3139/146.110573.
8. K Bochenek, M Basista, Advances in processing of NiAl intermetallic alloys and composites for high temperature aerospace applications, *Progress in Aerospace Sciences*, Volume 79, 1 November 2015, pages 136–146, https://doi.org/10.1016/j.paerosci.2015.09.003.
9. MT Jovanović, B Lukić, Z Mišković, I Bobić, I Cvijović, B Dimčić, Processing and some applications of nickel, cobalt and titanium-based alloys, *Journal of Metallurgy*, 2007, Association of Metallurgical Engineers of Serbia, Volume 13, pages 91–106.
10. PK Wright, M Jain, D Cameron, *High Cycle Fatigue in a Single Crystal Superalloy: Time Dependence at Elevated Temperature, Superalloys*, 2004, pages 657–666.
11. RF Decker, *Strengthening Mechanisms in Nickel—Base Super Alloys, Climax Molybdenum Company Symposium*, Pergamon press, United States, 1969.
12. RC Reed, *The Superalloys: Fundamentals and Applications*, Cambridge University Press, New York, 31 July 2008.
13. RG Davies, NS Stoloff, On yield stress of aged Ni-Al alloys, *Transactions of the Metallurgical Society of AIME*, Volume 233(Part 4), 1 January 1965, page 714.
14. JP Dennison, RJ Llewellyn, B Wilshire, The creep and fracture behaviour of some Dilute Nickel Alloys at 500 and 600 C, *Journal of the Institute of Metals*, Volume 94(Part 4), April 1966.
15. N Das, Advances in nickel-based cast superalloys, *Transactions of the Indian Institute of Metals*, Volume 63(Part 2), 2010, pages 265–274, https://doi.org/10.1007/s12666-010-0036-7.
16. DK Ganji, G Rajyalakshmi, Influence of alloying compositions on the properties of nickel-based superalloys: A review, recent, *Advances in Mechanical Engineering*, 2020, pages 537–555, https://doi.org/10.1007/978-981-15-1071-7_44.
17. PE Aba-Perea, T Pirling, PJ Withers, J Kelleher, S Kabra, M Preuss, Determination of the high temperature elastic properties and diffraction elastic constants of Ni-base superalloys, *Materials and Design*, Volume 89, 5 January 2016, pages 856–863, https://doi.org/10.1016/j.matdes.2015.09.152.
18. E Liu, J Jia, Y Bai, W Wang, Y Gao, Study on preparation and mechanical property of nanocrystalline NiAl intermetallic, *Materials and Design*, Volume 53, 1 January 2014, pages 596–601, https://doi.org/10.1016/j.matdes.2013.07.052.
19. HL Zhao, F Qiu, SB Jin, QC Jiang, High work-hardening effect of the pure NiAl intermetallic compound fabricated by the combustion synthesis and hot-pressing technique, *Materials Letters*, Volume 65(Part 17–18), 1 September 2011, pages 2604–2606, https://doi.org/10.1016/j.matlet.2011.05.091.
20. GX Yang, YF Xu, L Jiang, SH Liang, High temperature tensile properties and fracture behavior of cast nickel based K445 superalloy, *Progress in Natural Science: Materials International*, Volume 21(Part 5), pages 418–425, https://doi.org/10.1016/S1002-0071(12)60078-1.
21. Q Zhu, G Chen, C Wang, H Qin, P Zhang, Tensile deformation and fracture behaviors of a nickel-based superalloy via in situ digital image correlation and synchrotron radiation X-ray tomography, *Materials*, Volume 12(Part 15), 2 August 2019, page 2461. https://doi.org/10.3390/ma12152461.
22. C Guo, J Yu, J Liu, X Sun, Y Zhou, Tensile deformation and fracture behavior of nickel-based superalloy DZ951G, *Materials*, Volume 14(Part 9), 27 April 2021, page 2250.
23. https://doi.org/10.3390/ma14092250.
24. L Cui, H Su, J Yu, J Liu, T Jin, X Sun, The creep deformation and fracture behaviors of nickel-base superalloy M951G at 900 C, *Materials Science and Engineering: A*, Volume 707, 7 November 2017, pages 383–391, https://doi.org/10.1016/j.msea.2017.09.066.
25. W Hermann, HG Sockel, J Han, A Bertram, *Elastic Properties and Determination of Elastic Constants of Nickel-Base Superalloys by Free-Free Beam Technique, Superalloys*, September 1996, pages 229–238, https://doi.org/10.7449/1996/superalloys_1996_229_238.

26. MJ Donachie, SJ Donachie, *Superalloys: A Technical Guide*, 2nd edition, ASM International, 2002.
27. P Auburtin, Steve Cockcroft, A Mitchell, T Wang, Freckle formation in superalloys, conference: Superalloys, https://doi.org/10.7449/2000/Superalloys_2000_255_261.
28. P Caron, T Khan, Evolution of Ni-based superalloys for single crystal gas turbine blade applications, materials science, *Aerospace Science and Technology*, Volume 3(8), 1999, pages 513–523, https://doi.org/10.1016/S1270-9638(99)00108-X.
29. H Fecht, D Furrer, Processing of Ni-base superalloys for turbine engine disc applications, *Advanced Engineering Materials*, Volume 2, 2000, pages 777–787, https://doi.org/10.1002/1527-2648(200012)2:12<777: AID-ADEM777>3.0.CO;2-R.
30. Maurice Gell, DN Duhl, AF Giamei, The development of single crystal superalloy turbine blades, *Materials Science: Superalloys*, 1980, pages 205–214, https://doi.org/10.7449/1980/SUPERALLOYS_1980_205_214.
31. Niranjan Das, Advances in nickel-based cast superalloys, *Transactions of the Indian Institute of Metals*, Volume 63(2–3), 2010, pages 265–274, https://doi.org/10.1007/s12666-010-0036-7.
32. Atsushi Sato, Johan J Moverare, Magnus Hasselqvist, Roger C Reed, On the mechanical behavior of a new single-crystal superalloy for industrial gas turbine applications, *Metallurgical and Materials Transactions. Part A*, Volume 43(7), 2012, pages 2302–2315.
33. JS Benjamin, Dispersion strengthened superalloys by mechanical alloying, *Metallurgical and Materials Transactions. Part B*, Volume 1(10), 1970, pages 2943–2951, https://doi.org/10.1007/BF03037835.
34. Z Zhu, H Basoalto, N Warnken, RC Reed, A model for the creep deformation behaviour of nickel-based single crystal superalloys, *Acta Materialia*, Volume 60(12), 2012, pages 4888–4900. https://doi.org/10.1016/j.actamat.2012.05.023.
35. M Morinaga, N Yukawa, H Adachi, Alloying effect on the electronic structure of Ni_3Al (γ), *Journal of the Physical Society of Japan*, Volume 53(2), 1984, pages 653–663, DOI:10.1143/JPSJ.53.653.
36. Kyoko Kawagishi, Atsushi Sato, Toshiharu Kobayashi, Hiroshi Harada, Effect of alloying elements on the oxidation resistance of 4th generation Ni-base single-crystal superalloys, *Journal of the Japan Institute of Metals*, Volume 69(2), 2005, pages 249–252, https://doi.org/10.2320/jinstmet.69.249.
37. D Blavette, P Caron, T Khan, an atom probe investigation of the role of rhenium additions in improving creep resistance of Ni-base superalloys, *Scripta Metallurgica*, Volume 20(10), 1986, pages 1395–1400, https://doi.org/10.1016/0036-9748(86)90103-1.
38. Walter W Milligan, Stephen D Antolovich, Yielding and deformation behavior of the single crystal superalloy PWA 1480, *Metallurgical Transactions. Part A*, Volume 18(1), 1987, pages 85–95, https://doi.org/10.1007/BF02646225.
39. Heeman Choe, David C Dunand, Synthesis, structure, and mechanical properties of Ni–Al and Ni–Cr–Al superalloy foams, *Acta Materialia*, 52(5), 2004, pages 1283–1295, https://doi.org/10.1016/j.actamat.2003.11.012.
40. Seyed Abdolkarim Sajjadi, Said Nategh, Roderick IL Guthrie, Study of microstructure and mechanical properties of high-performance Ni-base superalloy GTD-111, *Materials Science and Engineering: Part A*, Volume 325(1–2), 2002, pages 484–489, https://doi.org/10.1016/S0921-5093(01)01709-9.
41. IS Kim, BG Choi, HU Hong, J Do, CY Jo, Influence of thermal exposure on the microstructural evolution and mechanical properties of a wrought Ni-base superalloy, *Materials Science and Engineering: Part A*, Volume 593, 2014, pages 55–63, https://doi.org/10.1016/j.msea.2013.11.031.
42. Q Wang, DL Sun, Y Na, Y Zhou, XL Han, J Wang, Effects of TIG welding parameters on morphology and mechanical properties of welded joint of Ni-base superalloy, *Procedia Engineering*, Volume 10, 2011, pages 37–41, https://doi.org/10.1016/j.proeng.2011.04.009.
43. JH Du, XD Lü, JL Qu, Q Deng, JY Zhuang, ZY Zhong, Microstructure and mechanical properties of novel 718 superalloy, *Acta Metallurgica Sinica (English Letters)*, 19(6), 2006, pages 418–422, https://doi.org/10.1016/S1006-7191(06)62082-6.
44. Vikash Chaudhary, Ajaya Bharti, Syed Mohd Azam, Naveen Kumar, Kuldeep K Saxena, A re-investigation: Effect of TIG welding parameters on microstructure, mechanical, corrosion properties of welded joints, *Materials Today: Proceedings*, Volume 45(Part 6), 2021, pages 4575–4580, https://doi.org/10.1016/j.matpr.2021.01.007.
45. Li Shuqi Zhuang Jingyun, Xie Xishan Li Bing, Effects of microstructure on crack propagation rate of GH169 alloy, *Journal of Materials Engineering*, 5, 1998, pages 26–27.
46. CY Cui, YF Gu, DH Ping, H Harada, Microstructural evolution and mechanical properties of a Ni-based superalloy, TMW-4, *Metallurgical and Materials Transactions. Part A*, Volume 40(2), 2009, pages 282–291, https://doi.org/10.1007/s11661-008-9746-4.

47. Y Yuan, YF Gu, ZH Zhong, T Osada, CY Cui, T Tetsui, T Yokokawa, H Harada, Creep mechanisms of a new Ni-Co-base disc superalloy at an intermediate temperature, *Journal of Microscopy*, Volume 248(1), 2012, pages 34–41, https://doi.org/10.1111/j.1365-2818.2012. 03647.x.
48. J Mao, Keh-Minn Chang, Wanhong Yang, K Ray, S Vaze, D Ferrer, Cooling precipitation and strengthening study in powder metallurgy superalloy U720LI, Volume 32(10), 2001, pages 2441–2452, https://doi.org/10.1007/S11661-001-0034-9.
49. Qiuzhi Gao, Yujiao Jiang, Ziyun Liu, Hailian Zhang, Chenchen Jiang, Xin Zhang, Huijun Li, Effects of alloying elements on microstructure and mechanical properties of Co–Ni–Al–Ti superalloy, *Materials Science and Engineering: Part A*, Volume 779, 2020, pages 139139, https://doi.org/10.1016/j.msea.2020.139139.
50. YC Lin, Jiao Deng, Yu-Qiang Jiang, Wen Dong-Xu, G Liu, Effects of initial δ phase on hot tensile deformation behaviors and fracture characteristics of a typical Ni-based superalloy, *Materials Science and Engineering. Part A*, Volume 598, pages 251–262, https://doi.org/10.1016/j.msea.2014.01.029.
51. Baig Gyu Choi, In Su Kim, Dae-Hyun Kim, ETA phase formation during thermal exposure and its effect on mechanical properties in Ni-Base superalloy GTD 111, *Conference: Superalloys*, 2004, pages 163–171, https://doi.org/10.7449/2004/Superalloys_2004_163_171.
52. L Garimella, PK Liaw, DL Klarstrom, Fatigue behavior in nickel-based superalloys: A literature review, Volume 49(7), 1997, page 67, docview/232557406/se-2.
53. Xuebing Huang, Yun Zhang, Zhuangqi Hu, Effect of small amounts of nitrogen on properties of a Ni-based superalloy, *Metallurgical and Materials Transactions. Part A*, Volume 30(7), 1999, pages1755–1761.
54. Winston O Soboyejo, TS Srivatsan, *Book: Advanced Structural Materials Properties, Design Optimization, and Applications*, 2006, pages 1–528, https://doi.org/10.1201/9781420017465.
55. A Sawant, Sammy Tin, JC Zhao, High temperature nanoindentation of Ni-Base superalloys, *Conference: Superalloys*, 2008, pages 863–871, https://doi.org/10.7449/2008/Superalloys_2008_863_871.
56. Haibo Long, Shengcheng Mao, Yinong Liu, Ze Zhang, Xiaodong Han, Microstructural and compositional design of Ni-based single crystalline superalloys —— A review, *Journal of Alloys and Compounds*, Volume 743, 2018, pages 203–220, https://doi.org/10.1016/j.jallcom.2018.01.224.
57. Qingqing Ding, Hongbin Bei, Xia Yao, Xinbao Zhao, Xiao Wei, Jin Wang, Ze Zhang, Temperature effects on deformation substructures and mechanisms of a Ni-based single crystal superalloy, *Applied Materials Today*, Volume 23, 2021, page 101061, https://doi.org/10.1016/j.apmt.2021.101061.
58. Suketa Chaudhary, PJ Guruprasad, Anirban Patra, Crystal plasticity constitutive modeling of tensile, creep and cyclic deformation in single crystal Ni-based superalloys, *Mechanics of Materials*, Volume 174, 2022, page 104474, https://doi.org/10.1016/j.mechmat.2022.104474.
59. Kinga A Unocic, Dongwon Shin, Xiahan Sang, Ercan Cakmak, Peter F Tortorelli, Single-step aging treatment for a precipitation-strengthened Ni-based alloy and its influence on high-temperature mechanical behavior, *Scripta Materialia*, Volume 162, 2019, pages 416–420, https://doi.org/10.1016/j.scriptamat.2018.11.045.
60. Qianying Guo, Michael Kirka, Lianshan Lin, Dongwon Shin, Jian Peng, Kinga A Unocic, In situ transmission electron microscopy deformation and mechanical responses of additively manufactured Ni-based superalloy, *Scripta Materialia*, Volume 186, 2020, pages 57–62, https://doi.org/10.1016/j.scriptamat.2020.04.012.
61. MV Nathal, LJ Ebert, The influence of cobalt, tantalum, and tungsten on the elevated temperature mechanical properties of single crystal nickel-base superalloys, *Metallurgical and Materials Transactions. Part A*, Volume 16(10), 1985, pages 1863–1870, https://doi.org/10.1007/BF02670373.
62. Nana Kwabena Adomako, Nima Haghdadi, Sophie Primig, Electron and laser-based additive manufacturing of Ni-based superalloys: A review of heterogeneities in microstructure and mechanical properties, *Materials and Design*, Volume 223, 2022, page 111245, https://doi.org/10.1016/j.matdes.2022.111245.
63. Md Shahwaz, Prekshya Nath, Indrani Sen, A critical review on the microstructure and mechanical properties correlation of additively manufactured nickel-based superalloys, *Journal of Alloys and Compounds*, Volume 907, 2022, page 164530, https://doi.org/10.1016/j.jallcom.2022.164530.
64. Guowei Wang, Lan Huang, Liming Tan, Zijun Qin, Chao Chen, Feng Liu, Yong Zhang, Effect of yttrium addition on microstructural evolution and high temperature mechanical properties of Ni-based superalloy produced by selective laser melting, *Materials Science and Engineering: Part A*, Volume 859, 2022, page 144188, https://doi.org/10.1016/j.msea.2022.144188.
65. JC Guo, SX Han, RN Yang, XW Lei, N Wang, Influence of γ/γ' morphology evolution on the mechanical properties of Ni-based single crystal superalloys formed by laser metal deposition, *Materials Science and Engineering: Part A*, Volume 836, 2022, page 142714, https://doi.org/10.1016/j.msea.2022.142714.

66. Reza Ghiaasiaan, Arun Poudel, Nabeel Ahmad, Muztahid Muhammad, Paul R Gradl, Shuai Shao, Nima Shamsaei, Room temperature mechanical properties of additively manufactured Ni-base superalloys: A comparative study, *Procedia Structural Integrity*, Volume 38, 2022, pages 109–115, https://doi.org/10.1016/j.prostr.2022.03.012.
67. Chuan Guo, Zhuoyu Li, Yuhe Huang, Minglin He, Gan Li, Qingqing Li, Yang Zhou, Fan Zhou, Gang Ruan, Wangqing Wu, Qiang Zhu, Heterogeneous structures in a Ni-based superalloy with tailoring mechanical properties produced by a twin laser powder bed fusion system, *Journal of Alloys and Compounds*, Volume 927, 2022, page 166983, https://doi.org/10.1016/j.jallcom.2022.166983.
68. Yu Li, Xiaogang Hu, Haifeng Liu, Qiulin Li, Qiang Zhu, High-throughput assessment of local mechanical properties of a selective-laser-melted non-weldable Ni-based superalloy by spherical nanoindentation, *Materials Science and Engineering: Part A*, Volume 844, 2022, page 143207, https://doi.org/10.1016/j.msea.2022.143207.
69. Lairong Xiao, Zhenwu Peng, Xiaojun Zhao, Xiaoxuan Tu, Zhenyang Cai, Qi Zhong, Sen Wang, Huali Yu, Microstructure and mechanical properties of crack - Free Ni - Based GH3536 superalloy fabricated by laser solid forming, *Journal of Alloys and Compounds*, Volume 921, 2022, page 165950, https://doi.org/10.1016/j.jallcom.2022.165950.
70. Pengcheng Wen, Chang Wang, Peishan Zhou, Mei Yang, Liang Wang, Bin Wang, Haiyang Wang, Effect of ultrasonic treatment on the microstructure and mechanical properties of a new Ni-base superalloy 4716MA0, *Journal of Alloys and Compounds*, Volume 922, 2022, page 166278, https://doi.org/10.1016/j.jallcom.2022.166278.
71. Stoichko Antonov, Kyle A Rozman, Paul D Jablonski, Martin Detrois, Implications of carbon content for the processing, stability, and mechanical properties of cast and wrought Ni-based superalloy Nimonic 105, *Materials Science and Engineering: Part A*, Volume 857, 2022, page 144049, https://doi.org/10.1016/j.msea.2022.144049.
72. Yayun Li, Pingwei Xu, Wei Jiang, Lei Zhou, Zihao Jiang, Yilong Liang, Yu Liang, Effect of gradient microstructure on elevated temperature mechanical properties of Ni-based superalloy ATI 718Plus, *Materials Science and Engineering: Part A*, Volume 843, 2022, page 143124, https://doi.org/10.1016/j.msea.2022.143124.
73. Zhongrun Xiao, Junyang He, Ji Gu, Bin Gan, Yu Hongyao, Zhongnan Bi, Jinhui Du, Min Song, Tensile properties and deformation mechanisms of a new Ni–Co base superalloy from room temperature up to 750 °C, *Intermetallics*, Volume 150, 2022, page 107697, https://doi.org/10.1016/j.intermet.2022.107697.
74. Amir R Khoei, Mehrdad Youzi, G Tolooei Eshlaghi, Mechanical properties and γ/γ' interfacial misfit network evolution: A study towards the creep behavior of Ni-based single crystal superalloys, *Mechanics of Materials*, Volume 171, 2022, page 104368, https://doi.org/10.1016/j.mechmat.2022.104368.
75. Hao Liu, Xinbao Zhao, Yong Yuan, Yingying Dang, Weiqi Li, Jiachen Xu, Yuan Cheng, Quanzhao Yue, Yuefeng Gu, Ze Zhang, Influence of thermal exposure on microstructural stability and tensile properties of a new Ni-base superalloy, *Journal of Materials Research and Technology*, Volume 21, 2022, pages 4462–4472, https://doi.org/10.1016/j.jmrt.2022.11.050.
76. Xingmao Wang, Yutian Ding, Yu Hongyao, Zhongnan Bi, Yubi Gao, Bin Gan, High strength and ductility in a novel Ni-based superalloy with γ' and nanotwins / stacking faults architectures, *Materials Science and Engineering: Part A*, Volume 847, 2022, page 143293, https://doi.org/10.1016/j.msea.2022.143293.
77. Mohsen Mehdizadeh, Hassan Farhangi, Effects of different elevated temperature and long-term exposure on microstructural evolution and mechanical characteristics of IN617 Ni-based superalloy, *Materials Science and Engineering: Part A*, Volume 841, 2022, page 143025, https://doi.org/10.1016/j.msea.2022.143025.
78. Feng Liu, Zexin Wang, Zi Wang, Jing Zhong, Lei Zhao, Liang Jiang, Runhua Zhou, Yong Liu, Lan Huang, Liming Tan, Yujia Tian, Han Zheng, Qihong Fang, Lijun Zhang, Lina Zhang, Hong Wu, Lichun Bai, Kun Zhou, High-throughput method–accelerated design of Ni-based superalloys, advanced functional materials, first published, 04 May 2022, https://doi.org/10.1002/adfm.202109367.
79. Chen Chen, Qing Wang, Chuang Dong, Yu Zhang, Honggang Dong, Composition rules of Nibase single crystal superalloys and its influence on creep properties via a cluster formula approach, *Scientific Reports*, 10, 2020, page 21621, https://doi.org/10.1038/s41598-020-78690-8.
80. Zijun Qin, Weifu Li, Zi Wang, Junlong Pan, Zexin Wang, Zihang Li, Guowei Wang, Jun Pan, Feng Liu, Lan Huang, Liming Tan, Lina Zhang, Hua Han, Hong Chen, Liang Jiang, High-throughput characterization methods for Ni-based superalloys and phase prediction via deep learning, *Journal of Materials Research and Technology*, Volume 21, 2022, pages 1984–1997, https://doi.org/10.1016/j.jmrt.2022.

16 Nanoindentation Studies on Physics of Deformation at Microstructural Length Scale of Metals

Sangeeta Rawal and Anoop Kumar Mukhopadhyay

16.1 INTRODUCTION

A large number of reports are available in our previous publication on the basic aspects and applications of the nanoindentation technique [1–3]. Therefore, the basics will not be reported once again for the sake of brevity. Further, a great deal of details are also provided in the companion Chapter 2, which focuses on the nanoindentation of various materials. Furthermore, the companion Chapter 5 also focuses especially on the deformation and fracture as well as various aspects of nanoindentation of ceramic materials. In addition, Chapter 18 provides how modern techniques like artificial intelligence (AI), machine learning (ML), and artificial neural networks (ANNs) could be utilized to decipher nanoindentation data of ceramics. However, the techniques are shown to be so general that they could be easily applied to the nanoindentation of metallic samples. Thus, the major focus of the current chapter is what has been already done in the case of nanoindentation of metals, what is the current literature scenario, and what are the most recent areas of emerging importance.

Nominally speaking, the nanoindentation technique provides experimental evaluation of the nanoscale/microstructural length scale mechanical properties of thin films and coatings as well as bulk samples of metals, ceramics, polymers, and composites. In these experiments, a continuous record of the applied load (P) versus depth (h) of penetration produced by (P) is recorded by using appropriate dedicated transducers. Since the tip typically has a radius of about 10–300 nm, the local stress is high enough to cause elastic and plastic deformation in a given metal. The experiments can be conducted in a conventional nanoindenter [3–6]. Most machines have a dedicated scanning probe microscope that provides high-resolution images of the nanoindents with typical ultralow loads in the range of 1–10,000 μN. The typical load sensing resolution in a commercial machine could be as good as 1 nN or better. Similarly, the depth sensing resolution could be as good as 0.04 nm or better. Depending on the nanoindenter tip shape, e.g., Berkovich, cube corner, Vickers, spherical, sphero-conical, and conical, different contact areas are produced on the sample surface. These contact areas are estimated by accurate measurement of the depth of penetration [1–4]. Thus, the nanohardness (H) is estimated as the ratio of load to the corresponding contact area. Similarly, from the experimental measurement of the stiffness of the sample and the nanoindenter which forms a series, Young's modulus (E) of the sample is estimated utilizing the known values of Poisson's ratio of the sample as well as the nanoindenter and Young's modulus of the nanoindenter. In most cases, a diamond nanoindenter is utilized. Young's modulus is almost globally accepted as 1000 GPa [4]. Many other important mechanical properties, such as fracture toughness (K_{Ic}) and residual stress can also be evaluated by the nanoindentation technique [1–3]. One of the most celebrated and globally accepted models utilized for the evaluation of (H) and (E) is due to the famous work of Oliver and Pharr [4]. Due to the large number of slip systems available, the metals can always deform by

DOI: 10.1201/9781003359364-20

plastic deformation. However, due to the intrinsic limitation on accurate measurement of (h), more often than not the projected area of contact is underpredicted. As the contact area is smaller than it should be at the corresponding ultralow loads, the ratio becomes relatively higher. This leads to an apparent enhancement in the nanohardness (H) at the ultralow depth in metals. However, at relatively higher loads, the depth of penetration increases. So the projected area of contact increases. Hence, the magnitude of (H) decreases. This apparent increase in (H) at ultralow load and depth and its stabilization to a relatively lower magnitude at corresponding higher load and depth, gives rise to what is most popularly known as the indentation size effect (ISE).

The most celebrated model that successfully explains the ISE, especially in metals, is the Nix and Gao model [5]. There are statistically stored dislocations (SSDs) available in metal to accommodate the strains generated due to thermal processes. When a nanoindentation is made, extra strain is produced in the metal. To accommodate this extra strain, geometrically necessary dislocations (GNDs) are created over and above the SSDs. These GNDs cause additional localized work hardening in metals over and above that caused by the SSDs. As a result, the highest change in the rate of change in strain at the lowest depth, e.g., the ultrasmall depth of penetration created by the ultralow load application. This process causes the maximum work hardening locally in the metal. Consequently, the magnitude of (H) increases. As the load is enhanced further, the magnitude of (h) is enhanced. At such relatively larger (h), the rate of change of strain with (h) becomes more reduced. This reduces the amount of extra work hardening. As a result, the magnitude of (H) drops with an increase in (P) and at higher depth. This is believed to be the mechanism of ISE in metals [5]. Further, it needs to be appreciated that there are many critiques of this simple picture. Therefore, many efforts are dedicated to developing a better understanding of the genesis of ISE in nanoindentation of metals [6–8]. Further, it is worth noting that there are three major ways in which nanoindentation experiments are done. The first type is done of course using a nanoindenter at room temperature and at high temperature. The second type is done using in situ nanoindentation experiments conducted inside a transmission electron microscope (TEM). The third type is done using atomic force microscopy (AFM).

Numerous works report on the nanoindentation of metals and metal matrix composites (MMCs). It is impossible to enumerate all of them within the smaller boundaries of this book chapter. Rather, it is decided to present a brief overview of some chosen work covering mainly the last decade to date. This is deliberately done to highlight the different varieties of angles in the structure–property relationship of metals that is being probed today by the nanoindentation technique.

16.2 NANOINDENTATION OF METALS: GLIMPSE OF THE JOURNEY DURING THE LAST DECADE

In a very interesting experiment reported [9] in 2014, in situ nanoindentation experiments are conducted inside a TEM. The results of these experiments reveal that inhomogeneous twin boundaries (ITBs) in nanotwinned (nt) aluminum can significantly affect the nanomechanical properties. It is shown that the ITBs put forward a very strong resistance against the pile-up of dislocations. As a result, a significant amount of work hardening happens. The higher the number of interactions between the ITBs and the dislocations, the higher the number of steps noted in the (P–h) plots. These steps indicate dislocation transmission across the ITBs. Thus, the ITBs play a major role in accommodating more localized plasticity at stress as high as 500 MPa [9]. Figure 16.1 shows a typical scenario of dislocation nucleation-dominated work hardening in nt Al under in situ nanoindentation experiments conducted inside a TEM.

Another related significant development in the same year reports [10] on the huge increase of about 31% in Young's modulus and a –7% reduction in nanohardness of Ni single crystal when a saturation magnetic field of about 2000 Oe is applied during the nanoindentation experiments (Figure 16.2). It is suggested that this happens due to the interaction between the dislocations and

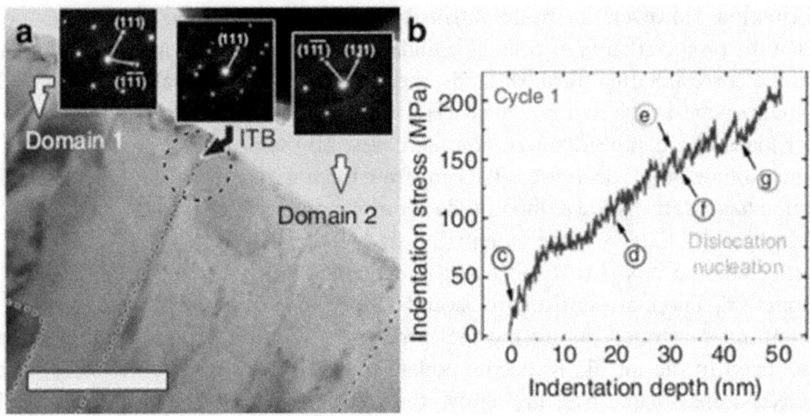

FIGURE 16.1 (a) Cross-section bright-field TEM micrograph (showing an overview of the area of interest before indentation) of nt Al, near Al [0-11] zone axis (scale bar, 200 nm). (b) Plot of indentation stress as a function of indentation depth for indentation cycle 1 [9].

the applied magnetic field. As a result, the plasticity of the Ni single crystal is intensified [10]. The importance of this lies in the fact that nanoindentation experiments can be utilized to study the interaction between the magnetic field and dislocations in magnetic metals.

Further, in situ nanoindentation experiments are reported [11] to be conducted in 2014 on Al-TiN multilayered nanocomposites. These experiments are conducted inside a TEM with a W tip. The

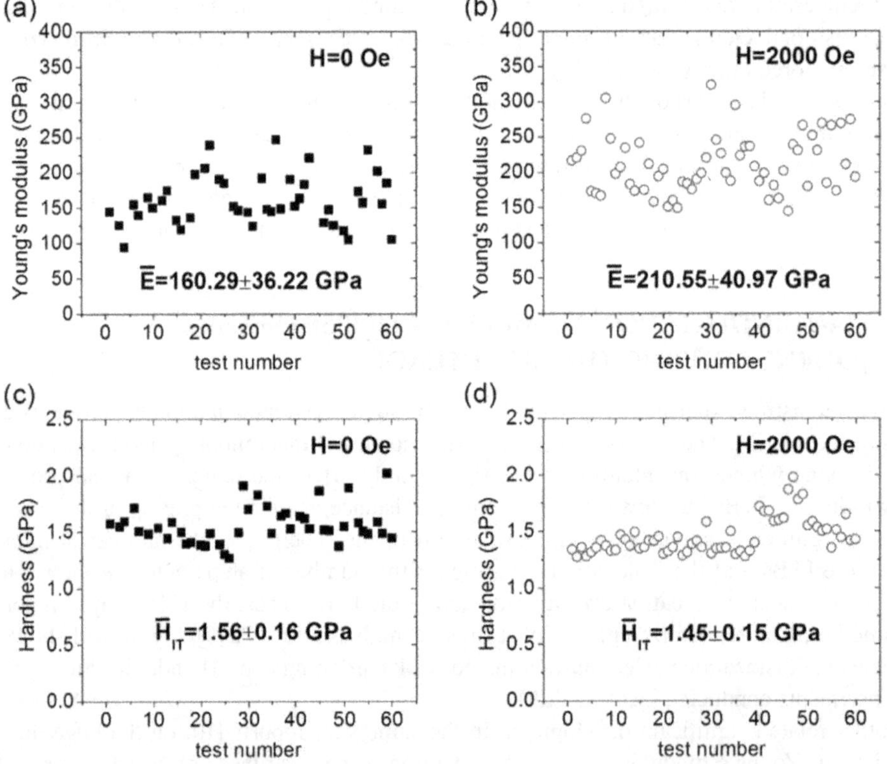

FIGURE 16.2 For Ni (111) single crystal, the variations of Young's modulus in magnetic field (a) 0 Oe, (b) 2000 Oe; and hardness in magnetic field (c) 0 Oe, (d) 2000 Oe [10].

results provide evidence that (Figure 16.3) under the high applied stress and due to the strong bonding, the otherwise plastically non-deformable ceramic layer of thickness 2.7 nm can plastically co-deform with the Al layer of the same thickness.

This is a very remarkable result that proves that to have significant plastic co-deformation, the thickness of both the Al layer and the TiN layer will need to be reduced to about 2.7 nm [11]. Furthermore, the intelligent combination of finite element modeling based on crystal plasticity theory and actual nanoindentation experiments reported in 2015 [12] establishes that the P–h plots (Figure 16.4) as well as the pile-up and the angles of lattice rotation can be reasonably well predicted for different surfaces, e.g., (001), (101), and (111) surfaces of Al single crystals.

A completely new recent approach [13] of nanoindentation with spherical nanoindenters with tip radii of 1, 10, and 100 μm is utilized to characterize the He ion irradiated W metal samples. This work was reported in 2017. This work shows that the spherical nanoindentation can provide elastic moduli during both loading and unloading. In addition, the nanoindentation-based yield strength can be estimated. Further, the post-yield strain response of W samples could be correlated with changes in the microstructure after irradiation. The microstructure is characterized by both EBSD (electron backscatter diffraction) and TEM. Furthermore, it provides new insight into the correlation of microstructure in different layers of the sample as obtained by EBSD and TEM with the corresponding nanomechanical response (Figure 16.5).

Another recent attempt reported in 2019 considers the pop-ins and elastic-to-plastic transition in Berkovich indentations of pure polycrystalline iron [14]. This work confirms that even a single P–h plot can be home to numerous pop-ins (Figure 16.6) induced by a multitude of mechanisms, e.g., dislocation nucleation, movement of previously pinned dislocations at interstitials, as well as interactions between the nanoindentation-induced stress and solute carbon atoms at the grain boundaries. These mechanisms can contribute either individually or in a combined manner. It is also reported that the random generation process of pop-ins in pure polycrystalline iron is characteristically stochastic, although it is relatively insensitive to crystallographic orientations.

From an important work reported recently in 2019 [15], it is found that the nanoindentation with a Berkovich nanoindenter can resolve the composition dependence of three high entropy alloys (HEAs) with at.% compositions of Fe40Mn20Co20Cr15Si5 (CS-HEA), Fe39Mn20Co20Cr15Si5Al1 (Al-HEA), and Fe38.5Mn20Co20Cr15Si5Cu1.5 (Cu-HEA) (Figure 16.7). Further, the annealing of

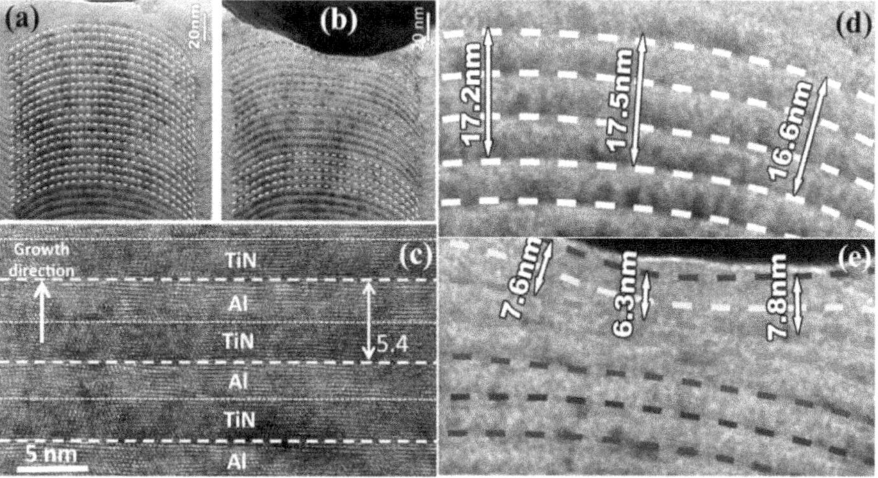

FIGURE 16.3 The TEM images of the 2.7 nm Al–2.7 nm TiN multilayer (a) before and (b) during the in situ nanoindentation by the W tip inside the TEM. (c–e) Dashed lines are Al-TiN interfaces. Following [11], the thickness reduction in the first three bilayers is from 17.5 nm to 6.3 nm [11].

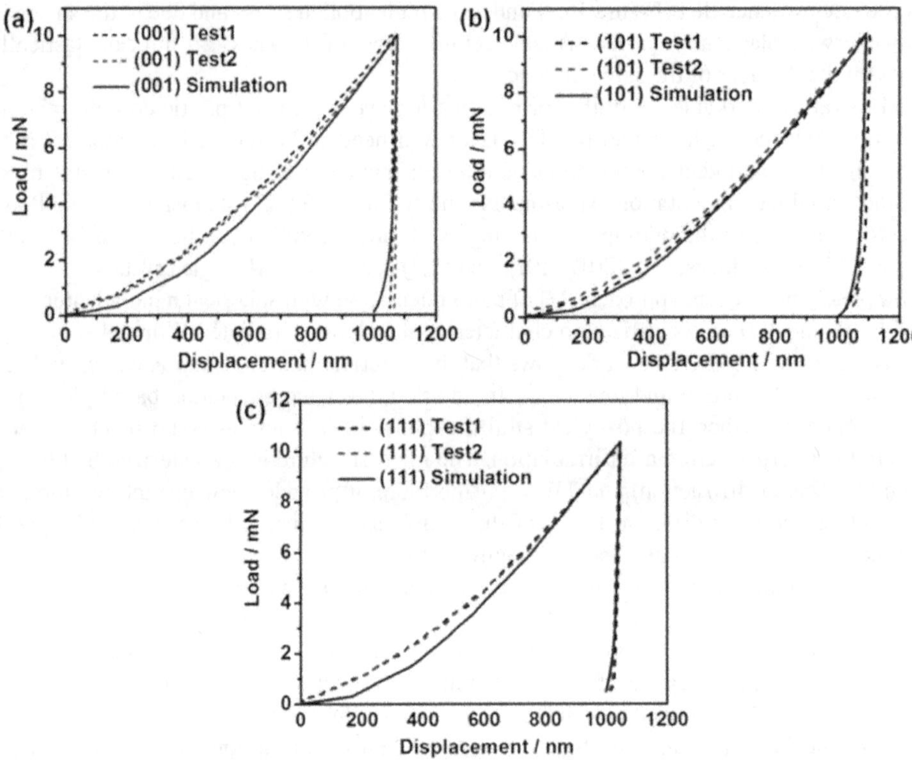

FIGURE 16.4 Comparison of experimentally measured and numerically predicted P–h plots in nanoindentation of Al single crystal surfaces. (a) (001), (b) (101), and (c) (111) [12].

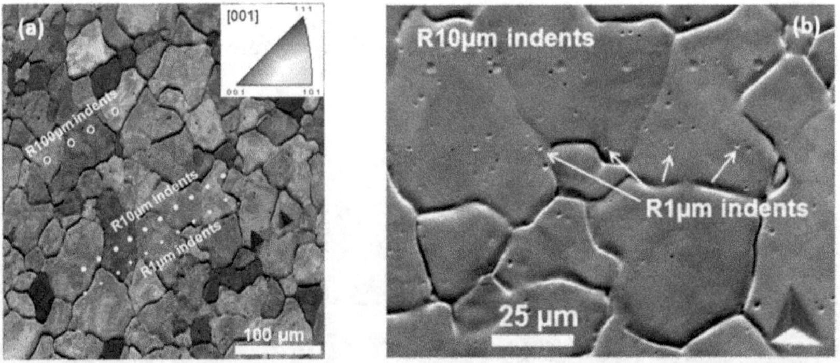

FIGURE 16.5 (a) EBSD image of nanoindents made with spherical nanoindenters with different tip radii of 1, 10, and 100 μm (artificially colored EBSD image with the inverse pole figure shown in the inset). (b) High magnification SEM images of the nanoindents shown in panel a with tip radii of 1 and 10 mm [13].

the as-cast Cu-HEA is shown [15] to provide the best combination of nanohardness (about 8.3 GPa) and Young's modulus (about 222 GPa).

Work reported in 2020 [16] proves that when indented at a maximum load of 1 mN by a Berkovich nanoindenter, a universal power law scaling exists between the stress drops, and corresponding loads measured for the second and subsequent pop-ins generated on (100) and (111) surfaces of body-centered cubic (BCC) Fe single crystals (Figure 16.8) and on (100) surface of face-centered cubic (FCC) Cu.

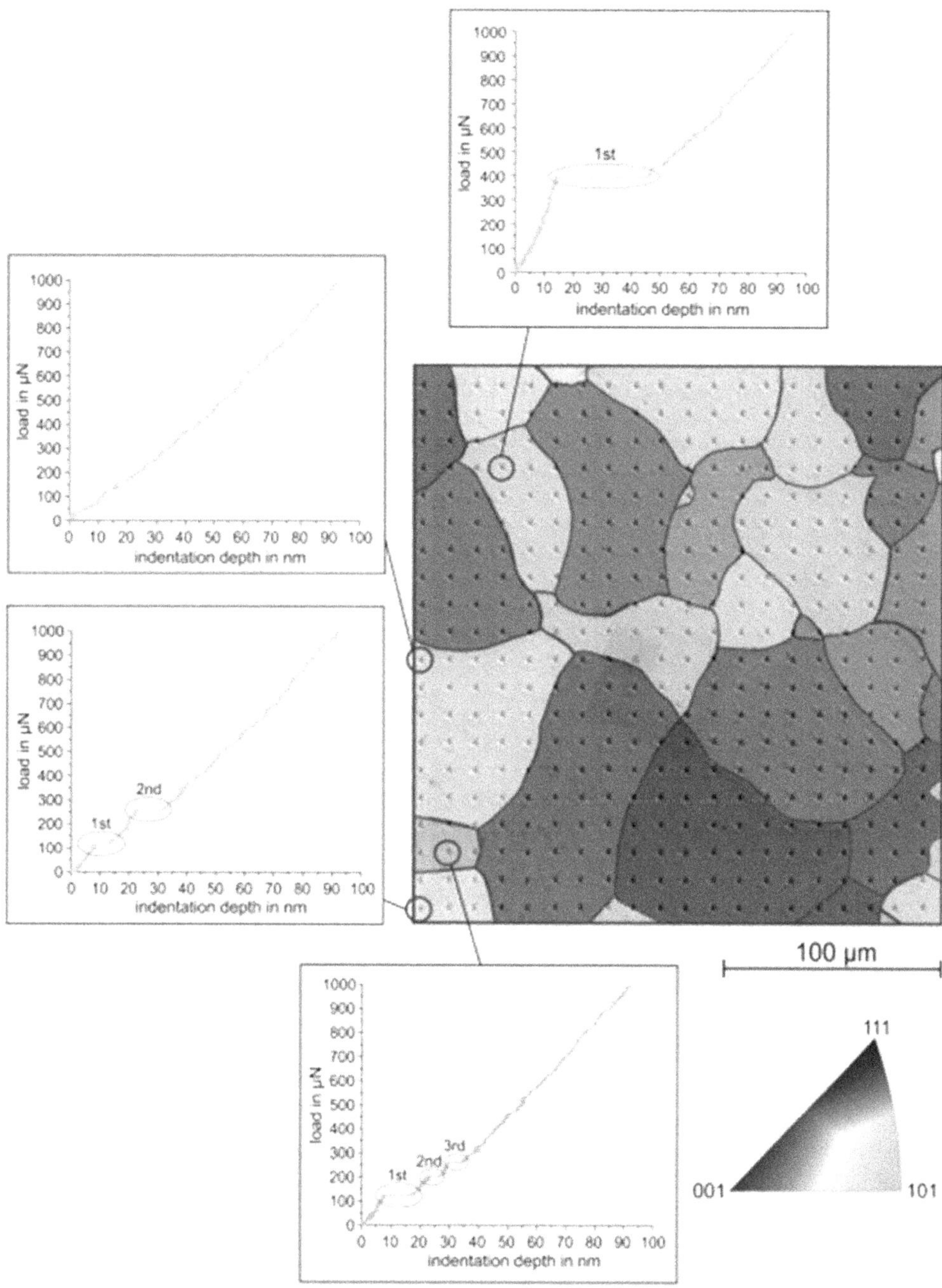

FIGURE 16.6 Array of Berkovich nanoindentation marks shown in pure iron along with an artificially colored and superimposed EBSD microstructure showing 0, 1, 2, and 3 pop-in events associated with the corresponding nanoindentation cavity formations on the sample surface [14].

The first pop-ins shown [16] to be linked to dislocation nucleation by MD simulation are noted to follow Gaussian-like statistics characterized by an average value, as expected. On the other hand, the second and subsequent pop-ins, which are shown to follow a power law dependency as mentioned earlier, are found by the MD simulation to be linked to dislocation network evolution leading to dislocation avalanche formation [16]. Finally, very recent work reported in 2022 [17] on additive

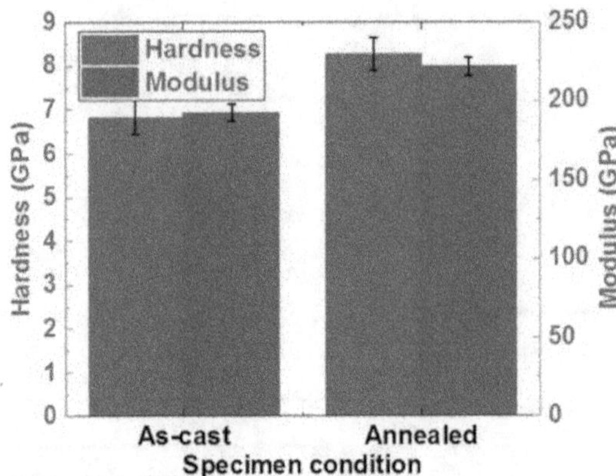

FIGURE 16.7 Variation of hardness and Young's modulus of the Cu HEA with processing conditions, e.g., as-cast and annealed [15].

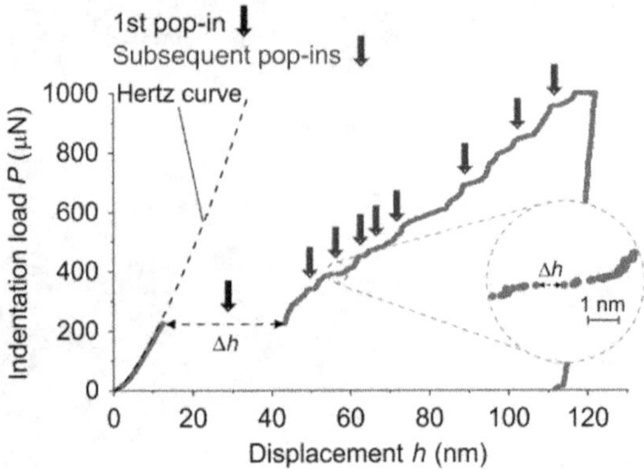

FIGURE 16.8 The occurrence of first and subsequent pop-ins in the P–h plot of (100) surface of BCC Fe single crystal during nanoindentation at a peak load of 1 mN [16].

manufacturing of metals shows that a very high value of nanohardness is achievable (Figure 16.9) when 3D hydrogels infused with metal precursors are calcined and ultimately reduced to convert the scaffolds to miniaturized, about 40 μm, metallic replicas.

Thus, it is evident from the preceding discussion that nanoindentation has been applied by a wide variety of means to explore the deformation of mainly metals and MMCs. Next, a brief survey of the literature is presented.

16.3 LITERATURE REVIEW ON NANOINDENTATION OF METALS

Nanoindentation is now established as a very useful tool for the measurement and analysis of mechanical properties of metals at small length scales, as mentioned earlier. With the help of the nanoindentation test phase transformations, dislocation source activation, hardness, fatigue, and other parameters can be found out. In certain studies, it is found that the surface contacts between many materials are highly dependent on their mechanical properties, and nanoindentation these days has become ubiquitous to measure these mechanical properties (such as elastic modulus and hardness) at the surface. The nanoindentation response of the material depends on the corresponding length scales, such as indentation depth, which is commonly termed the size effect. Nanoindentation can also be used for in situ and ex situ imaging, high-temperature testing, and acoustic emission

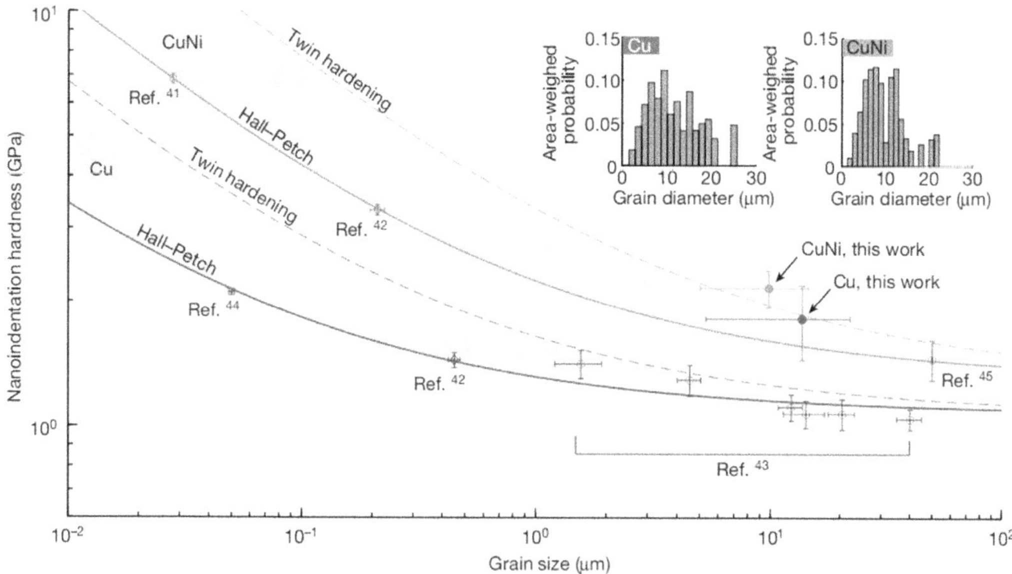

FIGURE 16.9 Variation of nanoindentation hardness as a function of grain size for Cu and Cu-Ni alloys additively manufactured by calcination and reduction of 3D hydrogels infused with metal precursors to convert the scaffolds to miniaturized, about 40 μm, metallic replicas [17]. Data from the literature are incorporated [17] for comparison only. Insets show grain size distributions for metallic Cu and Cu-Ni alloys.

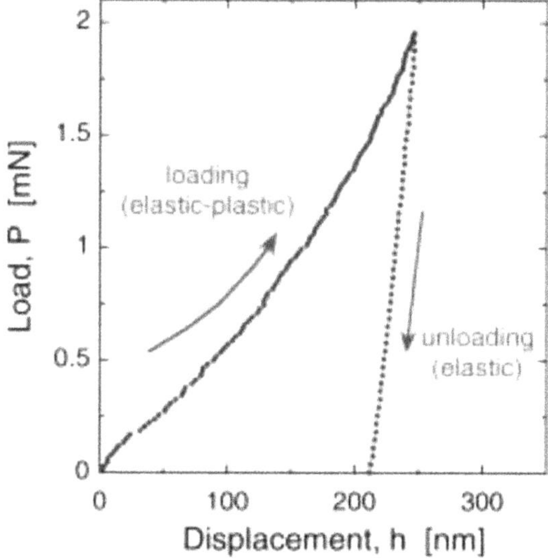

FIGURE 16.10 Example of a typical load–displacement curve (P–h curve) obtained from nanoindentation experiment of single-crystal Pt (100), an elastic-plastic material [18].

detection to analyze phase transformation, strain localization, and mechanical instability. The main components required in nanoindentation experiments are the specimen, actuators, and sensors, and indenter displacement and mechanical loads are measured. The indenter tip is generally made of diamond, and during a nanoindentation test, the indenter tip is pressed into the material's surface, and force and displacement are recorded. This process itself generates the experimental load (P) versus depth (h) plot (Figure 16.10).

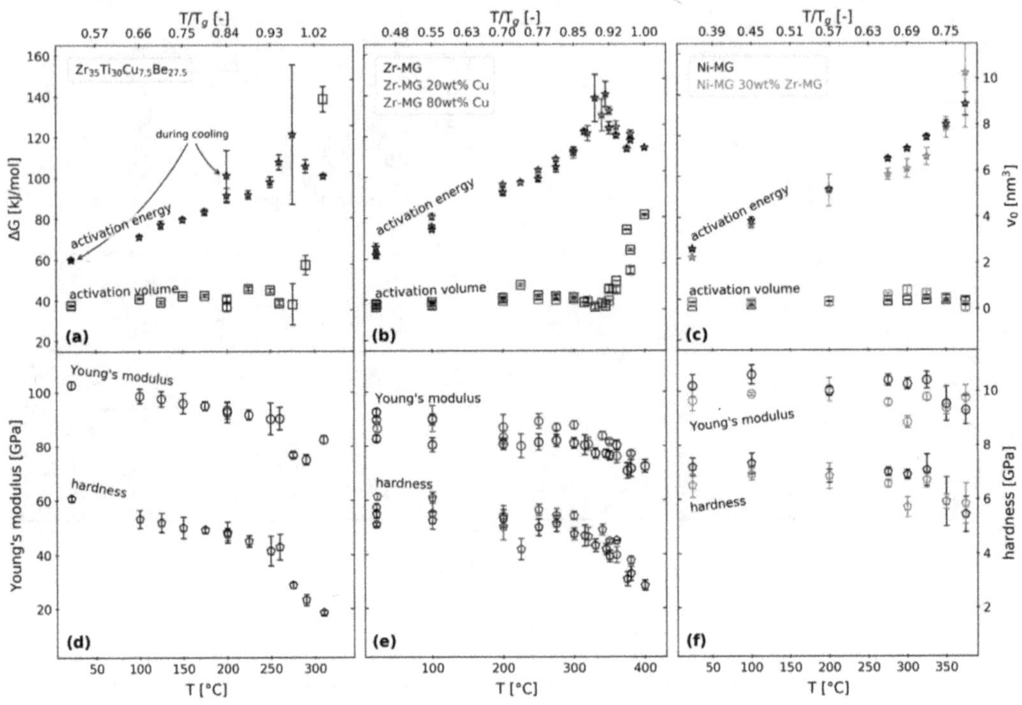

FIGURE 16.11 Plot of activation energy, Young's modulus, hardness, and activation volume, as a function of temperature for three different BMGs and BMGCs [26].

The P–h curve differs with material due to the different mechanical properties of the material. The P–h curve for indentation on single crystal Pt is shown in Figure 16.11 [18]. An important development during the last few decades is the development of bulk metallic glasses based on appropriately chosen metallic alloy compositions and advanced rapid cooling systems. They have huge application prospects. Since they are basically developed from metals, it is decided to show how their deformation is characterized by the nanoindentation system.

Bulk metallic glasses (BMGs) generally possess amorphous structures due to their short-range order. By changing these short-range orders by varying the annealing temperature and composition, certain intrinsic properties (such as toughness and hardness) of the BMGs [19–21] can be significantly improved. Nanoindentation is noted [22–26] to be an important tool for characterizing the small length scale deformation of various BMGs. The results of the nanoindentation strain rate jump tests [26] conclude that the activation energy and activation volume of these BMGs can be determined by the nanoindentation strain rate jump test. The Ni-MG and Zr-MG samples are reported to be [26] thermally cycled between room temperature and liquid nitrogen temperature. The respective homologous temperature for each plot is indicated on the top x-axis. Graphs display a decrease of Young's modulus and hardness with increasing temperature (more abruptly at T approaching T_g). The activation volume remains constant up to T_g and then shows a rapid increase. The activation energy increases until T_g is reached. Its magnitude drops at still higher temperatures.

It is noted further that [26] the activation volume and energy in these BMGs are not greatly influenced by the synthesis method, but the nanomechanical properties are greatly influenced by the temperature, especially near T_g (glass transition temperature). The variations of Young's modulus, activation energy, hardness, and activation volume with the number of thermal cycles [26] are displayed in Figure 16.12.

In nanoindentation experiments, the dependency of specimen hardness on the characteristic length is known as the size effect, and the mechanism for this effect depends on the nature of the

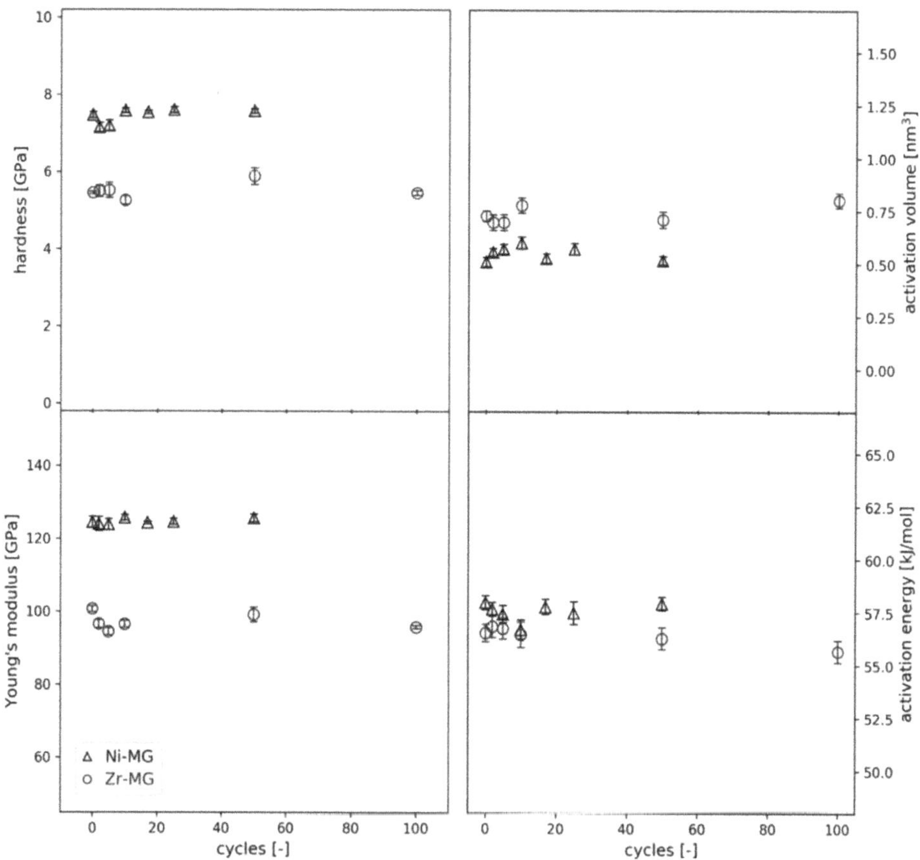

FIGURE 16.12 Plots of Young's modulus, activation energy, hardness, and activation volume as functions of the number of thermal cycles [26].

material. Such as for crystalline materials, the dislocation-based mechanism is responsible for the size effect, whereas for ceramics, polymers, amorphous metals, and semiconductor materials, the size effect is controlled by the phase transformation, shear transformation zones, non-dislocation mechanism, and cracking [27–30]. Various experimental studies have shown that the hardness of a material is dependent on the penetration depth (also referred to as indentation depth) and is known as the indentation size effect [31–34].

Some studies show that hardness increases with a decrease in indentation depth, whereas some literature studies have also shown a reverse size effect, that is, a decrease in hardness with a decrease in indentation depth (inverse size effect) [35, 36]. The inverse size effect could be due to instrumental errors or errors occurring during contact area measurement [37–39].

Besides these experimental techniques, various computer simulations are also very prevalent in the nanoindentation of metals. These include crystal plasticity [40], finite element [41], discrete dislocation dynamics [42], molecular dynamics [43], and quasi quantum method [44]. These techniques are also utilized to understand the physics of deformation at small length scales [45].

The hardness of a metal measured by the nanoindentation experiment is greatly influenced by the impurities present in the material. Recent work reports [46] on the effect of impurities in different FCC metals (Al, Cu, Ag, Ni, and Pb) on their hardness by the nanoindentation experiment. The hardness and Young's modulus of FCC metals are evaluated by the continuous stiffness method [46]. The hardness measured is displayed in Figure 16.13. It displays three distinct regions. It is found [46] that for Al and Pb, hardness decreases with an increase in indentation at high depths, and

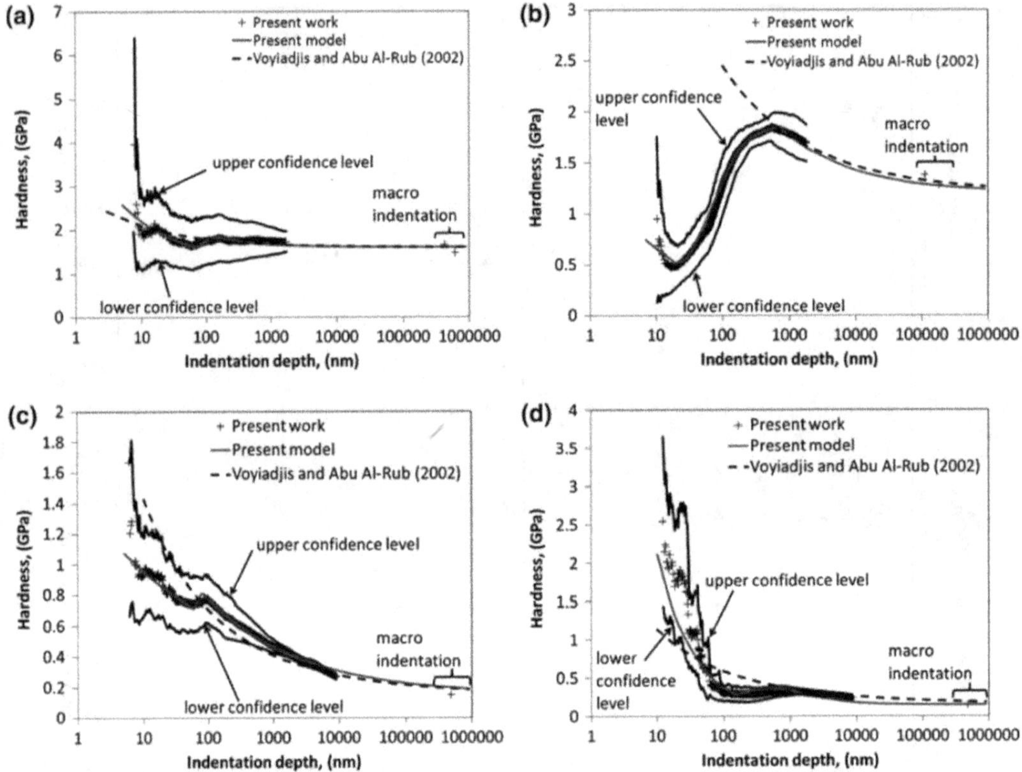

FIGURE 16.13 Hardness versus indentation depth plots for (a) copper (99.996% purity), (b) copper (99.9999% purity), (c) aluminum (99.99% purity), (d) aluminum (99.9999% purity), (e) silver (99.9% purity), (f) annealed silver (99.9% purity), (g) silver (99.9985% purity), and (h) nickel [45].

lower and upper confidence levels converge to the experimental data in region I because indentation depth can be reached up to 10 μm in these samples. The depth of 10 μm is not obtained for Cu, Ag, and Ni because these metals are comparatively harder than Al and Pb. The convergence of lower and upper confidence levels at higher indentation depth can be attributed to two reasons. First, at higher depths the material is homogeneous. Second, at higher depths the effect of crystal orientation and size is minimum.

However, at smaller indentation depths confidence levels diverge. This is attributed [46] to the pressing of the indenter on a grain boundary or on a grain far away from the grain boundary. In all the metals, except Pb, hardness decreases in regions I and III with an increase in indentation depth. It is interesting to note [46] that in region II where hardness increases with indentation, depth is not present in Al (purity 99.9999%) and Ag (purity 99.985%). Also, note that purity does not reduce in region II because a highly pure sample of Cu exhibits [46] clearer in region II in comparison to its extent in the low purity sample. Hence it can be concluded that region II in each case is mainly dominated by the grain boundaries, which causes a hindrance for the dislocation movement, and hence hardness increases with an increase in indentation depth. On the other hand, it is suggested [45] that in region I and region III, the strain gradient mechanism is the dominating factor and hardness decreases with an increase in indentation depth [46].

High-temperature nanoindentation of metals involves heating the sample and the indenter [47–49]. A recent study [49] uses the high-temperature nanoindentation approach to study the plasticity upon phase transformations on Co metal [49]. The bulk phase transformation and its effect on elasticity and plasticity are analyzed by the micromechanical methods in pure Co from RT to 873

K in both the HCP (hexagonal close-packed) and the FCC phases. The phase transformation in Co is confirmed by DSC (differential scanning calorimetry) and high-temperature X-ray diffraction around 700 K. It is observed that between RT and 473 K, plasticity is controlled mainly by the lattice friction. This enhances the strain rate sensitivity, and activation volumes increase with temperature (from about 40 b^3 to 80 b^3, where b is Burgers vector). This also leads to the increase of activation energy in the range of 1.5 to 2 eV for plastic deformation.

If the temperature is further increased, the strain rate sensitivity increases and activation energy decreases (0.8 to 1.0 eV) due to thermal-activation-initiated dislocation cross slip [49]. Deformation twinning (identified as the $\{10\bar{1}\,2\}$ $\langle\bar{1}2\ 10\rangle$ tension twin mode at RT) is noted [49] to exist at all temperatures in the HCP phase. If the temperature is raised to 523 K, a rising twinning activity is observed [49]. It happens due to a thermal activation of twin nucleation. With a further increase in temperature, twinning activity reduces due to a decrease in the number of dislocations. Further, it is reported [49] that a high-temperature FCC phase also exists. It has [49] high activation volumes of about 350 to 450 b^3, a low strain rate sensitivity (0.009), and activation energies in the range of about 3.5 to 4.0 eV. Furthermore, the presence of pop-ins is noted [50] to be present in nanoindentation experiments conducted on neutron-irradiated single crystals of Mo. Irradiation-induced defects result in the reduction of pop-in load.

Thus, the variation of nanoindentation hardness with depth at different irradiation doses is shown in Figure 16.14. The hardness (H) is measured by the relationship

$$H^2 = H_0^2\left(1 + \frac{h^*}{h}\right) \tag{16.1}$$

where H_0 represents the hardness at infinite depth and h^* represents a characteristic length scale. The relation of change in hardness with irradiation fluence is described by the following relation [50]:

$$\Delta H = A[1 - e^{-BD}]^c \tag{16.2}$$

where D is the fluence or dose, and A, B, C, and D are the fitting parameters. The defect size can also be evaluated from the pop-in test. As displayed in Figure 16.14c, the defect density rises rapidly in the lower dose regime. On the other hand, the irradiation-induced defects decrease with an increase in the irradiation dose. This is attributed [50] to the progress of defect clusters.

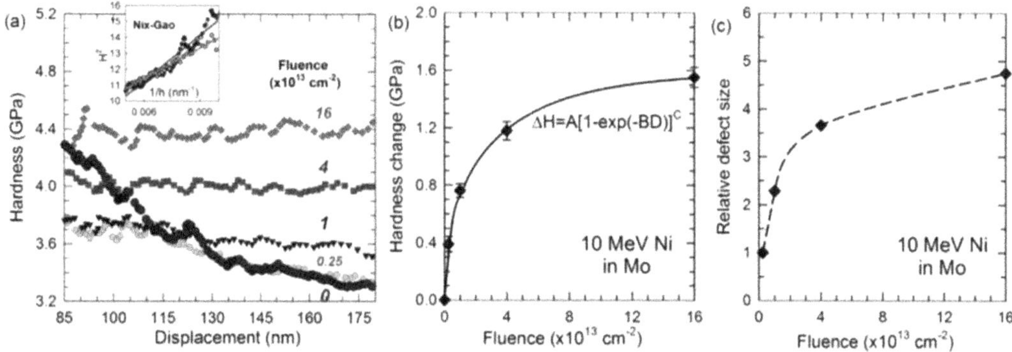

FIGURE 16.14 (a) Nanoindentation hardness versus depth after irradiation at different doses. Inset shows the Nix-Gao size effect for the pristine sample and corresponding to the lowest fluence. (b) Relation of nanoindentation hardness with irradiation fluence. (c) Change in relative defect size as a function of irradiation fluence [50].

16.4 FUTURE RESEARCH NEEDS

A very important area of emerging importance is high strain rate nanoindentation studies [51] on metallic and other materials. A recent review thus discusses [51] how a strain rate as high as $10^2 s^{-1}$ can be exploited in nanoindentation testing to derive the nanomechanical responses of metallic materials at high strain rates. Another upcoming area is the nanoindentation creep behavior studies [52] on metals, e.g., nanocrystalline (nc)-Ni and metallic alloys (e.g., Ni-20wt.% Fe alloys). Thus, the experimental and theoretical studies of significant importance reveal that in the transient regime, the creep resistance of Ni-Fe alloy is higher than that of nc-Ni. Further, it is suggested [52] that there are three regions of creep behavior. In region I it is mainly dislocation controlled. In region II it is controlled by both dislocation and grain boundary (GB) activities. In region III, however, it is mainly controlled by the GB activities. For the nc-Ni sample, only regions I and II are observed, while for the Ni-20wt.% Fe alloy only regions II and III are observed. This is an area where more future work will be desired given the fact that Ni and Ni-based alloys have tremendous industrial applications. It also needs to be appreciated that the knowledge base on nanoindentation creep of metals and metallic alloys is far from comprehensively developed. Thus, in future research, grain boundary engineering and microstructure engineering may be judiciously optimized to develop more creep-resistant metals and alloys.

An area of emerging importance is the correlation of crystal structures of various metallic materials [53], e.g., Cu, α–Fe, and α–Ti, with their nanoindentation characterizations. Recent work reveals that by using the Berkovich indenter and measuring the data in continuous stiffness mode (CSM), the extraneous influence of ISE can be eliminated. Further, it is also shown [53] that using the cylindrical flat punch indenter in load-controlled nanoindentation can also eliminate the undesired influence of ISE in the data.

Given the technological importance of various BMGs and shear-induced transformations (SITs) in them, this area is going to grow further in future research [54], thereby developing more of a fundamental understanding of the role of nanoindentation-induced localized plasticity governing the interaction between the shear transformation zones (STZs) and local microstructure. Recent studies [54] on the interaction between the serrated flow dynamics and STZs during nanoindentation experiments in BMGs, e.g., Co-, Fe-, Zr-, La-, and Al-based BMGs, reveal interesting insights. For instance, in soft Al- or Zr-based BMGs, the STZs are reported to be easily activated at low loading rates. It is noted [54] that in such situations, the shear stress exhibits a decreasing trend with a power law distribution. On the contrary, in the hard La-, Co-, and Fe-based BMGs, the STZs are formed at relatively higher loading rates. It is found [54] that in such situations the shear stress follows a Gaussian distribution. Thus, more work will be needed to better understand the genesis of such differences in shear-induced localized deformations in various BMGs.

Nanoindentation studies on metallic and metal alloy components developed by additive manufacturing (AM) technologies are one of the most important areas of future development [55]. Recent work on nanoindentation loading rate sensitivity analysis of functionally graded laser additive manufactured 316L and 410L SS shows [55] that for a predominantly austenitic microstructure, the gradients in grain size and morphology are not sensitive to loading rate. However, if the microstructure is predominantly martensitic, then the grains exhibit a sensitivity toward the local loading rate. Much more work will be needed to understand the scope of further microstructural tuning in such additively manufactured metallic materials and nanoindentation will continue to be a very useful tool for future research in such areas [55].

Further, nanotribology experiments are reported [56] to be conducted on martensitic stainless steel of nominal composition X190CrVMo20-4-1. The sintering in a nitrogen atmosphere and deep cryogenic heat treatment lead to superior scratch resistance. This superiority is ascribed to the presence of hard carbides and carbonitride phases, which take care of the transfer of load from the soft martensitic matrix. Thus, nanotribology of metals, HEAs, and MMCs is emerging as an important area of future research [56–58].

In recent times, major importance has been dedicated to the application of a wide variety of computational techniques for solutions of material-related and mechanical properties-related problems. Such computer-related techniques include the usage of machine learning, data mining, deep learning, artificial neural networks, and similar formalism (e.g., FEM) to predict the mechanical properties in both continuum and macroscale as well as in the microstructural length scale of various metals and alloys [59–62].

Finally, one of the most important emerging areas of research focus is the application of nanoindentation to characterize the AM 316L stents for biomedical application [–]. The stent is reported [63] to be developed by selective laser melting. The nanohardness of the AM stent is relatively higher than that of the commercially available 316L stent. This is attributed to the presence of a hierarchical microstructure, finer sub-grains, and enhanced dislocation density in the AM stent. Further, the nanohardness (H) and Young's modulus (E) exhibit grain orientation dependence, i.e., the grains near the (111) direction are reported to possess relatively higher magnitudes of (H) and (E). Thus, the application of the nanoindentation technique to biomedical metallic implant materials is going to grow. Thus, there are a host of emerging areas in the nanoindentation of metals and alloys.

16.5 SUMMARY AND CONCLUSIONS

This chapter focuses on the physics of deformation related to the nanoindentation response of metallic materials and alloys at the microstructural length scale. The first part of the chapter presents a brief introduction to nanoindentation and its importance for the characterization of metallic materials. This is followed by a very brief overview of the recent research and development scenario of the nanoindentation of metals and alloys. Next, the important results obtained from the huge literature on nanoindentation of metals are summarized. Finally, the emerging areas of importance in future research on nanoindentation of metals and metallic alloys are identified.

ACKNOWLEDGEMENTS

The authors acknowledge the kind support and encouragement received from various directors of the Council of Scientific and Industrial Research (CSIR)–Central Glass and Ceramic Research Institute (CGCRI), Kolkata, India, during the tenures of whom parts of the research work were carried out. They also gratefully acknowledge the infrastructural support received from various divisions of CSIR-CGCRI, Kolkata, India. The financial support received from Indian sponsoring agencies, including the CSIR, UGC, DST(SERB), DAE-BRNS, and IPR, of the Government of India (GOI) in terms of various project sponsorships and fellowships is gratefully acknowledged. Finally, the author AKM acknowledges the kind and continued support received from the authorities of the Sharda University, Greater Noida, Uttar Pradesh, India.

REFERENCES

1. Arjun Dey, A.K. Mukhopadhyay, *Nanoindentation of Brittle Solids*, CRC Press, Taylor and Francis Group, 2014, https://www.crcpress.com/Nanoindentation-of-Brittle-Solids/Dey-Mukhopadhyay/p/book/9781138076532.
2. Arjun Dey, A.K. Mukhopadhyay, *Microplasma Sprayed Hydroxyapatite Coatings*, CRC Press, Taylor and Francis Group, 2015.
3. https://www.crcpress.com/Microplasma-Sprayed-Hydroxyapatite-Coatings/Dey-Mukhopadhyay/p/book/9781138748866.
4. Arjun Dey, A.K. Mukhopadhyay, *Nanoindentation of Natural Materials with Hierarchical and Functionally Graded Microstructure*, CRC Press, Taylor and Francis Group.
5. https://www.crcpress.com/Nanoindentation-of-Natural-Materials-Hierarchical-and-Functionally-Graded/Dey-Mukhopadhyay/p/book/9781498784054.

6. W.C. Oliver, G.M. Pharr, Measurement of hardness and elastic modulus by instrumented indentation: Advances in understanding and refinements to methodology, *J. Mater. Res.*, 19(1), 2004, 3–20.
7. W.D. Nix, H. Gao, Indentation size effects in crystalline materials: A law for strain gradient plasticity, *J. Mech. Phys. Solids*, 46(3), 1998, 411–425.
8. Y. Huang, F. Zhang, K.C. Hwang, W.D. Nix, G.M. Pharr, G. Feng, A model of size effects in nano-indentation, *J. Mech. Phys. Solids*, 54(8), 2006, 1668–1686.
9. George M. Pharr, Erik G. Herbert, Yanfei Gao, The indentation size effect: A critical examination of experimental observations and mechanistic interpretations, *Annu. Rev. Mater. Res.*, 40(1), 2010, 271–292.
10. George Z. Voyiadjis, Mohammadreza Yaghoobi, Review of nanoindentation size effect: Experiments and atomistic simulation, *Crystals*, 7(10), 2017, 321, https://doi.org/10.3390/cryst7100321.
11. D. Bufford, Y. Liu, J. Wang, H. Wang, X. Zhang, In situ nanoindentation study on plasticity and work hardening in aluminium with incoherent twin boundaries, *Nat. Commun.*, 5, 4864, https://doi.org/10.1038/ncomms5864 | www. nature.com/naturecommunications.
12. Hao Zhou, Yongmao Pei, Daining Fang, Magnetic field tunable small-scale mechanical properties of nickel single crystals measured by nanoindentation technique, *Sci. Rep.*, 4, 2014, 4583, https://doi.org/10.1038/srep04583.
13. N. Li, H. Wang, A. Misra, J. Wang, In situ nanoindentation Study of Plastic Co-deformation in Al-TiN nanocomposites, *Sci. Rep.*, 4, 2014, 6633, https://doi.org/10.1038/srep06633.
14. Mao Liu, Cheng Lu, Kiet Anh Tieu, Ching-Tun Peng, Charlie Kong, A combined experimental numerical approach for determining mechanical properties of aluminum subjects to nanoindentation, *Sci. Rep.*, 5, 2015, 15072, https://doi.org/10.1038/srep1507.
15. Siddhartha Pathak, Surya R. Kalidindi, Jordan S. Weaver, Yongqiang Wang, Russell P. Doerner, Nathan A. Mara, Probing nanoscale damage gradients in ion-irradiated metals using spherical nanoindentation, *Sci. Rep.*, 7(1), 2017, 11918, https://doi.org/10.1038/s41598-017-12071-6.
16. Fabian Pöhl, Pop-in behavior and elastic-toplastic transition of polycrystalline pure iron during sharp nanoindentation, *Sci. Rep.*, 9, 2019, 15350, https://doi.org/10.1038/s41598-019-51644-5.
17. S. Sinha, R.A. Mirshams, T. Wang, S.S. Nene, M. Frank, K. Liu, R.S. Mishra, Nanoindentation behavior of high entropy alloys with transformation induced plasticity, *Sci. Rep.*, 9, 2019, 6639, https://doi.org/10.1038/s41598-019-43174-x.
18. Yuji Sato, Shuhei Shinzato, Takahito Ohmura, Takahiro Hatano, Shigenobu Ogata, Unique universal scaling in nanoindentation pop-ins, *Nat. Commun.*, 11, 2020, 4177, https://doi.org/10.1038/s41467-020-17918-7 | www. nature.com/naturecommunications.
19. Max A. Saccone, Rebecca A. Gallivan, Kai Narita, Daryl W. Yee, Julia R. Greer, Additive manufacturing of micro-architected metals via hydrogel infusion, Nature, 612(7941), 22/29 December 2022, 685.
20. Christopher A. Schuh, Nanoindentation studies of materials, *Mater. Today*, 9(5), 2006, 32–40, https://doi.org/10.1016/S1369-7021(06)71495-X .
21. F. Meng, K. Tsuchiya, I. Seiichiro, Y. Yokoyama, Reversible transition of deformation mode by structural rejuvenation and relaxation in bulk metallic glass, *Appl. Phys. Lett.*, 101(12), 2012, 121914, https://doi.org/10.1063/1.4753998.
22. X.D. Wang, Q.P. Cao, J.Z. Jiang, H. Franz, J. Schroers, R.Z. Valiev, Y. Ivanisenko, H. Gleiter, H. Fecht, Atomic-level structural modifications induced by severe plastic shear deformation in bulk metallic glasses, *Scr. Mater.*, 64(1), 2011, 81–84, https://doi.org/10.1016/j.scriptamat.2010.09.015.
23. Q. Cheng, E. Ma, Intrinsic shear strength of metallic glass, *Acta Mater.*, 59(4), 2011, 1800–1807, https://doi.org/10.1016/j.actamat.2010.11.046.
24. R. Yavari, A. LeMoulec, A. Inoue, N. Nishiyama, N. Lupu, E. Matsubara, W.J. Botta, G. Vaughan, M. Di Michiel, Å. Kvick, Excess free volume in metallic glasses measured by X-ray diffraction, *Acta Mater.*, 53(6), 2005, 1611–1619, https://doi.org/10.1016/j.actamat.2004.12.011.
25. Q.P. Cao, J.F. Li, Y.H. Zhou, A. Horsewell, J.Z. Jiang, Free-volume evolution and its temperature dependence during rolling of Cu60Zr20Ti20 bulk metallic glass, *Appl. Phys. Lett.*, 87(10), 2005, 101901, https://doi.org/10.1063/1.2037858.
26. M. Bletry, P. Guyot, Y. Brechet, J.J. Blandin, J.L. Soubeyroux, Homogeneous deformation of bulk metallic glasses in the super-cooled liquid state, *Mater. Sci. Eng. A*, 387–389, 2004, 1005–1011, https://doi.org/10.1016/j.msea.2004.02.085.
27. F. Jiang, M.Q. Jiang, H.F. Wang, Y.L. Zhao, L. He, J. Sun, Shear transformation zone volume determining ductile-brittle transition of bulk metallic glasses, *Acta Mater.*, 59(5), 2011, 2057–2068, https://doi.org/10.1016/j.actamat.2010.12.006.

28. Lisa Krämer, Verena Maier-Kiener, Yannick Champion, Baran Sarac, Reinhard Pippan, Activation volume and energy of bulk metallic glasses determined by nanoindentation, *Mater. Des.*, 155, 2018, 116–124, https://doi.org/10.1016/j.matdes.2018.05.051.
29. S.J. Bull, T.F. Page, E.H. Yoffe, An explanation of the indentation size effect in ceramics, *Philos. Mag. Lett.*, 59(6), 1989, 281–288, https://doi.org/10.1080/09500838908206356.
30. H. Li, R. Bradt, The microhardness indentation load/size effect in rutile and cassiterite single crystals, *J. Mater. Sci.*, 28, 1993, 917–926, https://doi.org/10.1007/BF00400874.
31. T.T. Zhu, A.J. Bushby, D.J. Dunstan, Size effect in the initiation of plasticity for ceramics in nanoindentation, *J. Mech. Phys. Solids*, 56(4), 2008, 1170–1185, https://doi.org/10.1016/J.JMPS.2007.10.003.
32. G.M. Pharr, E.G. Herbert, Y. Gao, The indentation size effect: A critical examination of experimental observations and mechanistic interpretations, *Annu. Rev. Mater. Res.*, 40(1), 2010, 271–292, https://doi.org/10.1146/annurev-matsci-070909-104456.
33. S.J. Bull, On the origins of the indentation size effect, *Z. Metallkd.*, 94(7), 2003, 787–792, https://doi.org/10.3139/146.030787.
34. T.T. Zhu, A.J. Bushby, D.J. Dunstan, Materials mechanical size effects: A review, *Mater. Tech.*, 23(4), 2008, 193–209, https://doi.org/10.1179/175355508X376843.
35. D. Kiener, K. Durst, M. Rester, A.M. Minor, Revealing deformation mechanisms with nanoindentation, *JOM*, 61(3), 2009, 14–23, https://doi.org/10.1007/S11837-009-0036-4.
36. K. Sangwal, On the reverse indentation size effect and microhardness measurement of solids, *Mater. Chem. Phys.*, 63(2), 2000, 145–152, https://doi.org/10.1016/S0254-0584(99)00216-3.
37. H. Bückle, Progress in microindentation hardness testing, *Metall. Rev.*, 4(1), 1959, 49–100, https://doi.org/10.1179/095066059790421746 .
38. N. Gane, The direct measurement of the strength of metals on a submicrometer scale, *Proc. R. Soc. Lond. A*, 317(1530), 1970, 367–391.
39. Jin Wang, T. Volz, S. Weygand, R. Schwaiger, The indentation size effect of single-crystalline tungsten revisited, *J. Mater. Res.*, 36(11), 2021, 2166–2175, https://doi.org/10.1557/S43578-021-00221-6.
40. G.P. Upit, S. Varchenya, I.P. Spalvin, Electromechanical effect in semi-metals, *Phys. Status Solidi (B) Basic Solid State Phys.*, 15(2), 1966, 617–621, https://doi.org/10.1002/PSSB.19660150222.
41. H. Westbrook, H. Conrad, *The Science of Hardness Testing and Its Research Applications; JH*, American Society for Metals, Metals Park, OH, 1973, Vol. 10, pp. 135–146.
42. Y. Wang, D. Raabe, C. Klüber, F. Roters, Orientation dependence of nanoindentation pile-up patterns and of nanoindentation microtextures in copper single crystals, *Acta Mater.*, 52(8), 2004, 2229–2238, https://doi.org/10.1016/j.actamat.2004.01.016.
43. N.A. Sakharova, J.V. Fernandes, J.M. Antunes, M.C. Oliveira, Comparison between Berkovich, Vickers and conical indentation tests: A three-dimensional numerical simulation study, *Int. J. Solids Struct.*, 46(5), 2009, 1095–1104, https://doi.org/10.1016/J.IJSOLSTR.2008.10.032.
44. M. Fivel, M. Verdier, G. Canova, 3D simulation of a nanoindentation test at a mesoscopic scale, *Mater. Sci. Eng. A*, 234–236, 1997, 923–926, https://doi.org/10.1016/S0921-5093(97)00362-6.
45. U. Landman, W. Luedtke, N.A. Burnham, R.J. Colton, Atomistic mechanisms and dynamics of adhesion, nanoindentation, and fracture, *J. Sci.*, 248(4954), 1990, 454–461, https://doi.org/10.1126/science.248.4954.454.
46. E.B. Tadmor, R. Miller, R. Philipps, M. Ortiz, Nanoindentation and incipient plasticity, *J. Mater. Res.*, 14(6), 1999, 2233–2250, https://doi.org/10.1557/JMR.1999.0300 .
47. J.B. Pethicai, R. Hutchings, W.C. Oliver, Hardness measurement at penetration depths as small as 20 nm, *Philos. Mag. A*, 48(4), 1983, 593–606, https://doi.org/10.1080/01418618308234914.
48. H. Amin, AlmasriGeorge Z. Voyiadjis, Nano-indentation in FCC metals: Experimental study, *Acta Mech.*, 209(1–2), 2010, 1–9, https://doi.org/10.1007/s00707-009-0151-x.
49. J.M. Wheeler, D.E.J. Armstrong, W. Heinz, R. Schwaiger, High temperature nanoindentation: The state of the art and future challenges, *Curr. Opin. Solid State Mater. Sci.*, 19(6), 2015, 354–366, https://doi.org/10.1016/j.cossms.2015.02.002.
50. V. Maier-Kiener, K. Durst, Advanced nanoindentation testing for studying strain-rate sensitivity and activation volume, *JOM*, 69(11), 2017, 2246–2255, https://doi.org/10.1007/s11837-017-2536-y.
51. Johann Kappacher, Michael Tkadletz, Helmut Clemens, Verena Maier-Kiener, High temperature nanoindentation as a tool to investigate plasticity upon phase transformations demonstrated on cobalt, *J. Mater.*, 6, 2021, 101084, https://doi.org/10.1016/j.mtla.2021.101084.
52. K. Jin, Y. Xia, M. Crespillo, H. Xue, Y. Zhang, Y.F. Gao, H. Bei, Quantifying early stage irradiation damage from nanoindentation pop-in tests, *Scr. Mater.*, 157, 2018, 49–53, https://doi.org/10.1016/j.scriptamat.2018.07.035.

53. P. Sudharshan Phani, B.L. Hackett, C.C. Walker, W.C. Oliver, G.M. Pharr, High strain rate nanoindentation testing: Recent advancements, challenges and opportunities, *Curr. Opin. Solid State Mater. Sci.*, 2023, 101054, https://doi.org/10.1016/j.cossms.2022.101054.
54. Weiming Sun, Yue Jiang, Zhihui Zhang, Zhichao Ma, Guixun Sun, Jiangjiang Hu, Zhonghao Jiang, Xiaolong Zhang, Luquan Ren, Nanoindentation creep behavior of nanocrystalline Ni and Ni-20 wt% Fe alloy and underlying mechanisms revealed by apparent activation volumes, *Mater. Des.*, 225, 2023, 111479, https://doi.org/10.1016/j.matdes.2022.111479.
55. Bowen Si, Zhiqiang Li, Xuexia Yang, Xuefeng Shu, Gesheng Xiao, Characterizations of specific mechanical behaviors for metallic materials with representative crystal structures using nanoindentation, *Vacuum*, 207, 2023, 111661, https://doi.org/10.1016/j.vacuum.2022.111661.
56. Yuexin Chu, Guishen Zhou, Shaoshan Wan, Yue Zhang, Fuyu Dong, Xiaoguang Yuan, Binbin Wang, Liangshun Luo, Yanqing Su, Weidong Li, Peter K. Liaw, Shear transformation zones and serrated flow dynamics of metallic glasses revealed by nanoindentation, *J. Alloys Compd*, 936, 2023, 168165, https://doi.org/10.1016/j.jallcom.2022.168165.
57. F. Khodabakhshi, M.H. Farshidianfar, A.P. Gerlich, A. Khajepour, V. Nagy Trembošová, M. Mohammadi, S.I. Shakil, M. Haghshenas, Nanoindentation plasticity and loading rate sensitivity of laser additive manufactured functionally graded 316L and 410L stainless steels, *Mater. Sci. Eng. A*, 862, 2023, 144437, https://doi.org/10.1016/j.msea.2022.144437.
58. P.K. Farayibi, J. Hankel, F. van gen Hassend, M. Blüm, S. Weber, A. Röttger, Tribological characteristics of sintered martensitic stainless steels by nano-scratch and nanoindentation tests, *Wear*, 512–513, 2023, 204547, https://doi.org/10.1016/j.wear.2022.204547.
59. S. Gonzalez, A.K. Sfikas, S. Kamnis, C.G. Garay-Reyes, A. Hurtado-Macias, R. Martínez-Sanchez, Wear resistant CoCrFeMnNi0.8V high entropy alloy with multi length-scale hierarchical microstructure, *Mater. Lett.*, 331, 2023, 133504, https://doi.org/10.1016/j.matlet.2022.133504.
60. E. Ahmadi, M. Goodarzi, Evaluation of nanomechanical and nanoscratch response of Inconel 718 reinforced with graphene nanoplates produced by combined ARB-GTAW processes, *J. Mater. Res. Technol.*, 22, 2023, 432–444, https://doi.org/10.1016/j.jmrt.2022.11.124.
61. F.E. Bock, R.C. Aydin, C.J. Cyron, N. Huber, S.R. Kalidindi, B. Klusemann, A review of the application of machine learning and data mining approaches in continuum materials mechanics, (2019), *Front. Mater.*, 6, 2019, 110, https://doi.org/10.3389/fmats.2019.00110.
62. S. Kossman, M. Bigerelle, Pop-in identification in nanoindentation curves with deep learning algorithms, *Materials (Basel)*, 14(22), 2021, 7027, https://doi.org/10.3390/ma14227027.
63. Yongju Kim, Gang Hee Gu, Peyman Asghari-Rad, Jaebum Noh, Junsuk Rho, Min Hong Seo, Hyoung Seop Kim, Novel deep learning approach for practical applications of indentation, *Mater. Today Adv.*, 13, 2022, 100207, https://doi.org/10.1016/j.mtadv.2022.100207.
64. Lu Lua, Ming Dao, Punit Kumar, Upadrasta Ramamurty, George Em Karniadakis, Subra Suresh, Extraction of mechanical properties of materials through deep learning from instrumented indentation, *PNAS*, 117(13) March 31, 2020, 7052–7062, www.pnas.org/cgi/doi/10.1073/pnas.1922210117.
65. E. Langi, L.G. Zhao, P. Jamshidi, M. Attallah, V.V. Silberschmidt, H. Willcock, F. Vogt, A comparative study of microstructures and nanomechanical properties of additively manufactured and commercial metallic stents, *Mater. Today Commun.*, 31, 2022, 103372, https://doi.org/10.1016/j.mtcomm.2022.103372.

17 Experimental Techniques to Study Physics of Deformation behavior in Glass at Microstructural Length Scale

Payel Bandyopadhyay and Anoop Kumar Mukhopadhyay

17.1 INTRODUCTION

As per conventional wisdom [1], the history of glassmaking can be traced back to 3500 BCE in Mesopotamia. Archaeological evidence suggests that the first true glass was made in coastal north Syria, Mesopotamia or Ancient Egypt. Even before that, obsidian, natural glass, was used as a tool. Today, the applications of glass span a truly spectacular range from simple household crockery to car windshields to gadget screens to even advanced applications. Higher quality bulletproof glass needs to be continuously developed as the requirement for protection level increases in proportion to newer threat levels that are being counter-invented. On the contrary, more transparent, scratch-resistant and impact-resistant glasses, e.g., Gorilla Glass, are being continuously developed by industries. The glass market also involves tablet computers, iPads, cameras, satellites, optical lenses and infrared glasses for remote sensing. Further advanced applications of glass encompass radiation shielding window glasses for nuclear technology, conventional laser glass, fiber lasers, biomedical instruments, transparent armors and advanced radio telescopes. Thus, we may conclude that the market size for glass in general and advanced glasses in particular will continue to grow and both conventional as well as advanced applications of glass will involve contact in one form or another. This is why and how the physics of deformation of glass assumes extraordinary scientific and technological importance today.

However, the most official genesis of classical contact mechanics is associated with Heinrich Hertz. In 1882, Hertz solved the contact problem of two elastic bodies with curved surfaces. These classical solutions are relevant even today. They provide the foundation stones for modern problems in contact mechanics. The Hertzian contact stress usually refers to the stress close to the area of contact between two spheres of different radii. Hertz first observed the cone-shaped fracture in glass in the 1880s [1]. Since then the breakthrough research works on fracture mechanics of materials in general and deformation and fracture of glass in particular by a host of seminal and pioneering attempts by Hill, Griffith, Lawn, Swain, Evans, Cook, Pharr, Fischer-Cripps, Bhushan and others have indeed focused brilliant lights in this field [2–12]. This age-old stream has its relevance even more brilliantly continuing today more than a century after Hertz, as there is much work ongoing on the contact mechanics of glass, whether the contact is static [2–12] or simply dynamic [13–15].

The pioneering work of Griffith gave the relation between the fracture strength (σ_c) and the flaw size (c) in an ideally brittle material [2, 17-19]. The total energy (U) of the system is given by [19] $U = U_M + U_S$, where U_M is the mechanical energy and U_S is the free energy expended in creating new crack surfaces, as the crack extends by an infinitesimal amount (dc) from the preexisting crack c. So, now the crack length is c + dc. But it needs to be appreciated that $dU_M/dc < 0$, as the mechanical

DOI: 10.1201/9781003359364-21

energy reduces due to propagation of the crack. But the surface energy term increases due to the extension of the crack as the cohesive force along (dc) must be overcome for crack propagation. So, we must have $dU_S/dc > 0$ for the crack to grow. In other words, the Griffith energy balance concept states that crack growth will occur only and only when the minimum energy required for the creation of two new fracture surfaces is provided by the process of crack growth that leads to the occurrence of the fracture process. The growth of a crack requires the creation of two new surfaces and hence an increase in the surface energy.

Using this procedure, Griffith found that $\sigma_c = (2E\gamma / \pi c)^{0.5}$, where E is the Young's modulus of the material and γ is the surface energy density of the material. Assuming $E = 62$ GPa and $\gamma = 1$ J/m^2 gives excellent agreement of Griffith's predicted fracture stress with experimental results for glass.

17.2 LITERATURE REVIEW

17.2.1 Theoretical Concepts and Important Parameters

The concept of stress intensity factors came into the picture after Griffith's work [2, 19]. The stress at the tip of the crack is quantified by the stress intensity factor. The crack generally propagates by displacement of the two surfaces newly created by generation of the crack. There are three modes of crack surface displacement. These are known as mode I, II and III types of loading. Mode I loading happens because of the separation of two crack surfaces due to tensile stresses. Mode II fracture happens due to loading normal to crack front i.e., due to in-plane, shear. Mode III loading happens due to transversal, i.e., out-of-plane, shear [19].

17.2.2 Concept of Critical Load

When the indenter penetrates, material stress is generated inside the material. The solution of stress fields beneath the spherical indenter was first provided by Hertz [1]. A Hertzian cone crack is formed beyond a critical load (P_c) for a brittle solid after the initial elastic field and due to this elastic field the maximum tensile stress occurs at the circle of contact [20]. The empirical relation $P_C \alpha$ r between the critical load and the indenter radius (r) was established by Auerbach [21]. Tillett [22] shows that for SLS glass there are two regions in the plot of P_C/r versus r. There is one region where P_C is proportional to r. In addition, there is another region where P_C is proportional to r^2 [22].

17.2.3 Concept of Preexisting Flaws

The cone crack is initially assumed to generate if the estimated tensile strength exceeds the limiting cohesive strength of the material. However, the paradox is that for usually conceivable practical magnitudes of r, the estimated tensile stress is generally much smaller than the limiting cohesive strength of the solid. Therefore, it cannot explain the genesis of the cone cracks. This is why it needs to be assumed that there are already flaws existing in glass and the cone cracks originate from such flaws.

In 1967, Frank and Lawn [4] introduced the already existing paradigm of Griffith–Irwin fracture mechanics into the Hertzian fracture problem. This seminal work reveals that the mechanics of cone crack initiation is much more complex than it was initially thought. Their treatment [4] of this problem involves two major steps. They assume that cone cracks start from flaws on the specimen top surface at (or just outside) the contact circle. It is this region where the tensile stresses are mostly concentrated. It is then argued that the embryonic cracks have the capacity to circumvent the contact circle subsequently. It does so as a shallow surface ring. It is shown [4] further that then the ring propagates both downward and sidewise outward, like an evolving out partial cone. It follows closely the σ_3 stress trajectories. Although close enough, these stress values are not exactly those of the σ_3 components.

As a consequence, at all points in space, these stress components emerge as nearly normal to the tensile stress components due to the σ_1 component in the prior stress field. In addition, it has been argued [4] that a stress intensity factor for the downward crack extension can be expressed uniquely in terms of the prior stress function. It is noted further [20] that the larger the temperature of contact (t_c), the smaller the critical load (P_c). Moreover, for a given temperature of contact (t_c), the critical load (P_c) is shown to decrease further with enhancement in humidity. Thus, this information provides knowledge on issues involved in the kinetics of crack growth in glass [20] when the environment changes in both humidity and temperature.

17.2.4 Crack Growth from Blunt and Sharp Contacts

It is important to bear in mind though that Hertzian cone cracks generally grow when the contact is due to blunt contact, e.g., a spherical indenter. In terms of practical situations, such blunted-out contacts do exist in, e.g., a worn-out grinding wheel. However, it may be plausible to argue that this bluntness grows over time due to repeated contact. It means the initial contact is not blunt, i.e., sharp. The excellent review work of Cook and Pharr [9] has provided knowledge about what types of cracks are generated on the surface of glass due to contact with a sharp indenter. There can be four types of cracks generated on the glass surface when it comes in contact with a sharp indenter. These are known as radial, median, lateral and half-penny [9]. The radial cracks generate [9] close to the surface originating from the edge of the plastic deformation zone. However, the median cracks generate [9] beneath the plastic deformation zone and remain parallel to the loading axis. The lateral cracks also generate [9] beneath the plastic deformation zone and evolve out in an effort to run parallel to the surface. The half-penny radial crack depicts an idealized sub-surface damage zone in glass [9]. Other researchers [17–19, 23–33] have also contributed significantly to the knowledge on cracks related to indentation and fracture in static contact of glass.

17.2.5 Other Issues in Static Contact-Based Fracture and Fatigue of Glass

Fracture lances are reported to form in soda–lime–silica glass during indentation with a spherical indenter [34]. Further, depending upon the load applied, the corresponding fractographs are comprised of four individual zones [35]. In addition, for SLS glass, both Vickers hardness and the lateral crack length decrease with an increase in temperature [35]. For SLS glass, up to 420°C, the radial crack length increases with temperature prior to a sharp fall with a further increase in temperature [35].

Recent research has also illustrated that crack tip shielding due to water penetration from the surrounding environment can have beneficial effects on both the strength and fracture toughness of silica glass [36]. The removal of large surface flaws and rounding of the crack tip due to hydrofluoric acid (HF) etching has a beneficial effect on the strength of SLS glass [25, 37]. The sensitivity of occurrence of the brittle-to-ductile transition (BDT) on the strain rates and the temperature has been also studied for SLS glass [38]. In addition, a model was developed that could successfully predict the temperature and strain rate dependency of BDT in the case of SLS glass [38].

It is also reported that the strength of HF-etched SLS glass increased with the depth of material etched away [39]. In addition, it is illustrated that when aged in water for 24 hours after etching, the strength of the soda–lime–silica glass is 30% higher than that of the as-received samples [39]. Other researchers show [40] that the plot of fracture strength of soda–lime glass exhibits an inverse dependency on indentation loads. Further, the plot has a slope of (–1/3) that matches with the theoretical prediction [40]. Moreover, the strength of soda–lime glass decreases with increasing surface length, and the depth of the indentation induces radial cracks [41].

A comparison of fatigue strength has been reported for soda–lime glass with and without a protective epoxy coating [41]. Considering the well-known environmentally assisted subcritical crack growth issue, in the absence of any protective coating, the fatigue strength of the bare soda–lime glass is reported to be the highest in liquid nitrogen, followed by those in dry nitrogen and distilled water environments [41]. As epoxy protects glass from water penetration, the fatigue strength of the coated soda–lime glass is reported to be the highest in distilled water followed by that in the dry nitrogen environment [41].

17.2.6 Deformation and Damage in SLS Glass Due to Dynamic Contact

A microfracturing process is involved during dynamic contact of glass [42]. The pattern of microfracturing observed in dynamic contact is reported to have a high degree of qualitative similarity with what happens in static contact. But there is a significant difference. In dynamic contact, the severity of the stress field is much more enhanced compared to that of static contact. This enhancement of the stress field happens due to the additional presence of the tangential load [42].

When the applied normal load (P) is very low ($P < 0.05N$), there is no cracking in the glass as the scratch track suffers from mainly plastic deformation [42]. At intermediate loads ($0.05N < P < 5N$), two characteristic features are noted in the scratch groove. The first one is a fully plastic scratch track that develops in the glass. In addition, well-developed median and lateral cracks are also noted in the scratch groove when scratched with a Vickers pyramid.

The scenario completely changes at relatively higher loads ($P > 5N$). At such high loads, the most characteristic feature is extensive crushing. This crushing of glass is reported [42] to occur mostly in and around the scratch groove that develops in the glass. Both the friction coefficient (μ) and the specific energy are shown to affect the grindability of brittle solids including glass [43–45]. It is found that the force ratio is independent of load, speed and environment for SA58 glass [44]. For this glass, the specific energy is proportional to Vickers hardness but is independent of the environment. In SA 58 glass, four types of cracks are found to be formed [44].

On the surface, the first type of cracks are found [44] to have propagated at an angle of about 30°. They occur in lateral directions with respect to that of the scratch direction. The second type of cracks are formed on the surface at higher loads (~3 N) but lower speeds (~400 $\mu m.s^{-1}$). These cracks follow a rounded path and are perpendicular to the first type of cracks. The most interesting observations happen on the subsurface region where two types of cracks are reported to be observed. One of them is parallel to the surface and the other is found to be perpendicular to the surface [44].

On the other hand, the concept of critical load (P_c) becomes important when a hard steel sphere slides over the flat surface of glass [46]. It is reported [46] that when the effective force (P_{eff}) is greater than the critical load (P_c), usually the classical Hertzian ring cracks appear on the glass surface, especially at positions located at the rear of the sphere. It has been proposed that the magnitude of the critical load could be as small as ~2N for a steel ball of radius 3 mm in dynamic contact with soda–lime glass [46]. The initiation of such cracks in the glass subjected to dynamic contact is explained [46] on the basis of Griffith's energy balance criterion [2, 19] and the inhomogeneous stress field [47]. Further, it is reported that for a coefficient of friction (μ) greater than 0.02 [46], the P_c is proportional to r^2, where r stands for the radius of the steel ball (i.e., r = 3 mm). When the coefficient of friction reaches 0.1, the load required to create a cone fracture is predicted to reduce by a factor of 100 times or more [46].

17.2.7 Concept of Limiting Radius

There is also independently carried out work on a glass scratched by a steel ball [48]. The ball has a radius (r) larger than the limiting radius (r_c). The results of this study [48] match with the earlier results [46] that as long as μμis greater than 0.02, P_c is proportional to r^2. For instance, up to an r_c of 7 μm [49] when μμ is about 0.15, P_c becomes proportional to r. However, the proportionality relation (P_c proportional to r) is practically absent when μ becomes as high as about 0.5 [49]. In such a situation, however, P_c is found to be proportional to r^2 [49]. What all the aforesaid information [2, 19, 42–48] amounts to is that both the contact load and the contact radius are factors of paramount importance as far as the dynamic contact deformation of glass in general and SLS glass in particular are concerned.

17.2.8 Role of Friction and Speed

An interesting question that arises from this work [48] is what factors control the coefficient of friction in dynamic contact of glass. Any comprehensive work covering the load range of from a few micronewtons to a few newtons is yet to emerge for soda–lime–silica glass exposed to dynamic contact. However, for a given applied normal load of 0.1 N, the friction coefficient of glass is reported to be independent of scratching speed in the range of 2 to 200 $\mu m.s^{-1}$ [49]. Nevertheless, the most interesting observation is that the higher the scratching speed, the lower the crack density in glass [49], thereby suggesting that speed has an important role to play in scratch deformation of glass and that the information available for soda–lime–silica glass is far from complete.

17.2.9 Role of Load Dependent Transition in Damage Initiation

There is no reason to think though that the role of load is not so important in the scratching of glass. In fact, it is one of the most important [42, 46] factors. The single-pass scratch test on SLS glass is conducted using a Rockwell C indenter with progressive loads up to applied normal loads of as high as 100 N [50]. This experimental work [50] identifies a critical load (P_c) of 23 N for SLS glass beyond which chipping occurs. The repeated scratch test on SLS glass at a progressive load up to ~25 N [50] is also conducted. In this case of repeated scratch tests, three distinct zones are observed.

These zones are the (1) groove with only plastic ploughing, (2) the appearance of the first crack and (3) the chip formation. The critical load (P_c) for cracking (i.e., transition from only ploughing to initiation and formation of partial cone cracks) is 6–1 N depending upon the number of cycles, e.g., 1–10 [50]. The magnitude of P_c is high (e.g., 6N) when the number of cycles is low (e.g., 1 or so). However, the magnitude of P_c becomes low (e.g., 1N) when the number of cycles is large (e.g., 10). In a similar fashion, the critical load (P_c) for chipping (i.e., transition from only initiation and formation of partial cone cracks to chip formation) is reported to be 23–12 N depending upon the number of cycles (e.g., 1–10) [50]. As explained earlier, in this case also, the critical load decreases with an increase in the number of cycles [50].

It is interesting to note further that both single and double scratch experiments have been conducted on BK7 optical glass using a sharp Vickers indenter at 30, 90, 150 and 300 mN applied normal loads [13]. For the single-pass scratch test, both lateral and radial cracks are observed for applied normal loads of 90–300 mN, but chipping is not observed to have happened even at applied normal loads up to as high as 300 mN [13]. Most surprisingly, however, for paired scratches made at 30 mN applied normal load, the distance between neighboring scratches is reported to have had a null effect on normalized volume removal. This information suggests that there is a critical minimum value of load below which such effects do not manifest.

The scenario changes drastically, however, at much higher applied normal loads of 90, 150 and 300 mN. The normalized volume of BK7 glass removed is maximized at a particular separation distance of 25 μm for the pair of scratches made at 90 mN load. A pair of scratches made at distances less than 25 μm or a pair of scratches made at 90 mN load, do not necessarily maximize the normalized volume of BK7 glass removed. Similarly, the critical separation distances are identified to be 35 and 100 μm for paired scratches made at still higher loads of 150 and 300 mN, respectively [13]. These data would suggest that the higher the normal load, the larger could be the critical separation distance between the pair of neighboring parallel scratches to achieve the maximum normalized volume of BK7 glass removed.

In the case of obsidian, a glass of archaeological importance, three distinct zones are identified from the results of incremental scratch tests conducted at various applied normal loads up to 50 N [51]. The first zone is called the micro-ductile zone. No damage occurs in this zone except the formation of a plastically deformed groove formation. The corresponding critical magnitude of the applied normal load is ~1 N. From the graphical data given in [51] the second zone, called the micro-cracking zone, appears to have occurred in the applied normal load range of greater than

~1–4 N. It is reported that subsurface cracks appear in this zone, but it happens without any chipping in the obsidian.

However, there is a third zone located near the end of the scratch groove. This zone is called the micro-abrasive zone. From the graphical data given in [51], this zone appears to have occurred in the applied normal load range of greater than about ~4 N. This zone is reported to have extensive chipping that occurs along with the consequent debris formation and crack evolution in the obsidian glass. Depending on the applied normal loads, the existence of the three zones, i.e., micro-ductile, micro-cracking and micro-abrasive zones, is also observed for scratch tests conducted on SLS glass at applied normal loads up to as high as 50 N [52]. It is indeed observed that no cracks are generated up to applied normal loads of 0.3 N. Beyond a 0.3 N applied normal load, the median and lateral cracks are reported to have intersected the SLS glass surface, and finally at an applied normal load greater than 0.8 N scratch grooves are filled with crushed materials [52].

The lateral crack generated during scratching never reached the surface when the tests were conducted under absolute dry conditions with an insignificantly small humidity level. Further, it is observed that no radial cracks are formed in SLS glass under such experimental conditions [52]. Further, SLS glasses with devitrite phase (i.e., it contains more Na_2O and CaO) are reported to be more prone to chipping, while glasses with only a silica network-like structure are found to be more resistant to chipping [52]. Similarly, it has been noted [53] that in alcohol-lubricated scratch experiments, the scratch hardness of SLS glass is a sensitive function of the chemical structure of the alcohol, e.g., the chain length.

17.2.10 Scratch Test Parameters

It is indeed very interesting to note that a wide variety of test conditions are employed by researchers worldwide during scratch tests conducted on different types of glass, including SLS glass [13, 42, 44, 48–53, 54]. For instance, in the case of blunt indenters, the tip radius could be as low as 3 μm [49] to as high as 9500 μm [48]. Similarly, in the case of blunt indenters, the applied normal load for single-pass scratch tests could be as low as 0.049 N [49] to as high as 20 N [48]. On the other hand, for multiple-pass scratch tests, the applied normal load may be as low as 0.1 N to as high as 25 N [50]. However, in the case of a ramping single-pass scratch test, the maximum applied normal load could be as high as 100 N [50]. Interestingly, the critical load (P_c) could be 392.4–2746.8 N in correspondence to the tip radius of ~1500–9500 μm [48]. In the case of blunt indenters, the scratching speed for single-pass scratch tests could be as low as 2 $\mu m.s^{-1}$ [49] to as high as 660 $\mu m.s^{-1}$ [48].

However, in the case of sharp indenters, the applied normal load for single-pass scratch tests could be as low as 0.01 N [44] to as high as 4 N [42]. In an analogous manner, the scratching speed for single-pass scratch tests could be as low as 0.04 $\mu m.s^{-1}$ [45] to as high as 1000 $\mu m.s^{-1}$ [42]. The radii of the tip of the sliding indenters are not mentioned in most of the cases [13, 42, 52–53], except as to be 0.3 μm in [44].

It is found for soda–lime glass that the erosion rate depends on air pressure, stand-off distance, nozzle diameter with a positive exponent and abrasive mass flow rate with a negative exponent [55]. The formation of four zones along the depth direction and the role of low nozzle speed and high jet pressure for higher depth of cut and good surface quality during the water jet milling have been reported by some researchers for soda–lime glass [56]. Dynamic contact deformation studies are still very important.

The material removal processes due to lapping for borofloat, BK7 and quartz have been studied. It is found that the removal rate is lowest for BK7 and highest for borofloat, and the removal rate not only depends on the hardness but also on the silica content of the glass which softens the surface of glass by forming a hydration layer [57]. It is reported that severe cracks are formed parallel to the grinding direction. These are the cracks from which the fracture occur in the case of tensile loading for glass bars [58].

17.3 DEFINITIVE MICROSCRATCH TESTING

A commercially available scratch tester (Model TR-102-M3, Ducom, Bangalore, India) equipped (Figure 17.1) with a Rockwell C diamond indenter with a ~200 μm tip radius was also used recently [52–56] to perform the scratch tests following [1, 4–6]. The scanning electron microscopy (SEM) image of the scratch grooves made at applied normal loads of 15 N at scratching speeds of 100, 200, 500 and 1000 μm.s^{-1} are shown in the inset of Figure 17.1a. The indenter, the stage used to hold the SLS glass sample and the acoustic emission (AE) sensor arrangements of the scratch tester are shown in Figure 17.1b–d. The scratch experiments were reported [52–56] to be conducted in air under ambient laboratory conditions (~30°C and relative humidity 70%) at four different constant applied normal loads (P) (2, 5, 10 and 15 N) with scratching speeds (v) of 100, 200, 500 and 1000 μm.s^{-1}. The load cell could measure both normal load (P) and lateral (F) forces in the range of 20–200 N with a resolution of ±0.1 N.

These results [52–56] show that the coefficient of friction, wear volume and wear rate increase with load. The wear volume and wear rate follow a power law dependency with the increasing speed of the scratch in microstructural length scale. At a typical low value of normal load (2 N), both complete and partial Hertzian tensile cracks are found [52–56] to be adjacent yet separated, thereby limiting the interaction among them. In addition, there are very few edge cracks oriented at an angle of about 30° to the direction of the scratch, emanating from the periphery of the Hertzian tensile cracks. At a slightly higher normal load of 5 N, the extent of interaction between the Hertzian tensile cracks increases significantly. Another interesting observation is the presence of much larger "intra-Hertzian tensile-crack" cracks about 15–20 μm at 5 N at the periphery of the Hertzian tensile cracks and their growth both toward and away from the central region of the Hertzian tensile cracks. Such features are reported [52–56] to be present throughout the length of the scratch made at 5 N.

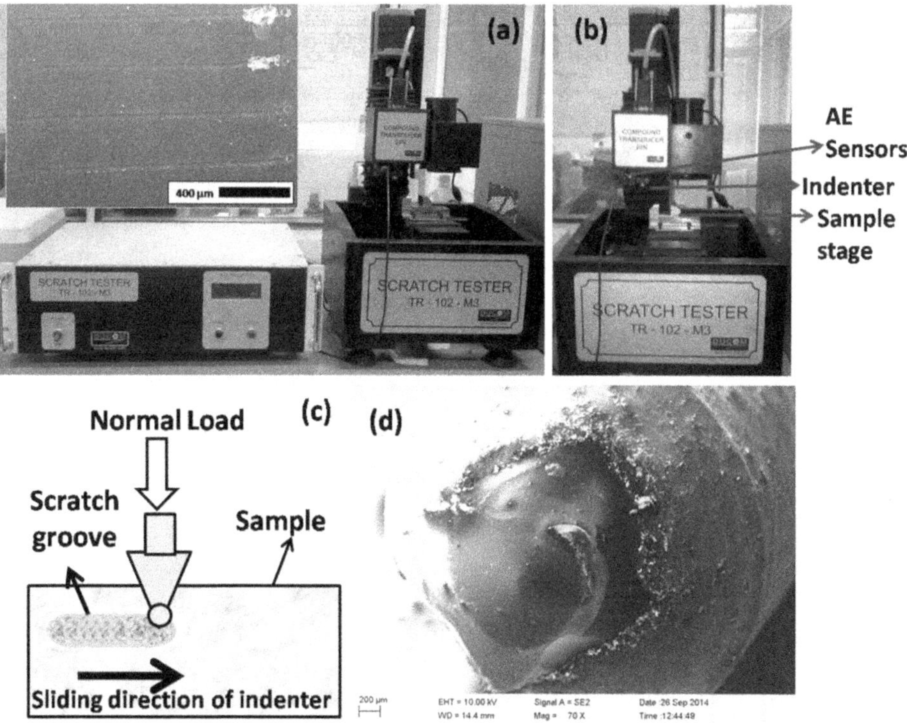

FIGURE 17.1 (a) The scratch tester. The inset shows the SEM image of scratch groove made at 15 N applied normal load. (b) The exploded view of the scratch tester. (c) The schematic of the scratch testing. (d) The Rockwell C indenter used in the study.

At a higher scratching speed of 500 μm.s^{-1} a crack is reported [52–56] to occur with two branches generated perpendicularly from the main crack.

At a relatively higher normal load of 10 N, the separation between the Hertzian tensile cracks decreases manifold leading to more interaction among the Hertzian tensile cracks. It is also an important observation [52–56] that now the consecutive Hertzian tensile cracks are regularly joined and intersected by the edge cracks. At the highest applied normal load of 15 N, the additional cracks that appear at 10 N normal load manifest more abundantly with a much higher spatial density. The severity of interaction among the consecutive Hertzian tensile cracks increases so much at 15 N normal load that they almost overlap and intrude into each other's growth territory. As a consequence of this intense overlapping and interaction, possibly a separate series of inclined cracks, which are termed [52–56] as "secondary cracks" evolve nearly parallel to each other behind the trailing edge of the indenter and located at the edge of the scratch groove. The angle of orientation of these new set of cracks (e.g., ~8.23° for 10 N and ~12.08° for 15 N) are [52–56] completely different from those (e.g., ~30° for 2 N and ~14° for 5N) of the edge cracks. It is also observed that the extent of the whitish zone becomes gradually enlarged as the applied normal load is increased from 2 to 15 N. This observation suggests the possible presence of subsurface damages. The interactions among the Hertzian tensile cracks create zones [52–56] that are about to chip out from the rest of the materials. The deformation and damages are reported [52–56] to decrease with increasing scratching speed for all the applied normal loads.

17.4 FUTURE ISSUES

However, there are still many unresolved issues that will require further investigation. For instance, a portion of the scratch groove made at the applied normal load of 10 N in the present SLS glass is depicted in the FESEM photomicrographs given in Figures 17.2 and 17.3. It shows another very

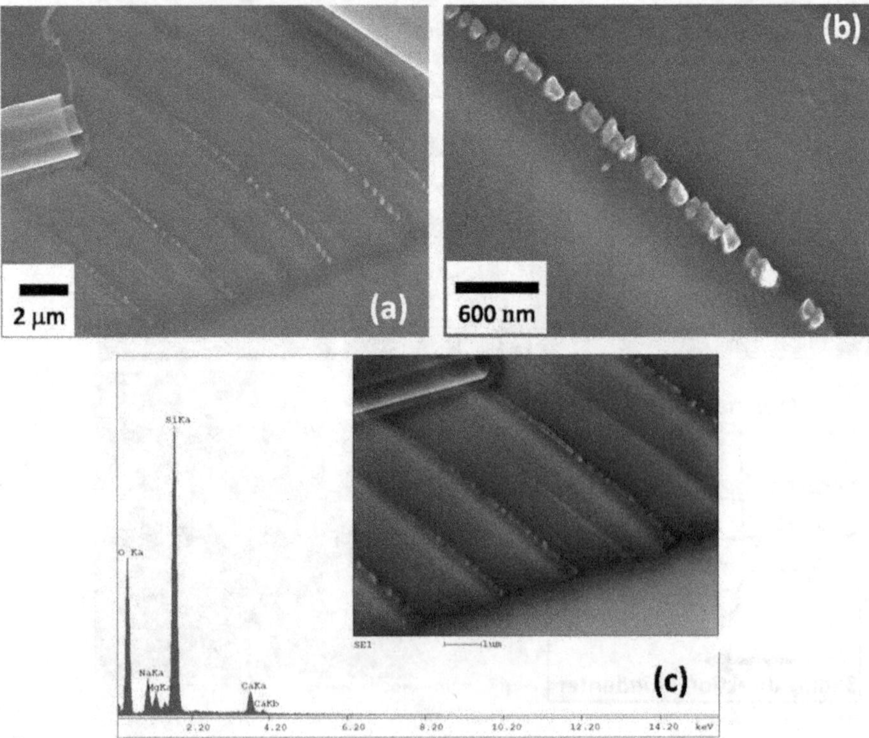

FIGURE 17.2 (a) FESEM photomicrograph of the scratch grooves. (b) Magnified view of panel a. (c) The EDAX data of the marked region.

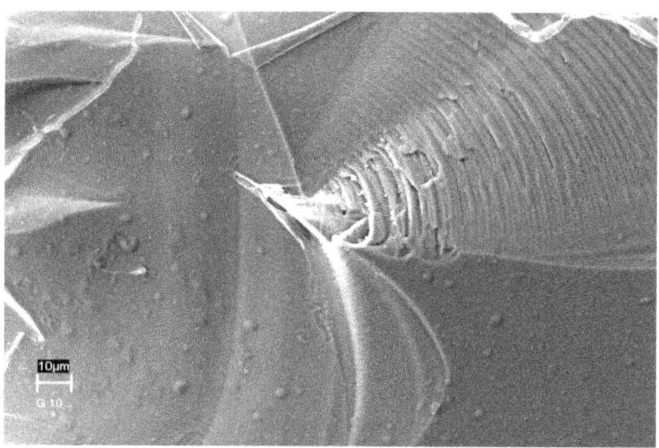

FIGURE 17.3 FESEM photomicrograph of the damage zone.

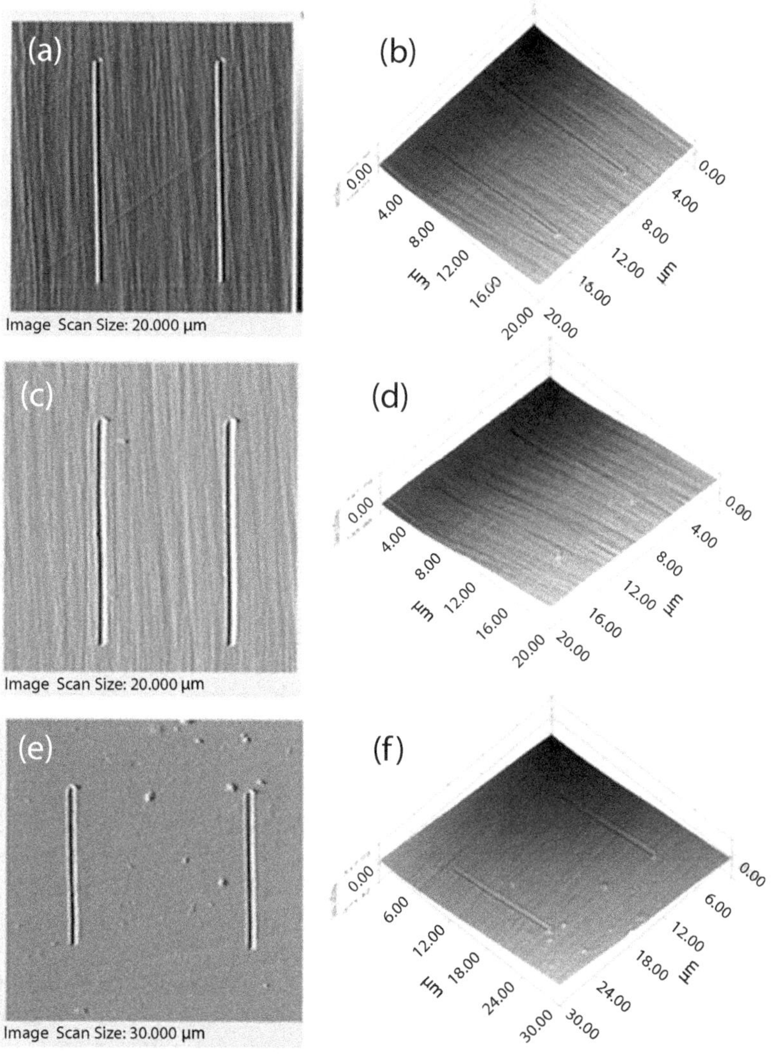

FIGURE 17.4 SPM photomicrographs of the nanoscratches. (a, c, e) Top view. (b, d, f) Topography (Reprinted with permission from Reference [50]).

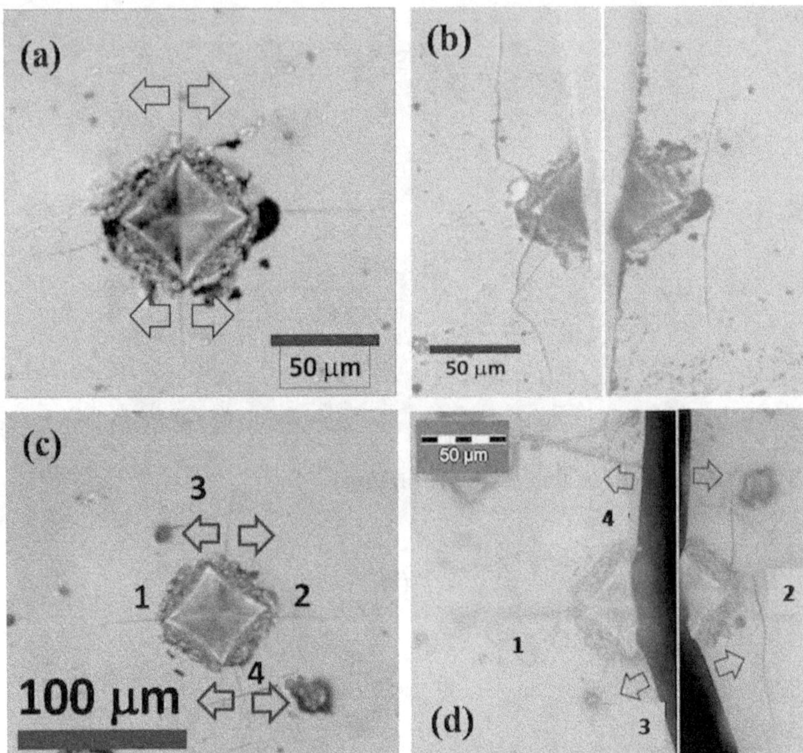

FIGURE 17.5 Optical microscopy of Vickers indentation. (a and b) Before and after testing for strain rate 0.001%.s^{-1}. (c and d) Before and after testing for strain rate 0.05%.s^{-1}.

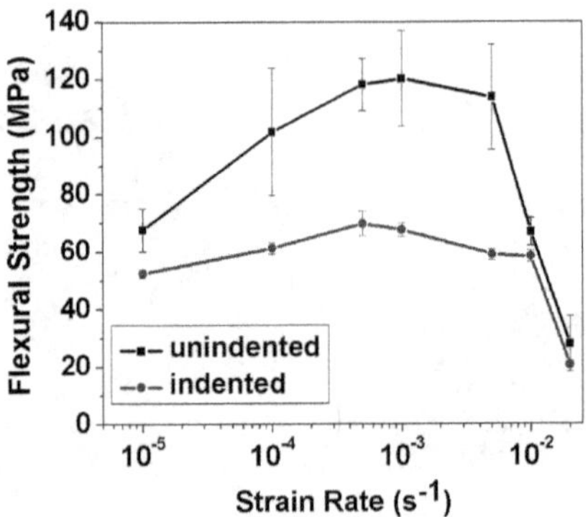

FIGURE 17.6 Variation of flexural strength of SLS glass, with and without artificial indentation.

peculiar, interesting feature. The surface of the glass inside the scratch groove forms a structure like the wave propagation inside the water. It also shows extensive regions of micro-shear and micro-shear-induced localized fractures. It is suggested that their genesis could be linked to the propagation of stress waves inside the SLS glass material [13–16]. However, further studies are needed to clarify the validity of this suggestion for the present SLS glass.

Further, the nanoscratch tests were conducted using a nanoindenter (Triboindenter, Ubi 700, Hysitron Inc., Minneapolis, Minnesota, USA) on the same SLS glass at 100 μN to 1000 μN and the scanning probe microscope (SPM) results (Figure 17.4) show that brittle material like SLS glass could also deform, at least at the nanoscale, by plastic grooving mechanism [59].

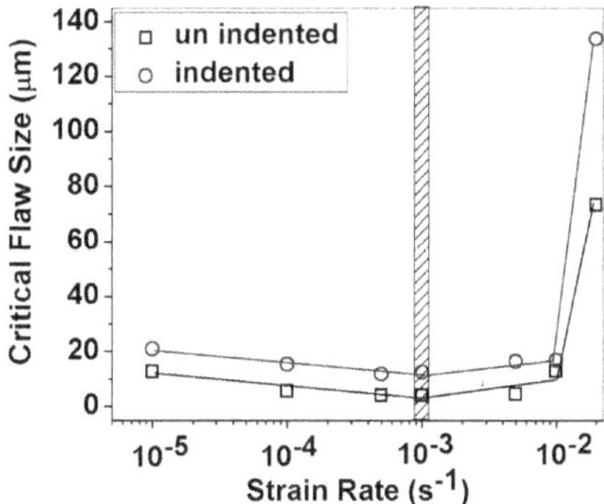

FIGURE 17.7 Critical flaw size as a function of strain rate.

In addition, Vickers microindents simulating grinding flaws are deliberately introduced in the same SLS glass, followed by three-point bend fractured at different increasing strain rates of 10^{-5} s^{-1} to 0.5×10^{-1} s^{-1} (Figure 17.5). The comparison of failure strength of the same SLS glass with and without these artificial flaws (Figure 17.6) exhibits a strong sensitivity on strain rate. The strength of the indented SLS glass is lower than that of the unindented SLS glass at any given strain rate, as expected. In both cases, there is an initial enhancement in the flexural strength at different rates with an increase in the applied strain rate up to $10^{-1}s^{-1}$ for the unindented or 10^{-2} s^{-1} for the indented glass. However, beyond these critical strain rates, the flexural strength abruptly decreases (Figure 17.6). These results are commensurate with the increase in critical flaw size (Figure 17.7), assuming a fracture toughness of about 0.72 $MPa.m^{0.5}$ for the SLS glass.

However, it must be admitted that more systematic and thorough experimentation and surface as well as subsurface fractographic confirmation will be certainly required to confirm the validity of the current work and hence may definitely constitute a very important area of future research in dynamic contact of commercially available SLS glass in particular and glass in general.

ACKNOWLEDGEMENTS

The authors acknowledge the kind support and encouragement received from various directors of the Council of Scientific and Industrial Research (CSIR)–Central Glass and Ceramic Research Institute (CGCRI), Kolkata, India, during the tenures of whom parts of the research work were carried out. They also gratefully acknowledge the various infrastructural support received from various divisions of CSIR-CGCRI, Kolkata, India. The financial support received from Indian sponsoring agencies, including CSIR, UGC, DST(SERB), DAE-BRNS, and IPR, of the Government of India (GOI) in terms of various project sponsorships and fellowships is gratefully acknowledged by the authors. Finally, the author AKM acknowledges the kind and continued support received from the authorities of the Sharda University, Greater Noida, Uttar Pradesh, India.

REFERENCES

1. http://en.wikipedia.org/wiki/History_of_glass
2. A. A. Griffith, The phenomena of rupture and flow in solid, *Philos. Trans. R. Soc. Lond. A* 221 1920 163.
3. R. Hill, *The Mathematical Theory of Plasticity*, Oxford University Press, 1950.
4. B. R. Lawn, F. C. Frank, On the theory of Hertzian fracture, *Proc. R. Soc. Lond. A* 299(1458) 1967 291.
5. B. R. Lawn, T. R. Wilshaw, Indentation fracture: Principles and applications, *J. Mater. Sci.* 10(6) 1975 1049.
6. B. R. Lawn, M. V. Swain, Microfracture beneath point indentations in brittle solids, *J. Mater. Sci.* 10(1) 1975113.

7. A. G. Evans, E. A. Charles, Fracture toughness determinations by indentation, *J. Am. Ceram. Soc.* 59(7–8) 1976 371.
8. G. M. Hamilton, Explicit equations for the stresses beneath a sliding spherical contact, *Proc. Inst. Mech. Eng.* 197C(1) 1983 53.
9. R. F. Cook, G. M. Pharr, Direct observation and analysis of indentation cracking in glasses and ceramics, *J. Am. Ceram. Soc.* 73 1990 787.
10. A. C. Fischer-Cripps, Elastic–plastic behaviour in materials loaded with a spherical indenter, *J. Mater. Sci.* 32(3) 1997 727.
11. Andreas Momber, Fracture features in soda-lime glass after testing with a spherical indenter, *J. Mater. Sci.* 46(13) 2011 4494.
12. Manuel L. B. Palacio, Bharat Bhushan, Depth-sensing indentation of nanomaterials and nanostructures, *Mater. Char.* 78 2013 1.
13. W. Gu, Z. Yao, X. Liang, Material removal of optical glass BK7 during single and double scratch tests, *Wear* 270(3–4) 2011 241.
14. K. Dadkhahipour, T. Nguyen, J. Wang, Mechanisms of channel formation on glasses by abrasive water jet milling, *Wear* 1 2012 292–293.
15. Byoung-Jun Cho, R. Hyuk-MinKim, R. Manivannan, Deog-Ju Moon, Jin-Goo Park, On the mechanism of material removal by fixed abrasive lapping of various glass substrates, *Wear* 302(1–2) 2013 1334.
16. Brian Lawn, *Fracture of Brittle Solids*, 2nd Edition, Cambridge University Press, 1993.
17. R. Brian Lawn, Indentation of ceramics with spheres: A century after hertz, *J. Am. Ceram. Soc.* 81 1998 1977.
18. F. Auerbach, Measurement of hardness, *Ann. Phys. Chem* 43 1891 61.
19. J. P. Tillett, Fracture of glass by spherical indenters, *Proc. Phys. Soc. Lond., B* 69(1) 1956 47.
20. Andreas Momber, Fracture features in soda-lime glass after testing with a spherical indenter, *J. Mater. Sci.* 46(13) 2011 4494.
21. Trevor E. Wilantewicz, E. James, R. Varner, Vickers indentation behavior of several commercial glasses at high temperatures, *J. Mater. Sci.* 43(1) 2008 281.
22. Sheldon M. Wiederhorn, Theo Fett, Gabriele Rizzi, Michael J. Hoffmann, Jean-Pierre Guin Water, Penetration—Its effect on the strength and toughness of silica glass, *Metall. Mater. Trans. A* 44a 2013 1164.
23. C. Symmers, J. B. Ward, B. Sugraman, Studies of the mechanical strength of glass, *Phys. Chem. Glasses* 3 1962 76.
24. Tanguy Rouxel, Jean-Christophe Sangleboeuf, The brittle to ductile transition in a soda-lime-silica glass, *J. Non-Cryst. Solids* 271 2000 224.
25. E. K. Pavelchek, R. H. Doremus, Fracture strength of soda-lime glass after etching, *J. Mater. Sci.* 9(11) 19741803.
26. G. Scott Glaesemann, Karl Jakus, John E. Ritter, Jr., Strength variability of indented soda-lime glass, *J. Am. Ceram. Soc.* 70(6) 1987 441.
27. J. E. Ritter, M. R. Lin, Effect of polymer coating on the strength and fatigue behavior of indented soda-lime glass, *Glass Technol.* 32 1991 51.
28. M. V. Swain, Microfracture about scratches in brittle solids, *Proc. R. Soc. Lond. A* 366(1727) 1979 575.
29. K. W. Peter, Densification and flow phenomena of glass in indentation experiments, *J. Non-Cryst. Solids* 5(2) 1970 103.
30. A. Broese van Groenou, N. Maan, J. C. B. Veldkamp, Single point scratches as a basic for understanding grinding and lapping, in *The Science of Ceramic Machining and Surface Finishing II*, eds., B. J. Hockey, R. W. Rice, National Bureau of Standards Special Publication 562, U.S. Government Printing Office, Vol. II, 43–60, 1979.
31. J. B. D. Veldkamp, N. Hattu, V. A. C. Snijders, *Crack Formation during Scratching of Brittle Materials, Fracture Mechanics of Ceramics*, eds., R. C. Bradt, D. P. H. Hasselman, F. F. Lange, Plenum Press, Vol. 3, 273–300, 1978.
32. B. R. Lawn, Partial cone crack formation in a brittle material loaded with a sliding spherical indenter, *Proc. R. Soc. Lond. A* 299(1458) 1967 307.
33. G. M. Hamilton, L. E. Goodman, The stress field created by a circular sliding contact, *J. Appl. Mech.* 33(2) 1966 371.
34. D. R. Gilroy, W. J. Hirst, Brittle fracture of glass under normal and sliding load, *Phys. D, Appl. Phys.* 2 1969 1784–1787.
35. K. Li, Y. Shapiro, J. C. M. Li, Scratch test of soda-lime glass, *Acta Mater.* 46(15) 1998 5569.
36. F. Petit, C. Ott, F. Cambier, Multiple scratch tests and surface-related fatigue properties of monolithic ceramics and soda lime glass, *J. Eur. Ceram. Soc.* 29(8) 2009 1299–1307.

37. W. Gu, Z. Yao, X. Liang, Material removal of optical glass BK7 during single and double scratch tests, *Wear* 270(3–4) 2011 241.
38. H. B. Abdelounis, K. Elleuchb, R. Vargiolu, H. Zahouani, A. L. Bot, On the behaviour of obsidian under scratch test, *Wear* 266(7–8) 2009 621–626.
39. V. L. Houerou, J. C. Sangleboeuf, S. Deriano, T. Rouxel, G. Duisit, Surface damage of soda–lime–silica glasses: Indentation scratch behavior, *J. Non-Cryst. Solid* 316(1) 2003 54–63.
40. S. Yoshida, T. Hayashi, T. Fukuhara, K. Soeda, J. Matsuoka, N. Soga, *Scratch Test for Evaluation of Surface Damage in Glass, Fracture Mechanics of Ceramics*, eds., R. C. Bradt, D. Munz, M. Sakai, K. W. White, Springer, Vol. 14, 101–111, 2003.
41. Payel Bandyopadhyay, Arjun Dey, Sudakshina Roy, Anoop K. Mukhopadhyay, Effect of load in scratch experiments on soda lime silica glass, *J. Non-Cryst. Solids* 358(8) 2012 1091.
42. J. M. Fan, C. Y. Wang, J. Wang, Modelling the erosion rate in micro abrasive air jet machining of glasses, *Wear* 266(9–10) 2009 968.
43. K. Dadkhahipour, T. Nguyen, J. Wang, Mechanisms of channel formation on glasses by abrasive water jet milling, *Wear* 292–293 2012 1.
44. Byoung-Jun Cho, R. Hyuk-Min, Kim Manivannan, Deog-Ju Moon, Jin-Goo Park, On the mechanism of material removal by fixed abrasive lapping of various glass substrates, *Wear* 302 2013 1334.
45. John J. Mecholsky, Jr., S. W. Freiman, Roy W. Rice, Effect of grinding on flaw geometry and fracture of glass, *J. Am. Ceram. Soc.* 60(3–4) 1977 114.
46. Xu Nie, Weinong W. Chen, Douglas W. Templeton, Dynamic ring-on-ring equibiaxial flexural strength of borosilicate glass, *Int. J. Appl. Ceram. Technol.* 7(5) 2010 616.
47. Xu Nie, Weinong W. Chen, Andrew A. Wereszczak, Douglas W. Templeton, Effect of loading rate and surface conditions on the flexural strength of borosilicate glass, *J. Am. Ceram. Soc.* 92(6) 2009 1287.
48. J. Schneidera, S. Schulaa, W. P. Weinhold, Characterisation of the scratch resistance of annealed and tempered architectural glass, *Thin Sol, Films* 520(12) 2012 4190.
49. J. Zhanga, F.-Z. Xuana, F. Yang, Effect of surface scratches on the characteristics of nonlinear Rayleigh surface waves in glass, *J. Non-Cryst. Sol.* 378(15) 2013 105.
50. C. Bernarda, V. Keryvina, J.-C. Sangleboeufa, T. Rouxela, Indentation creep of window glass around glass transition, *Mech. Mater.* 42(2) 2010 196.
51. P.-L. Larsson, On representative stress correlation of global scratch quantities at scratch testing of elastoplastic materials, *Mater. Des.* 49 2013 536.
52. P. Bandyopadhyay, A. Dey, S. Roy, A. K. Mukhopadhyay, Effect of load in scratch experiments on soda lime silica glass, *J. Non-Cryst. Sol.* 358(8) 2012 1091–1103.
53. Payel Bandyopadhyay, Arjun Dey, Ashoke K. Mandal, Nitai Dey, Sudakshina Roy, Anoop K. Mukhopadhyay, Effect of scratching speed on deformation of soda–lime–silica glass, *Appl. Phys. A* 107(3) 2012 685–690.
54. Payel Bandyopadhyay, Arjun Dey, Ashoke K. Mandal, Nitai Dey, Anoop K. Mukhopadhyay, New observations on scratch deformations of soda lime silica glass, *J. Non-Crystal. Solids* 358(16) 2012 1897–1907.
55. Payel Bandyopadhyay, Arjun Dey, Sudakshina Roy, Nitai Dey, Anoop Kumar Mukhopadhyay, Nanomechanical properties inside the scratch grooves of soda–lime–silica glass, *Appl. Phys. A* 107(4) 2012 943–948.
56. Payel Bandyopadhyay, Arjun Dey, Anoop K. Mukhopadhyay, Novel combined scratch and nanoindentation experiments on soda-lime-silica glass, *J. Appl. Glass Sc.* 3(2) 2012 163–179.
57. Liu Ming, Wu Jianan, Gao Chenghui, Sliding of a diamond sphere on K9 glass under progressive load, *J. Non-Cryst. Sol.* 526 2019 119711.
58. Jiacheng Fu, Hongtu He, Weifeng Yuan, Yafeng Zhang, Jiaxin Yu, Towards a deeper understanding of nanoscratch-induced deformation in an optical glass, *Appl. Phys. Lett.* 113(3) 2018 031606.
59. P. Bandyopadhyay, A. K. Mukhopadhyay, Unique observations in nanoscale dynamic contact of glass, *Mater. Today Proc.* 1 2021 4.

Part 5

Future Research Directions

18 Future Directions
Applications of Artificial Intelligence in Material Deformation and Fracture

Ashok Kumar, Payel Maiti, and Anoop Kumar Mukhopadhyay

18.1 INTRODUCTION

Artificial intelligence (AI) is shaping our current world in significant ways, and its increasing impact in every sphere of our lives is likely to be felt much more strongly in the coming years. There have been many path-breaking applications of artificial intelligence in computer vision, medical diagnosis, speech recognition, optical character recognition and natural language processing. Currently, there are two major popular branches known as traditional machine learning (ML) [1] and deep learning (DL) [2]. Applications of AI in science and engineering are continuously increasing at an accelerated rate.

The scientific and technical development of a society is critically dependent upon materials with desirable properties. But development of such materials often requires a large number of time-consuming, labor-intensive and expensive trial-and-error-based experimentation. The situation is equally problematic with performing theoretical calculations or numerical simulations on a case-by-case basis. In large measure, such hard endeavors are ultimately ineffective. One important help that artificial intelligence can provide to a scientist or engineer is to reduce the number of time-consuming and expensive experiments, calculations or simulations by narrowing the search space and suggesting a much smaller number of options that could save a lot of time and expense. Artificial intelligence can provide crucial and valuable help in developing a focused and efficient experimental, theoretical or numerical strategy that would be more likely to bear fruit. This can make the exploration effort more focused and informed.

If an AI system is trained on a large number of samples, it can figure out a general nonlinear function between the inputs and the outputs (although the number of parameters can be very large). Hornik [3–5], in a series of papers, established that a feedforward artificial neural network with at least one hidden layer with an arbitrary number of units can approximate any Borel measurable function between inputs and outputs of finite dimensions. In real-life situations, such a model could be used for forecasting output results for newer input conditions, provide useful trends for extrapolation and interpolation, and help significantly narrow the search space for useful scientific and technical parameters.

18.2 INTRODUCTION TO ARTIFICIAL INTELLIGENCE

Artificial intelligence refers to intelligent behavior demonstrated by machines. A measure of whether a machine could be considered intelligent or not is given by the famous Turing test [6],

DOI: 10.1201/9781003359364-23

first proposed by Alan Turing in 1950. In that famous paper he posed the question "Can machines think?" That paper is often considered to originate the concept of artificial intelligence, although the actual term was introduced later. Turing proposed a hypothetical machine called the Turing machine that could in principle perform any mathematical or logical computation by manipulating some finite number of symbols such as the binary digits 0 and 1. This insight played a great role in the development of modern computers. The development of digital computers led to great expectations of machines surpassing other capabilities of humans just as they had surpassed ordinary computational capabilities of humans. But despite many efforts by many great minds, old AI programs were not very fruitful. Ordinary human capabilities such as speech recognition and face recognition turned out to be much more difficult for computers. It was eventually realized that a different approach was required. Believing that one needed to learn how the human brain functions to make any progress, Rosenblatt [7] in 1958 proposed a computational unit that was a simplified artificial neuron called a perceptron. Mutually connecting such perceptrons in layers led to artificial neural networks (ANNs). A famous algorithm called the backpropagation algorithm was proved in 1986 by Rumelhart et al. [8], which showed that feedforward ANNs could be trained by the backpropagation of errors. Although this was a major conceptual step in showing that ANNs were perhaps useful, their real glory had to await the advent of computers with higher computing capabilities, which occurred around 2010. ANNs can be trained to recognize very complex patterns of data and are widely used in the AI field.

Since human intelligence can be applied in all fields of human activity, so can AI. The current state of AI is called narrow artificial intelligence, where AI has proven to be quite successful and useful in narrowly defined tasks. AI systems can learn from examples. An AI model usually has a large number of trainable parameters and other parameters called hyperparameters that are kept constant during the training. After each batch of input data, error or loss is computed, which is then used to revise the AI model's trainable parameters. This revision of trainable parameters is called learning. In ANNs, which are a very popular and highly effective AI model type, the trainable parameters are called the weights that link one artificial neuron to another and their biases. These weights are analogous to synaptic strengths between biological neurons in the human brain that are supposed to be updated following learning.

Training an AI system by repetitively presenting a large number of examples and minimizing a loss or error function computed from the mismatch of AI-derived answers and correct or ground truth answers is called supervised learning. There are other kinds of learning possible. Such as unsupervised and reinforcement learning, where ground truth answers are not known a priori.

18.3 MACHINE LEARNING

Early successes for artificial intelligence in the 21st century were based on traditional machine learning where features had to be designed by humans in a process called feature engineering. For example, if one had to make an AI system to determine a type of vehicle, then one would manually define specific features that would be important in classifying images of vehicles such as a car, a bus, a truck, a motorbike, a train, etc., into classes. Then based on these manually defined features, either an ANN or other machine learning algorithm could be employed to classify the images into respective classes. A popular and free machine learning library is Scikit-learn [9] (https://scikit-learn.org/) available for the PYTHON programming language, which provides a uniform interface for implementing a large number of machine learning algorithms such as support vector machines, random forest, k-means, Gaussian mixture models, and naïve Bayes. PYTHON, which is a modern object-oriented open-source programming language with a large number of free packages for scientific and technical purposes as well as for AI and ML, is available in many distributions. One popular distribution is called ANACONDA (https://www.anaconda.com/) and it provides many pre-installed PYTHON packages for scientific work.

18.4 DEEP LEARNING

Traditional ML algorithms may perform better than deep learning for smaller-sized datasets. But as the data size increases, the performance of traditional ML algorithms tends to plateau. For larger datasets, another paradigm called deep learning has been very successful in the last decade. Deep learning is based on ANNs with a large number of hidden layers (hence deep).

The second decade of the 21st century saw the advent of deep learning, which led to many remarkable successes for AI. One major difference between traditional machine learning and deep learning is that the features are not required to be defined by humans. But the AI system itself figures out the important features, provided a large number of examples are given to it. One particular deep learning paradigm is called convolutional neural networks (CNNs). These networks were designed based on the human visual pathway and were extremely successful in computer vision tasks such as image classification and segmentation [10–12]. There are other network designs called recurrent neural networks (RNNs) and long short-term memory (LSTM) networks that are very useful for various tasks where the current state of the system depends on its previous states. A popular platform to implement DL-based networks is using TensorFlow [11] (https://www.tensorflow.org/) and Keras (https://keras.io/) packages in the PYTHON programming language. In its latest versions, Keras comes bundled with TensorFlow. Another very popular PYTHON-based package for deep learning is PyTorch (https://pytorch.org/). AI-based systems can also be implemented in C++, MATLAB or MATHEMATICA, but PYTHON-based implementations are most popular.

There have been many more advances in deep learning, such as those leading to networks called transformers, that use mechanisms of attention that have proved very fruitful in natural language processing (NLP). Other generative kinds of networks such as generative adversarial networks can generate a new sample after being trained on a large number of examples. These can be used in creating new music or art or even deep fakes, in which a computer-generated photo, video or audio is created that appears to be genuine but is not. In fact, the developments in the AI field have been so rapid with so many widespread implications that no field is left unaffected, and the importance of AI in other hitherto less invaded fields is only going to rapidly increase.

18.5 AI IN MATERIALS DEFORMATION AND FRACTURE RESEARCH

In cases where large amounts of data are not available, usually machine learning algorithms are employed that can perform well with even smaller amounts of data. But due to the requirement of feature engineering, they need more human intervention and often do not generalize well to new data. Deep learning on the other hand can learn important features of data by itself, but requires much larger amounts of data to train than traditional machine learning. Next, we present a few examples of researchers employing AI methods in materials research, especially as applicable to deformation and fracture.

Picklum and Beetz [13] note that technical innovations demand the development of materials with novel specified properties. But developing such materials with the required properties is expensive in terms of cost, labor and time. They developed a system based on machine learning and semantic knowledge to model relationships between materials, processes and properties. Tang et al. [14] working on high-entropy ceramics mentioned that although they have highly desirable properties, due to the nonexistence of systematic design procedures for producing them with desired mechanical properties, the usual trial-and-error experiments and isolated simulations are ineffective and costly, while machine learning methods could accelerate the development of such ceramics with desirable properties. Couet [15] presented a review of predictive models for high-dose irradiation and corrosive effects of molten salt on compositionally complex alloys and observed that the irradiation and corrosion response of a large number of such alloys can be tested at small cost and in a relatively short time, which could be used to establish a large database for utilizing machine learning methods to narrow the search space for compositions of such alloys and provide useful trends.

Liang et al. [16] applied several machine learning algorithms for predicting creep behavior in concrete. Machine learning algorithms used by them included random forest, extreme gradient boosting (XGBoost) and light gradient boosting. They obtained high accuracy for training data with coefficients of determination (R^2) higher than 0.95. For the testing data, their ML models show significantly higher performance than a numerical model.

Vranjes-Wessely et al. [17] studied high throughput nanoindentation mapping of rocks that are rich in organic matter and that are difficult to study for nanoindentation using ML techniques. They focused on hardness and elastic constant measurements and used the ML algorithm of k-means clustering to separate representative values of these physical quantities. Vignesh et al. [18] also used k-means clustering in their deconvolution experiments for determining individual phase properties from the spatial map of properties of multiphase thermal barrier coatings. Konstantopoulos et al. [19] applied ML-based classification on nanoindentation mapping data to detect the mechanism of reinforcement in the interface between the fiber and matrix of carbon fiber-reinforced polymer composites and obtained classification accuracies of about 67%. They used ANNs and ML algorithms such as random forest (RF), k-means, stochastic gradient boosting and support vector machines (SVMs) and found SVMs to be more adept at their task. They emphasize that often one has to employ many AI–ML algorithms to see which one performs best in a certain task.

Zhang et al. [20] working on high-entropy alloys developed a database using molecular dynamics simulations describing the relationship between alloy composition and mechanical properties that was used for training the AI–ML models. They considered SVMs, kernel-based extreme learning machines, and deep neural networks (DNNs) and found that the DNNs performed best for the binary classification of yield stress. Some studies, such as that by Samaei and Choudhury [21], use finite element analysis (FEA) combined with generative adversarial (GANs) for determining the elastic properties of heterogeneous materials where the GAN was used to generate microstructures with varying pore shapes, porosities and pore distributions.

This is just a panoramic view of only a few kinds of applications of AI in materials science and material deformation and fracture. In fact, the scope of applications is quite wide and increasing every day.

18.6 EXAMPLE APPLICATION OF AI: NANOINDENTATION OF ALUMINA

18.6.1 Nanoindentation Basics

In various applications, the behavior of materials under varying stress is an important piece of required information. For many materials, nanoindentation experiments can provide useful data and estimates of physical quantities like hardness and elastic constants [22–28]. For nanoscale and microscale materials, this technique is especially useful. Even for large-scale materials, nanoindentation studies are useful, especially since even larger size indentations start as nanoindentations.

In a nanoindentation experiment, an instrument applies a steadily increasing load up to a maximum and then steadily decreases the load back to zero. During the experiment, the applied load (P) and corresponding depth (h) of the indenter tip is recorded. The most common indenter tips are made of diamond and have a tetrahedral-shaped tip called a Berkovich indenter. A typical simplified indentation geometry is illustrated in Figure 18.1.

From the P versus h plots several useful quantities could be determined as shown in Figure 18.2.

18.6.2 Details of Nanoindentation Experiments

The data presented here were originally obtained from experiments performed by the second author under the supervision of the third author. They consisted of from 16 up to 25 repeated indentation experiments on an alumina substrate with maximum loads of 10, 30, 50, 70, 100, 200, 300, 500, 700, 900 and 1000 mN. The applied load versus time curves are shown in Figure 18.3, and the depth

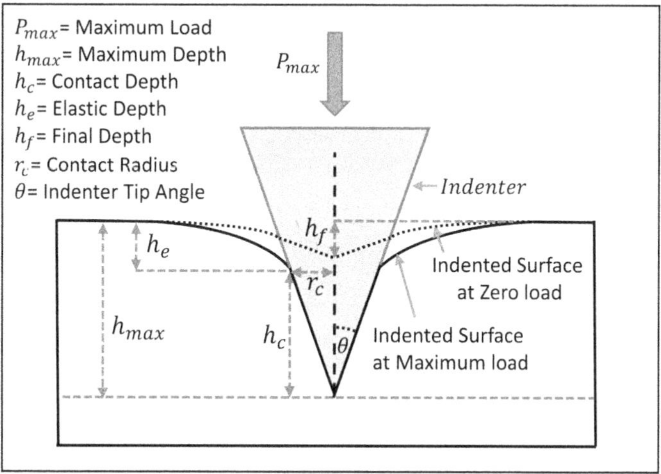

FIGURE 18.1 A simplified nanoindentation geometry with various quantities defined.

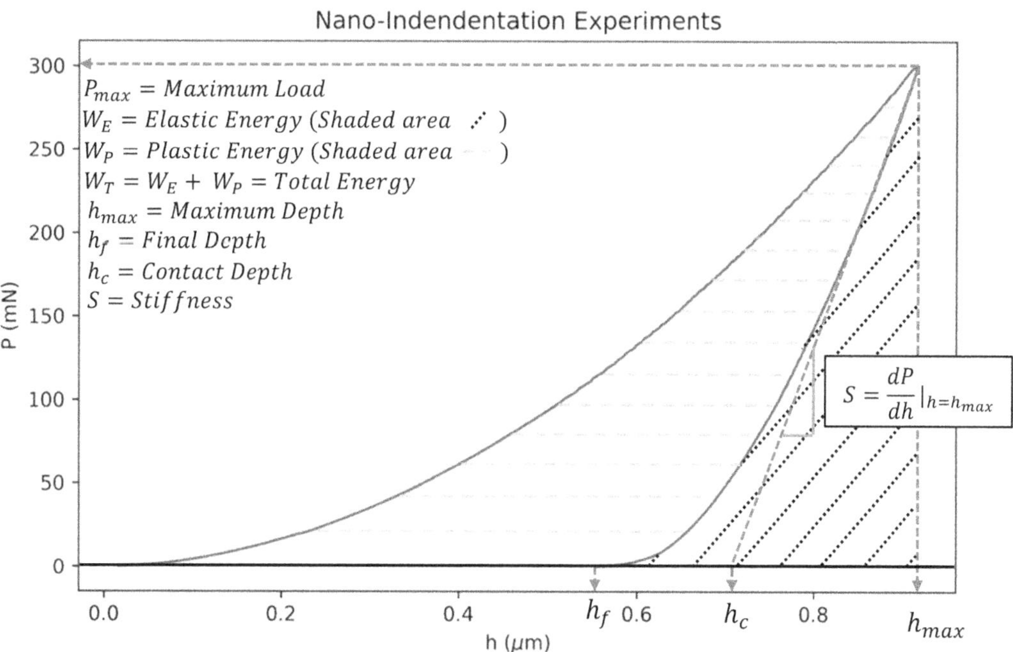

FIGURE 18.2 An experimentally obtained load versus depth, or P–h curve for alumina at 300 mN max load. Several useful quantities can be derived from the P–h curve, and some of them are shown in the figure.

of indentation versus time curves are shown in Figure 18.4 and the derived load versus depth (P versus h) curves are shown in Figure 18.5.

18.6.3 Power Law Fits

Oliver and Pharr [22, 29] argued that the loading curves follow a power law of form

$$P_{loading} = Ah^m \tag{18.1}$$

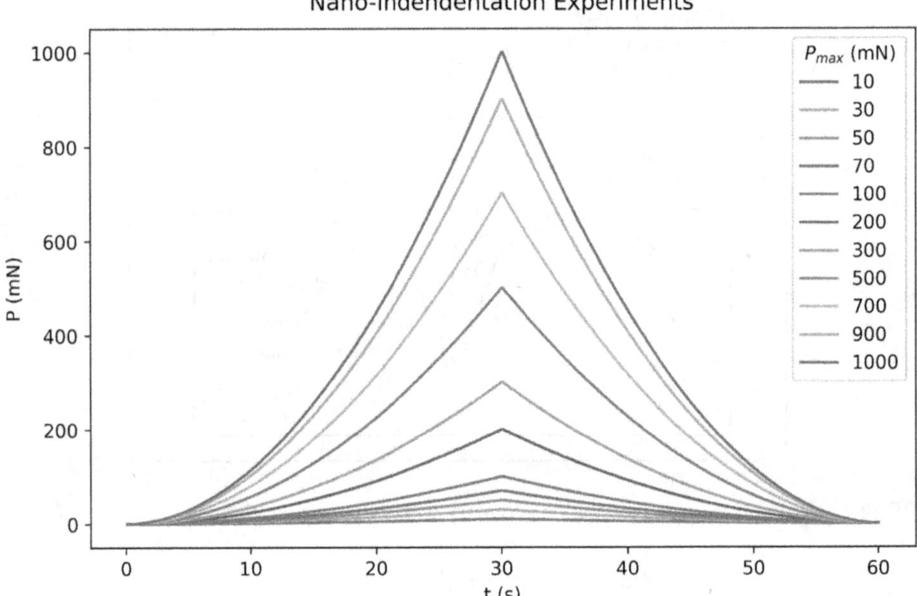

FIGURE 18.3 Applied load versus time curves for different maximum loads. Each curve is obtained from an average of 16 to 25 experiments.

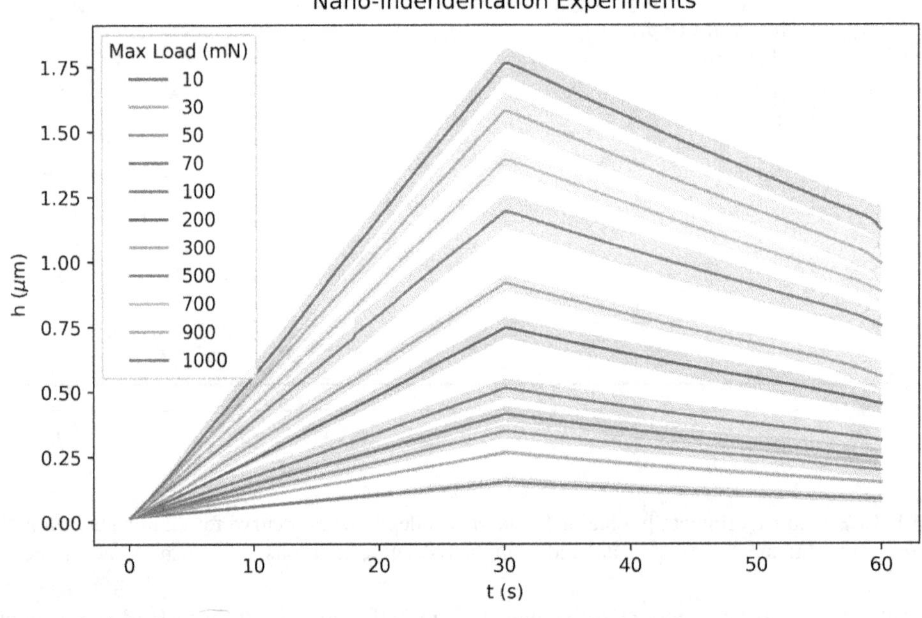

FIGURE 18.4 Depth of indentation versus time curves for different maximum loads. Each curve is obtained from an average of 16 to 25 experiments. The shaded regions show the standard deviation interval around the mean shown by the solid curves.

where $P_{loading}$ is the applied load during increasing load, h is the depth of indentation, and A and m are constants to be determined from fit. The power law fits to the loading curves are shown in Figure 18.6.

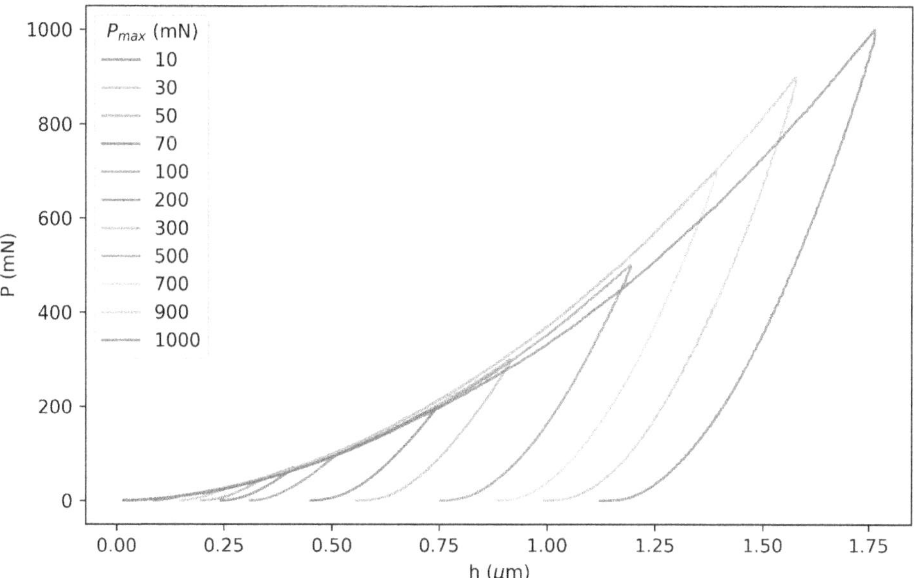

FIGURE 18.5 Applied load versus depth of indentation curves for different maximum loads. Each curve is obtained from an average of 16 to 25 experiments.

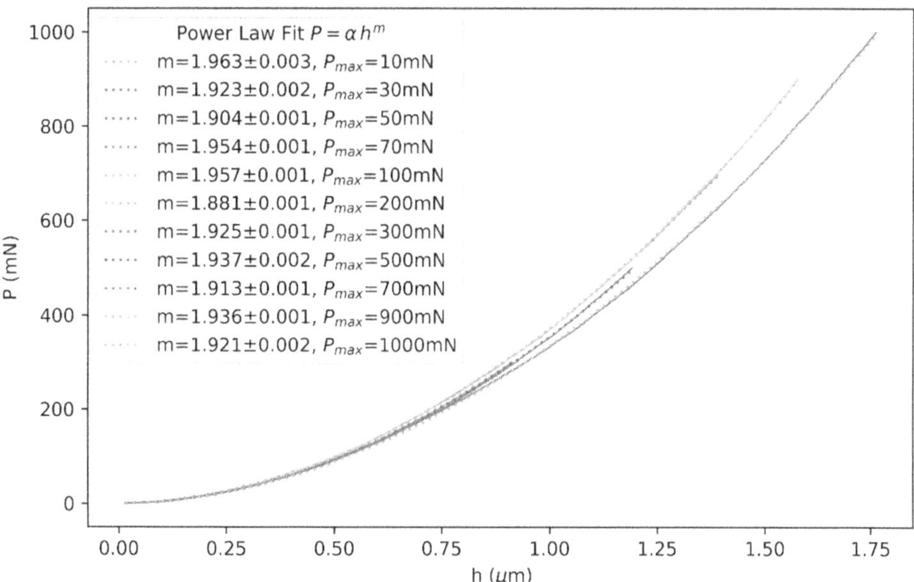

FIGURE 18.6 Applied load versus depth of indentation curves for different maximum loads during increasing load. Each curve is obtained from an average of 16 to 25 experiments. A power law fit was performed on each curve that shows the exponent ranges between 1.881 and 1.963.

Similarly, the unloading curves also follow a power law of form [22, 29]

$$P_{unloading} = B\,(h - h_f)^m \tag{18.2}$$

where $P_{unloading}$ is the applied load during decreasing load, h is the depth of indentation, and B and m are constants to be determined from fit. The power law fits to the unloading curves are shown in Figure 18.7.

18.6.4 Normalized Power Law Fits

The power law equations shown in Equation 18.1 and Equation 18.2 suffer from a drawback. The equations are not written in a dimensionless form, which leads to awkward units for the fit coefficients A and B, especially if m happens to be nonintegral. This also creates a problem when converting from one set of units to another. Oliver and Pharr [22] did show a normalized form for the unloading curve in one of their plots of the form

$$P = B'\left(\frac{h-h_f}{h_{max}}\right)^m. \tag{18.3}$$

Kossman et al. [26] rewrote Equation 18.3 in a dimensionless form as

$$\frac{P}{P_{max}} = B\left(\frac{h-h_f}{h_{max}}\right)^m \tag{18.4}$$

where coefficient B would be dimensionless and may be easier to compare between different expressions.

18.6.5 Proposed Modification in the Form of Power Law Fits

We propose the following fully normalized forms for the loading and unloading power law functions such that the fit coefficients are dimensionless and physically reasonable:

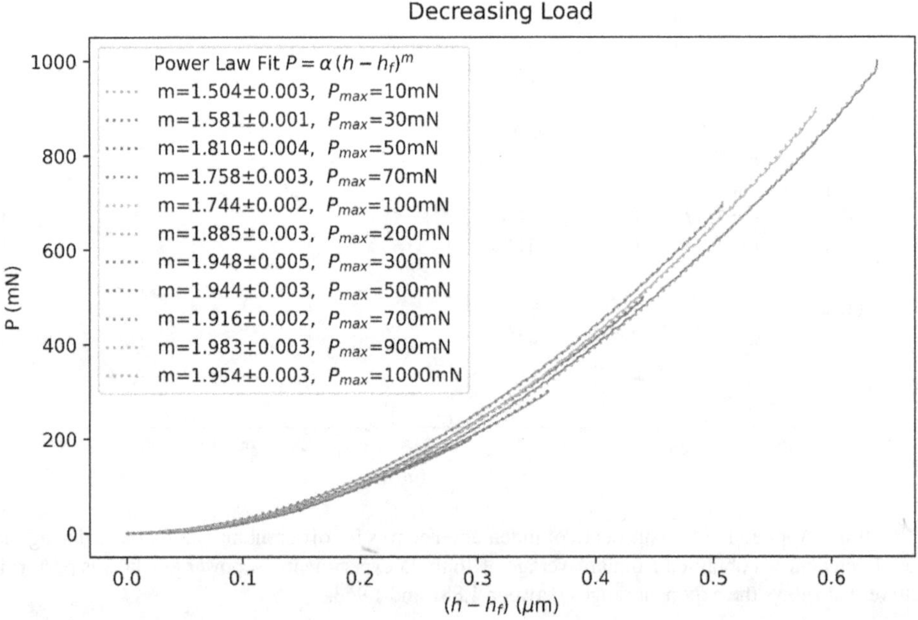

FIGURE 18.7 Applied load versus depth of indentation curves for different maximum loads during decreasing load. Each curve is obtained from an average of 16 to 25 experiments. A power law fit was performed on each curve that shows the exponent ranges between 1.504 and 1.983.

$$\frac{P_{loading}}{P_{max}} = \alpha \left(\frac{h}{h_{max}} \right)^{m_l}, \tag{18.5}$$

$$\frac{P_{unloading}}{P_{max}} = \beta \left(\frac{h - h_f}{h_{max} - h_f} \right)^{m_u}, \tag{18.6}$$

where m_l and m_u are power law exponents during loading and unloading, respectively. Since at $P = P_{max}$ we have $h = h_{max}$, we must have

$$\alpha = \beta = 1 \tag{18.7}$$

It shows that one of the fit coefficients is superfluous. In fact, with the knowledge of P_{max}, we can rewrite the proposed fit equations as

$$P_{loading} = P_{max} \left(\frac{h}{h_{max}} \right)^{m_l} \tag{18.8}$$

$$P_{unloading} = P_{max} \left(\frac{h - h_f}{h_{max} - h_f} \right)^{m_u} \tag{18.9}$$

These new proposed forms behave as expected in the limiting case of $h = h_{max}$, $P = P_{max}$, and the unloading form performs as expected in the limiting case of $h = h_f$, $P = 0$. The normalized fit plots

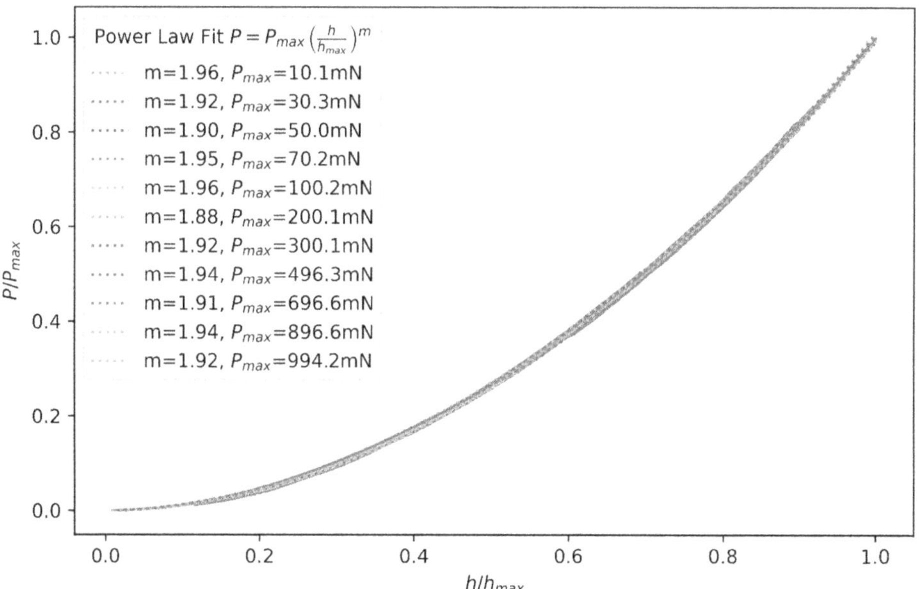

FIGURE 18.8 Applied normalized load versus normalized depth of indentation curves for different maximum loads during increasing load. Each curve is obtained from an average of 16 to 25 experiments. A power law fit was performed on each curve that shows that the exponent ranges between 1.88 and 1.96. In the normalized form it is clear that all loadings very closely follow the same curve. Also, notice that P_{max} treated as a fit parameter is very close to the actual maximum load.

where P_{max} was treated as a fit parameter are shown in Equations 18.8 and 18.9. It is evident that the fitted values of P_{max} are quite close to the actual values.

18.7 ANN MODEL FOR PREDICTING W_P, W_E, H_F AND H_{MAX} AS FUNCTIONS OF P_{MAX}

As an example application, we present a simple fully connected artificial neural network to predict values of plastic energy W_p, elastic energy W_e, maximum depth h_{max} and final depth h_f as functions of P_{max} only, since for a given target material, indenter and loading–unloading pattern, these quantities can depend only upon P_{max}.

A fully connected artificial neural network for a general nonlinear regression between input values and output values was implemented using TensorFlow (version 2.9) and Keras (version 2.9) packages of PYTHON (version 3.9) programming language. The feature vector consists of only the maximum load value P_{max}, which is used as input to the neural network, therefore the input layer of the network has a single unit neuron or perceptron. The output has four values, therefore, the output layer has four units or neurons. There are two hidden layers in the model with 128 and 64 neurons, respectively. Since it is a fully connected model, every neuron in a layer is connected to every neuron in the next layer. Before the training, the weights of the links between the neurons are initialized with random values drawn from a normal distribution. The training proceeds according to the feedforward backpropagation algorithm with the ADAM (Adaptive Moment Estimation) optimizer with a learning rate of 0.001. The error function or loss function used was the mean square error function computed between the actual expected values and outputs of the network. The activation function used was the rectified linear unit (ReLU), except for the output layer where a

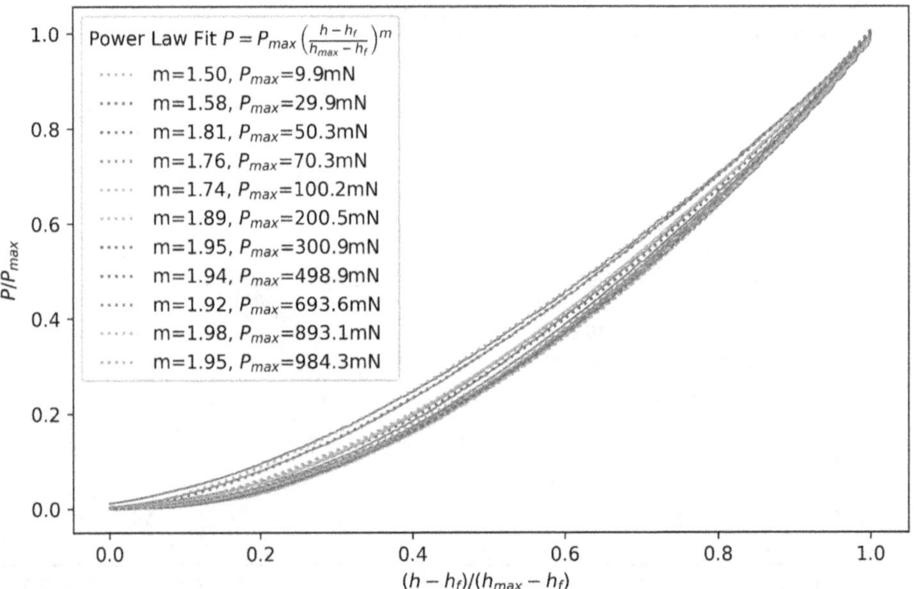

FIGURE 18.9 Applied normalized load versus normalized depth of indentation curves for different maximum loads during decreasing load. Each curve is obtained from an average of 16 to 25 experiments. A power law fit was performed on each curve that shows that the exponent ranges between 1.50 and 1.98. In the normalized form it is clear that the unloading curves show more variation than the loading curves. There is a trend of increasing exponent with load. Also, notice that P_{max} treated as a fit parameter is very close to the actual maximum load.

linear activation was used since this was a regression problem. Network link weights are adjusted only after a batch is passed through the network, and completing an epoch means presenting all the training data to the network in batches. The batch size used was 32 and epochs used for training were 100. The model details and training parameters are shown in Figure 18.10. Prior to training the ANN, a normalization on the data was performed using the Scikit-learn package's sklearn.preprocessing.MinMaxScaler class. The MinMaxScaler scales the data to lie between 0 and 1. After the ANN is trained, data predictions from the network need to be scaled back using the inverse function provided by the MinMaxScaler function.

The performance of the network during the training phase is shown in Figure 18.11. The mean square error or loss function initially decreases and eventually flattens, indicating completion of training.

After completion of the training process, the values of h_{max}, h_f, W_t and W_p predicted by the trained network for given P_{max} values were plotted against the actual experimental values in Figures 18.12 and 18.13. A linear regression was performed for each case between predicted and actual values and the coefficient of performance (R^2) and the slope of the resulting line were computed, demonstrating a good performance by the network.

18.8 SOME MACHINE LEARNING METHODS TO PREDICT THE LOAD–DISPLACEMENT (P–H) CURVE: PREDICTING M_L, M_U, H_{MAX} AND H_F AS FUNCTIONS OF P_{MAX}

As another example application, we applied a few ML algorithms to predict the full load–displacement (P–h) curve. For a given material and a given indenter, we expect that the load–displacement curve should depend only upon P_{max}. As seen from Equation 18.8 and Equation 18.9, the loading part just has P_{max}, power law exponent m_l and maximum depth h_{max} as parameters, and the unloading part has P_{max}, unloading power law exponent m_u, maximum depth h_{max} and final depth h_f as parameters. Therefore, determining the P–h curve implies determining m_u, m_l, h_{max} and h_f as functions of P_{max}.

Fully Connected Artificial Neural Network Model

Layers	Units	Type	Activation	Initializer	Parameters
Input Layer	1	Dense	Rectified Linear Unit	Normal	2
Hidden Layer 1	128	Dense	Rectified Linear Unit	Normal	256
Hidden Layer 2	64	Dense	Rectified Linear Unit	Normal	8256
Output Layer	4	Dense	Linear	Normal	260
				Total Parameters	8774
				Total Trainable Parameters	8774

Model Training Parameters	
Optimizer	ADAM (Adaptive Moment Estimation)
Loss Function	Mean Square Error (MSE)
Learning Rate	0.001
Batch Size	32
Epochs	100

FIGURE 18.10 Details of the fully connected artificial neural network (ANN) model and associated training parameters that were used for training the network on the nanoindentation data.

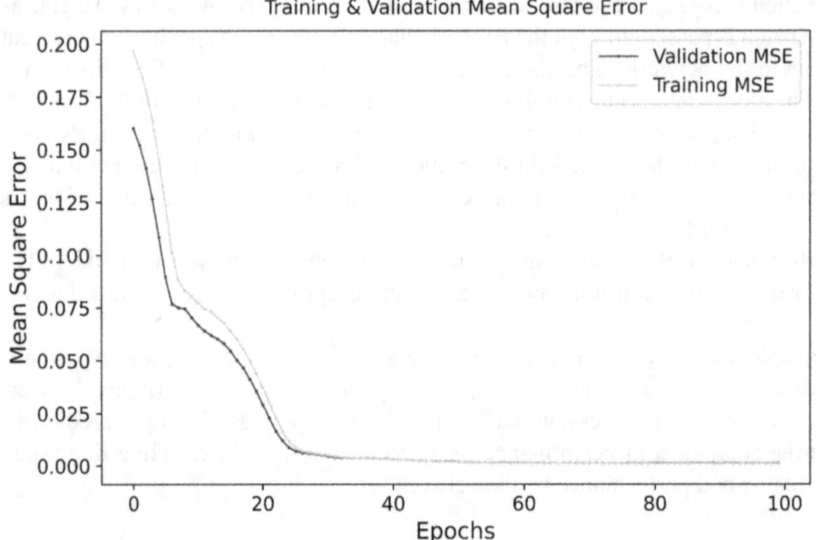

FIGURE 18.11 Training and validation mean square error (MSE) plots for the artificial neural network model. The minimum MSE after 100 epochs of training was 0.0011 for training and 0.0012 for validation. The training curves and validation curves converge together demonstrating no overfitting. The mean square error drops and eventually flattens indicating that the network has been trained.

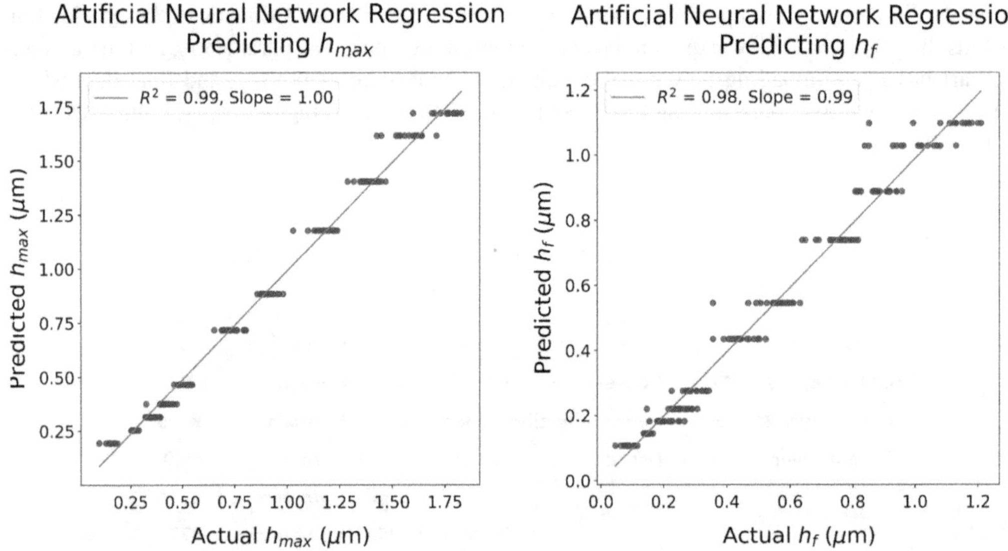

FIGURE 18.12 The comparison between the actual h_{max} and h_f values and those predicted by the trained neural network. A linear regression between the actual and predicted values shows a good performance with the coefficient of performance R^2 and slope of the line indicated.

Since all these quantities (m_u, m_l, h_{max}, h_f and P_{max}) are continuous variables, the problem is regression rather than classification. Since we will train the machine learning models using experimentally determined m_u, m_l, h_{max} and h_f for each value of P_{max}, this will be a case of supervised learning.

The Scikit-learn machine learning library [9] provides PYTHON language implementations of a large number of ML algorithms including dozens of regression models. For an illustration, we will

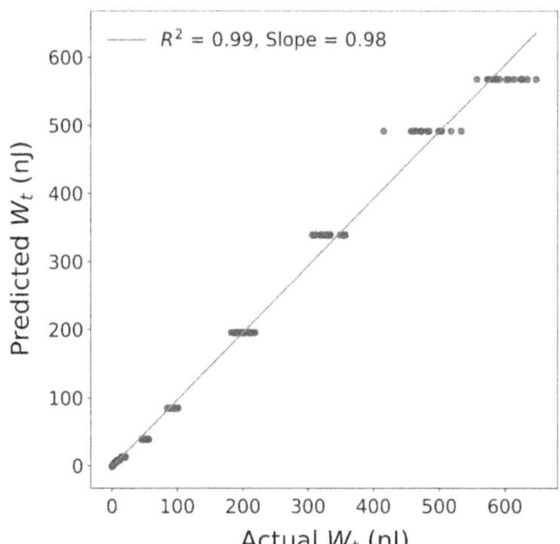

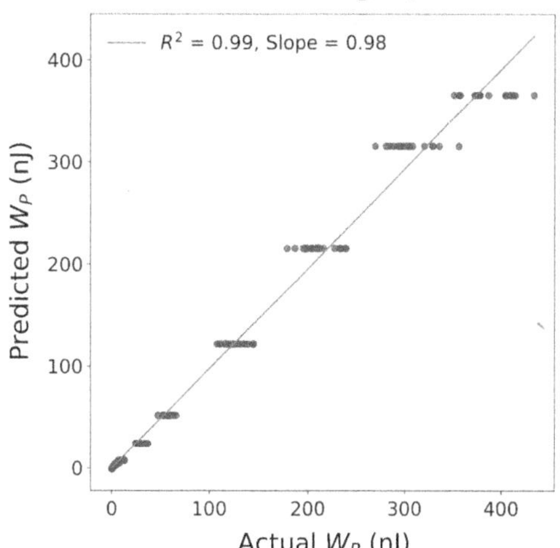

FIGURE 18.13 The comparison between the actual W_t and W_p values and those predicted by the trained neural network. A linear regression between the actual and predicted values shows a good performance with the coefficient of performance R^2 and slope of the line indicated.

be showing only three nonlinear regression approaches. One will be SVM [30–32] with radial basis function (RBF)-based regressors. Another is based on gradient boosting (GB) and XGBoost [33]. Lastly, since an ANN can approximate any nonlinear function between inputs and outputs due to the universal approximation theorem [4], we will also implement an ANN-based regressor.

SVM regressors are implemented in the SVM subpackage of Scikit-Llearn with the name SVR (sklearn.svm.SVR class). The main parameters chosen for the SVM radial basis function regression were kernel = 'rbf', gamma = 8.185×10^{-6} and regularization parameter c = 2. Parameter gamma was chosen with the default 'scale' option, which computes gamma as 1/ (Number of features × Variance of feature vector). The full set of parameters including all the default parameters for SVR-RBF are:

```
svr_rbf params= {'C': 2, 'cache_size': 200, 'coef0': 0.0, 'degree': 3, 'epsilon': 0.1, 'gamma': 'scale', 'kernel': 'rbf', 'max_iter': -1, 'shrinking': True, 'tol': 0.001, 'verbose': False}
```

Gradient boosting regressors are implemented in the SVM subpackage of Scikit-learn with the name GradientBoostingRegressor (sklearn.ensemble.GradientBoostingRegressor class). The main parameters chosen for gradient boosting regression (GBR) were learning_rate = 0.03 and subsample = 0.8. These were chosen to reduce the overfitting tendency of GBR. As shown by Hastie et al. [34], overfitting in GBRs can be controlled by shrinkage (keeping the learning rate less than 1) and subsampling (keeping sampling less than 1). The full set of parameters including all the default parameters for GBR are:

```
gbr params= {'alpha': 0.9, 'ccp_alpha': 0.0, 'criterion': 'friedman_mse', 'init': None, 'learning_rate': 0.03, 'loss': 'squared_error', 'max_depth': 3, 'max_features': 1, 'max_leaf_nodes': None, 'min_impurity_decrease': 0.0, 'min_samples_leaf': 1, 'min_samples_split': 2, 'min_weight_fraction_leaf': 0.0, 'n_estimators': 100, 'n_iter_no_change': None, 'random_state': None, 'subsample': 0.8, 'tol': 0.0001, 'validation_fraction': 0.1, 'verbose': 0, 'warm_start': False}
```

The ANN-based regressor was based on a fully connected ANN with two hidden layers. The details of the ANN model are given in Figure 18.14.

Fully Connected Artificial Neural Network Model For Predicting Load-Displacement Curve

Layers	Units	Type	Activation	Initializer	Parameters
Input Layer	1	Dense	-	-	2
Hidden Layer 1	256	Dense	Rectified Linear Unit	Normal	512
Hidden Layer 2	128	Dense	Rectified Linear Unit	Normal	32,896
Output Layer	4	Dense	Linear	Normal	516
				Total Parameters	33,924
				Total Trainable Parameters	33,924

Model Training Parameters	
Optimizer	ADAM (Adaptive Moment Estimation)
Loss Function	Mean Square Error (MSE)
Learning Rate	0.001
Batch Size	16
Epochs	50

FIGURE 18.14 The fully connected artificial neural network (ANN) model and associated training parameters that were used for training the network for predicting the load–displacement curve, predicting loading exponent (m_l), unloading exponent (m_u) and maximum depth (h_{max}) for given values of P_{max}.

The dataset was split into training and validation sets in the ratio of 80/20 using the train_test_split function from the Scikit-learn's sklearn.model_selection class. Figure 18.15 shows the mean square error (MSE) for training data as well as the validation data plotted against epoch number during training of the ANN model. The MSE value for both training and validation data eventually approaches a constant minimum value of 0.0077 and 0.0079, respectively, after 50 epochs of training, indicating a successful training with negligible overfitting.

18.8.1 Predicting H_{MAX} as Function of P_{MAX}

The results of predicting the maximum depth h_{max} as a function of maximum load P_{max} are shown in Figure 18.16, where plots for SVM-RBF regression, GBR, ANN-based prediction and power law fits are shown together. We can see that all the methods perform well with ANN, GBR and power law fits performing better than SVM-RBF.

The power law fit where the coefficient and exponent were both varied gives:

$$h_{max} = 0.044\, P_{max}{}^{0.53}, \tag{18.10}$$

where h_{max} is in the units of microns and P_{max} is in the units of mN.

The power law exponent of 0.53 is quite close to 0.5 or to the square root function. If the exponent is fixed to be exactly 0.5, then the fit gives

$$h_{max} = 0.053\sqrt{P_{max}}\,. \tag{18.11}$$

By squaring and rearranging we obtain

$$\frac{P_{max}}{h_{max}{}^{2}} = 356\ GPa. \tag{18.12}$$

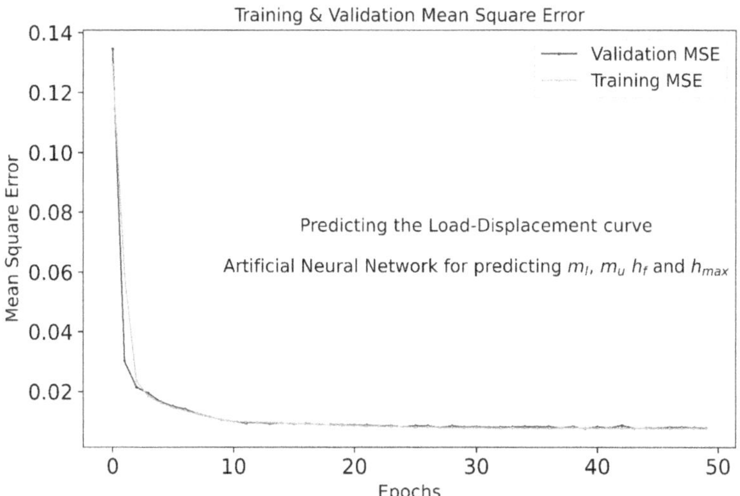

FIGURE 18.15 Training and validation mean square error (MSE) plots for the artificial neural network model for predicting loading exponent (m_l), unloading exponent (m_u) and maximum depth (h_{max}). The minimum MSE after 50 epochs of training was 0.0098 for training and 0.0096 for validation. The training curves and validation curves converge together demonstrating no overfitting. The mean square error drops and eventually flattens indicating that the network has been trained.

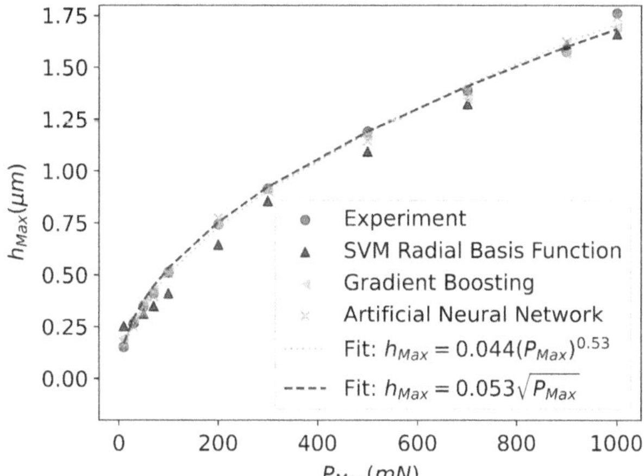

FIGURE 18.16 Predicting maximum depth h_{max} as a function of maximum load $P_{max.}$ The experimental data are shown as circles, the results of support vector machines–radial basis function (SVM-RBF) regression are shown as upright triangles, the results of gradient boosting regression (GBR) are shown by left-pointing triangles and the results of ANN-based prediction are shown as crosses. A power law fit with two parameters (a multiplicative coefficient and a power law exponent) was also performed, which is shown by dots. The exponent of 0.53 is quite close to 0.5. Another fit with the exponent fixed at 0.5 (square root function) is shown with dashes. We can see that all the methods perform well, with ANN, GBR and power law fits performing somewhat better than SVM-RBF.

This value of 356 GPa is close to Young's modulus for high-purity alumina, which is the substrate material in this case.

18.8.2 Predicting M_L as Function of P_{MAX}

The results of predicting loading power law exponent m_l as a function of maximum load P_{max} are shown in Figure 18.17 where plots for SVM-RBF regression, GBR, ANN-based prediction and

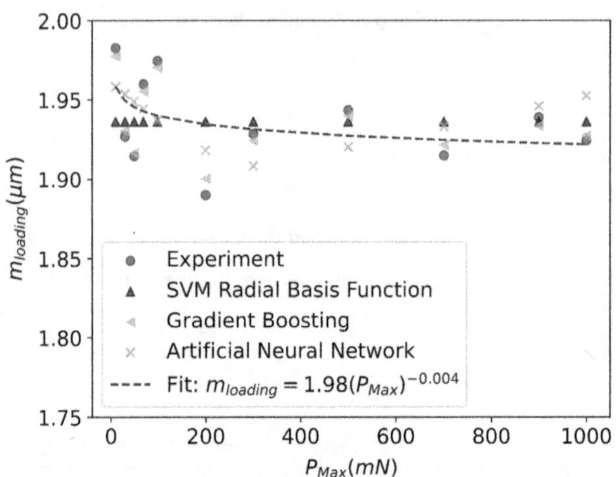

FIGURE 18.17 Predicting power law exponent m_l of the loading part of the load–displacement curve as a function of maximum load $P_{max.}$ The experimental data are shown as circles, the results of support vector machines–radial basis function (SVM-RBF) regression are shown as upright triangles, the results of gradient boosting regression (GBR) are shown by left-pointing triangles and the results of ANN-based prediction are shown as crosses. A power law fit was also performed, which is shown with dashes. We can see that loading exponents are confined in a narrow range, which becomes even narrower for larger loads and appears to have a very small downward trend with increasing maximum load. ANN regression follows the power law fit most closely, and the SVM-RBF gives a constant value prediction. Gradient boosting with the chosen parameters fits the fluctuations of the experimental data more closely, indicating some overfitting.

power law fits are shown together. The power law fit where the coefficient and exponent were both varied gives:

$$m_l = 1.98 P_{max}^{-0.004}. \tag{18.13}$$

The values of m_l show a slight decreasing trend with P_{max}, starting from around 1.98 at zero max load and approaching a slightly lower value of 1.93 for larger loads. We can see that ANN follows the power law fit more closely, while SVM-RBF predicts a constant value close to 1.94. GBR follows the experimental fluctuations more closely indicating possible overfitting. The predictions of different methods show more variation at lower P_{max} values, while at larger P_{max} values they tend to converge.

18.8.3 Predicting M_U as Function of P_{MAX}

The results of predicting unloading power law exponent m_u as a function of maximum load P_{max} is shown in Figure 18.18 where plots for SVM-RBF regression, GBR, ANN-based prediction and power law fits are shown together. The power law fit where the coefficient and exponent were both varied gives:

$$m_u = 1.39 P_{max}^{0.053}. \tag{18.14}$$

The values of m_l show an increasing trend with P_{max}, starting from around 1.5 at zero P_{max} and approaching a higher value of 2.0 for larger loads. We can see that ANN follows the power law fit more closely, while SVM-RBF matches well at lower loads but deviates for larger loads and predicts a constant value close to 1.88. GBR follows the experimental fluctuations more closely indicating possible overfitting. The predictions of different methods are close together at lower P_{max} values, while at larger P_{max} values they show more divergence.

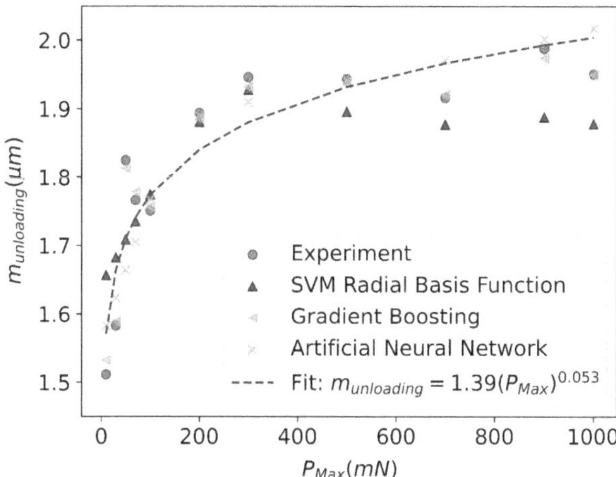

FIGURE 18.18 Predicting power law exponent m_u of the unloading part of the load–displacement curve as a function of maximum load $P_{max.}$ The experimental data are shown as circles, the results of the support vector machines–radial basis function (SVM-RBF) regression are shown as upright triangles, the results of gradient boosting regression (GBR) are shown by left-pointing triangles and the results of ANN-based prediction are shown as crosses. A power law fit was also performed, which is shown with dashes. We can see that, unlike loading exponents, the unloading exponents have a clear increasing trend with increasing maximum load. The ANN regression follows the power law fit most closely, while the SVM-RBF deviates from others and approaches a constant value prediction as the maximum load increases. Gradient boosting with the chosen parameters fits the fluctuations of the experimental data more closely indicating some overfitting.

18.8.4 Predicting H_f as Function of P_{MAX}

The results of predicting maximum depth h_f as a function of maximum load P_{max} are shown in Figure 18.19 where plots for SVM-RBF regression, GBR, ANN-based prediction and power law fits are shown together. We can see that all the methods perform well, with ANN, GBR and power law fits performing better than SVM-RBF.

The power law fit where the coefficient and exponent were both varied gives:

$$h_{max} = 0.023P_{max}^{0.56}, \tag{18.15}$$

where h_{max} is in units of microns and P_{max} is in units of mN.

18.9 CONCLUSIONS

We have given a brief overview of AI and its applications in the science of material deformation and fracture. We presented a simple example of predicting parameters and quantities related to P–h curves as functions applied maximum load using a fully connected artificial neural network. We then presented several ML methods such as ANN regression, SVM-RBF regression and gradient boosting regression for predicting the whole P–h curve as a function of applied maximum load. Although methods that can be employed for similar problems are large in number, including many machine learning algorithms and neural network architectures, we think that simple examples that demonstrate some of the power and methodology of applying artificial neural networks and other ML algorithms to such problems would be more useful to the wider audience. For actual usage, one would have to go through actual implementations in the computational package of one's choice; we have indicated some of the popular choices in this chapter. Applications of AI methods are likely to be quite fruitful and likely to be quite widespread in the near future in the study of materials science

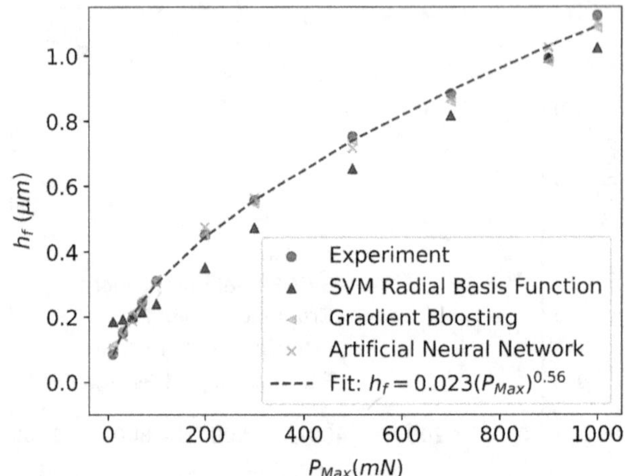

FIGURE 18.19 Predicting final depth h_f as a function of maximum load P_{max}. The experimental data are shown in circles, the results of support vector machines–radial basis function (SVM-RBF) regression are shown as upright triangles, the results of gradient boosting regression (GBR) are shown by left-pointing triangles and the results of ANN-based prediction are shown as crosses. A power law fit with two parameters (a multiplicative coefficient and a power law exponent) was also performed, which is shown by dots. The exponent of 0.56 is close to 0.5. Another fit with the exponent fixed at 0.5 (square root function) is shown with dashes. ANN, GBR and power law fit perform better than the SVM-RBF.

in general and in the study of deformations and fractures in particular. Therefore, we hope that this chapter will serve as a stepping stone for many readers new to artificial intelligence.

ACKNOWLEDGEMENTS

All the authors acknowledge the kind support and encouragement received from various directors of the Council of Scientific and Industrial Research (CSIR)–Central Glass and Ceramic Research Institute (CGCRI), Kolkata, India, during the tenures of whom parts of the research work were carried out. They also gratefully acknowledge the various infrastructural supports received from various divisions of CSIR-CGCRI, Kolkata, India. The financial support received from Indian sponsoring agencies, including the CSIR, UGC, DST(SERB), DAE-BRNS, and IPR, of the Government of India (GOI) in terms of various project sponsorships and fellowships is gratefully acknowledged. Finally, the author AKM acknowledges the kind and continued support received from the authorities of Sharda University, Greater Noida, Uttar Pradesh, India.

REFERENCES

1. J. VanderPlas, *Python Data Science Handbook*, vol. 53, no. 9, O'Reilly Media, Inc., Sebastopol, CA, 2019.
2. I. Goodfellow, Y. Bengio, and A. Courville, *Deep Learning*, MIT Press, 2016.
3. K. Hornik, M. Stinchcombe, and H. White, "Universal approximation of an unknown mapping and its derivatives using multilayer feedforward networks," *Neural Netw.*, vol. 3, no. 5, pp. 551–560, 1990.
4. K. Hornik, M. Stinchcombe, and H. White, "Multilayer feedforward networks are universal approximators," *Neural Netw.*, vol. 2, no. 5, pp. 359–366, 1989.
5. K. Hornik, "Approximation capabilities of multilayer feedforward networks," *Neural Netw.*, vol. 4, no. 2, pp. 251–257, 1991.
6. A. M. Turing, "Computing machinery and intelligence," in Epstein, R., Roberts, G., and Beber, G. (Eds.), *Parsing the Turing Test*, Springer, 2009, pp. 23–65.

7. F. Rosenblatt, "The perceptron: A probabilistic model for information storage and organization in the brain," *Psychol. Rev.*, vol. 65, no. 6, pp. 386–408, 1958, doi: 10.1037/h0042519.
8. D. E. Rumelhart, G. E. Hinton, and R. J. Williams, "Learning representations by back-propagating errors," *Nature*, vol. 323, no. 6088, pp. 533–536, 1986.
9. F. Pedregosa et al., "Scikit-learn: Machine learning in Python," *J. Mach. Learn. Res.*, vol. 12, pp. 2825–2830, 2011.
10. Y. LeCun, P. Haffner, L. Bottou, and Y. Bengio, "Object recognition with gradient-based learning," in Forsyth, D. A., Mundy, J. L., di Gesu, V., and Cipolla, R. (Eds.), *Shape Contour Grouping Computer Vision*, Springer, pp. 319–345, 1999.
11. K. Fukushima, "A self-organizing neural network model for a mechanism of pattern recognition unaffected by shift in position," *Biol. Cybern*, vol. 36, pp. 193–202, 1980.
12. K. Fukushima, "Neocognitron: A hierarchical neural network capable of visual pattern recognition," *Neural Netw*, vol. 1, no. 2, pp. 119–130, 1988.
13. M. Picklum and M. Beetz, "MATCALO: Knowledge-enabled machine learning in materials science," *Comput. Mater. Sci.*, vol. 163, pp. 50–62, December 2018, 2019, doi: 10.1016/j.commatsci.2019.03.005.
14. Y. Tang, D. Zhang, R. Liu, and D. Li, "Designing high-entropy ceramics via incorporation of the bond-mechanical behavior correlation with the machine-learning methodology," *Cell Rep. Phys. Sci.*, vol. 2, no. 11, p. 100640, 2021, doi: 10.1016/j.xcrp.2021.100640.
15. A. Couet, "Integrated high-throughput research in extreme environments targeted toward nuclear structural materials discovery," *J. Nucl. Mater.*, vol. 559, p. 153425, 2022, doi: 10.1016/j.jnucmat.2021.153425.
16. M. Liang, Z. Chang, Z. Wan, Y. Gan, E. Schlangen, and B. Šavija, "Interpretable Ensemble-Machine-Learning models for predicting creep behavior of concrete," *Cem. Concr. Compos.*, vol. 125, September 2021, 2022, doi: 10.1016/j.cemconcomp.2021.104295.
17. S. Vranjes-Wessely et al., "High-speed nanoindentation mapping of organic matter-rich rocks: A critical evaluation by correlative imaging and machine learning data analysis," *Int. J. Coal Geol.*, vol. 247, p. 103847, September 2021, doi: 10.1016/j.coal.2021.103847.
18. B. Vignesh, W. C. Oliver, G. S. Kumar, and P. S. Phani, "Critical assessment of high speed nanoindentation mapping technique and data deconvolution on thermal barrier coatings," *Mater. Des.*, vol. 181, p. 108084, 2019, doi: 10.1016/j.matdes.2019.108084.
19. G. Konstantopoulos, E. P. Koumoulos, and C. A. Charitidis, "Classification of mechanism of reinforcement in the fiber-matrix interface: Application of Machine Learning on nanoindentation data," *Mater. Des.*, vol. 192, p. 108705, 2020, doi: 10.1016/j.matdes.2020.108705.
20. L. Zhang, K. Qian, J. Huang, M. Liu, and Y. Shibuta, "Molecular dynamics simulation and machine learning of mechanical response in non-equiatomic FeCrNiCoMn high-entropy alloy," *J. Mater. Res. Technol.*, vol. 13, pp. 2043–2054, 2021, doi: 10.1016/j.jmrt.2021.06.021.
21. A. Samaei and S. Chaudhuri, "Mechanical performance of zirconia-silica bilayer coating on aluminum alloys with varying porosities: Deep learning and microstructure-based FEM," *Mater. Des.*, vol. 207, p. 109860, 2021, doi: 10.1016/j.matdes.2021.109860.
22. W. C. Oliver and G. M. Pharr, "An improved technique for determining hardness and elastic modulus using load and displacement sensing indentation experiments," *J. Mater. Res.*, vol. 7, no. 6, pp. 1564–1583, 1992.
23. M. R. VanLandingham, "Review of instrumented indentation," *J. Res. Natl. Inst. Stand. Technol.*, vol. 108, no. 4, p. 249, 2003.
24. P. Sudharshan Phani, W. C. Oliver, and G. M. Pharr, "Measurement of hardness and elastic modulus by load and depth sensing indentation: Improvements to the technique based on continuous stiffness measurement," *J. Mater. Res.*, vol. 36, no. 11, pp. 2137–2153, 2021, doi: 10.1557/s43578-021-00131-7.
25. C. Su, E. G. Herbert, S. Sohn, J. A. LaManna, W. C. Oliver, and G. M. Pharr, "Measurement of power-law creep parameters by instrumented indentation methods," *J. Mech. Phys. Solids*, vol. 61, no. 2, pp. 517–536, 2013, doi: 10.1016/j.jmps.2012.09.009.
26. S. Kossman, T. Coorevits, A. Iost, and Di. Chicot, "A new approach of the Oliver and Pharr model to fit the unloading curve from instrumented indentation testing," *J. Mater. Res.*, vol. 32, no. 12, pp. 2230–2240, 2017, doi: 10.1557/jmr.2017.120.
27. W. C. Oliver and G. M. Pharr, "Measurement of hardness and elastic modulus by instrumented indentation: Advances in understanding and refinements to methodology," *J. Mater. Res.*, vol. 19, no. 1, pp. 3–20, 2004, https://doi.org/10.1557/jmr.2004.19.1.3.

28. H. Huang and H. Zhao, "Determination of residual indentation depth in incomplete or irregular unloading curves," *Meas. Sci. Technol.*, vol. 25, no. 8, 2014, doi: 10.1088/0957-0233/25/8/087003.
29. W. C. Oliver and G. M. Pharr, "Nanoindentation in materials research: Past, present, and future," *MRS Bulletin*, vol. 35, no. 11, pp. 897–907, 2010, doi: 10.1557/mrs2010.717.
30. D. A. Pisner and D. M. Schnyer, "Support vector machine," in *Machine Learning*, Elsevier, 2020, pp. 101–121, doi: 10.1016/B978-0-12-815739-8.00006-7.
31. M. A. Hearst, S. T. Dumais, E. Osuna, J. Platt, and B. Scholkopf, "Support vector machines," *IEEE Intell. Syst. Their Appl.*, vol. 13, no. 4, pp. 18–28, 1998.
32. W. S. Noble, "What is a support vector machine?," *Nat. Biotechnol.*, vol. 24, no. 12, pp. 1565–1567, 2006.
33. C. Bentéjac, A. Csörgő, and G. Martínez-Muñoz, "A comparative analysis of gradient boosting algorithms," *Artif. Intell. Rev.*, vol. 54, no. 3, pp. 1937–1967, 2021.
34. T. Hastie, R. Tibshirani, J. H. Friedman, and J. H. Friedman, *The Elements of Statistical Learning: Data Mining, Inference, and Prediction*, vol. 2, Springer, 2009.

Index

D

G

H

N

O

P

R

S

T

U

V

For Product Safety Concerns and Information please contact our EU
representative GPSR@taylorandfrancis.com
Taylor & Francis Verlag GmbH, Kaufingerstraße 24, 80331 München, Germany

www.ingramcontent.com/pod-product-compliance
Lightning Source LLC
LaVergne TN
LVHW081314110826
845149LV00006B/1501

* 9 7 8 1 0 3 2 4 1 7 0 7 3 *